A Dictionary of the Natural Environment

A Dictionary of the Natural Environment

F. J. Monkhouse and John Small

Edward Arnold

© John Small and the Estate of the late F. J. Monkhouse 1978
First published 1978 by
Edward Arnold (Publishers) Ltd
25 Hill Street, London W1X 8LL

The first edition of F. J. Monkhouse's Dictionary of Geography
© the Estate of F. J. Monkhouse 1965
and the second edition
© the Estate of F. J. Monkhouse 1970
were published by
Edward Arnold (Publishers) Ltd
in 1965 and 1970 respectively and have been revised
for inclusion in part of this publication.

ISBN: 0 7131 5957 x Cloth Edition
0 7131 5958 8 Paper Edition

All rights reserved. No part of this publication may be reproduced, stored in a retrieval system, or transmitted in any form or by any means, electronic, mechanical, photocopying, recording or otherwise, without the prior permission of Edward Arnold (Publishers) Ltd.

This book is published in two editions. The paperback edition is sold subject to the condition that it shall not, by way of trade or otherwise, be lent, re-sold, hired out, or otherwise circulated without the publisher's prior consent in any form of binding or cover other than that in which it is published and without a similar condition including this condition being imposed upon any subsequent purchaser.

Printed in Great Britain by
Richard Clay (The Chaucer Press), Ltd,
Bungay, Suffolk

Preface

When Professor Monkhouse died early in 1975, he had nearly completed a full-scale revision of *A Dictionary of Geography*, in an attempt to produce, as he put it, 'a final definitive version'. In the decade after the first appearance of the *Dictionary* in 1965, Geography had continued to experience vast and in some respects traumatic changes. Among these was the rapid expansion in the use of statistical techniques and models, both in the physical and human fields, and far-reaching developments in such branches of human geography as urban and social geography. It was inevitable that, to cope with the increasingly specialized and sophisticated terminology used by modern geographers, the *Dictionary* should expand commensurately. Professor Monkhouse himself made, for the 1966 and subsequent editions, substantial modifications and additions; and the revisions that he had hoped to incorporate in his final version would have enlarged it still more.

In these circumstances the Publishers felt it right to investigate the case for restricting the field covered by the *Dictionary*. After full discussions I was entrusted with the task of producing *A Dictionary of the Natural Environment*, to be based essentially on 'physical' material already assembled by Professor Monkhouse. In this I was at an early stage forced to make decisions that may strike the reader as somewhat arbitrary. However, what does or does not constitute a 'physical' definition is sometimes surprisingly difficult to decide. Moreover, I quickly realized that it is often useful to have readily available in a dictionary material that is not related to the natural environment in the strictest sense, but which may nevertheless be referred to frequently by physical geographers and environmental scientists (such as SI scales, methods of field survey, and even on occasions map projections).

As for new material, I have been able to include 465 additional definitions (approximately half of which had been assembled by Professor Monkhouse himself), together with 18 new line diagrams and—a wholly new departure—a selection of appropriate photographs. It is intended that the result will be a 'new' book, in terms of extended coverage of the natural environment and some aspects of presentation, but that the spirit of Professor Monkhouse's original *Dictionary* will be retained as far as is possible. To this end I have deleted very little of his own material; and in producing my own definitions I have tried to emulate his method and style.

I know from many conversations with him that Professor Monkhouse derived immense pleasure both from the task of compiling *A Dictionary of Geography* and from its subsequent wide popularity. My hope is that the present volume would have met with his approval.

R. J. Small

Southampton, October 1976

Abbreviations

Note. Where the listed word is repeated in the text entry, it is indicated by its initial.

abbr.	abbreviation	L.	Lake
adj.	adjective	Lat.	Latin
alt.	alternative	l.c.	lower case
app.	appertaining	lb(s)	pound(s)
approx.	approximate(ly)	lit.	literally
avge	average	m	metre
c.	about, with numbers such as dates	max.	maximum
C	Centigrade, Celsius	mb.	millibar
cf.	compare	mi.	mile
cm	centimetre	min.	minimum
cm^3	cubic centimetre	mm	millimetre
$cm^3\ s^{-1}$	cubic centimetre per second	mtn	mountain
ct.	contrast	N.	north
cu ft	cubic foot, feet	N.E.	northeast
E.	east	N.W.	northwest
e.g.	for example	occas.	occasionally
=	equals	opp.	opposite
esp.	especially	orig.	originally
estim.	estimated	oz.(s)	ounce(s)
excl.	excluding	partic.	particularly
[*f*]	figure attached	p.h.	per hour
[*f* word]	figure attached to word cited, with some reference to present entry	pl.	plural
		pron.	pronounced
F	Fahrenheit	R.	River
Fr.	French	ref.	reference
ft	foot, feet	rel.	relative, relatively
fthm(s)	fathom, fathoms	resp.	respectively
gen.	generally, general	S.	south
Germ.	German	S.E.	southeast
Gk.	Greek	SI	*Système Internationale (d'Unités)*
g	gram(me)	sing.	singular
Gt.	Great	Sp.	Spanish
I.	island, isle in proper name	specif.	specifically
i.e.	that is	sq.	square
incl.	including	S.W.	southwest
in.(s)	inch, inches	syn.	synonymous
It.	Italian	vb	verb
kg m²	kilogram per square metre	W.	west
km	kilometre	yd(s)	yard(s)

If not otherwise indicated, place names refer to the United Kingdom.

A Dictionary of the Natural Environment

aa (Hawaiian) (pron. *ah-ah*) A LAVA-flow solidified into irregular block-like masses of a jagged, clinkerous, angular appearance, the result of gases escaping violently from within the lava and the effects of the drag of still molten material under the hardening surface-crust. E.g. on the slopes of Mauna Loa (4171 m, 13 680 ft), Hawaiian National Park. Ct. PAHOEHOE, PILLOW LAVA.

abime (Gk.) A deep vertical shaft in limestone country (KARST), opening at the bottom into an underground passage. Lit. 'bottomless'.

ablation The wasting or consuming of snow and ice from the surface of an ice-sheet or glacier. It involves: (i) *melting*, caused by solar radiation acting esp. by conduction by way of solid debris on the surface or neighbouring rock walls, by rel. warm rainfall, and by melt-water streams; (ii) *sublimation*, the direct transference of water from the solid to gaseous state, depending on wind, temperature and humidity; (iii) ABRASION, caused by powerful winds blowing hard ice-particles along the surface (esp. in Polar regions); (iv) the calving of icebergs where the ice-margins reach tide-water. A *factor*: the rate at which the snow- or ice-surface wastes. A is sometimes used for the removal of rock debris by wind, but deflation is better. Ct. ALIMENTATION. See pl. 14, 45.

ablation till An assemblage of unstratified and unsorted clay, stones and boulders, contained in and upon stagnant ice, then let down as the ice melts and decays *in situ*, usually in the presence of much water, thus causing slumping within the mass. Finer particles are commonly removed by this active melt-water, so that it tends to be coarser than LODGEMENT-TILL.

Abney level A surveying instrument, comprising a spirit-level mounted above a sighting-tube, the bubble being reflected in the eye-piece. Used to measure the angle of inclination of a line joining an observer to another point. If the linear distance between the points is measured, their difference in height can be calculated using the tangent ratio. This affords a rapid and convenient field-survey method when an accuracy of $\pm$ a half-degree is acceptable.

abrasion The mechanical or frictional wearing-down, specif. of rock by material (e.g. quartz-sand) which forms the abrasive medium, transported by running water, moving ice, wind and waves (which supply the energy or momentum) (cf. sand-blast). A is the result of the process of CORRASION of rock. A of an ice-surface can be caused by wind-blown ice-particles.

abrasion platform A nearly smooth rock p worn by forces of ABRASION, as along the coast (wave-cut p).

absolute age In GEOCHRONOLOGY, the dating of rocks in actual terms of years; ct. RELATIVE A. Various tables of dating exist; e.g.:

(beginning millions of years ago)	
Quaternary	2
Tertiary	65
Cretaceous	136
Jurassic	190
Triassic	225
Permian	280
Carboniferous	345
(in USA) {Pennsylvanian / Mississippian}	
Devonian	395
Silurian	440
Ordovician	500
Cambrian	570

Note: Other estimates variously date the beginning of the Cambrian at between 500 and 600 million years ago.

absolute drought A period of at least 15 consecutive days, each with less than 0·25 mm (0·01 in) of rainfall (British climatology). The record duration of **a d** in Britain is 60 days during spring, 1893 in Sussex.
Note: In USA a criterion of 14 days without measurable rain is called a DRY SPELL (which has a different definition in UK).

absolute flatland map A **m** in which all areas with slopes below a selected critical value are outlined and distinctively shaded.

absolute humidity The mass of water-vapour per unit volume of air, expressed in g m^{-3}. A body of air of a given temperature and pressure can hold water-vapour up to a limited amount, when it becomes saturated (at the DEW-POINT). Cold air has a low **a h**, warmer air has a higher figure; e.g. air at 10°C (50°F) can contain 9·41 g m³; at 20°C (68°F), 17·117 g; at 30°C (86°F), 30·036 g. A **h** over the land is highest near the Equator, lowest in central Asia in winter. The alt. term VAPOUR CONCENTRATION is now favoured by the Meteorological Office. Ct. RELATIVE H, SPECIFIC H, MIXING RATIO.

absolute instability The state of an air mass with an ENVIRONMENTAL LAPSE-RATE greater than the DRY ADIABATIC LAPSE-RATE, and which is therefore unstable. A **i** is regardless of moisture content, in ct. to CONDITIONAL I, where there is a dependence on moisture content.

absolute stability The state of an air mass in which the ENVIRONMENTAL LAPSE-RATE is less than the SATURATED ADIABATIC LAPSE-RATE, and which is therefore stable.

absolute temperature A scale of **t** based on Absolute Zero ($= -273\cdot 15°C$), the point at which thermal molecular motion ceases; i.e. 0° Absolute or KELVIN (K). The Kelvin degree has the same value as the CENTIGRADE degree; it is sufficiently accurate to use a scale obtained by adding 273°C to the observed **t** e.g. $-10°C = 263$ K (the ° is now officially omitted). This scale is esp. valuable in that there are no negative quantities; it is used in meteorology in expressing upper air **t**.

absorption (i) The physical process by which a substance retains radiant energy (heat- and light-waves) in an irreversible form of some other kind of energy, as opposed to reflecting, refracting or transmitting it; ct. a dull black surface which absorbs a high proportion (e.g. a BLACK BULB THERMOMETER) with ice and snow (which absorb little), and burnished silver (which absorbs only 5%). (ii) The process whereby the energy of electromagnetic radiation is taken up by a molecule and changed into another form of energy.

abstraction Used by some geomorphologists as syn. with river CAPTURE. Strictly **a** involves lateral widening of the master-stream; capture involves HEADWARD EROSION by the master-stream.

abyss (Lat.) Lit. bottomless (cf. ABIME), hence indicating something of very great, almost unfathomable, depth (a chasm, ocean DEEP).

abyssal App. to ocean depths between 2200 m and 5500 m (1200 and 3000 fthms); some authorities use loosely for depths of only 1800 m (1000 fthms) or even 900 m (500 fthms) and even gen. as the ocean-floor. *A plain:* undulating deep-sea plain. *A zone:* area of accumulation of PELAGIC marine deposits, notably ooze. See SUBMARINE RIDGE. [*f*]

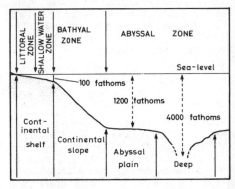

accelerated erosion An increase in the natural rate of erosion due to the activities of man. Formerly regarded as syn. with soil erosion, **a e** is now applied more broadly e.g. to the modification of the channel geometry of streams as a result of urbanization. This produces rapid and increased stream discharge, and scour and bank erosion of stream channels ensue, because there has been no compensating increase of sediment load, the soil surface having been 'frozen' by spreads of concrete, tarmac etc.

accessory (and **supplementary**) **clouds** In the *International Cloud Atlas* a series of 9 additional features to the major **c** classification; e.g. *arcus*, a dense horizontal roll of dark **c**, arched in front of CUMULONIMBUS; *incus*, the 'anvil' above a cumulonimbus; *mamma*, hanging protuberances on the underside of a **c**; *tuba*, a cone or column protruding from a **c** base (*funnel c*).

accessory minerals Varied ms widely distributed in rel. small quantities in an igneous rock, whose absence would not alter its essential nature.

accident, climatic Used by W. M. Davis and adopted partic. by C. A. Cotton to indicate interruptions to the 'normal' CYCLE OF EROSION which are the result of marked changes in climate. The 'normal' cycle was conceived as taking place under humid temperate conditions, from which aridity or glaciation might provide accidental departures. The term has become virtually obsolete since the idea of a humid temperate climate being 'normal' no longer holds credence, in view of increased knowledge of geomorphological processes elsewhere in the world.

accidented relief A rugged, 'broken' or highly dissected physical landscape.

accordant drainage A systematic relationship apparent between rock-type and structure on the one hand, and surface d pattern on the other. [*f*]

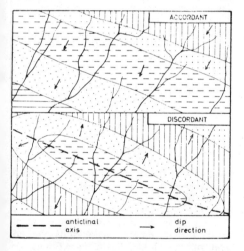

accordant junction (of rivers) A tributary which joins a main river along a course leading normally to a junction at that level. First enunciated in 1802 by J. Playfair as PLAYFAIR'S LAW.

accordant (or **concordant**) **summit levels** Where ss of hills or mountains rise to approx. the same elevation; this accordance may be explained either: (i) by assuming the existence of a former upland with a plane surface, much dissected; or (ii) the result of uniform denudation of an area of evenly spaced valleys, where erosion and weathering reduce hilltops uniformly in height. A summits may be plotted by drawing a SUPERIMPOSED PROFILE from a contour map. [*f, opposite*]

accretion (i) The accumulation of material; e.g. sediment on a FLOOD-PLAIN. (ii) In meteorology, the growth of an ice particle by collision with a drop of water.

accumulated temperature The sum or 'accumulation' of DEGREE-DAYS above a basic critical value (e.g. 5·5°C (42°F) as a base **t** for the growth of grass) over a period of time. The concept was introduced in 1855 by A. de Candolle, who used 6°C as his critical temperature. If on a given day the **t** is above the critical value for h hours and the mean **t** during that period exceeds the datum value by m degrees, the accumulated **t** for that day above the datum is hm degree-hours or $hm/24$ degree-days. To cut down the calculations necessitated by using daily figures, estimates based on monthly means may be made. If 5·5°C (42°F) is the datum, and 7°C (45°F) is the mean for a specific month, it will count (in a 31-day month) as $1·5 \times 31 = 46·5$ degree-days (C) or $3 \times 31 = 93$ degree-days (F) towards the final total.

acid lava A mass of molten igneous material, flowing slowly from a volcanic vent, stiff and viscous, rich in silica, and with a high melting-point (about 850°C). Hence it solidifies rapidly and does not flow far, forming a steep-sided dome; e.g. Mount Lassen, Cascades, California. Some **a l**s solidify in a fine-crystalled state as rhyolite or dacite, others in a glassy form as obsidian. A **l** is obsolescent, though still widely used; see ACID ROCK.

acid rock An igneous **r** with over 10% free quartz, or consisting of minerals rich in silica. Many common **r**-forming minerals are silicates, or compounds of silica with metallic oxides. Examples of **a r**s are granite, rhyolite, obsidian. The term **a r** (which arose from the

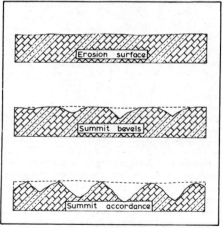

concept of silica as an acidic oxide) is, however, considered to be misleading, and is becoming obsolete.

acid soil A s with a hydrogen-ion value (pH) below 7·2, base-deficient and sour. In cool, moist areas, percolating ground-water leaches out the soluble bases, partic. calcium, from the A-HORIZON. The s gradually becomes lime-deficient; i.e. increasingly acid. See PODZOL.

acre A unit of English measure of area

1 a = 4840 sq. yds
640 as = 1 sq. mi.
1 a = 0·4047 hectares (ha)
2·4711 as = 1 ha
247·11 as = 1 km²

acre-foot Used in USA, esp. in irrigation engineering, signifying the amount of water required to cover an a. of land to the depth of 1 ft. This amounts to 43 560 cu. ft = 1219·2 m³.

actinometer An instrument for measuring the intensity of RADIATION, esp. from the sun.

active fault A f-line along which repeated earth-movements are currently in progress; e.g. San Andreas F, California.

active layer, in the soil The thickness of soil frozen in winter, that thaws out in summer. Syn. with *mollisol*. Ct. PERMAFROST.

actual isotherm An i for which values plotted are actual means, not reduced to sea-level by a correction for the altitude of the station; the pattern of is therefore closely resembles that of a contour map.

adiabatic App. to the change in temperature of a mass of gas (notably air), undergoing expansion (cooling) or compression (heating) without actual loss or gain of heat from outside. This commonly occurs within an ascending or descending air-mass, since on expansion individual molecules are more widely diffused, on compression more closely packed. On expansion the lowering temperature may reach DEW-POINT, causing condensation and precipitation. Adverb *adiabatically*. See DRY and SATURATED A. LAPSE-RATE

adobe A hard-baked clayey deposit in desert basins of USA. Probably of wind-blown origin, but possibly reworked and redeposited by running water.

adolescence A stage in the CYCLE OF EROSION, following youth and preceding maturity; the features characteristic of the latter are as yet only slightly developed. There has been much recent criticism of the use of such evocative language in descriptive analysis.

adret (Fr.) A hill-slope, esp. in the French Alps, which faces S. or S.W. and so receives the max. available amount of sunshine and warmth, in ct. to the N.-facing shady side (UBAC). Germ. *Sonnenseite*, It. *adretto* or *adritto*, Turkish *günvey*.

adsorption The concentration, penetration, and physical adhesion of particles of one substance on the surface of another, as molecules of gases or a material in solution held on the surface of a solid, though not involved in chemical combination. In sediments, it includes films of water surrounding individual particles. In soils, it includes base-salts in colloidal form surrounding mineral particles.

advanced-dune A sand-d formed ahead of a larger d accumulating round some obstacle, kept distinct from the main d by eddy motion of the wind. [*f* DUNE]

advection The movement of air, water and other fluids in a horizontal direction; ct. CONVECTION, movement in a vertical direction. In the case of air, it may result in the transfer of heat and of water-vapour, as from lower to higher latitudes, or from warm sea to cooler land (and vice-versa).

advection fog A f formed when a warm, moist air-stream moves horizontally over a cooler land or sea surface, thus reducing the temperature of the lower layers of the air below the DEW-POINT; e.g. near the Grand Banks of Newfoundland, where warm air from over the Gulf Stream, moving N. from the Florida Channel, passes over the waters of the Labrador Current; this is 8° to 11°C (15° to 20°F) cooler, since it brings melt-water from disintegrating PACK-ICE farther N., thus forming dense f on 70 to 100 days in the year. Also **a f** occurs on an avge. of 40 days in the year off the Golden Gate, San Francisco, caused by warm air moving E. over cold offshore currents.

adventive cone A parasitic or subsidiary c, which breaks out on the flanks of a volcano; e.g. on Mt Etna.

aegre See BORE.

aeolian App. to the effects of wind, esp. on relief; e.g. in deserts where the unconsolidated surface is unprotected by vegetation, or along a sandy sea-coast. *A erosion* (akin to sand-blast) may produce ZEUGEN, YARDANGS, DEFLATION hollows; *a transport* and *deposition* may produce DUNES, LOESS. Often spelt eolian, esp. in USA. Derived from Aeolus, Gk. god of the winds.

aeon, eon (i) A vague and indefinite term for a very long period of time in Earth history. (ii) More precisely defined as 10^9 years, on which basis the Earth is about 4·7 as old. (iii) One of the 2 major divisions of Earth time, CRYPTOZOIC, PHANEROZOIC A. Ct. EOZOIC.

aerial photograph A p, vertical or oblique, of the Earth's surface from an aircraft. Used for mapping (PHOTOGRAMMETRY) and for gen. study, esp. of landforms and archaeology. A ps are taken in strips (*sorties*) of overlapping prints, and used to make a *mosaic* (in USA '*print lay-down*'). The scale of an **a p** is the relation between the height of the aircraft and the focal length of the camera-lens; e.g. with a 100 cm camera at 10 000 m height, RF = 1/10 000.

aerobic In the biological sense, referring to organisms living in the presence of free oxygen, specif. with ref. to those in the soil. Ct. ANAEROBIC.

aerology Scientific study of the ATMOSPHERE throughout its vertical extent, in ct. to CLIMATOLOGY, confined to the atmosphere adjacent to the Earth's surface. Adj. *aerological*, hence the UK Meteorological Office's *Daily Aerological Record*.

aeroplankton Minute organisms (spores, etc.) which float freely in the atmosphere.

aerosphere The entire gaseous envelope surrounding the Earth, including the TROPOSPHERE and STRATOSPHERE. Used because some authorities limited ATMOSPHERE to the lowest layer only of the gaseous envelope, but the latter is more usually accepted for the whole.

affluent Obsolescent term for the tributary of a river. Sometimes restricted to a small stream joining a larger, ct. 2 of more or less equal size, which are CONFLUENTS.

after-glow A faint and diffuse arch of radiance occasionally visible in the W. sky after sunset, when the sun is 3° or 4° below the horizon, probably caused by the scattering effect on light of dust-particles in the atmosphere. Ct. ALPINE GLOW.

aftershock Vibrations of the Earth's crust after main earthquake waves have passed, originating at or near the same SEISMIC FOCUS, caused by minor adjustments of rocks after the main rupture. These may go on for hours, days or even months, and may cause structural damage to buildings weakened by a main shock.

agglomerate A mass of angular fragmental material ejected by a volcano, and cemented in an ash or tuff matrix to form a PYROCLAST. See BRECCIA.

agglomeration The growth of water particles by collision of minute droplets in the process of cloud formation, resulting in either *coalescence* (both droplets water) or *accretion* (one an ice crystal).

aggradation The building-up of the land-surface by the deposition and accumulation of solid material derived from denudation by a river. It is also applied to the development of a marine beach. Vb., to *aggrade*. See pl. 8, 71.

aggradational ice A form of underground **i**, developed as the upper surface of PERMAFROST gradually rises, thus incorporating existing **i** lenses at the base of the ACTIVE LAYER. The **i** appears as white horizontal bands with intervening dirt.

aggregate (i) A structural unit in the soil, in which individual particles are held together. (ii) An **a** of mineral particles is one definition of a rock. (iii) Gravel and sand used for concrete.

agonic line A l joining the Earth's magnetic poles, along which MAGNETIC DECLINATION is zero; i.e. a magnetic needle points to True North. This l seems to be moving slowly in a W. direction. [*f*]

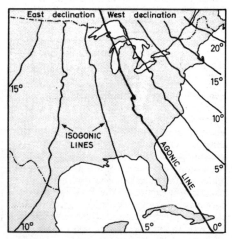

agricultural climatology Gen., c in its application to agriculture. Sometimes called AGROCLIMATOLOGY. See MICROCLIMATE.

A-horizon The top zone in the SOIL PROFILE [*f*], immediately below the surface, from which soluble salts (esp. bases) and colloids have been leached. Usually black or grey in colour, and contains some HUMUS. Recent practice is to denote the sub-surface ELUVIAL horizon (A2) as E.

aiguille (Fr.) Prominent needle-shaped rock-peak, esp. in the Mont Blanc massif; e.g. A. du Midi, Charmoz, Grépon, Verte.

airglow The faint light of the night sky, even when moonless.

air-mass A largely homogeneous mass of air, sometimes extending over hundreds of km (though applied to more limited local phenomena), with marked characteristics of temperature and humidity, bounded by FRONTS, and originating in a specif. source-region. An **a-m** may travel great distance, transporting its orig. characteristics, though gradually modified. On a basis of temperature, an **a-m** is *Polar* or *Tropical*; on a basis of humidity *Maritime* (having crossed oceans and so moist), or *Continental* (originating over continents and so dry). In combination, **a-m**s are Polar Maritime (*Pm* or in USA and by the UK Meteorological Office *mP*), Polar Continental (*Pc*), Tropical Maritime (*Tm*), and Tropical Continental (*Tc*). Other categories include **a-m**s from the Arctic Ocean (*A*), and Antarctic continent (*AA*), and equatorial oceans (*E*). An indication of warming through equatorward movement is given by adding *W* (warm) (e.g. *TcW*); of cooling through poleward movement by adding *K* (Germ. *kalt*) (e.g. *TcK*). If monsoonal in character, the suffix *M* is added (e.g. *PcM* and *TmW*(*M*)). Considerable modification in temperature or humidity since an **a-m** left its source-region is denoted by prefix *N* (e.g. *NTm*); stable or unstable conditions by suffix *S* or *U* respectively; and source-region by suffix initials, *NP* (N. Pacific Ocean), *SI* (S. Indian Ocean). A mixed **a-m** may be indicated by *X*. *A-m climatology* is concerned with the frequencies and properties of **a-m**s affecting the region under consideration.

air-meter A simple form of ANEMOMETER, consisting of a wheel (with a calibrated dial) which rotates in the wind. Useful for measuring winds of low velocity, but tends to continue rotating by its own momentum for some time after a gust ceases, thus giving inaccurate readings.

air-stream A moving current of air; a wind.

ait (eyot) A small island in a river; e.g. Chiswick E., R. Thames.

Aitoff's Projection A p based on ZENITHAL EQUIDISTANT P, in which horizontal distances from the central meridian are doubled; it resembles MOLLWEIDE P, but parallels (except the Equator) and meridians (except the central) are curves, and there is less distortion at the margins. Cf. HAMMER P. [*f*]

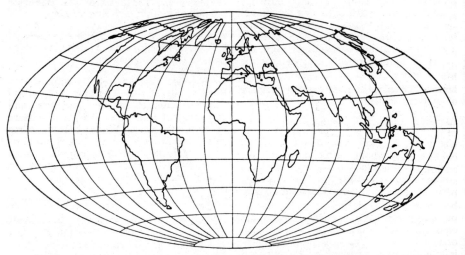

---

alas, -es A large THERMOKARST depression, with steep bounding walls and a flat floor, sometimes occupied by a shallow lake. As are well developed in Siberia (notably central Yakutia, where they are up to 40 m deep and 15 km in diameter). As are formed by localized melting of the PERMAFROST, sometimes following destruction of forest cover by fire. ICE-WEDGES thaw to give hummocky terrain (incl. BAYDJARAKHS); as the ice-wedge troughs develop, collapse of intervening ridges creates a lowland and further destroys forest cover. Accumulation of meltwater forms lakes, and PINGOS may grow and collapse. As individual as grow, they merge to form *a valleys*, which may be tens of km long. In Siberia, where up to half the surface area may be occupied by as, formation has occurred during the Post-Glacial period. [*f*]

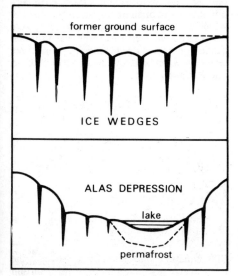

albedo The reflection coefficient, or ratio between total solar electromagnetic RADIATION falling (incident) upon a surface and the amount reflected, expressed as a decimal or percentage. The Earth's average a is about 0·4 (40%); i.e. 4/10 of solar radiation is reflected into space. It varies from 0·03 for dark soil to 0·85 for a snow-field. Water has a low a (0·02) with near-vertical rays, but a high a for low-angle slanting rays; grass about 0·25.

Albers' Projection A CONICAL **p** of equal area properties, with 2 standard parallels along which scale is correct. Parallels are concentric circles spaced more closely to N. and S. of the standard parallels; meridians are radiating straight lines, spaced equally, with true scale on the standard parallels, the scale too great between them and progressively smaller outside. Meridian and parallel scales are constructed in inverse proportion to each other to obtain the equal area property. There is little distortion of shape on a continental scale. Used for maps of USA, since the greatest scale error for that area is only 1·25%.
[*f, opposite*]

alcove An arcuate, steep-sided seepage cavity in a rock face, due to several causes. (i) As in the Snake R. valley of Oregon, Washington and Idaho, cut into lava sheets by HEADWARD EROSION, the result of SPRING SAPPING. (ii) On the flanks of sandstone canyons in the Colorado Plateau, formed by percolating ground water emerging as a seepage on a cliff face, thereby removing the cementing calcium carbonate and causing disaggregation of the sandstone. As the cavity enlarges, roof collapse may occur, stimulated by winter frost action.

alcrete A DURICRUST with a high proportion of aluminium hydroxide, a near-syn. for BAUXITE. See also CUIRASS.

Aleutian 'low' The mean sub-polar atmospheric low pressure area in the N. Pacific Ocean, most marked in winter. It is not a very intense stationary 'low', but an aggregate of rapidly moving individual 'lows', interrupted by occasional anticyclones. Cf. ICELANDIC LOW.

Algonkian Derived from the Algonkian Indians, between the Great Lakes and Hudson Bay, a stratigraphic name denoting the youngest Precambrian sedimentary rocks.

alidade (i) A rule with peep-sights at either end, or (on more elaborate instruments) with a telescope mounted exactly parallel to the rule. Used in conjunction with a PLANE-TABLE to draw lines of sight on a distant object when surveying. (ii) The index of any graduated survey instrument, such as a SEXTANT.

aligned sequence A series of glacial OVERFLOW CHANNELS, crossing several interfluves, which exhibit a distinct alignment.

alimentation In glaciology, the accumulation of snow on a FIRN-field, through direct snowfall, the contribution of avalanches, and refreezing of melt-water. This may nurture an outflowing glacier. When a near the source of a glacier exceeds wastage (ABLATION) at its end, it 'advances'; when these are the same it is stationary; when a is less than wastage the glacier shrinks, 'recedes' or 'retreats'.

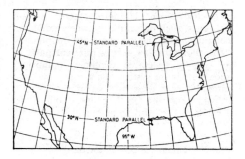

8 ALKALI

alkali In chemistry, the soluble hydroxide of metal, esp. of sodium, potassium and calcium, which reacts with an acid to form a salt and water. Specif. applied to soils which have a pH reaction of above 7·2, found usually in dry areas where soluble salts have not been washed or leached away. In a SOIL PROFILE alkaline earths (calcium carbonate, magnesium carbonate, calcium sulphate or gypsum, sodium chloride, potassium carbonate) accumulate in the B-HORIZON, washed down from the A-HORIZON. *A-flat*: plain of sediment with a high proportion of **a** salts, often crusted, formed by evaporation of a former lake; e.g. in Jordan Valley (Israel); near Great Salt Lake (USA). Ct. PLAYA.

alkaline rocks Igneous rocks rich in sodic and potassic feldspars, micas, sodic amphiboles and pyroxenes; e.g. the riebeckite-microgranite of Ailsa Craig, W. Scotland.

Allerød (Danish) A Late-Glacial interstadial phase, dating from approx. 11 000 B.P. At this time the climate became rel. warm, and birches spread in S. Britain. Also referred to as Zone II of the LATE-GLACIAL period.

allocthon, -ous Something which has been transported; ct. AUTOCTHON, -OUS. Used specif. in connection with an overthrust and far-travelled rock-mass, i.e. a NAPPE. It also refers to COAL MEASURES formed from transported vegetation, and to any EXOTIC feature.

allogenic Having an origin elsewhere. Applied to: (i) Streams which derive their water-supply from outside the immediate area, as those crossing a desert or area of limestone country; (ii) Constituents of certain sedimentary rocks, orig. part of other rocks, which have been transported, redeposited and compacted; e.g. pebbles in CONGLOMERATES. (iii) A plant succession developing as a result of changing physiographic conditions; e.g. in a DELTA, with silt accumulation.

allometric growth The development of stream networks, such that ratios between individual components of the networks are maintained. 'The allometric law relates different parts of an organic system in dynamic equilibrium such that, as the system as a whole grows, the ratios between each part and the whole ... remain constant' (R. J. Chorley and B. A. Kennedy). E.g. as the number of stream segments within a network increases, the proportions of the segments falling into each STREAM ORDER remain approx. unchanged. It is suggested that all 'laws' of MORPHOMETRY (stating the relationships between stream order, stream length, basin area etc.) derive from the **a g** of stream systems.

alluvial App. to ALLUVIUM.

alluvial cone A form of A. FAN, though with a higher angle of slope, in which the mass of material is thick and coarse, and its surface steep, as in semi-arid areas, where deposits are carried by short-lived torrents. Sometimes termed a *dejection c*. E.g. in Arizona, S. Utah and S. California. [*f*]

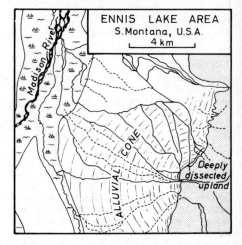

alluvial fan A fan-shaped mass of sand and gravel, with apex pointing upstream and with a convex slope, deposited by a stream where it suddenly leaves a constricted course for a main valley or an open plain. Fs are produced where the constriction of a valley abruptly ceases, not solely where there is a change of gradient, though this contributes. E.g. the upper Rhône valley in Switzerland, where torrents from the Bernese and Pennine Alps leave gorges for the open floor of the main valley.

alluvial flat A near-horizontal **a** area near river, on which ALLUVIUM is deposited in time of flood; on a larger scale it becomes an *a plain*. [*f, page 9*]

alluvial terrace Following REJUVENATION a river cuts down its channel, leaving at a higher level portions of an A. FLAT, which may be PAIRED on either side as terraces. Some authorities regard **a** ts as syn. with river ts, but this is only possible if '**a**' is taken to include coarse sands and gravels as well as fine-grained deposits. [*f* A. FLAT]

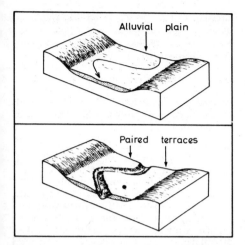

alluvium, adj. **alluvial** (i) In a broad sense, all unconsolidated fragmental material laid down by a stream as a cone or fan, in its bed, on its flood-plain, and in lakes, deltas and estuaries; comprising silt, sand, gravel. (ii) In a restricted sense, fine-grained silt and silt-clay, as on maps of the Geological Survey of UK. *A soils*, fine, well mixed rock waste, frequently replenished in flood with a high mineral content, and near the river responsible for deposition (therefore irrigation is possible), are often of great agricultural value; e.g. valleys of Nile, Indus, Ganges, Mekong, Yangtse, Hwangho. The chief disadvantage of an alluvial plain is its liability to destructive flooding. See pl. 20, 72.

almwind Local name for a wind of FÖHN type which blows from the S. across the Tatra Mtns, descending into the foreland of S. Poland. It may be strong and blustery, and can raise the temperature very rapidly, causing avalanches in late winter and spring.

alp A high-lying gentle slope, bench or 'shoulder' in the mountains, esp. in Switzerland, commonly above a U-shaped glaciated valley, at the level where a marked change of slope occurs. Though snow-covered in winter, it provides summer pasture to which animals are driven. [*f, opposite*]

Alpine App. to the Alps; used more widely (with l.c.) of any high mountains, their characteristic relief (esp. glacial features), climate and flora. Specif. the last series of major mountain-building movements (*A Orogeny*) in mid-Tertiary times, responsible for the main ranges of Europe and Asia (sometimes known as *Alpides*).

alpine glacier See GLACIER.

alpine glow (Germ. *Alpenglühen*) A short-lived pinkish tint on mountain peaks, esp. when snow-covered, just after sunset and before sunrise. The **g** may start when the rim of the sun is about 2° above the horizon. Colours in the morning have a purplish tint, in the evening orange. Ct. AFTER-GLOW.

Altaides Ranges of mountains uplifted during the *Altaid Orogeny* of Upper Carboniferous-Permian times, of which the Altaid range in Asia is the type-example. See ARMORICAN, HERCYNIAN, VARISCAN.

altimeter An instrument used in aircraft or by surveyors to show height above sea-level, based on fall in atmospheric pressure with height, which avges 34 mb. (1 in. of mercury) for each 300 m; i.e. an ANEROID BAROMETER calibrated for height. In surveying, an **a** should be used only to determine heights rel. to a near-by BENCH-MARK or SPOT-HEIGHT, since absolute determinations may give a large error. For accurate work, tables are available which make allowances for latitude and air temperature. Modern types of *radio a* use electronic techniques. An *altigraph* is a self-recording **a**.

altimetric frequency graph A method of analysing relief by graphing as a HISTOGRAM the **f** of occurrence of specif. heights above sea-level, esp. useful when seeking to recognize and correlate PLANATION surfaces. The **f** of occurrence may be obtained by: (i) counting summit spot-heights; (ii) covering the area with a grid of small squares, and in each square noting the highest point, or the mean of the highest and lowest points, or the

$$\frac{\text{height of each corner} + 4\,(\text{height at centre})}{8}$$

On the graph, plot heights above sea-level in m on the horizontal scale, the % frequency on the vertical scale. Heights are grouped in

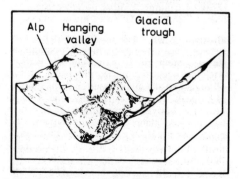

altitude class intervals; e.g. 25-m groups (0–25, 26–50, 51–75). [*f*]

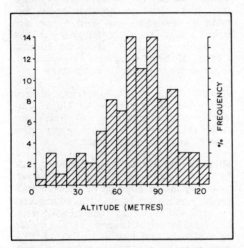

altiplanation A levelling process caused by MASS-WASTING, which may under certain conditions produce terrace-like forms and flattened summits. Some are accumulation features of rock-material, but planation of solid rock may be caused by FREEZE-THAW processes under PERIGLACIAL conditions. The recognition of a terraces of this kind in S.W. England (e.g. Dartmoor) has been claimed. (See GOLETZ TERRACE).

altiplano A high-lying plateau-basin in the Andes, specif. in Bolivia, between the E. and W. Cordillera, at about 3700–4500 m (12–15 000 ft) above sea-level. Covered with sheets of alluvium and glacial drift. L. Titicaca (fresh water) and L. Poopo (saline) are foci of inland drainage on this plateau.

Altithermal The period approx. 8000–3000 B.P. in the W. USA when July mean temperatures reached values up to 4°F above the present, and when 'drought' conditions prevailed. During the A phase, headward erosion of many tributary valleys provided the sediment for the aggradation of major valleys.

altitude (i) The height above a chosen datum surface (mean sea-level) in a topographical survey system. (ii) In surveying, the vertical angle between the horizontal plane of the observer and any higher point; e.g. the summit of a peak. (iii) In astronomy, the a of a heavenly body is the angle of elevation between the plane of the horizon and the body, measured along a GREAT CIRCLE through the body and the ZENITH. A is used in conjunction with AZIMUTH to fix the position of a heavenly body. [*f* AZIMUTH]

altocumulus (*Ac*) A fleecy cellular cloud, in bands or waves of globular masses, gen. separated by blue sky, though they may be so close together that their edges join. They usually consist of supercooled droplets. They occur at middle altitudes, about 2400–6000 m (8–20 000 ft), and are usually, though not invariably, a sign of fair weather. The sun and moon, when seen through these clouds, may be surrounded by a CORONA. See pl. 1.

altostratus (*As*) A greyish uniform sheet-cloud, of wide extent, through which it is possible to see the sun 'as through ground-glass', with a 'watery look'. It usually heralds rain, since it is associated with the approach of a WARM FRONT, and is formed from the thickening of CIRROSTRATUS. It occurs at middle altitudes, about 2400–6000 m (8–20 000 ft).

alveolate relief A surface dominated by cupola-shaped hills of gen. low relief. Identified by J. Hurault in rain-forest areas of W. Africa, mainly in granite terrains where chemical decomposition is rapid. Rejuvenation of streams results in some cupolas being transformed into domed INSELBERGS, partic. where mass movements strip the regolith to reveal fresh rock which subsequently resists further decay.

aluminium (in USA, **aluminum**) A metal of remarkable lightness, high strength cf. weight (esp. in alloys), great resistance to corrosion, and high electrical conductivity.

amber A yellowish translucent substance, fossilized resin from coniferous trees, occurring in Oligocene strata along the Baltic coast of Samland (formerly East Prussia, now USSR); of value for ornaments and jewellery.

ambient temperature The t in any part of the atmosphere immediately surrounding a specif. entity such as a cloud or THERMAL.

amorphous The state of a mineral with no definite crystalline structure; e.g. limonite (Fe_2O_3).

amphibole One of a group of ferromagnesian silicate minerals, of which the most widespread is hornblende.

amphidromic system A unit area in the sea, within which the surface is set oscillating by the tide-producing forces, with a period related to the dimensions of the unit. In addition, a gyratory movement is produced by the Earth's rotation (CORIOLIS FORCE), so that high water rotates around the *nodal* (or *amphidromic*) *points* in an anticlockwise direction in the N. hemisphere. At these points

Plate 1. ALTOCUMULUS CLOUDS. (*Royal Meteorological Society*)

water level remains approx. the same (zero range), while CO-TIDAL LINES radiate outwards, along which the tidal range increases. In the English Channel a KELVIN WAVE is set up, and the **a** point is theoretically located on the land (actually in Wiltshire), known as a *degenerate a point*. [*f*]

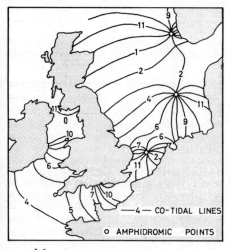

amygdale A VESICULE, a gas- or steam-escape bubble cavity in a volcanic rock, spherical or oval, filled with a secondary mineral such as quartz. Hence *amygdaloidal*. E.g. some of the Borrowdale Volcanic Series, English Lake District.

anabatic wind A local w blowing up-valley during the afternoon, esp. in summer. The air on mountain slopes is heated by conduction to a greater extent than air at the same level above a valley floor. This causes convectional rising from above the slopes, hence air moves up from the valley to take its place.

anabranch, -es An individual stream segment within a BRAIDED RIVER system. See DISTRIBUTARY.

anaclinal Opposed to the DIP of the surface rocks; applied partic. to streams and valleys trending against the dip of beds which they cross. Introduced by J. W. Powell in 1875, together with CATACLINAL, but now rarely used. Ct. OBSEQUENT STREAM.

anaerobic In the biological sense, organisms living in the absence of free oxygen. An *a soil* is in an airless state, notably when waterlogged.

anafront A COLD FRONT in which a warm AIR-MASS is for the most part rising over a wedge of cold air. Ct. KATAFRONT.

anaglyph A method of visualizing relief in 3 dimensions. 2 photographs printed side by side in red and green are viewed through lenses tinted green and red resp. Used effectively in some textbooks on landforms.

analemma A graph which plots the sun's DECLINATION for each day of the year on a vertical scale, and the EQUATION OF TIME plotted on a horizontal scale, with sun times fast (in ct. to MEAN SOLAR TIME) to the left of the centre axis, sun times slow to the right. An **a** is sometimes printed on a globe. [*f*]

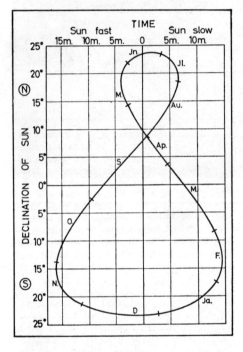

analogue Used specif. in long-range meteorological forecasting, involving the deduction of a series of future repeating patterns of weather situations by analysing and comparing synoptic patterns and sequences of previous years. See also PERIODICITY, SINGULARITY.

anamolistic cycle The c of tides which results from the varying distance of the Earth from the moon; PERIGEE to perigee = 27·5 days.

anastomosis (In physiology, the cross-connections between arteries and veins in the body.) In geomorphology, applied to streams which are braided into a network of small water-courses (*anastomosing streams*), the result of extensive deposition on the stream-bed of part of a heavy traction load.

anchor ice I formed on and frozen to the bed of a stream or other moving water-body, when the rest of the water is not frozen. This may be in the form of FRAZIL I. on submerged objects around which cold water is flowing. A **i** is not GROUND I, according to some authorities, though others regard it as syn.

andesite A fine-crystalled, light grey, VOLCANIC igneous rock of INTERMEDIATE composition, in the form of tiny crystals embedded in a glassy ground-mass. While gen. extrusive in origin, it may be found also in minor intrusions. First studied in the Andes of S. America, hence the name. It forms massive crags and buttresses in the English Lake District, Snowdonia and the Ben Nevis group.

Andesite Line A petrological boundary in the Pacific Ocean, traced from Alaska, Japan, the Marianas, the Bismarck Archipelago, Fiji and Tonga to E. of New Zealand. W. of the **A L** the rocks are INTERMEDIATE (andesite, dacite, rhyolite); to E. they are basic (basalt, trachyte, olivine). On the E. of the Pacific the **A L** runs close to the coast of the Americas. Its significance is that inside the **A L** is the true ocean-basin, floored with basic rocks; outside are continental types of rock. [*f*]

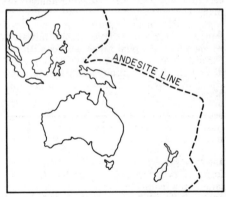

anemometer An instrument for indicating, and, in more elaborate forms, for automatically recording (*anemograph*) direction and velocity of the wind. Types used: (i) *Pressure-plate a*, a simple device with a wooden base and an upright support holding a metal plate suspended from a knife-edge, placed at right-angles to the wind; the angle to which the plate blows is observed, and the velocity is read off from tables. (ii) *Cup-a*, with cups mounted on horizontal cross-arms attached to a vertical rotating spindle; in some types the rotating cups generate a voltage which registers on a dial; (iii) *Pressure tube*, which depends on the difference in pressure between

2 pipes, one facing the wind, the other connected to a system of suction holes. Pressure difference is communicated to a scribing device on a revolving drum, while the vane on the top of the tube indicates changes of direction, which are also recorded (now obsolescent). A continuous record is an *anemogram*.

aneroid barometer An instrument for measuring atmospheric pressure, consisting of a metallic box, almost exhausted of air, whose flexible sides expand and contract with changing air pressure; the magnified movements are communicated by a spring to a needle on a calibrated circular dial (ct. BAROGRAPH). Invented by Lucien Vidie about 1843. While the instrument is light, portable and convenient, it is subject to slight errors. It gained use in geomorphology for field determination of heights, giving readings accurate to $\pm 1 \cdot 5$ m under favourable conditions. Very accurate 'precision models' are produced for meteorological and aircraft use.

angle of rest The max. slope at which a mass of moving unconsolidated rock (SCREE) becomes stable. This may be upset by the high pore water pressure induced by heavy rain, earth-tremors, and the passage of animals or man.

Anglian The ante-penultimate glacial period in Britain, equivalent to the ELSTER GLACIATION of N.W. Europe. Represented typically by glacial deposits at Corton, Suffolk.

angular unconformity An u in which older underlying rock strata dip at a different angle, usually steeper, from that of younger overlying ones. Ct. DISCONFORMITY, NON-SEQUENCE, UNCONFORMITY.

angulate pattern (of drainage) A modified form of TRELLIS DRAINAGE, in which tributaries join main streams at acute or obtuse angles. It reflects the influence of major jointing in the rocks, sometimes of faulting.

anhydrite Anhydrous calcium sulphate, i.e. without its water of crystallisation. Specif. gravity 2·89–2·98. From Gk. *anhydros*, waterless. See GYPSUM.

annular drainage A pattern of SUBSEQUENT d in which streams follow arcuate courses as discontinuous portions of what appear to be near concentric circles. This occurs partic. around dissected domes, where subsequent streams are eroding valleys in less resistant strata; e.g. Black Hills of S. Dakota, with a central crystalline dome rising to 2207 m (7242 ft) (Harney Peak), round which the Cheyenne and its tributaries have developed a patterns in weak shales. [*f*]

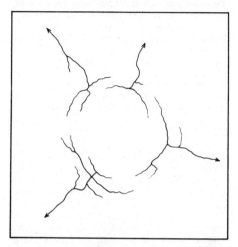

anomaly The departure of any element or feature from uniformity, or from a normal state, or from a long-term avge value, used partic. in meteorology in connection with temperature, in oceanography, and in connection with gravity. (i) A *temperature a* is the difference in degrees between the mean temperature (reduced to sea-level) of a station, and the mean temperature for all stations in that latitude. The result is either a *positive* (higher than average) or a *negative* (lower than average) *a*. Points with equal a may be plotted and joined with ISANOMALOUS LINES. (ii) A *gravity a* is the difference between observed gravity and that computed for an idealized globe. If allowance is made for height above sea-level (thus distance from the Earth's centre), a *free-air a* is obtained. If the attraction of other masses is allowed for, the *Bouguer a* is obtained. The *isostatic a* allows for disturbances due to ISOSTATIC movements. (iii) A *salinity a* is the difference between observed salinity at any point and the mean salinity for the entire oceans.

Antarctic App. to S. Polar regions, opp. to ARCTIC. Strictly an adjective, as *A land-mass* (*Antarctica*) and *A ice-sheet*. Also used as a noun to describe that portion of the earth's surface lying within the A CIRCLE (66° 32′ S.). The area of the ice-sheet is about 13 million km² (5 million sq. mi.) and its max. thickness about 2300 m (7500 ft), though near the coast ranges of mountains with rock-peaks (NUNATAKS) project above the ice. Also loosely used to describe features or conditions similar to those found in the A region.

Antarctic Circle The parallel of latitude 66° 32′ S., along which on about 22 Dec. (summer SOLSTICE in S. Hemisphere) the sun does not sink below the horizon, and on about 21 June (winter solstice) it does not rise above it. S. of this line the number of days without sun in winter increases until at the S. Pole six months of darkness are succeeded by six months of daylight. Ct. ARCTIC C.

antecedent drainage, antecedence A river system originating before a period of uplift and folding of land as a result of Earth-movements. The river continues to cut down its valley at approx. the same rate as uplift, and so maintains its gen. pattern and direction. This is one type of INCONSEQUENT D. E.g. the Indus in Kashmir, the Brahmaputra where it crosses from Tibet into Assam, and the Ganges and its headstreams; in each case deep gorges have been cut across lofty mountains. The R. Colorado has eroded a canyon across the plateaus of S.W. USA; the term **a d** was coined by J. W. Powell (1875) in specif. connection with this river. [*f*]

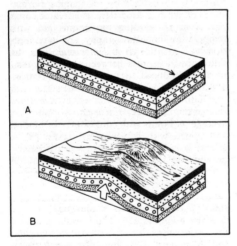

anteposition Used to denote a combination of ANTECEDENT and SUPERIMPOSED DRAINAGE; e.g. R. Colorado.

anthodite A delicate, flower-like formation of white CALCITE, found on the roof of a limestone cave; e.g. Skyline Caverns, Virginia, USA.

anthracite A type of hard, shiny coal, which contains (according to various classifications) 85–98%+ of carbon, and 2–8% volatile matter, and which burns smokelessly with great heat. It is formed by the large-scale decomposition and compaction of buried deposits of coal-forests and swamps of Carboniferous times. Theophrastus, a philosopher-pupil of Aristotle, knew of coal and called it *anthrax*, from which the word **a** is derived.

anthropogeomorphology The study of man-made landforms; e.g. opencast pits, quarries, waste-tips, lakes.

anticentre The point at the ANTIPODES of the EPICENTRE of an earthquake.

anticline An upfold caused by compressive forces in the earth's crust; the strata dip outwards, forming limbs on either side of the AXIS or central line. The **a** may be overfolded so that it appears to lie on its side. If the axis of an **a** begins to dip, the fold is said to PITCH in that direction. See pl. 12. [*f*A]

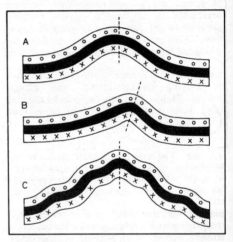

anticlinorium A complex ANTICLINE, on which minor upfolds and downfolds are superimposed; e.g. the Weald of S.E. England. [*f* ANTICLINE C]

anticyclone An area of atmospheric pressure high in relation to its surroundings, diminishing outwards from the centre, indicated on a synoptic chart by a series of roughly concentric, usually widely spaced, closed isobars. In the N. hemisphere air moves clockwise around an **a**, in the S. hemisphere it moves anticlockwise; see BUYS BALLOT'S LAW. An **a** system is slow-moving, winds are light, variable or absent near the centre, and the weather is usually settled: dry, warm and sunny in summer, and either cold, frosty and clear, or foggy ('anticyclonic gloom') in winter. The tendency is for horizontal convergence of air at high levels, slow subsidence within it, and slow divergence at low levels. First introduced by Francis Galton in 1861.

The word *'high'* is now commonly used as a noun. An **a** may be warm or cold. A *warm a*, usually large-scale, has a high TROPOPAUSE and a cold STRATOSPHERE, occurring mainly in the subtropics; e.g. the Azores High; it may form a BLOCKING HIGH in high latitudes. A *cold a*, with a low tropopause and rel. warm stratosphere, is shallower, and occurs both on a small and local scale moving in mid-latitudes between depressions, and on a larger continental scale in winter, almost stationary over N. America and central Asia. [*f*]

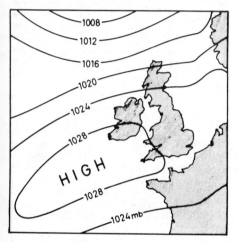

antidune A deposit of sand on a stream bed, in which the steeper slope faces upstream. 'Normal' dunes form at rel. low velocities in streams, but as velocity increases these are planed. With further increase of velocity, **as** (themselves marked by surface waves in the stream) develop and migrate upstream, through the addition of sediment on the steeper slope and the removal of sediment from the gentler lee slope. As are transitory features, continually forming, being destroyed and reforming. [*f, opposite*]

antimeridian A MERIDIAN which is 180° of longitude from a given **m**.

antimony A metal, derived from the ore *stibnite* (Sb₂S₃), used as an alloy-metal.

antipleion Syn. with *meion*, indicating a climatological station with a high negative temperature ANOMALY. If ISANOMALS are drawn, areas of high negative anomaly (conventionally tinted blue) stand out; e.g. continental interiors and E. coasts of continents in high latitudes.

antipodal bulge The tidal effects at a point **a** to the 'tidal bulge' on the side of the Earth nearest the moon (where lunar attraction is greatest); at the **a b**, 13 000 km (8000 mi.) further from the moon, lunar attraction is weakest and in effect the crust is 'pulled away', to produce an apparent **a b** in ocean waters.

antipodes, adj. **antipodal** Points at either end of a diameter of the Earth, i.e. 180° of longitude apart, one as many degrees of latitude S. of the Equator as the other is N. Antipodes Island is the nearest point **a** to Britain (actually to the Channel Islands), situated S.E. of New Zealand at 49° 42′ S., 178° 50′ E. Note the broadly **a** arrangement of the world's landmasses and oceans, with a few exceptions (Patagonia is **a** to N. China, New Zealand is **a** to Spain). As Australia and New Zealand lie on the other side of the world from Britain, they are loosely referred to as 'the A.'

anti-trades Winds in the upper atmosphere (above 1500 m, 6000 ft) above the surface TRADE WIND belt, blowing in a gen. W. direction; also called *Counter trades* (now obsolete). Formerly used to describe surface W. winds in high latitudes, so liable to some confusion. In modern usage only the first meaning is employed. They are not high altitude 'return currents', transporting to higher latitudes rising air at the INTER-TROPICAL CONVERGENCE ZONE, but represent part of the troposphere Westerlies.

anvil cloud The flattening of the top of a large convective cloud (CUMULONIMBUS) when it reaches the base of the STRATOSPHERE, spreading in the direction of high-level winds. Ice and snow crystals falling below the spreading layer produce the characteristic wedge-shaped cloud.

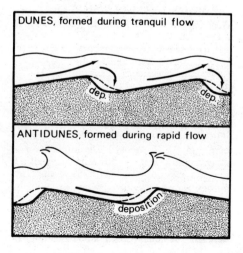

aphelion The furthest point in the orbit of a heavenly body from the sun; the Earth at **a** on 4 July is 152 million km (94·5 million mi.) away; 1·5% greater than the annual mean distance. The velocity of the Earth in its orbit is least at **a**. Ct. PERIHELION.

aphotic zone Those parts of the seas and oceans in which light is insufficient for photosynthesis by algae, below a depth of about 60–150 m (200–500 ft). Ct. PHOTIC.

apogean tide When the moon is at its furthest distance (APOGEE) from the Earth, its gravitational attraction is less; high **t**s are lower and low **t**s are higher than usual, and the tidal range is less.

apogee (i) The point in the orbit of a planet when its max. distance from the Earth; ct. PERIGEE. In **a** the moon is 398 579 km (247 667 mi.), from the Earth. The term had far greater significance when the Earth was believed to be the centre of the universe. Now used with specif. reference to the moon. (ii) The meridional altitude of the sun on the longest day of the year, when it reaches its greatest altitude (at midday).

apparent dip The DIP of a stratum in a section which is not at right-angles to the STRIKE; i.e. the angle of **a d** is between 0° and that of TRUE D.

apparent time (or local solar time) The **t** of day indicated by the apparent movement of the sun, as shown on a sun-dial; *a* noon is when the sun's centre crosses in transit the local meridian (i.e. the highest point of its apparent diurnal course). The interval between two successive diurnal transits is not the same, because the sun's orbit is an ellipse and is inclined to the Equator; hence the need for MEAN SOLAR TIME, as calculated from the EQUATION OF T. See also ANALEMMA.

Appleton Layer A **l** in the IONOSPHERE at about 300 km (190 mi.) above the Earth's surface, which reflects short radio-waves (which have penetrated the HEAVISIDE-KENNELLY LAYER) back to Earth. Now known as the *F2 layer*.

applied climatology 'The scientific analysis of a statistical collective of individual weather conditions directed toward a useful application for operational purposes'. (H. E. Landsberg and W. C. Jacobs.) The application of climatic information towards agriculture, industry, medical problems, house design, fuel economy, military operations, pest control, transportation, and all other aspects of human, cultural and economic relationship to man's physical environment.

apse line A **l** joining the points of PERIHELION and APHELION, which moves slowly around the Earth's orbit in the same direction as its orbital movement. The times of perihelion and aphelion are progressively later, about $1\frac{1}{4}$ seconds per annum.

aquatic environment Used of a *continental e* in ct. to a *terrestrial e*, in which water is a fundamental presence, though not marine. Includes: fluvial, estuarine, deltaic, paludal (swampy), lacustrine (limnol) and lagoonal es.

aquiclude In a strict sense, a rock stratum which is porous and may hold much water (e.g. clay), but the pores are filled with water held by surface tension, so sealing the rock against the downward movement and free passage of water. Some writers use the term more widely as syn. with AQUIFUGE. Neither is much employed in UK, more commonly in USA.

aquifer (alt. **aquafer**) A stratum of PERMEABLE rock, such as chalk and sandstone, which can hold water in its mass and will allow it to pass through. If underlain by an IMPERMEABLE stratum, it will act as a storage reservoir for GROUND WATER. The chief **a**s in S.E. England are the Chalk and Lower Greensand; in S.W. Lancashire the Triassic sandstones underlie more than 1·2 million km (0·5 million sq. mi.). The Dakota Sandstone on the E. flanks of the Rockies in USA allows water to move underground from the mountain slopes in an E. direction into the Great Plains. The **a** may be either *confined*, bounded on its upper surface by a layer of IMPERMEABLE rock, or *unconfined*, with no such overlying layer.
[*f* ARTESIAN BASIN]

aquifuge Little used in UK, more pop. in USA, an IMPERMEABLE rock stratum which will neither hold water in its mass nor allow it to pass through; e.g. shale, granite, quartzite. Ct. AQUICLUDE. [*f* ARTESIAN BASIN]

aragonite A crystalline form of CALCIUM CARBONATE.

arbitrary projection A somewhat vague term to indicate a **p** which is neither equidistant, conformal or equivalent. Sometimes called an *aphylactic p.*

arc A line of islands in the form of a curve; e.g. Aleutians; island as of E. Asia. Usually it lies along a line of folding in the Earth's crust (see PLATE TECTONICS), with distinct crustal weakness, and associated with earthquakes and volcanoes. Ocean deeps commonly lie close and parallel to an **a** on its outer (oceanic) side.

arc, of the meridian A geodetic measurement along a meridian, made to determine accurately the shape and size of the Earth, esp. with ref. to its oblateness. The first as measured were by French expeditions in Peru (1735–43) and Sweden (1736–7).

arch A natural opening through a rock mass, caused: (i) by collapse of a limestone cavern, leaving a portion of the 'roof'; e.g. Gordale Scar, Malham (Yorkshire); Marble Arch, the valley of the R. Cladagh, near Enniskillen (N. Ireland); (ii) on the coast, where caves are worn by marine erosion penetrating a rock projection, e.g. Durdle Door (Dorset); Needle Eye, near Wick (N. Scotland); see STACK [f]; (iii) where a river forms a very acute meander, and then breaks through the narrow 'neck', abandoning the meander and flowing through the a; e.g. Rainbow Bridge (Utah); (iv) by the weathering of the weak part of a mass of rock, finally forming a hole right through; e.g. Arches National Monument and Natural Bridge, Bryce Canyon (Utah). See pl. 64.

Archaean (i) In gen. usage, syn. with PRE-CAMBRIAN, i.e. referring to the oldest rocks. (ii) By some authorities, the rocks of the ARCHAEOZOIC era, the 2nd of 3 eras of Precambrian time. (iii) In USA and Canada, the earlier of 2 eras of Precambrian time.

Archaeozoic (i) By some authorities, the middle of the 3 eras into which Precambrian time is divided, and the associated groups of rocks. The preceding era is the Eozoic, the following one the Proterozoic. (ii) By others, all time before the beginning of the Palaeozoic era.

archipelago (It.) A group of islands scattered in near proximity about a sea. Orig. meant 'chief gulfs' (e.g. Aegean Sea), but this meaning is now obsolete. Later applied to a sea over which numerous islands are scattered, and now to the islands themselves; e.g. Sporades and Cyclades in the Aegean Sea. Some groups are called A; e.g. the Tuamotu A in S. Pacific Ocean.

Arctic Of or pertaining to N. Polar regions, strictly that part of the Earth's surface lying N. of the A CIRCLE (66° 32′ N.) Also used to describe features or conditions of climate, landscape, animals and plants characteristic of A regions. Though strictly an adjective, A is also used as a substantive (*the A*), esp. in Canada and USA. Some climatologists define an A climate as one where the mean temperature for the coldest month is below 0°C (32°F), for the warmest month below 10°C (50°F).

Arctic air-mass A very cold AIR-MASS orig. over the A Ocean, denoted by symbol A. Note possible confusion with POLAR A-M, coined orig. merely to denote temperature contrast with TROPICAL A-M; A was introduced later, after Polar had become established, and not all authorities distinguish between them.

Arctic Circle The parallel of latitude 66° 32′ N., along which about 21 June (summer solstice in the N. hemisphere) the sun does not sink below the horizon, and about 22 Dec. (winter solstice) it does not rise above it. N. of this line the number of days without sun in winter increases until at the N. Pole there are 6 months of darkness, succeeded by 6 months of daylight. Ct. ANTARCTIC C.

Arctic Front A semi-permanent frontal zone (see FRONTOGENESIS) along which (in the N. hemisphere) cold air from the A region meets moderately cool air in latitudes 50°N. to 60°N. Though shortage of observations prevents any detailed knowledge of its location, it probably extends from N. of Iceland along the N. coast of Eurasia, across the N. Pacific and N. Canada; i.e. it lies N. of the POLAR FRONT in the Atlantic and Pacific Oceans. Not a very active frontal zone, as temperature contrasts between A and POLAR AIR-MASSES are not very marked.

'Arctic smoke' A type of fog formed in high latitudes, when cold air passes over a warmer water surface and moisture, evaporated from the water surface, condenses, so that the water appears to 'smoke'. The fog is usually shallow, and is typically found in bays in the Arctic Ocean. Some regard A s as syn. with STEAM-FOG, but strictly the former develops over salt water, the latter over fresh.

arcuate delta A DELTA with a rounded arcuate, convex-outwards margin; e.g. Nile, Danube. [*f* DELTA]

area The extent of a surface, measured in square units:

$$144 \text{ sq. ins.} = 1 \text{ sq. ft}$$
$$9 \text{ sq. ft} = 1 \text{ sq. yd}$$
$$30\tfrac{1}{4} \text{ sq. yds} = 1 \text{ sq. pole}$$
$$40 \text{ sq. poles} = 1 \text{ rood}$$
$$4 \text{ roods} = 4840 \text{ sq. yds}$$
$$= 1 \text{ acre}$$
$$640 \text{ acres} = 1 \text{ sq. mi.}$$

English	Metric
1 sq. mi.	$= 2 \cdot 58999 \text{ km}^2$
0·386103 sq. mi.	$= 1 \text{ km}^2$ (100 hectares) (ha)
1 acre	$= 0 \cdot 40468 \text{ ha} = 4050 \text{ m}^2$
2·47106	$= 1 \text{ ha}$ (100 ares)
1 sq. in.	$= 645 \cdot 16 \text{ mm}^2$
1 sq. ft	$= 0 \cdot 0929 \text{ m}^2$
1 sq. yd	$= 0 \cdot 83613 \text{ m}^2$
1·19599 sq. yds	$= 1 \text{ m}^2$

To convert units per sq. mi. to units per km², multiply by 0·3861; to convert units per km² to units per sq. mi., multiply by 2·58999.

area-height diagram A d indicating relationship between area and altitude, comprising a graph with a vertical scale of heights and horizontal scale of either: (i) areas between each pair of selected contours; or (ii) percentage of the total area occupied by the area between each pair.

arenaceous A rock composed largely of cemented quartz-grains and other small particles; e.g. sandstone, gritstone, greywacke, arkose.

arenization The formation of deep sandy REGOLITHS, e.g. from the deep weathering of granitic rocks. A occurs widely in the tropics (owing to intense chemical weathering), but is also found in the mid-latitudes (where deep weathering probably operated during warmer, moist periods of the Tertiary era).

arête (Fr.) (i) A steep-sided rocky ridge, specif. the crest between two adjacent CIRQUES; e.g. Striding and Swirral Edges on Helvellyn, English Lake District. Also known, esp. in USA, as *combe-ridge*. (ii) More gen. any clean-cut ridge in high mountains; e.g. Brouillard and Peuteret As. on the S. face of Mont Blanc; Zmutt, Hörnli, Fürggen and Italian As. on the Matterhorn; A. du Diable on Mont Blanc du Tacul. In Germ., *Grat*; e.g. Viereselsgrat on the Dent Blanche in the Pennine Alps. See pl. 5, 34.
[*f* PYRAMIDAL PEAK]

aretic drainage Syn. with INTERNAL or inland d.

argillaceous A rock largely composed of clay minerals; e.g. mud, clay, mudstone, shale, marl.

arid, noun aridity Lit. dry, parched or deficient in moisture. Various definitions include: (i) less than 250 mm (10 ins.) of rainfall per annum (USA); (ii) insufficient rainfall to support vegetation in any quantity; (iii) insufficient rainfall to support agriculture without irrigation; (iv) where total EVAPORATION potential exceeds actual precipitation. Various formulae have been devised to define the boundary of an a climate, mostly empirical, involving relationships between temperature, precipitation and/or evaporation; values are plotted and isopleths are drawn. Formulae include:

(i) $I = \dfrac{P}{T+10}$; where I is the index of aridity, T is the mean annual temperature in °C, and P is the mean annual rainfall in mm (E. de Martonne).

(ii) Rain (or moisture) factor $= P/T$ (R. Lang)

(iii) Rain mainly in winter: $R = t$
Rain evenly distributed through the year: $R = t + 7$
Rain mainly in summer: $R = t + 14$
where t is the mean annual temperature in °C; if the annual rainfall (R) in cm is less than $2t$ and greater than t, the climate is regarded as a (W. Köppen and R. Geiger). (iv) The PRECIPITATION-EFFICIENCY INDEX of C. W. Thornthwaite (1931), on which an a climate has an index of less than 16.

arkose A coarse-grained sandstone or grit containing a high percentage of fragments of FELDSPAR, usually with a siliceous cement. The feldspar fragments are little altered by weathering, which suggests rapid disintegration of the parent granite or gneiss.

Armorican orogeny Applied by E. Suess (1888) to the Late Palaeozoic (Carbo-Permian) mountain-building period of W. Europe, corresponding to the VARISCAN of central Europe. Named after Armorica, that part of N.W. France usually known as Brittany. Also applied to features of nature; e.g. A times, A trend-lines. The resultant mountains are sometimes regarded as part of the ALTAIDES. The gen. usage is to regard A mountains as W. representatives of the whole HERCYNIAN system, Variscan as E. representatives.

arroyo (Sp.) A stream-bed, usually dry, in an arid area, occas. carrying a short-lived torrent after intensive rain; used esp. in Latin America and S.W. USA. Cf. WADI, NULLAH.

artesian basin A b in the earth's crust, sometimes of great extent, in which one or more AQUIFERS are enclosed above and below by impermeable strata; e.g. London B. Formerly water here saturated both the Chalk and the layer of Lower Eocene sandstones, lying in a syncline below the London Clay and above the Gault Clay; when wells were sunk water rose to the surface. But great withdrawals (at present about 1200 million litres a day) have so reduced the orig. supply that consumption now exceeds intake on the N. Downs' and Chilterns' CATCHMENT areas. The WATER-TABLE and hydrostatic pressure are falling, and water has to be pumped up. The greatest **a bs** in the world are in Australia, where Jurassic sandstone aquifers underlie 1·3 million km² (0·5 million sq. mi.), deriving intake from rain falling on the E. Highlands. Many of the 9000 wells are deep, as much as 1·6 km (1 mi.), but much water is slightly saline and used for watering stock rather than irrigation. Some experts consider that the amount withdrawn each year exceeds annual intake from rainfall; i.e. part of the water is derived from accumulation during the past, which is not being replaced. [*f*]

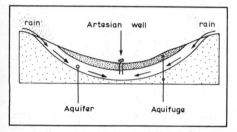

artesian well A boring put down into an AQUIFER in an **a** BASIN; if the outlet of a **w** in the centre of the basin is at a lower level than the WATER-TABLE within the aquifer around the edges of the basin, water will rise in the well under hydrostatic pressure. Named from ws of this type in Artois, N.E. France. Also loosely applied to any **w** in which water rises under pressure, though not to the surface, sometimes called *sub-artesian*; e.g. a deep **w** was sunk near Wool in Dorset through Tertiary beds into Chalk to 221 m (726 ft), and water rose to within 28 m (93 ft) of the surface. Pumping is required in a sub-**a**-**w**.

[*f* ARTESIAN BASIN]

asbestos A highly fibrous group of AMPHIBOLE minerals, the most important being *chrysotile* (fibrous serpentine), which occurs attached to walls of veins traversing metamorphic rocks.

ash Fine material ejected from the crater of a volcano during an eruption; really a misnomer, for the **a** is not a product of combustion, but consists of finely comminuted particles of lava. Specif., coarse **a** has grains 0·25–4·0 mm in diameter, fine **a** less than 0·25 mm. The **a** is so fine that it may be carried very great distances. After the eruption of Krakatoa in the Soenda Straits in 1883, **a** was carried twice round the world. When Aniakchak in Alaska blew up in 1912, 0·3 m thickness of **a** fell on Kodiak, 80 km away. In March 1963, the volcano Irazu in Costa Rica began to pour out **a** across 650 km² of country; this lay 'like a deep fall of black snow'.

ash cone A small volcanic **c** of **a**; its shape depends on the nature of the material, but usually concave due to the spreading outwards of material near the base, and less steep than a CINDER C. Many **a cs** in Iceland; e.g. a group of ninety, each 35–45 m high, at Rauholar, near Reykjavik. Monte Nuovo, W. of Naples in Italy, grew in three days to over 140 m as the result of a single eruption. In 1937 an **a c**, Vulcan, accumulated rapidly at Blanche Bay near Rabaul in the Bismarck Archipelago, growing to 180 m during the first day and 226 m in three days. A few major **cs** are entirely of **a**; e.g. Volcano de Fuego in Guatemala, 3350 m (11 000 ft) in height.

ash flow See NUÉE ARDENTE.

aspect The direction in which a slope faces, partic. with ref. to possible amounts of sunshine; see ADRET, UBAC. **A** has marked effects on siting of settlements, vegetation and cultivation. Ct. S.-facing sides (with villages, farms, orchards) and N.-facing sides (with coniferous forests) of Alpine valleys which trend W. to E. *A angle*: in the N. Hemisphere the angle between the line of max. downslope and S., measured positive E.-ward, negative W.-ward.

asphalt A naturally occurring viscous hydrocarbon; e.g. Trinidad Pitch Lake, Athabasca Tar-sands.

assimilation In petrology, material incorporated into an igneous rock through melting and solution; a marginal feature of an INTRUSION; see AUREOLE.

association A common word given a specialized ecological meaning, as by botanists and geographers. One of the hierarchy of plant-groupings, an assemblage of plants living in close inter-dependence, with similar growth and habitat requirements, and with one or more DOMINANT species used to denote it; e.g. oak forest. Rather more complex definitions of an **a** are to be found in botanical literature, including its relationship to a CLIMAX COMMUNITY.

Asteroids A belt of about 1500 small heavenly bodies (*planetoids*), each less than c. 800 km (500 mi.) in diameter (the largest is *Ceres*), revolving in the Solar System between Mars and Jupiter. Possibly result from the break-up of a single former planet.

asthenosphere A zone of weaker and hotter rocks in the upper MANTLE underlying the LITHOSPHERE, a plastic substratum beneath the more rigid crust, occurring from 100–200 km below the surface. Thought to coincide largely with the *Gutenberg Channel* (not to be confused with the GUTENBERG DISCONTINUITY), and is characterized by a marked reduction in shear-wave velocity as a result of its low strength and rigidity. Possibly the scene of horizontal convection flow, associated with movement of the lithospheric PLATES.

astrolabe An instrument formerly used: (i) in fixing latitude by observing the apparent transit of the sun across the meridian at midday; (ii) for measuring the altitude of any heavenly body.

asymmetrical fold An ANTICLINE [fB] or SYNCLINE wherein one limb dips more steeply than the other; e.g. Hampshire Basin syncline. In extreme form it becomes an OVERFOLD.

asymmetrical valley A valley with the slopes on one side steeper than those on the other. The degree of asymmetry may be expressed by an *A-index* (the ratio between the max. angles of the two valley sides). A vs may be determined by geological structure (where streams experience UNICLINAL SHIFT, and so undercut one valley side), or may be related to climatic influences. Many **a** vs in Britain and Europe are regarded as RELICT LANDFORMS, developed under past periglacial conditions. Differential exposure to the sun's rays and prevailing winds probably resulted in differential frost action and solifluction. The 'active slopes' (on which these periglacial processes were most effective) were either steepened or reduced in angle more rapidly than 'inactive slopes', and asymmetry resulted. In the Chilterns the steeper slopes face towards the S.W., but in the Marlborough Downs the asymmetry is often 'reversed'. Thus in different areas and at different times, differential exposure to sun and wind has varied in its effects. [*f, opposite*]

at-a-station analysis, -es The study of changes in the HYDRAULIC GEOMETRY of a particular stream channel cross-section as discharge varies through time. Changes in discharge are accommodated mainly by adjustments of channel width, channel depth and stream velocity. The basic relationships are expressed in the formulae

$$w = aQ^b$$
$$d = cQ^f$$
$$v = kQ^m$$

in which w is width, d is depth, v is velocity, Q is discharge, and a, b, c, f, k and m are constants. Values for b, f, and m are very important. For the Seneca Creek at Dawsonville, Maryland, calculated values are $f = 0.52$, $m = 0.30$, $b = 0.18$ (note that these add up to unity, indicating the interdependence of width, depth and velocity). These figures imply that increases in discharge are accommodated mainly by increases in stream depth and only to a small extent by increases in width. This can occur only if the channel is narrow with steep banks rather than wide and dish-shaped (when accommodation mainly by increased width would have occurred).

Atlantic Polar Front The POLAR FRONT between POLAR MARITIME and TROPICAL MARITIME AIR-MASSES in the N. Atlantic Ocean.

Atlantic stage, of climate A climatic phase, *c.* 5500 to 2500 B.C., when the climate of W. Europe (and probably elsewhere) was milder, cloudier and damper, with temperatures about 2° to 3°C above those of the present; hence 'climatic optimum'. In Gt. Britain there was extensive growth of mixed oak forest, widespread peat formation, and a gen. rise of sea-level of about 3 m as a result of the return of water from melting ice-sheets. The formation of the Strait of Dover occurred *c.* 5000 B.C. (the accepted start of the Upper Holocene). Sometimes called the *Megathermal Period*.

Atlantic type, of coastline A type of coastline which develops where the trend of mountain ridges and the 'grain' of the relief gen. are at right-angles or oblique to the coastline; e.g. S.W. Ireland, N.W. France, N.W. Spain, Morocco. Ct. Pacific (CONCORDANT) type.

atmosphere A thin layer of odourless, colourless, tasteless gases surrounding the Earth. It

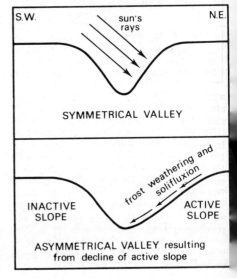

consists of nitrogen (78·0%), oxygen (20·95%), argon (0·93%), carbon dioxide (0·03%), and small proportions of neon, krypton, helium, methane, xenon, hydrogen, etc., with amounts of water-vapour varying from 0 to 4·0%. The **a** is held to the earth by gravitational attraction. A very small amount of OZONE (O_3), the allotropic form of oxygen, is present, with its max. density at 25–30 km. Half the **a** lies within 5·6 km of the earth's surface, 75% within 11 km, 90% within 16 km, and 97% within 27 km. The 'weather-making' layers are limited to a few km, esp. as more than half the water-vapour is below 2300 m. The **a** is divided into a number of 'layers' by temperature (TROPOSPHERE, STRATOSPHERE, MESOSPHERE, THERMOSPHERE); by physico-chemical properties (HOMOSPHERE, HETEROSPHERE): and by electromagnetic properties (IONOSPHERE, MAGNETOSPHERE). Total weight of the **a** 5.9×10^{15} tons. In the atmosphere properties typical of a gas really cease to exist at 600 km.

atmospheric circulation The gen. c and movement of air, in the form of pressure-cells and wind-systems, near sea-level and in the upper **a**. See PLANETARY WINDS. [*f*]

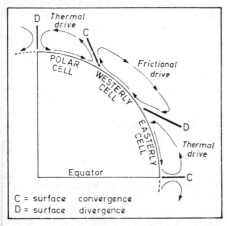

C = surface convergence
D = surface divergence

atmospheric pressure The p exerted by the atmosphere as a result of its weight on the surface of the Earth, expressed in millibars (1000 mb. = 1 bar = 1 million dynes per cm²). In SI terms, 1 bar = 10^5 newtons per m², or 10^5 pascals (since 1971). The avge **p** over the Earth's surface at sea-level is 1013·25 mb., equivalent to the weight of a column of 760 mm (29·92 ins) of mercury at 0°C or to a weight of air of 1033·3 g cm^{-1} (14·66 lb. per sq. in.) = 1.01078×10^5 pascals. Lowest mean sea-level **a p** recorded in UK = 925 mb., in world 877 mb.; highest in UK = 1055 mb., in world 1079 mb. See also BAROMETER, BAROGRAPH.

atoll A CORAL REEF of circular, elliptical or horse-shoe shape, enclosing a lagoon. Found most commonly in W. and central Pacific; e.g. Gilbert and Ellice Is., Cook Is., Marshall Is. These reefs pose problems both of explaining their shape and also that coral extends downward to depths at which, under present conditions, it is unable to grow. Charles Darwin's subsidence theory postulated that rings of coral grew around islands subjected to slow gradual submergence. These have sunk gradually beneath the sea (or the sea has risen), while the reef has maintained itself through upward growth. However, some reefs occur where insufficient submergence is evident. R. A. Daly suggested that coral formation could be related to a rise in sea-level associated with the return of melt-water to the oceans after the Quaternary glaciation. Sir J. Murray put forward the idea that coral reefs could grow up on banks of their own debris, where growth would be most vigorous on the outside, thus causing a coral ring to grow outwards. Recent research includes the determination of ocean temperatures during the Pleistocene from organic remains deposited on the ocean floor; exact measurement of coral growth rate and determination of a reef calcium budget; dating of core material drilled from the ocean floor; seismic profiling of reefs. Darwin's concepts involving submergence seem to be justified; Daly was correct in principle in stressing the importance of the Pleistocene but wrong in his concept of mechanisms involved, notably that there was inadequate time for intertidal marine planation during glacial periods; much erosional modification is now attributed to subaerial karstic processes. The gen. conclusion is that 'modern' reefs (formed within the last 5000–6000 years) are but a thin veneer of 10 m or so of coral, resting on great thicknesses of old reefs (BIOHERMS), some even of Eocene age.

[*f*]

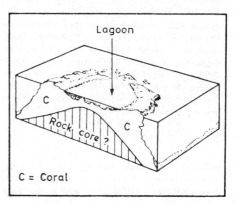

C = Coral

attached-dune A sand-DUNE, either TAIL-DUNE or HEAD-DUNE, accumulating as a sand-drift around a rock or other obstacle in the path of wind-blown sand in desert lands. [*f* DUNE]

attitude App. to the disposition of a rock-stratum, which may be horizontal or tilted. Includes the relationship of a BEDDING-PLANE to the horizontal, given in terms of DIP and STRIKE.

attrition The constant wearing-down into progressively more rounded, finer particles of a load of rock-material through frictional grinding during transport by running water, the wind and waves. A does not include wearing away of river-beds and banks and of sea-cliffs; this is ABRASION, leading to CORRASION.

aufeis (Germ.) A surface accumulation of ice, formed in periglacial environments (i) where streams overtop banks and inundate surrounding areas during periods of freezing; (ii) where ground water emerges at springs; (iii) where freezing builds up hydrostatic pressure in an unfrozen AQUIFER and causes the water to burst through overlying layers to the surface. A melts during the spring thaw, except where covered by insulating debris that may cause its incorporation with PERMAFROST.

aureole A zone of alteration (METAMORPHISM) of rocks in contact with an igneous intrusion, so that complex mineralogical changes were caused by high temperatures. Can be traced around the margins of a BATHOLITH, e.g. Dartmoor, and is usually associated with metal ores. [*f* BATHOLITH]

aurora The luminous effect of electromagnetic phenomena in the IONOSPHERE, visible in high latitudes as red, green and white arcs, 'draperies', streamers, rays and sheets in the night sky, best developed at a height of about 100 km. Probably the result of magnetic storms and electrical discharges from the sun during periods of sun-spot activity, mainly of electron particles, funnelled into the earth's magnetic field and accelerated to the high energies necessary, causing ionization of gases. Called *A Borealis* (or 'N. Lights') in the N. hemisphere, *A Australis* in the S. Occas. seen in England, but more common in N. Scotland, Orkneys and Shetlands, and presents a magnificent spectacle in N. Scandinavia and N. Canada. See pl. 2.

australite A small 'button' of silica found in the interior of Australia, believed to be debris formed *c*. 5000 years ago when some meteorites came into collision with the moon. A form of TEKTITE.

Plate 2. AURORA BOREALIS, forming an 'auroral curtain' with a clearly defined 'ray' structure which continuously changes its shape, as if moved by the wind. (*Popperphoto*)

autochthon, -ous (i) Strata that have been shifted little by earth-movement from their orig. sites, though often strongly faulted or folded; e.g. Helvetides on N. side of the Alpine ranges in Switzerland and Austria. Crystalline massifs, fragments of the old Hercynian continent which have not moved horizontally, though involved within the folding (e.g. Pelvoux, Mont Blanc, Aar-Gotthard), are also **a**. (ii) The formation of coal from *in situ* vegetation. (iii) Any non-transported feature, formed essentially *in situ*; e.g. CONGLOMERATE formed by the break-up and cementation of an underlying rock; an EVAPORITE. Ct. ALLOCTHON, -OUS.

autoconsequence A theory that rivers flowing over their own alluvial deposits may be superimposed on to the underlying solid rocks by reason of a fall in sea-level. Possibly the Belgian rivers (Scheldt, Scarpe, Lys, etc.) are of this nature.

autogenic App. to a plant succession (PRISERE) resulting from modification and alteration of the environment by the vegetation itself; e.g. on a sand-dune, in a bog. Ct. ALLOGENIC.

autometamorphism METAMORPHIC changes (METASOMATISM) during the cooling of an igneous rock as a result of PNEUMATOLYSIS or HYDROTHERMAL processes.

autumn Astronomically, the transitional period between the autumnal EQUINOX and the winter SOLSTICE; in the N. hemisphere this extends from *c*. 21 Sept. to 22 Dec., in the S. hemisphere from *c*. 21 March to 21 June. Popularly, a in UK comprises the months of Sept. and Oct.; in America ('the Fall') it comprises Sept., Oct. and Nov.

available relief The vertical distance between the height of a land-surface undergoing dissection and the height of valley-floors of adjacent dissecting streams. This may be used in quantitative statements about landscape geometry and in morphometric calculations.

avalanche The fall by gravity of a mass of material down a mountainside; it can be used for rock (better LANDSLIDE or ROCK-FALL), but is usually restricted to masses of snow and ice. On steep slopes as may occur both in winter, when newly fallen non-coherent snow slides off older snow surface, and in spring when wet, partially thawed, masses of enormous size fall down valley slopes, and where a crust compacted by the wind (WIND-SLAB) breaks away suddenly, often through the passage of a skier. Other falls come from the margins of a HANGING GLACIER. An a can be very destructive; during spring 1951 a sudden thaw after heavy winter snowfall caused widespread as in Switzerland and Austria, with much loss of life and destruction of property. Probable a tracks are known, and partic. dangerous areas are avoided. Villages, roads, and railways are carefully sited, natural a breaks such as rock spurs and thick pine-woods are utilized, and steel a-sheds are erected at critical points to protect roads (e.g. Great St Bernard road-tunnel) and approaches to railways. The section of the Trans-Canada Highway through Rogers Pass in British Columbia is protected by many a-sheds. See pl. 3.

avalanche wind The rush of wind in front of an AVALANCHE produced by the falling mass; this may be very destructive, and can cut a swathe through a forest or a village.

aven (Fr.) (i) A deep shaft-like hole in limestone country (syn. PONOR), orig. derived from *patois* word in the Causses region of Central Massif of France; e.g. the A. Armand in the Causse Méjan, 198 m deep, leading to a vast series of caves. (ii) Used more specif. in England to describe enlarged vertical joints in the roof of a cave which narrow upward (the result of carbonation-solution), and sometimes, though not always, open out into depressions on the surface.

Plate 3. A massive AVALANCHE which discharged huge quantities of rocky debris onto the surface of the Sherman Glacier, Alaska, in August 1964. (*Austin Post, Geological Survey*)

axial plane A p bisecting the upper (or lower) angle between the limbs of a fold on each side of the crest-line of an anticline, or the trough-line of a syncline. Inclinations of this plane from the vertical characterize different types of fold. [*f* AXIS, OF FOLD]

axis, of Earth The diameter between the N. and S. Poles, about which the earth rotates in 24 hours, tilted at an angle of about $66\frac{1}{2}°$ to the plane of the ECLIPTIC, i.e. at an angle of about $23\frac{1}{2}°$ from a line perpendicular to that plane. The length of the axis is 12 714 km (7900 mi.). [*f*]

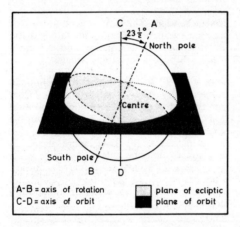

axis, of fold The central line of a FOLD (crest or trough) from which the strata dip away (as in an ANTICLINE: the crest) or rise (as in a SYNCLINE: the trough) in opposing directions. See AXIAL PLANE. [*f*]

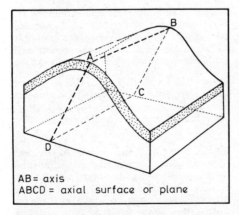

ayala A local wind in the Central Massif of France, strong, sometimes violent, and very warm; similar to the MARIN.

azimuth (i) In astronomy, the angle intercepted between the meridian plane of an observer and the vertical plane passing through a heavenly body, measured the nearest way from the meridian plane. Ct. ALTITUDE. (ii) In surveying, a bearing read clockwise from true N.; i.e. the **a** of a point due E. is 90°, of W. is 270°. By the US Coast and Geodetic Survey, **a** is measured clockwise from zero = S.; i.e. an **a** of 90° = W., 270° = E. It may be (a) *magnetic a* (measured from MAGNETIC N.); (b) *true a* (measured from TRUE N.); or (c) the *grid a* (measured from GRID N.). In a geodetic survey the **a** is used to fix the orientation of the system of TRIANGULATION. [*f*]

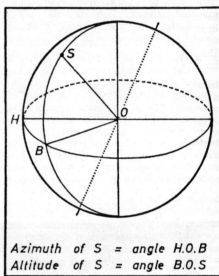

Azimuth of S = angle H.O.B
Altitude of S = angle B.O.S

azimuthal projection Syn. with ZENITHAL P.

Azoic An obsolescent term from Gk., 'without life', the meaning of which has undergone change. It has been used: (i) for all PRECAMBRIAN stratified rocks; (ii) for the earliest part of Precambrian times, irrespective of the nature of the rocks. It is a misnomer in that remains of life have been found in later Precambrian rocks. See CRYPTOZOIC.

azonal soil A 'young' s, not sufficiently long under the effects of soil-forming processes, agencies and influences to develop mature characteristics. A ss include: (i) *mountain* or SCREE ss, on unstable slopes of rock fragments; (ii) *alluvial s*s, derived from ALLUVIUM deposited by running water, which provides the material for a good agricultural s (with well-mixed, fine texture, good mineral content, and replenishment in times of flood); (iii) *marine s*s derived from mud-banks, sandbanks and dunes, often saline; (iv) *glacial* ss, developed on DRIFT; (v) *wind-blown* ss, including sand and esp. LOESS (LIMON) as parent-materials; (vi) *volcanic* ss, derived from recent volcanic parent-materials (LAVA, ASH, CINDER, PUMICE), which because of high mineral content form good soils when weathered.

Azores 'high' The subtropical anticyclone in the atmosphere, situated gen. over the E. side of N. Atlantic Ocean. More extensive and continuous during N. hemisphere summer, and may extend N.E. to affect W. Europe and UK.

backing A change of direction of the wind in an anti-clockwise direction; e.g. from S.W. to S., to S.E. In USA thus used for the N. Hemisphere, but in the reverse (i.e. clockwise) for the S. Cf. VEERING.

backset bed Used mainly in USA to denote a deposit of sand on the gentler slope to windward of a dune, often trapped by tufts of sparse vegetation.

backshore The area extending from the average high-water line to the cliffline, covered with BEACH deposits. Largely corresponding with BERM. [*f* COASTLINE]

back-slope The gentler slope of a CUESTA, compared with the steeper ESCARPMENT. Most authorities use DIP-SLOPE in connection with a cuesta, but as the slope is seldom the same as the dip of the rock, this can be misleading, and **b-s** is strictly more correct. [*f* CUESTA]

back-swamp Marshy areas in a low-lying FLOOD-PLAIN behind the LEVEE bordering a river; used specif. in USA in respect of the Mississippi flood-plain.

back-wall The steep rock wall at the back of a CIRQUE, rising to an ARÊTE enclosing it. In cirques in the British mountains, the **b-w** may be from 60 to 180 m high. The huge Walcott Cirque in Antarctica has a **b-w** 3000 m high.

backwash A mass of water running back down the slope under the influence of gravity, after a wave breaks on a beach; receding movement of a wave. See also LONGSHORE DRIFT [*f*].

backwater An area of virtually stagnant water, still joined to a stream, but rel. unaffected by its current. It develops readily where a stream splits its channel (see BRAIDED RIVER) or forms such an acute MEANDER that the current cuts through its neck and flows on a more direct course, leaving its old channel as a **b**; see OXBOW.

backwearing A term describing the recession of slopes with little or no loss of steepness (cf. PARALLEL RETREAT OF SLOPE). It may be a dominant process in tropical environments e.g. where sheets of LATERITE are incised by streams into plateaus and mesas, whose hard laterite cap-rocks are undermined by wash and creep affecting the underlying weak MOTTLED and PALLID ZONES. [*f*]

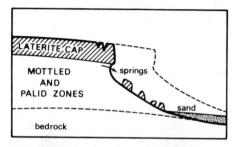

badlands Orig. applied to an area of semi-arid climate in S. Dakota (called '*les terres mauvaises à traverser*' by the French, because it was difficult to cross). Now a National Monument of 5200 km^2, with a spectacular landscape of gullies and 'saw-tooth' ridges, cut in vari-coloured Oligocene shales and limestones. Applicable to any such erosional landscape, where a maze of ravines and valleys dissects plateau-surfaces. **B** formation may result from severe SOIL EROSION, esp. in an area of sparse and intermittent rainfall; the first stage is caused by the destruction of a protective turf mat, sometimes by overgrazing.

baguio A tropical storm in the Philippine Is., experienced esp. from July to November. During July 1911, a **b** resulted in a rainfall of 1170 mm (46 ins.), one of the wettest days (24 hours) ever recorded on the earth's surface.

Bai-u The season of heaviest rain in parts of China and Japan, in late spring and summer. **B** rains are sometimes called '*plum rains*', since they occur during the plum-ripening season.

bajada (Sp.), anglicized to *bahada* A continuous, gently sloping fringe of angular scree, gravel and coarse sand around the margins of an inland basin, or along the base of a mountain range, in a semi-arid region. Formed by the coalescence of a series of adjacent ALLUVIAL CONES, each deposited by a torrential, usually intermittent, stream where it leaves a constricted valley; e.g.

Arizona, Nevada, Mexico and the Atacama Desert of central Chile. [*f* PLAYA]

ball lightning A rare form of L, in which a luminous ball, sometimes moving, sometimes apparently stationary, is seen. The cause or substance of this effect is not known.

balloon-sonde, ballon-sonde (Fr.) A balloon, hydrogen-filled, liberated into the atmosphere, carrying self-recording meteorological instruments. It is thus possible to obtain data at great altitudes. The **b** rises, expanding, and ultimately bursts. The remains of the **b** and carefully protected instruments return to earth by parachute. From recorded measurements, a profile of changes of temperature and pressure with altitude is obtained. Ct. RADIOSONDE.

bamboo A genus of giant grasses (*Bambusoidea*), growing widely within tropical latitudes (esp. S.E. Asia), and in favoured areas even well outside the tropics (as in S.W. Ireland and Cornwall). It grows profusely and extremely rapidly, ranging from a few cm to 36 m high, and up to 20 cm in diameter. Some **b**s are climbers, over 60 m in length. It is so plentiful that it is one of the commonest and cheapest media for housing, scaffolding and construction work in S.E. Asia.

band A long ridge-like hill or SPUR, esp. in the English Lake District; e.g. the Band, Bowfell.

bank (i) A vague term for the margin of a river. In the cases of a mountain torrent or a river in a CANYON, which usually occupies the entire valley-floor, or of some wide FLOOD-PLAINS, **b**s are hardly distinguishable. It is best marked where a channel with distinct sides has been cut in the floor of a valley by a stream with powers of vertical erosion. At high-water the space between the **b**s is full (BANK-FULL stage) and water then overflows on to the FLOOD-PLAIN; at low-water beds of gravel and alluvium appear. (ii) An area of mud, sand or shells (not rock) in the sea, covered with fairly shallow water, though deep enough not to be a hazard to navigation. The waters above are usually good fishing-grounds; e.g. Grand Bs. of Newfoundland, Dogger B. (iii) In the N. of England and Scotland a hillside; e.g. 'Ye bs and braes o' bonnie Doon'.

bank-caving On the outside of a curve in a river's course, the force of the current impinges against the bank. Undercutting and undermining result, causing slumping of masses of clay, sand and gravel into the river, where it is swept away as part of its load.

banket (Afrikaans) A pebble-conglomerate containing gold, found in the Witwatersrand of the Republic of S. Africa and in Ghana. Individual beds are known as REEFS.

bankfull The state of river flow when the space from bank to bank is completely filled with water; beyond this, FLOOD-STAGE is reached. At **b** stage the velocity of the river is thought to be constant along its whole length.

banner-cloud A **c** which streams out on the lee-side of a peak in a clear sky, the result of condensation within a rising air-current produced as a forced up-draught. Descent and re-warming of air occurs down-wind, and the 'banner' ends as water droplets evaporate. While this flow of air continues, the position of the **c** remains fixed. E.g. 'Table Cloth' over Table Mountain (Republic of S. Africa). A **b-c** may be partic. pronounced in the case of a prominent isolated peak; e.g. Matterhorn. See LEE-WAVE.

bar (i) A unit of atmospheric pressure. At 45°N. latitude, at sea-level, and at a temperature of 0°C, 1 **b** = 29·5306 ins. (750·062 mm) of mercury. It is divided into 1000 millibars = 1 bar = 10^3 millibars = 10^6 dynes per cm^2 = 10^5 newtons per m^2 (10^5 Nm^{-2}) = 10^5Pa (Pascal). (ii) A bank of mud, sand and shingle, deposited in water offshore parallel to the coast (*offshore b*), across the mouth of a river, across the exit to a harbour (*harbour b*), across a bay between 2 headlands (BAY-B), and between an island and the mainland (TOMBOLO). A **b** may be either an upstanding and exposed feature, or a submerged ridge covered at least at high tide. More specif. BARRIER BEACH or BARRIER ISLAND should be used for the former to avoid confusion. (iii) Deposits of alluvium, sand and gravel found in stream-channels (see RIFFLE). [*f*]

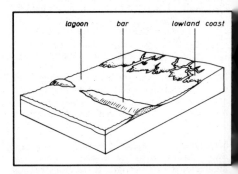

barbed drainage pattern A p of d in which tributaries form obtuse angles with the main stream; they appear to point up-river. This

results from CAPTURE, which has caused the main stream to flow in a direction opposite to its orig. [*f*]

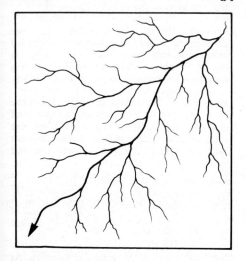

barchan See BARKHAN.

barkhan (barchane, barchan) (Turki) A crescentic sand-dune, lying transversely to the wind direction, with 'horns' trailed downwind. It forms when the wind direction remains constant. It varies in height to over 30 m with a gentle slope on the windward side, steeper on the 'slip-face' or lee side where eddy motion assists in maintaining a concave profile. If the supply of sand is continuous, the dune may advance as a result of endless movement of sand up the windward slope and over the crest, sometimes constituting a threat to oases. A dune may be found as an individual hill, but usually they occur in groups, a chaotic ever-changing 'sand sea'. The main problem is how it originates, since it appears to occur readily on fairly level open surfaces; possibly patches of pebbles, a shrub, even a dead animal, or a low bump in the ground may cause the accumulation of a heap of sand, from which features of a characteristic **b** can develop. [*f*]

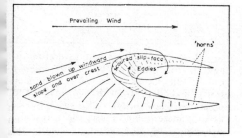

bar-and-swale In an area of MEANDER growth and downstream sweep of the meander belt, with formation of OX-BOWS, embankments of sand and alluvium (**bs**) alt. with troughs (**ss**) are formed.

baroclinic The state of a fluid or gas in which surfaces of constant pressure intersect surfaces of constant temperature; a measure of *baroclinicity* (sometimes called *baroclinity*) may be derived for their relationship. This implies in meteorology large-scale atmospheric INSTABILITY and the presence of a frontal zone. [*f*]

barograph A self-recording ANEROID BAROMETER; differences of atmospheric pressure are transmitted to a needle which traces an inked line on a moving drum, thus giving a continuous record (*barogram*). In another form, the position of a mercury meniscus is recorded photographically.

barometer An instrument invented by E. Torricelli (1643) for measuring ATMOSPHERIC PRESSURE, by balancing the weight of a column of mercury against that of a column of the atmosphere. Modern **b**s (e.g. KEW or FORTIN patterns) have various adjustments for applying corrections, and vernier-scales for exact readings. Corrections are applied for: (i) *latitude*, standardized to 45°N.; (ii) *temperature*, standardized to 12°C; (iii) *altitude*, an avge decrease of 33·9 mb. (1 in. of mercury) for each 274 m (900 ft) for the lowest 1000 m, above which the rate of decrease becomes progressively less; see ALTIMETER; (iv) *instrumental individuality*, as compared with a standard instrument. Another type is the ANEROID B. See also BAROGRAPH and BAR.

barometric gradient The amount of change in atmospheric pressure between 2 points, as indicated by the distance apart on a level surface of the ISOBARS on a synoptic chart, i.e. a 'steep' g involves considerable difference in pressure, with closely spaced isobars, while a 'gentle' g has only slight difference, with widely spaced isobars. The direction of max. g is at right-angles to the isobars. A steep g is associated with strong winds. In a tropical storm (e.g. HURRICANE or TYPHOON), there may be a difference of 40 mb. over only 30 km. Also known as *pressure g*.
[*f* GEOSTROPHIC FLOW]

barometric tendency The *nature* (increasing or decreasing) and the *amount* of change of atmospheric pressure during a specified period, gen. 3 hours.

barotropic In a fluid or gas, where surfaces of constant pressure are parallel to surfaces of constant temperature; this **b** state is equivalent to zero BAROCLINICITY. The noun is *barotropy*. [*f* BAROCLINIC]

barranca (Sp.) A steep-sided, water-eroded ravine, specif. in Mexico and S.W. USA, partic. the deep gashes on the flanks of a volcanic cone.

barren lands, barrens Applied to much of N. Canada, characterized by bare rock, swamp, TUNDRA, permanently frozen subsoil (PERMAFROST) and a short intense summer. The term has fallen into disuse, partly because the areas are not as barren as once believed (esp. in respect of mineral wealth).

barrier-beach, -island A sandy bar above high tide, parallel to the coastline, and separated from it by a lagoon; it may be sufficiently above high tide, with dunes lying on it, to be a *b-island*. From New Jersey along the E. coast of USA to Florida, and along the shores of the Gulf of Mexico as far as the Mexican border, extends a string of **b-b**s, where a flat coastal plain is bordered by shallow water. Some probably started as offshore bars of sand far out from the mainland, but have been pushed gradually shoreward by wave action forming lagoons (SOUNDS); e.g. Albemarle and Pamlico Sounds behind Cape Hatteras (N. Carolina). Some b-islands represent portions of former SPITS. Others are fragments of a breached dune-line; e.g. Frisian Is. off the mainland coast of Netherlands, Germany and Denmark. Many b-islands are seaside resorts; e.g. Miami Beach, Palm Beach (Florida); Lido at Venice.

barrier-lake A l produced by the formation of a natural dam across a valley. Types include: (i) Is ponded behind a land-slide or avalanche; e.g. Earthquake L., Montana, formed after a rock-fall in August 1959; (ii) Is enclosed by deposition of alluvium in a DELTA; e.g. Étang de Vaccarès at the mouth of the R. Rhône; (iii) Is formed by the deposition of a TERMINAL MORAINE across a glaciated valley; e.g. Is of the English L. District (in part), Finger Ls. of New York State; (iv) Is formed by ice-dams; e.g. Vatnsdalur in Iceland; (v) Is ponded by vegetation dams; e.g. in the heathlands of W. Europe; (vi) Is ponded behind a dam of calcium carbonate; e.g. L. Plitvicka, on the Crna R. in Yugoslavia; (vii) Is ponded behind a lava dam; e.g. Lac d'Aydat in Central Massif, France.

barrier reef A CORAL **r** parallel to the coast, but separated by a lagoon of considerable depth and width. The Great **B. R** of Australia extends from Torres Strait at 9°S. to 22°S., over 1900 km, varying in width from 30 to 50 km off Cairns to 160 km in S. Individual rs, crowned with islands, are spread over an irregular platform covered with shallow water, possibly the result of depression of a denuded surface below sea-level to a depth which allowed reef-building organisms to flourish. Smaller examples are found around Pacific islands (e.g. Aitutaki in the Cook group). For theories of development of coral reefs, see ATOLL. [*f*]

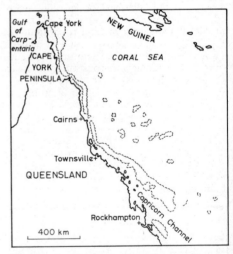

barysphere Used loosely with several meanings. (i) The whole of the Earth's interior beneath the CRUST (*lithosphere*), i.e. incl. MANTLE and CORE; syn. with *centrosphere*, and sometimes (incorrectly) called *bathysphere*. (ii) In a more limited sense, the core only, 3476 km (2160 mi.) radius, of nickel-iron under great pressure. (iii) By some writers. used syn. with mantle and in contrast to both surface crust and inner core. The consensus of opinion favours (i).

barytes A whitish or colourless mineral, an ore of barium ($BaSO_4$), found in veins; used in the manufacture of paint, glass, pharmaceuticals etc.

basal complex The ancient rocks in the 'shield' areas of the Earth.

basal conglomerate A CONGLOMERATE found in the lowest part of strata resting unconformably on older rocks. It may appear above any erosional break, typically as a beach deposit spread over a former land-surface.

basal platform Used by D. L. Linton to describe the unweathered granite 'surface' above which Dartmoor tors rise. The **p** was considered to represent the former surface of the water-table, at a time when the overlying granite was being decomposed into GROWAN with CORESTONES by vadose water (see TOR). It is now believed that weathering can operate effectively *below* the water-table, and that the weathered material lying above the **b p** will itself support PERCHED WATER-TABLES. **B p** is now regarded as syn. with BASAL SURFACE OF WEATHERING. See pl. 58.

basal sapping Intense 'erosion' of the lower parts of slopes, owing to concentrated chemical weathering, wash processes, spring and seepage action etc. **B s** is esp. active in tropical environments (i) where laterite-capped tablelands are undermined by spring sapping, which removes weathered clay and causes collapse and CAMBERING of the laterite margins; (ii) around INSELBERGS, where an abundance of moisture (derived partly from run-off over the bare rock of the inselberg) promotes intense rock decomposition. **B s** is considered to produce sharp changes of slope profile ('predmont angles') or MARGINAL DEPRESSIONS. Alt. the disintegration of rocks along the back-wall of a CIRQUE, the result of melt-water making its way down the BERGSCHRUND, and enabling alternate freezing (by night) and thawing (by day) to shatter the rocks.

basal-slip A sliding movement of basal ice in a glacier over the underlying rock-floor, caused by the solid character of the ice, its weight higher up, and the gradient of its floor.

basal surface of weathering The lower limit of active weathering, marking an abrupt change from weathered to sound rock. In the tropics the **b s of w** often lies at depths of 30 m or more (over much of tropical Africa—e.g. in southern Uganda—the surface lies at a depth of over 50 m), indicating a protracted period of decomposition of granitic and gneissic rocks of the pre-Cambrian 'Basal Complex'. However, in some areas climatic changes and/or regional uplifts have led to stream rejuvenation, and the stripping and exposure of parts of the **b s of w**. Some pediments and low inselbergs represent little modified parts of the **b s of w**. See pl. 58.

basalt A fine-grained, usually dark-coloured, igneous rock, belonging to the basic group (over half plagioclase feldspars and the rest ferro-magnesian silicates), extruded from a fissure in the Earth's crust, hence a VOLCANIC ROCK. It flows readily, forming extensive sheets, and may solidify in hexagonal columns; e.g. the Giant's Causeway, N. Ireland, and Fingal's Cave, Staffa. It makes up over 90% of the igneous rocks. Most large SHIELD VOLCANOES (Mauna Loa, Hawaii) and extrusive plateaus (Columbia-Snake in USA, N.W. Deccan in India, Antrim in N. Ireland, Iceland) consist of **b**.

basal wreck An alt. name for a CALDERA.

base-flow recession curve A graph showing the discharge of a stream fed only by springs under conditions of nil water percolation (i.e. the springs are continually draining the regional AQUIFER). As the water-table falls, and hydraulic gradients decline, spring discharge is progressively reduced, but at a decelerating rate. Thus a very long and severe drought (such as that experienced in Britain during 1976) is required before springs dry up completely. In streams fed partly by springs and partly by direct run-off the properties of discharge derived from springs is referred to as the '*base-flow component*'.

base-level The lowest level to which a river can erode its bed (or to which a land-surface can be reduced by running water), normally assumed to be sea-level (*ultimate b-l*), though there may be temporary b-ls, a lake or resistant stratum of rock (*local b-l*). While b-l is thought of as a horizontal surface, some writers, recognizing that streams require a gradient in order to flow, regard it as an inclined or curved surface, the theoretical limit to stream erosion. Such a surface is practically impossible to define.   [*f* REJUVENATION]

base-line (i) A carefully measured line, the initial stage in a triangulation survey. From its ends a system of triangles is projected by angular measurement as an accurate framework for detailed topographical surveying.

The UK Ordnance Survey has measured numerous bases (e.g. on Hounslow Heath and Romney Marsh); the current triangulation depends on: (*a*) Ridgeway base, 11 260·1931 m in length, from White Horse Hill in Berkshire to Liddington Castle in Wiltshire; and (*b*) Lossiemouth base (7170·7234 m) near the shores of the Moray Firth. A **b-l** is measured with every refinement, using an INVAR tape, and various corrections, as for temperature, are applied, so that an accuracy of 1 in 300 000 is attained. (ii) See LAND SURVEY SYSTEM (USA).

basement complex A mass of ancient igneous and metamorphic rocks, usually (though not always) Precambrian in age and complex in structure, which underlies stratified sedimentary rocks.

basic grassland A type of **g** occurring mainly on chalk and limestone terrain, the dominant grasses sheep's fescue (*Festuca ovina*) and red fescue (*F. rubra*).

basic lava A mass of molten igneous material on the surface of the earth, rich in iron, magnesium and other metallic elements, and rel. poor in silica. It has a low melting-point and flows readily for a considerable distance before solidifying. Usually its flow from the vent of a VOLCANO is unchecked and free from explosive activity. The **b l** may form a large flattish cone (SHIELD VOLCANO) or, if it flows from fissures, an extensive plateau. **B l** congeals usually as basalt, sometimes known as *flood basalt*. It is estimated that **b l** comprises 90% of all **l**s.

basic rock Rather loosely used for a quartz-free igneous **r**, with more than 45% basic oxides (aluminium, iron, calcium, sodium, magnesium, potassium); e.g. basalt, dolerite, gabbro. (Formerly simply the antithesis of now obsolete *acid*, though **b** is conveniently current.) Sometimes used loosely to indicate an igneous **r** composed of dark-coloured minerals. The term is apt to be misleading, even inaccurate, and is obsolescent. In USA it has been suggested that **b r** should be replaced by *subsilicic* (in respect of its silica content), *mafic* (in terms of its base content), or *melanocratic* (in respect of the dark-coloured minerals). **B** rocks grade into INTERMEDIATE as the sodium content of the feldspar increases, and into ULTRABASIC by decrease in the feldspar itself.

basin (i) A large-scale depression on the Earth's surface, occupied by an ocean, hence *ocean b*. (ii) The area drained by a single river-system, hence *river b*. (iii) A shallow structural downfold in the Earth's crust; e.g. Hampshire, London, Paris **B**s. (iv) A shallow downfold containing *Coal Measures*, hence *coal b*; e.g. Saar B. (v) An area enclosed by higher land, with or without an outlet to the sea; e.g. Great **B** of USA, Tarim **B** (central Asia). (vi) Syn. with CIRQUE in parts of W. America. (vii) A depression resulting from the settlement of the surface through removal in solution of underground deposits of salt or gypsum, naturally or by man's action.

basin-and-range A series of tilted FAULT-BLOCKS, forming asymmetrical ridges alternating with **b**s. Applied as a proper name ('Basin and Range country') to part of S.W. USA between the Sierra Nevada and Wasatch Mountains, where the **r**s have steep E. faces and more gentle W. ones.

basisol A tropical soil formed by the weathering of basalt under conditions of heavy rainfall and high temperature; its blackness is due to titanium. Characterized by low organic content and commonly by a zone of calcium carbonate concretions. In the dry season it forms a black soil which crumbles into dust, or bakes hard if not ploughed; in the wet season it may be plastic or sticky. Occurs widely in N.W. Deccan, India (used for cotton growing, hence *Black Cotton Soil*). Similar soils are in Kenya (also there known as Black Cotton Soil), Morocco, N. Argentina, and in parts of the west Indies.

'basket-of-eggs' relief See DRUMLIN.

batholith (bathylith) A large mass of igneous rock, usually granite, formed by the deep-seated intrusion of MAGMA. The dome-like upper surface may be exposed by prolonged denudation to form uplands; e.g. Dartmoor, Bodmin Moor in S.W. England; Wicklow and Mourne Mountains in Ireland; and British Columbia Coast Range **b**, 2400 km long by 160 km wide (the last now known to be a series of separate **b**s, intruded at different times). Many **b**s are elongated along the axes of fold-ranges; e.g. Aar-Gotthard, Mont Blanc and Pelvoux massifs along the axes of the Alps; Montagnes Noires and M. d'Arrée along the Armorican axes of Brittany. The mechanism by which **b**s are emplaced is not wholly understood. One theory is of EMPLACEMENT, by which a vast block of country-rock foundered and sank into the underlying magma; another is that the country-rock has been absorbed by, or transformed into, magma by GRANITIZATION. A **b** may be either sharply defined or surrounded by a mineralized AUREOLE. [*f*]

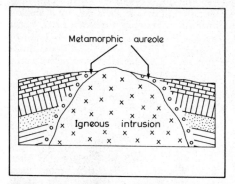

bathyal App. loosely to that part of the ocean between 180 m and 1800 m (100 and 1000 fthms), broadly the CONTINENTAL SLOPE, partic. to benthonic life (BENTHOS) and types of deposit found. Hence **b** deposits of Blue, Red, Green and Coral muds.
[*f* ABYSSAL ZONE]

bathymetric Relating to the measurement of the depth of a water-body; a **b** map of an ocean, sea or lake, with contours (*isobaths*) related to the datum of the mean water-surface depicting the relief of its floor; e.g. *Carte générale bathymétrique des Océans*, published by the International Bureau of Hydrographics in Monaco. Hence *bathymetry*.

bathy-orographical Applied to a map depicting both altitude of land and depths of sea, usually by layer-colouring, with water conventionally in shades of blue, land in green, yellow, brown.

bathythermograph A self-registering thermometer used for measuring great depths in the ocean.

battue ice Large thick masses of sheet-**i** during winter on the St Lawrence, a major impediment to navigation.

bauxite An amorphous mass of clay with a high content of aluminium hydroxide ($Al_2O_3 . 2H_2O$), usually with ferruginous and other matter; the chief ore of ALUMINIUM, the third most important element in the earth's crust (after oxygen and silicon), comprising about 15% by volume, 8% by weight. Occurs widely in feldspars and other silicates. Under tropical weathering, feldspars break down into clay-minerals, leaving **b**, varying in colour from off-white to red and brown.

Baventian An early pre-PASTONIAN cold phase in the Pleistocene history of Britain, represented by marine deposits at Easton Bavents, Suffolk.

bay (i) A wide, open, curving indentation of the sea or a lake into the land; e.g. Hudson **B**, **B** of Fundy, **B** of Biscay, Georgian **B** (L. Huron). Some authorities use **b** in a hierarchy of coastal openings, larger than a COVE, smaller than a GULF; cf. BIGHT, EMBAYMENT. A **b** is defined precisely in delimiting TERRITORIAL WATERS. A straight line is drawn between natural promontories on either side of the indentation, and the water area so delimited is considered to be a **b** if its area is as large as, or larger than, a semicircle whose diameter is equal to the delimiting line. (ii) A shallow elongated depression in the coastal plain of E. USA (Carolina **B**s). (iii) A translation of Germ. *Bucht*, an extension of lowland into an upland along a river valley; e.g. *Kölnische Bucht*, Cologne 'Bay', Leipzig 'Bay'. [*f*]

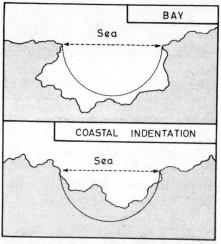

bay-bar (in USA, **baymouth-bar**) A bank of sand, mud or shingle, extending wholly across a bay, linking two headlands, and straightening off the coast. Where only one end of such a feature is attached to the coast, it is a SPIT. A **b** linking the mainland with an island is a TOMBOLO. **B-b**s are produced: (i) by the convergence of two spits from opp. sides of a **b**; (ii) by a single spit extending in a constant direction across the **b**; (iii) as a development from an offshore BAR driven onshore. E.g. Loe B. (Cornwall); b-b across Corpus Christi Bay (Texas); NEHRUNGEN on the E. Baltic coast.

Baydjarakh A conical hillock in Arctic regions, resulting from the melting of the ice-wedges of ice-wedge polygons. A common feature of THERMOKARST.

Plate 4. A small sandy BAY-HEAD BEACH at Coldbackie, Sutherland, Scotland. Note the effects of WAVE REFRACTION on the wave pattern. (*Aerofilms*)

bay-head beach, pocket beach A crescent of sand and shingle at the head of a small cove between 2 headlands; e.g. Pembroke and Gower coasts of S. Wales; Cornwall; Maine and Oregon coasts of USA. See pl. 4.

bay-head delta A d formed by a heavily laden stream depositing much of its load at the head of a bay.

bay-ice Sea ice, formed in a bay during a series of winters, thick enough to impede navigation.

bayou A marshy creek or sluggish, swampy backwater (OX-BOW), specif. along the lower Mississippi and its delta, and along the Gulf coast of USA.

beach Tee gently sloping accumulation of material (shingle, sand) along the coast between low-water spring-tide line and the highest point reached by storm-waves. The most typical b has a gently concave profile; the landward side is backed by sand-dunes, succeeded by shingle, then an area of sand, and rocks covered with seaweed at or about the low-tide mark. Some bs may comprise an extensive area of sand uncovered at low tide; e.g. Morecambe Bay (Lancashire); Long B. (Washington State); Florida bs; coast of New Jersey (Atlantic City, Ocean City). See also RAISED B.

beach cusp A projection of sand and shingle, alt. with rounded depressions, along a b. The projections are of open cone-shape, apexes pointing seawards, giving a kind of scalloped pattern. They are the result of a powerful SWASH and BACKWASH, esp. when waves break at or near right-angles to the coast. It is difficult to explain their initiation, though once started the eddying swash scours the depression, moving coarser material on either side on to the c, and progressively emphasizing it.

beach profile A p transverse (at right-angles) to the coastline, usually concave upwards, since it is steeper above high water and gentler below. The b p is in equilibrium when the amount of material accumulated is more or less balanced by the amount removed, representing the net result of the avge set of conditions obtaining along that stretch of coast. This is easily disturbed by strong onshore winds, esp. with exceptionally high tides, producing destructive storm-waves which comb down and remove material. The b p of equilibrium is part of the overall shore p.

beaded drainage A series of pools and small ponds linked by short streams, developed by surface melting in PERMAFROST regions. The pools may result from the thawing of ground ice at the intersection of ice-wedge polygons, and the streams may develop along troughs formed by the thawing of ice-wedges.

beaded esker A winding ridge of gravel and sand of fluvioglacial origin, with larger humps of material at intervals, the results of periods of pause in the retreat of the glacier nurturing the stream which deposited the e material. E.g. behind Flamborough Head, Yorkshire. [f ESKER]

beaded valley A **v** in which a narrow section alternates with a wider, more open one; e.g. upper Yosemite V., California.

bearing The horizontal angular difference between the meridian and a point viewed by an observer, measured in degrees clockwise from the meridian (*true b*); e.g. a true **b** of 270° is due W., of 180° is due S. If measured from magnetic N., it is a *compass* or *magnetic b*; in this context syn. with AZIMUTH.

Beaufort Notation A code of letters devised by Admiral Sir Francis Beaufort, FRS, in the early 19th century, to indicate the state of the weather: *b* blue sky; *c* cloudy; *o* overcast; *g* gloom; *u* ugly threatening sky; *q* squalls; *kq* line squall; *r* rain; *p* passing showers; *d* drizzle; *s* snow; *rs* sleet; *h* hail; *t* thunder; *l* lightning; *tl* thunderstorm; *f* fog; *fe* wet fog; *z* haze; *m* mist; *v* unusual distant visibility; *e* wet air, but no rain falling; *y* dry air; *w* dew; *x* hoar-frost. Refinements include capital letters to indicate intensity (R = heavy rain); double capitals for continuity (RR = continued heavy rain); suffix $_o$ for slightness (r_0 = slight rain); *i* for intermittence (*ir* = intermittent rain). On weather maps, this code is now replaced by an international system of meteor signals; more than 50 in all; listed in the *International Cloud Atlas* (WMO) e.g. ● rain, ❜ drizzle, ▲ hail, ✴ snow, ⦤ lightning, ⚲ dew.

Beaufort Wind Scale A scale of wind force, devised by Beaufort in 1805, modified in 1926; it is related to the descriptions of wind effects and estimated velocity at 10 m above the ground.

Scale No.	Wind	Force (m.p.h.)	($m\ s^{-1}$)	Observed effects
0	calm	0	0	Smoke rises vertically
1	light air	1–3	0·3–1·5	Wind direction shown by smoke drift, but not by vane
2	light breeze	4–7	1·6–3·3	Wind felt on face; leaves rustle, vane moves
3	gentle breeze	8–12	3·4–5·4	Leaves and small twigs in motion; a flag is extended
4	moderate breeze	13–18	5·5–7·9	Raises dust; small branches move
5	fresh breeze	19–24	8·0–10·7	Small trees sway; small crests on waves on lakes
6	strong breeze	25–31	10·8–13·8	Large branches in motion; wind whistles in telephone wires
7	moderate gale	32–38	13·9–17·1	Whole trees in motion
8	fresh gale	39–46	17·2–20·7	Breaks twigs off trees
9	strong gale	47–54	20·8–24·4	Slight structural damage to houses
10	whole gale	55–63	24·5–28·4	Trees uprooted. Considerable structural damage
11	storm	64–75	28·5–32·6	Widespread damage
12	hurricane	above 75	over 32·7	Devastation

beck A small stream in N. England, with rapid flow, winding course, and irregular bed.

bed (i) A layer or stratum of rock, usually a feature of the deposition of sedimentary rocks, divided from layers above and below by well-defined BEDDING-PLANES. Used in 2 ways: (*a*) in a gen. context, as simply syn. with layer; (*b*) in a hierarchical sense, within an order of increasing magnitude of LAMINATION, laycr, **b**, STRATUM, FORMATION. By contrast, some writers consider that a **b** is larger than a stratum, consisting of several strata. (ii) Applied to a layer of PYROCLASTS (e.g. an ash-**b**), and to individuals in a sequence of lava flows; e.g. in N. Skye, where 25 distinct **b**s of basalt have an overall thickness of 300 m; in the N.W. Deccan where one boring revealed 29 **b**s. (iii) The floor of an area covered with water (river, lake, sea), usually permanently, though it may dry out temporarily.

bedding The pattern of layers of sedimentary rocks, sometimes of distinctive banding, colour, thickness, composition and texture, separated above and below by well-defined planes. When the beds are very clearly defined, the rock is *well-bedded*. Also sometimes applied to the tendency of certain metamorphic and igneous rocks to split in well-defined planes. See also CURRENT B. [*f*].

bedding-plane The surface or plane separating distinctive layers of sedimentary rock, indicating the end of one phase of deposition and the beginning of another. Sometimes referred to as part of the STRATIFICATION.
[*f* STRIKE]

bed-load, of river The solid material carried along the bed of a river by SALTATION and gravity; i.e. a *traction load*, the particles pushed and rolled along the stream bed. It contrasts with the load carried in suspension; the b-l consists of coarser, larger material. See SIXTH-POWER LAW.

bed-rock (i) Solid unweathered rock underlying the superficial layer of top-soil, sub-soil and unconsolidated material (the weathered or residual mantle). (ii) Sometimes used specif. of solid rock underlying PLACER deposits of gold or tin.

Beestonian An early pre-CROMERIAN cold phase in the Pleistocene history of Britain, represented by silts and fluviatile deposits at Beeston, Norfolk.

beheading, of river See CAPTURE, RIVER.

Behrmann's projection Modified version of the CYLINDRICAL EQUAL AREA P, published 1910, in which the principal scale is preserved along 2 standard parallels, 30°N. and 30°S.

belt (i) A zone of distinctive character, sometimes elongated, sometimes concentric; e.g. of rock outcrops (B. Series of metamorphic rocks in N.W. USA); of climate (TRADE-WIND b); of vegetation (CONIFEROUS FOREST b). (ii) A narrow strip of water; e.g. Great B., Little B., in the Baltic Sea. (iii) An elongated area of PACK-ICE.

belted outcrop plain An EROSION SURFACE upon which a series of rocks outcrop in parallel strips. CUESTAS and lowlands may develop from such areas; e.g. coastal plain of S.E. USA, lying E. of the FALL-LINE.

belt of no erosion The upper part of a hill-slope, extending from the crest for some distance down-slope, on which the surface accumulation of water from precipitation is insufficient to cause 'erosion'. The concept was devised by the American engineer, R. E. Horton (1945), who demonstrated that beyond a certain distance from the crest surface wash would become sufficiently powerful to pick up soil particles, but that close to the hill-top surface run-off would lack the volume and velocity to transport such particles. The distance from the crest to the point at which running water attains 'erosion velocity' is referred to as the 'critical distance' (x_c). Horton showed that x_c is not constant, but varies according to lithology, vegetation, climate etc. In an area of bare, impermeable clay, under a semi-arid climate where rainfall is rare but intense, x_c will be very short. Conversely, under a temperate humid climate where rainfall is more protracted but less intense, a grassed chalkland slope will experience no run-off; thus x_c extends to the slope foot.

ben A Scottish peak, derived from the Gaelic *beinn* or *beann*. Used as a prefix to many Scottish mountains; e.g. B. Nevis, Lomond, More, Cruachan.

bench A terrace, step or ledge, usually narrow, backed and fronted by a perceptible steepening of slope, produced by denudational processes (e.g. WAVE-CUT B), structural movements (STEP-FAULT), or artificial excavation (e.g. quarrying or open-cast mining).

bench-mark (BM) A surveyor's defined and located point of reference. The UK Ordnance Survey uses a broad arrow with a bar across its apex cut in solid rock, walls and buildings, with a specif. height related to ORDNANCE DATUM at Newlyn, Cornwall. The US Geological Survey, US Coast and Geodetic Survey and US Corps of Engineers use an engraved brass disc, fixed in solid rock or concrete, related to the Sea-Level Datum of 1929 (there is a $250 fine for disturbing a **BM**).

Benioff zone Linear earthquake-active z indicated by island-arc systems, esp. in the Pacific Ocean, named after H. Benioff.

benthos Plant and animal organisms living on the sea-floor, in ct. to free-floating life (PLANKTON) and swimming life (NEKTON). B may be divided into: (i) *littoral* (between high-water spring tide mark and about 200 m depth); and (ii) *deep-sea* groups. They may live in a fixed position, or be capable of crawling or burrowing. See also NERITIC and PELAGIC.

berg Germ. and Afrikaans word as an element in proper names: (i) a single hill or mountain in Germany (ct. *Gebirge*, a range);

e.g. Feldberg, Hesselberg; (ii) a range in S. Africa; e.g. Drakensberg, Groot Zwartberg. The element **b** is used in combinations for physical features; e.g. BERGSCHRUND, BERG WIND, ICEBERG.

bergschrund (Germ.) (Fr. *rimaye*) A crack around the head of a FIRN-filled CIRQUE, separating the steep ice-slope on the BACK-WALL above from the main snow-field. It represents the point where the moving ice-mass is drawing away from the enclosing walls of the cirque. Ct. RAND-KLUFT. [*f* CIRQUE]

berg wind Lit. 'mountain wind', specif. a warm, dry, sometimes gusty, w blowing mainly in winter down from the plateau of S. Africa towards the coast, thus warming ADIABATICALLY.

berm (i) A narrow terrace, shelf or ledge of shingle thrown up on the beach by storm-waves. (ii) A remnant of a partially eroded valley-terrace, indicating an interruption of the CYCLE OF EROSION of a river, leaving portions of the earlier valley-floor above present river-level.

bevel Used to describe a surface that has been planed-off, such as the crest of a hill-top or a CUESTA.

bevelled cliff A sea-cliff in which the lower part comprises a steep or vertical free-face, and the upper part a gentler slope, often rectilinear in profile and covered by rock fragments and soil. In parts of Devon and Cornwall, the upper bevel may include separate slope segments, indicating a complex origin. **B cs** develop in two ways: (i) where the lower cliff is developed in a massive rock formation (e.g. the Portland Stone of the Isle of Purbeck) which is directly attacked by waves, and the upper cliff is formed by a weaker stratum (e.g. the Purbeck Beds) which has been weathered subaerially to a gentle slope; (ii) where, owing to a fall of sea-level during the Pleistocene, a sea-cliff has been 'abandoned' and reduced in angle by frost weathering and solifluction, only to be 'revived' in its lower part by renewed wave attack during a subsequent rise of sea-level. [*f, opposite*]

B-horizon The layer in the soil beneath the A-HORIZON, where much material (colloids, bases and mineral particles) washed down is deposited or precipitated; sometimes referred to as the zone of ILLUVIATION, or *illuvial horizon*, it may be marked by the presence of a HARDPAN. Usually sub-divided in detailed soil-studies into B_1, B_2, B_3, of which the 1st and 3rd are transitional to horizons above and below.

bifurcation ratio The proportion between the number of streams of one order of magnitude to the number of larger streams of the next order of magnitude which they create through their confluence. In hydrological work streams are divided into orders, the highest being the main river or 'trunk stream'; see STREAM ORDER. If in a particular drainage basin 150 rivulets of order 1 unite to form 50 streams of order 2, the **b r** is 3·0. The term is somewhat of a misnomer, since **b** would imply division, not uniting.

bight (i) A wide gentle curve or indentation of the coast, commonly between two headlands; e.g. B. of Benin, B. of Biafra, Great Australian B. (ii) Used specif. of a curved re-entry in an ice-edge in Polar regions.

bill A long narrow promontory, ending in a prominent 'beak' or spur; e.g. Portland B. Extended to include other coastal projections; e.g. Selsey B.

billabong (Austr. aboriginal) A stagnant backwater in a stream that flows only temporarily.

binodal tidal unit An AMPHIDROMIC tidal system with two nodes, rather than the usual one.

biochore (i) A region or unit-area with a distinctive plant or animal life. (ii) The part of the earth's surface which can support life

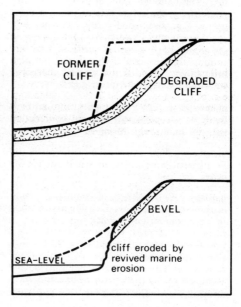

(W. Köppen). 3 major **bs** are recognized by plant geographers: (i) forest; (ii) grassland; (iii) scrub and desert.

bioclimatology The study of climate in relation to organic life, incl. human beings, other animals and plants. This may involve esp. questions of human habitability: housing, clothing and other health requirements depending on climatic conditions.

biodegradation The consumption (or 'scavenging') of waste matter (e.g. hydrocarbons) by organisms, such as seaborne microbes which consume oil-spillage at sea.

biogenic sediment Sedimentary, organically derived deposits, created by the life processes of animals and plants; e.g. chalk, coral limestone.

biogeography The study of geographical aspects of the distribution of plant (PHYTOGEOGRAPHY) and animal (ZOOGEOGRAPHY) life, esp. in terms of reasoned distributions.

bioherm (i) An old CORAL REEF (ct. BIOSTROME). (ii) The sedentary organisms (corals, gasteropods, foraminifera) which have contributed to its formation. Sometimes a distinction is drawn between the mound-like form of a **b** and the more horizontally bedded BIOSTROME.

biomass The amount of chemical energy contained in a group of growing plants at any given time, expressed in gm of dry matter per m^2. This should not be confused with *productivity*, the actual rate at which a plant grows. E.g. a redwood forest has a large **b**, small productivity, while PHYTOPLANKTON in the ocean have a low **b** (because they are continually consumed by predators) but high productivity. *Yield* is the amount of energy stored in the desired fruit or grain.

biome A complex of plants and animals living as a sociological unit. In conjunction with the habitat this forms an ECOSYSTEM.

biosphere The surface zone of the earth and its adjacent atmosphere, in which organic life exists.

biostasy A condition of biological equilibrium (ct. RHEXISTASY, denoting a disruption of this equilibrium), which may obtain in tropical rain-forests. It is suggested that such rain-forests promote rapid rock decomposition, but hold the products of that decomposition *in situ*. The supply of carbon dioxide and chelating agents to the soil by forest growth and decay is a major factor in rock weathering; and many of the weathered products are recycled by the deep-rooted trees, rather than being removed from the 'system' altogether. 'Under such conditions most energy is used in the development of the plant community, and not in erosion and removal of debris by streams' (I. Douglas).

biostrome A coral reef currently in course of formation, with a more or less horizontal structure. Ct. BIOHERM.

biotic Of or app. to life. **B** factors reflect the influence of living organisms, esp. in development of soil and vegetation (bacteria, earthworms, ants, termites, moles, field-mice; in USA also prairie-dogs, ground-squirrels and gophers), in ct. to climate, EDAPHIC and physiographic factors.

biotic complex 'The interacting complex of soils, plants and animals which, in response to climatic and other environmental conditions, forms a varied covering over much of the earth' (S. R. Eyre).

biotite A common rock-forming mineral in both IGNEOUS and METAMORPHIC rocks, consisting of ferromagnesian silicates. It ranges in colour from brown to dark green, with a glassy or pearly lustre, and is transparent and translucent. One of the MICA group of minerals.

bird's foot (or **birdfoot**) **delta** A **d** which extends into the sea with DISTRIBUTARIES bordered by sediment; e.g. Mississippi **d**. [*f*]

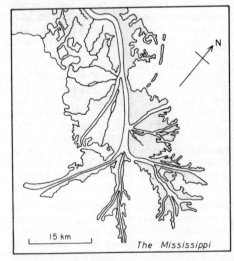

The Mississippi

'**biscuit-board**' **relief** The result of CIRQUE erosion cutting into several sides of a mountain massif; e.g. Snowdon. See pl. 5.

Plate 5. BISCUIT BOARD RELIEF on Deeply glaciated uplands near Blosseville Hyst, Greenland. Note the pattern of CIRQUES, ARETES and incipient HORNS, and the large OUTLET GLACIER to the left. (*Geodetic Institut, Copenhagen*)

bise, bize (Fr.) A dry, rather cold N. or N.E. wind blowing across parts of Switzerland, N. Italy and S. France, similar in nature to MISTRAL and TRAMONTANA. Occurs most frequently in spring, when it may bring bright sunny weather, and also common in winter when it may bring snow (*b noire*). It can cause harmful frosts in spring.

bitumen A group of solid or liquid hydrocarbons soluble in carbon disulphide. Applied gen. to pitch, tar, asphalt; e.g. the substance obtained from the Pitch Lake, Trinidad.

bituminous Applied to a free-burning coal containing 12–35% volatile matter; includes house-coal and gas-coal. The name was applied in error long ago under the impression that these coals contained BITUMEN.

blackband A layer of bedded carbonaceous ironstone or *siderite* ($FeCO_3$) commonly occurring in the Coal Measures, containing about 30% of iron.

black box approach A type of 'systems approach', in which the whole system is treated as a unit, without any consideration of its internal structure, and attention is directed solely to the character of the outputs which result from identified inputs (R. J. Chorley and B. A. Kennedy). Mass balance studies of glaciers, in which inputs of snow and ice are calculated, and outputs of water determined by ablation and discharge measurements, are an example of the b b a, since there is no consideration of the structure of the glacier, its rate of movement etc. (ct. GREY BOX and WHITE BOX APPROACHES).

black bulb thermometer A mercury t with its b blackened, mounted horizontally in a glass tube from which air has been removed, and exposed to the sun's rays, thus giving 'sun temperatures'. This instrument is now rarely used.

black cotton soil See REGUR.

black-earth See CHERNOZEM.

black ice A pop. name for GLAZED FROST on a road.

blanket-bog A BOG formed under conditions of high rainfall, swathing the whole land-surface where it is rel. horizontal; e.g. Moor of Rannoch in Inverness and Perthshire.

blind valley A v in limestone, dry or occupied by a stream, enclosed at the lower end by a rock-wall, at the base of which the stream disappears underground. This may be the result of: (i) the collapse of the roof of an underground stream-course; (ii) the grading of a surface stream to a progressively falling BASE-LEVEL, forming a horizontal cave-passage through which the stream goes underground. E.g. Cladagh R., Marble Arch Cave, near Enniskillen (N. Ireland); Bonheur, a tributary of the Dourbie (thence the Tarn) in the Grands Causses in the Central Massif of France.

blizzard A very strong, bitterly cold, wind accompanied by masses of dry powdery snow or ice-crystals, with poor visibility (WHITE-OUT), under polar or high-altitude conditions. Derived from N.W. gales experienced in USA in winter, but now used widely, esp. in Antarctica.

block-diagram A drawing in either 1-point or 2-point (true) perspective, giving a 3-dimensional impression of a landform; in effect a sketch of a relief model. Geological sections can be appended to the sides of the diagram.

block disintegration The break-up of well-bedded, jointed rocks by mechanical means, notably frost action. When water freezes in rock interstices, its volume increases by about 10%, exerting great pressure and shattering the rocks along lines of weakness, producing rectangular blocks; e.g. Cambrian quartzites of the Canadian Rockies. See pl. 6, 23.

block-faulting The division of a section of the earth's crust into individual blocks by FAULTING, some of which may be raised,

Plate 6. BLOCK DISINTEGRATION in large boulders of granitoid gneiss, released by the weathering of JOINT-PLANES, at Lukenya Hill, near Nairobi, Kenya. (*R. J. Small*)

others depressed, others tilted. See FAULT-BLOCK, TILT-BLOCK. [*f*]

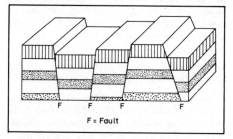

block-field, -spread See FELSENMEER.

'blocking high' An area of atmospheric high pressure (an ANTICYCLONE), which remains relatively stationary by comparison with approaching depressions, and thus blocks their passage across its particular location.

block lava Usually syn. with AA, but sometimes restricted to truly angular, non-clinkerous masses of congealed silica-rich lava.

block mountain A FAULT-BLOCK (i.e. a block bordered and outlined by faults), standing up prominently because of either: (i) its elevation by earth-movements; or (ii) the sinking of the surrounding area. E.g. Harz Mtns, Vosges, Black Forest. See HORST, TILT-BLOCK.

block-slumping MASS-MOVEMENT down steep faces of escarpments and sea-cliffs, esp. where a stratum of clay underlies more massive but well-jointed rocks such as chalk and limestone. The action of water (esp. SPRINGS) undermines the upper strata, and lubricates the clay. This b-s involves shearing of rocks, tearing away of a soggy mass of material, usually with a distinct rotational movement on a curved plane (ROTATIONAL SLIP), leaving a scar on the hillside, while the material slumps downwards; e.g. Cotswold scarp between Gloucester and Cheltenham.
[*f* ROTATIONAL SLIP]

blood rain Rain-drops containing fine particles of reddish dust, transported by the wind from deserts, and washed down from the atmosphere during precipitation; e.g. in S. Italy, Sicily, Malta, as far as the Canary Is.; plants are often covered with a reddish powder. It was even experienced in England on 1 July 1968.

blow-hole A near-vertical cleft linking a sea-cave with the surface inland of a cliff-edge; spray is thrown out of the hole by the compressional force of a wave surging into the cave below. Formed by erosion concentrated along a well-marked JOINT or FAULT extending vertically from the surface into the roof of the cave. A b-h is formed in well-jointed hard rocks; e.g. Old Red Sandstone of the Orkneys, Isle of Soay (off Skye), and coast of Caithness (Mermaid's Kirk); and in Carboniferous Limestone (Huntsman's Leap, near St. Gowan's Head, S. Dyfed). Sometimes called *gloup* or *gloap* in Scotland. [*f* STACK]

blow-out A hollow in a sandy terrain, esp. among dunes in deserts, heathlands and along coasts, or in an arid plain, formed by wind-eddying (DEFLATION), partic. where protective vegetation has been destroyed. In a rock-desert, a b-o may form from a break in a resistant surface-layer (e.g. by faulting or localized weathering), which the wind enlarges by eddying into a deflation-hollow. The floor of an arid intermont basin (e.g. in Arizona, S. California and New Mexico) is partic. subject, esp. after a temporary salt-lake has dried out. A b-o can occur in an area of peat; e.g. on Millstone Grit plateaus of the Pennines where wind-erosion of peat is occurring.

blow-well, blowing well An artificial w or natural spring, esp. in E. Yorkshire and E. Lincolnshire, which, working on the ARTESIAN principle, creates fountains or flows of water above ground. The chalk AQUIFER is sometimes covered by a clay AQUICLUDE, which when penetrated leads to a water-flow of this kind.

blue-band A distinctive layer of blue ice in a GLACIER, free from air-bubbles, probably due to the freezing of melt-water in crevasses.

Blue John A blue ornamental variety of FLUORITE (CaF_2), found in Derbyshire.

Blue Mud A widely occurring m on the CONTINENTAL SLOPE, of terrigenous origin, earthy and plastic, deriving its colour from the presence of iron sulphide. Individual particles are fine (less than 0.03 mm in diameter).

blue sky The apparent colour of the cloudless s during daytime, the result of scattering of sunlight by molecules of air. The short waves at the blue-violet end of the solar spectrum are scattered more readily by finer molecules at high altitudes, where the s appears deep blue. The shade of blue can be measured on a scale devised by F. Linke, with 14 shades ranging from white to ultramarine.

bluff A steep prominent slope. Used specif. of a valley-slope cut by lateral erosion of a river on the outside of a MEANDER (*river-cliff*). As a river extends its lateral erosion, it creates a wide FLOOD-PLAIN, bounded by low bs on either side, cut back wherever a meander impinges. In a large river (e.g. lower Mississippi), the bs may lie back more from the channel. A continuous series of bs is the *bluffline*.

Bodden (Germ.) An irregularly shaped inlet along the S. Baltic coast, produced by a rise of sea-level over a former uneven lowland surface. The island of Rügen (off the coast of E. Germany) consists of a few former irregular islands, now linked by sand-spits and enclosing the Wieker B., Jasmünder B., Kubitzer B., Greifswalder B. [*f*]

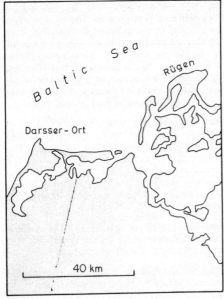

bog (i) Waterlogged, spongy ground, with a surface layer of decaying vegetation, esp. sphagnum and cotton-grass, ultimately producing highly acid PEAT. (ii) The vegetation complex thus associated. E.g. Dartmoor, gritstone plateaus of the Pennines, W. Scotland (Moor of Rannoch), Offaly (B. of Allen), Connemara and Mayo (Ireland). Widespread bs occur in high latitudes in Canada (MUSKEG), Scandinavia, Finland and USSR, where an uneven surface of impermeable rocks with a partial glacial drift cover causes waterlogging. See also BLANKET-BOG, RAISED BOG.

bogaz plur. **bogazi** (Serbo-Croat) An elongated trench-like chasm in limestone country, notably in the KARST of Yugoslavia. It has developed along a clearly defined joint through CARBONATION-SOLUTION.

bog burst When a b has developed into a bed higher than the surrounding country, a period of exceptional rainfall may cause oversaturation and mass instability, so that a stream of black organic matter flows away for some distance; e.g. in Finland, Canadian Shield.

boghead coal A close-grained black or dark brown type of c, containing algal remains.

bog-ore A layer of hydrated iron oxide (usually *limonite*–$2Fe_2O_3.3H_2O$), found in peat bogs or shallow lakes, probably precipitated by the agency of bacterial organisms.

bolson A basin of inland drainage among the high plateaus of S.W. USA in Arizona, New Mexico and S. California, commonly containing a salt-lake, with sheets of rock-salt or gypsum, and rimmed by mountains. Should a permanent or temporary stream flow across the basin, it is a *semi-b*.

bomb, volcanic A mass of lava thrown into the air during a volcanic eruption, solidifying in globular masses of rock before reaching the ground. Included in the class of PYROCLASTS.

bone bed A stratum containing fragments of fossil bs, teeth and scales of vertebrates, esp. fishes. The concentration of this accumulation may be the result of some rapid catastrophe; e.g. a submarine earthquake which killed all life simultaneously. The b is rich in calcium phosphate. E.g. Ludlow B.B. (Silurian). When cemented it forms *bone breccia*.

Bonne, Bonne's Projection An EQUAL AREA P, of modified CONICAL type. The chosen central meridian (*c.m.*) is divided truly (i.e. 10° lat. = $\frac{2\pi R}{36}$, where R is the radius of the globe to scale), and is crossed by a standard parallel (*s.p.*). Each *c.m.* and *s.p.* is selected as centrally as possible rel. to the area to be mapped; e.g. for N. America, *c.m.* 90°W., *s.p.* 40°N. The *s.p.* is drawn using the formula $r = R \cot \theta$ (where θ is *s.p.*). All other parallels are concentric to the *s.p.*, whose function is merely to control their curvature. Each parallel is divided truly; e.g. 10° of meridian interval $= \frac{2\pi R . \cos lat}{36}$. The other meridians are curves drawn through the division points on each parallel. Each quadrangle formed by intersecting parallels and meridians has its base and height true to scale, and therefore the p is equal area, but scale and shape distortion increase rapidly to W. and E. Commonly used for compactly shaped countries in midlatitudes (e.g. official maps of France on a scale of 1/80 000), and for such continents as

N. America and Eurasia. The limiting case of B, using the Equator as the *s.p.*, is the SANSON-FLAMSTEED (SINUSOIDAL). [*f*]

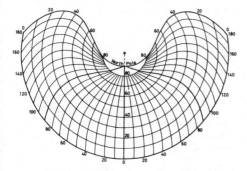

bora (It.) A bitterly cold, squally wind, blowing violently down the mountains from N. to N.E. into the N. Adriatic Sea, from a high-pressure area over central Europe towards the rear of a depression over the Mediterranean, strengthened by gravity flow (cf. MISTRAL). Most common in winter, but known in a weaker form in summer as the *borino*. Wind velocities are at times very high, when it is sometimes called the *boraccia*; 137 km h^{-1} has been recorded at Trieste, with gusts up to 209 km h^{-1}. Its strength is partic. marked when the trend of bays, islands and ridges coincides with wind direction, thus funneling it. It can be a great danger to navigation and to transport on coastal roads. The **b** is sometimes assoc. with cloud and precipitation when the cold air undercuts the moist warm air of the depression to the S. ('the rainy **b**'), but usually the depression lies more distant, the air is dry and the skies clear ('the dry **b**').

bore A 'wall' of broken water moving upstream in a shallow-water estuary with an appreciable tidal range; caused by the progressive constriction of the advancing flood at spring tide, retarded by friction at its base and by a powerful opposing river current. The **b** gradually diminishes in height and ultimately dies out. E.g. Severn **b**, sometimes a metre high; *eagre* (also *egre*, *aegir*) on the Trent, and *mascaret* on the Seine (both much reduced by dredging for navigation); the Hooghly in the Bay of Bengal. On the Tsientang-kiang in N. China (near Hangchow), the front of the wave attains a height of 3–4·5 m and advances upstream at 16 km h^{-1}. The tide rises so markedly in the Bay of Fundy

Plate 7. A massive granitic BORNHARDT near Kaduna, Nigeria. Note the large-scale EXFOLIATION of the dome summit. (*R. J. Small*)

that the term **b** is commonly applied, though not in the strict sense.

Boreal (i) A climatic zone in which winters experience snow, and summers are short. (ii) App. to the N., specif. the northern coniferous forests. (iii) A climatic period from about 7500 to 5500 B.C., gen. dry, with cold winters and warm summers; indicated by development of a pine-hazel flora.

bornhardt (Germ.) A residual hill rising abruptly from the surface of a plain; cf. INSELBERG, of which term W. Bornhardt was the originator. Some writers use it with a precise meaning, implying a core of massive unjointed granite which resisted sub-surface chemical disintegration under conditions of high temperature and periodic humidity, but was exhumed through the removal of surrounding rotted regolith. See pl. 7.

boss A small BATHOLITH with a more or less circular plan, having a surface area (when exposed by denudation) of about 100 km^2; e.g. N. Arran (W. Scotland); Shap (Cumbria). Usually regarded as syn. with STOCK, though some regard a **b** as circular, a stock as more irregular.

bottom (i) The low-lying alluvium-covered FLOOD-PLAIN along a river valley, esp. in USA; e.g. Mississippi Bottomlands. (ii) The floor of a lake, sea or ocean. (iii) A dry valley in chalk country; e.g. Rake B., near Butser Hill, in Hampshire. (iv) The former head of a lake in a U-shaped valley, now infilled with sediment deposited by inflowing streams; e.g. Warnscale B. in Buttermere valley, English Lake District.

bottomset beds In a DELTA, the fine material carried out to sea and deposited in advance of the main delta; over them in due course FORESET BEDS are laid down as the delta advances seaward. [*f*]

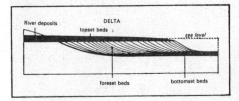

Bouguer Anomaly A phenomenon first observed by P. Bouguer (1735), who made gravitational observations near Chimborazo in the Andes, and discovered that the plumb-line deviation towards that peak was less than calculated. The same discrepancy was found during surveys in the Indo-Gangetic Plain, where the estimated deviation resulting from the mass of the Himalayas was 15″ of arc, but the measured deflection of a plumb-line was only 5″. This **a** results from deficiency of mass in the crust underlying mountain ranges. In calculating the **a**, the observed gravity is compared with a formula 'corresponding to a spheroidal surface with an avge value of the ellipticity' (H. Jeffreys), corrected for height above sea-level.

boulder A large individual fragment of rock, exceeding 200 mm in diameter (*Brit. Stand. Inst.*), or 256 mm (*US Wentworth Scale*). An example of a common word given a precise meaning.

boulder-clay An unstratified mass of clay, containing stones of all shapes and sizes, often striated, their nature depending on source of origin. The product of glacial erosion and deposition, without any water transport involved. Many local names are applied; e.g. Chalky **B-c** and Cromer Drift of E. Anglia; Hessle and Purple **B-c** of Lincolnshire, Humberside and E. Yorkshire. TILL is now preferable, since it does not imply any definite or specif. constitution.

boulder controlled slope A concept introduced by K. Bryan, in which it is envisaged that slope angles under desert conditions are determined by the angle of repose of the boulders resting upon them. The latter (whose size is controlled by joint spacing) are released by rock weathering but remain on the slope, although subsequently reduced by GRANULAR DISINTEGRATION. In time a layer of such boulders will accumulate, resting on a bedrock slope whose angle equals that of the boulder layer. Providing no change in joint spacing occurs, the size of the boulders released by weathering will remain constant, as will the angle of slope i.e. a **b c s** experiences PARALLEL RETREAT OF SLOPE. [*f*]

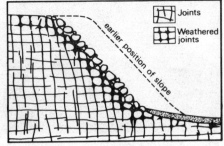

boulder-field See FELSENMEER.

boulder-train A series of ERRATICS deposited by a glacier; usually worn from the same identifiable bedrock source, carried forward by moving ice, and deposited either in a broad fan-pattern, with the apex pointing to the origin of the rock, or in a more or less straight line. They can be used to map the exact movement of former ice; e.g. a train of dark Silurian boulders can be traced across the limestone Craven district in W. Yorkshire.

boundary current Part of the circulation of oceanic water at depth, along the W. side of oceans: (i) flowing S. in the Atlantic from high latitude source of deep cold water off S. Greenland; (ii) flowing N. in the Pacific from S. of New Zealand; and (iii) a less marked N. flow from S. of Africa along the E. coast of that continent into the Arabian Sea. It has been suggested by H. Stommel that (i) meets in latitude 35°S. another **b c** flowing N. from the Weddell Sea, where the water turns E. and produces a great flow around the Antarctic continent. The edges of currents are strongly marked by sudden changes in temperature and salinity, hence 'boundary'.

Bourdon tube Thermometer or barometer consisting of a closed curved tube, elliptical in section, containing spirit; its curvature alters as temperature or pressure changes.

bourne A temporary or intermittent stream which may flow in a DRY VALLEY in chalk country after heavy rainfall, partic. in winter, when the WATER-TABLE rises above the level of the valley-floor. A common place-name element in chalk country; e.g. Ogbourne in Wiltshire. A considerable time-lag occurs between the period of heavy rain and the b-flow; water percolates slowly through closely jointed chalk towards the saturated layer. Also known as 'winter bournes', 'woebournes', 'nailbournes', 'gypseys' and 'lavants'.

B.P. Before the present day. A widely used form of dating, esp. for Late-Glacial and Post-Glacial events (e.g. the ALLERØD phase dates from approx. 11 000 B.P.).

brackish Slightly saline, with salt content less than that of sea-water; sometimes defined as containing 15–30 parts of salt per thousand.

brae (Scottish) A hillside; the BROW of a hill. See BANK.

braided river A r whose course consists of a tangled network of inter-connected diverging and converging shallow channels, with banks of alluvial material and shingle between. These bars may be revealed at times of low water. They are formed esp.: (i) in a heavily laden though shallow **r**; (ii) on the surface of an ALLUVIAL FAN; and (iii) by a melt-water stream as it flows across FLUVIOGLACIAL deposits. Braiding is encouraged when the banks of the main stream consist of easily erodible alluvium, sands and gravels. E.g. Spree near Berlin; Rhône between St Maurice and L. Geneva. See pl. 8.

braiding terrace Produced by a B RIVER, transitional in character between PAIRED and

Plate 8. The course of the Eyra Fjordur, a BRAIDED RIVER, northern Iceland. The valley-floor shows the influence of extensive AGGRADATION. (*Eric Kay*)

MEANDER TS, the result of spasmodic erosion rel. to the growth and breakdown of detrital fans.

brash (i) Broken rubbly rock beneath the soil and sub-soil, partially weathered *in situ*. Appears in Cornbrash, a Jurassic rock noted for the 'brashy' soils it has produced, and long known as good for corn growing. (ii) Loose masses of small ice fragments floating in the sea.

Brave West Winds Nautical term for ws of the S. hemisphere, blowing from a W. or N.W. direction over the ocean between latitudes 40°S. and 65°S., moving N. and S. with the seasonal change of world pressure belts. Characterized by strength and persistence, and by stormy seas, overcast skies, and damp, raw weather. Seamen call the latitudes in which they blow the 'ROARING FORTIES'.

breached anticline The development of a drainage pattern on an **a** may result in a valley along its axis, so removing the overlying rocks and exposing older ones. The result is an anticlinal valley with infacing escarpments. This is common since: (i) the axis of an **a** is structurally weaker, having been subjected to tension, than a neighbouring syncline; and (ii) it is higher above BASE-LEVEL and thus more readily attacked. E.g. breached Fernhurst **a** in the Weald; Vale of Pewsey. On a large scale, the Bow R. in W. Alberta has cut a deep valley along the crest of a major **a** trending N.N.W. to S.S.E., leaving remnants of the W. limb as the main crest of the Canadian Rockies. [*f*]

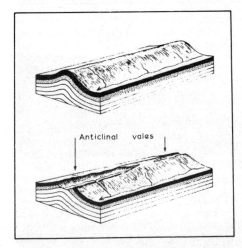

breadcrust bomb A volcanic **b** with a glassy crust seamed with cracks, the result of shrinkage following cooling and congealing.

breakaway The upper parts of slopes showing retreat by undermining processes e.g. tropical laterites, which often form bold freefaces rising above gentler slopes developed in weak weathered material. Retreat of the latter produces vertical fissuring of the unsupported laterite, large masses of which break away to litter the slopes beneath.

breaker A mass of turbulent broken water and foam rushing up a beach, formed when a wave travelling across deep water passes into shallow water; its crest steepens, curls over and breaks. Gen. this oversteepening occurs when the ratio of wave-height to wavelength exceeds 1/7. Waves can also break in deep water when generated by strong local winds (sometimes known as 'white horses' or 'whitecaps'), or if a mass of rock (e.g. REEF) lies near the surface. Some authorities classify them into: (i) *spilling b*s; (ii) *plunging b*s; and (iii) *surging b*s. (i) and (iii) tend to be *constructive*, i.e. move shingle and sand up the beach; (ii) is *destructive*, i.e. 'combs down' the beach and moves material seawards.

break of slope A marked or sudden change of inclination in a **s**. In the bed of a river, such a **b** may represent a KNICKPOINT or a transverse resistant band of rock. In a U-shaped valley, the **b** of **s** in a cross-profile may represent the max. height of glacial erosion. Elsewhere it may indicate the edge of an EROSION SURFACE.

break-point bar An offshore BAR formed along the line where waves first break.

breccia (It.) A rock composed of angular CLASTIC fragments mixed with finer material. **B** may be: (i) a SCREE deposit; (ii) a sedimentary rock cemented during consolidation; (iii) a 'volcanic **b**', composed of PYROCLASTIC material (AGGLOMERATE); (iv) 'intrusion **b**', formed of pieces of country rock embedded in material of magnetic origin; (v) FAULT-**b**; and (vi) formed of specific fragments, such as 'bone **b**' found in caves from the remains of animals. Derived from It. *breccia*, broken wall material. In English a distinction is made between **b** (angular constituents) and CONGLOMERATE (rounded constituents).

breckland (i) An area of heathland, with thickets and bracken. (ii) An area of land taken in from heathland for temporary cultivation. (iii) Proper name of an area on the borders of Norfolk and Suffolk, largely under coniferous plantations of the Forestry Commission.

breeze A wind between Force 2 (light breeze, 5 knots) and Force 6 (strong breeze, 28 knots) on the BEAUFORT WIND SCALE.

brickearth (i) Orig. any loamy clay used for brick-making. (ii) Specif., a fine-textured deposit found on river-terraces, resulting from wind-blown material which has been re-worked, re-sorted and re-deposited by water; e.g. on the Thames terraces. Soils derived from **b** are usually fertile and easy to work.

brickfielder A dry, dusty, squally wind in Victoria, Australia, sometimes with temperatures exceeding 38°C (100°F), blowing from the continental interior in a S. direction in front of a depression.

brig, brigg A headland of hard rock, specif. where oolitic limestone outcrops along the coast of Yorkshire; e.g. Filey B.

brigalow A type of SCRUB in semi-arid areas of Australia, on the borders of the MULGA; consists mainly of a species of *acacia* (*A. harpophylla*).

Brillouin Scale A logarithmic linear s produced by L. Brillouin (1964), ranging from 10^{-50} to 10^{30}, and thus including the largest (10^{27}) and smallest (10^{-13}) units yet measured by man. The equatorial circumference of the earth is taken as 0. Cf. G-SCALE.

brine A solution containing a higher proportion of common SALT than sea-water, often occurring as a spring.

broad A sheet of reed-fringed, fresh water, forming part of, or joined to, a slow-flowing river near its estuary. Typically found in Norfolk and extends into Suffolk, here used as a proper name. This area was a shallow bay in Roman times, in which peat accumulated. There has been much discussion about the origin of the **B**s, but the consensus of opinion is that cutting and removal of peat has been a major factor. One category of **b** is separated from rivers by narrow 'washlands' (*ronds*) (e.g. Salhouse), linked to the rivers by artificial cuts; the other is an actual broadening of the river; e.g. Thurne B. Breydon Water is a portion of the old estuary, consisting of mud-flats crossed by creeks and the main river channel at low tide, a sheet of water 6 by 0·8 km at high tide.

Brockenspectre The apparent shadow of an observer standing on a mountain summit with the sun behind him, projected as a diffraction effect on to a cloud or fog-bank beyond. Round this shadow are rings of coloured light. So called after the Brocken, a summit in the Harz Mtns (Germany). *Glory* and *anticorona* are sometimes used for this phenomenon.

brockram A sedimentary rock, angular BRECCIA, found notably in the Permian.

Bronze Age A major phase in the development of Man's culture, succeeding the PALAEOLITHIC, MESOLITHIC and NEOLITHIC. In the early stages, copper was used in its pure form for objects of adornment ('the Copper Age'), starting in Mesopotamia *c.* 4000 B.C. Its use spread during the next millennium. About 3000 B.C. the alloy of copper and tin (**b**) was discovered, and the **B A** reached its height in the 2nd half of the 2nd millennium; the gradual superseding of bronze brought in the IRON AGE. In Britain **B A** lasted from about 2000 B.C. to 6th century B.C.

brook A small stream. Used loosely and poetically in England (e.g. Lord Tennyson's 'Brook'). In USA regarded as smaller than a CREEK.

brow The upper part of a hill. Cf. Scottish *brae*.

brown-coal A brown fibrous deposit, intermediate in development, form and character between peat and coal proper.

brown forest soil (brown earth) One of the main ZONAL SOIL-groups, characteristic of areas in middle latitudes formerly covered with deciduous woodland, rich in organic matter derived from accumulation and decay of leaves. The surface A_1-horizon is usually a slightly acid humus layer, since the climate is humid and some leaching occurs. The A_2-horizon is a grey brown, somewhat leached, zone (but less so than in a PODZOL). The B-horizon is thick, dark brown, and contains colloids and bases carried down from the A-horizon. Found notably in N.E. USA, N. China, central Japan, and N.W. and central Europe. In the last, most **s** has been cultivated for centuries, and the orig. forest has been long removed.

brown steppe soil See PRAIRIE SOIL.

brunizem See PRAIRIE SOIL.

Brückner cycle A **c** of climatic change, with an avge (though irregular) periodicity of about 35 years. Postulated by E. Brückner in 1890, who examined rainfall and temperature records back to the 18th century. The actual amplitude of oscillations is not great: for temperature less than 1°C, and for precipitation only from 9% above normal to 8% below. Individual **c**s may range from 25 to 50 years. Variation between cold and humid, and warm and dry periods may be reflected in glacier fluctuation, level of the Caspian and other inland seas, date of grape harvests, patterns of tree-rings, and other phenomena.

Buchan spell An unseasonable s of cold (summer) or warm (winter) weather, postulated in 1867 by A. Buchan on the basis of 50 years' observations, which appear to occur at 9 specif. times of year. The statistical reliability of these dates is doubtful.

bunch grass A coarse tufted g, separated by bare sandy or stony ground, which grows e.g. in the semi-arid parts of the American Midwest and on intermont plateaus of the Western Cordillera. It provides extensive grazing, but for only a small number of head per unit of area.

Bunter The lowest formation of the Triassic system, consisting mainly of variegated red sandstones and pebble-beds. Probably laid down under arid or semi-arid conditions, sometimes resting UNCONFORMABLY on rocks below, sometimes on the Permian with little indication of a break. Hence New Red Rocks, which may embrace the Permo-Trias boundary.

buran A strong N.E. wind experienced in central Asia at all seasons, though occurring particularly as a fierce, bitterly cold blizzard ('white b') in winter.

burn (Scottish) A small stream in Scotland, N. England and N. Ireland.

bush An area of uncleared shrubby vegetation, sometimes with scattered trees; in gen., wild and unsettled, as opposed to cultivated, terrain. In S.W. USA applied to areas of low bushes (e.g. creosote) adapted to semi-arid conditions.

bushveld (Afrikaans, *bosveld*) A type of 'tall grass—low tree' SAVANNA, occurring in subtropical and tropical Africa. The trees may be scattered to form open parkland, or so dense as nearly to form forest.

butte A small, though prominent, flat-topped hill, usually capped with a resistant rock-stratum, remaining after the partial denudation of a plateau in semi-arid areas. It may be a small isolated portion of a MESA. E.g. in Arizona and New Mexico. It is flanked by slopes of angular SCREE, the result of desert-weathering. Isolated hills formed by other means are sometimes also regarded as bs; e.g. igneous intrusions which remain upstanding, or a hill protected by an extrusive cap of basalt; examples of the last can be seen along the Front Range of the Rockies, esp. near Denver, Colorado. [*f* MESA]

butte témoin (Fr.) (Germ. **Zeugenberge**) A flat-topped OUTLIER, beyond the edge of a plateau or an ESCARPMENT of which it was once part, with its surface in broadly the same plane. Most true buttes are in fact **b** ts. *t* = witness, i.e. evidence of a former extension of the plateau.

buttress A protruding or outstanding mass of rock on a mountainside; e.g. Kern Knotts B., Bowfell B., Gillercombe B. in English Lake District; Northeast, Tower and Observatory Bs. on Ben Nevis.

Buys Ballot's Law A law put forward in 1857 by C. H. D. Buys Ballot of Utrecht, that if an observer stands with his back to the wind in the N. hemisphere, pressure is lower on his left hand than on his right; in the S. hemisphere pressure is lower on his right. This is a result of the CORIOLIS FORCE (SEE FERREL'S LAW) on the movement of a body on the surface of the earth; air tends to move along the isobars of a pressure system, in the N. hemisphere clockwise round ANTICYCLONES, anti-clockwise round DEPRESSIONS. See GEOSTROPHIC FLOW.

bysmalith A large IGNEOUS INTRUSION, in the rough shape of a cylinder, which has arched up the country rock above it. Its surface characteristics are similar to a LACCOLITH, but its sides plunge steeply down for a vast distance, and it is commonly associated with faulting round its margins. Sometimes it may be exposed as an upland mass by denudation; e.g. Mt Hillers in Henry Mtns (Utah), Mt Holmes in Yellowstone National Park (Wyoming).

caatinga (Portuguese) (i) A type of tropical woodland, growing in N.E. Brazil; XEROPHILOUS plants which can withstand long periods of aridity but take advantage of seasonal rains. It forms a thick, virtually impenetrable mass of thorny jungle, from which rise occasional taller trees: giant cacti, wax-palms, acacias and euphorbias. For 7 months it is a grey, leafless, dead-looking tangle, but when short rains occur the vegetation bursts into intensive life: trees develop brilliant blossom, herbs and bulbous plants flower. After 3 to 4 months, the vegetation reverts to its dormant state. (ii) Used in the Rio Negro valley, Brazil, for low evergreen forest, where there is no marked dry season.

cacimbo Low cloud and fog, sometimes with drizzle, along the coast of Angola, resulting from the cooling effect of air moving inland from over the cold offshore Benguella Current.

Cainozoic, Cenozoic, Kainozoic From Gk. 'recent life', indicating the 3rd of the eras of geological time subsequent to the Precambrian (the last 60–70 million years) and the rock-groups deposited during that time. There has been much confusion and difference of interpretation in its use. (i) Commonly in the past, used as syn. with TERTIARY, now obsolescent. (ii) In USA, continental Europe and increasingly in UK (as by A. Holmes and Geological Society of London), C is divided into 2 periods/systems, Tertiary and Quaternary, including all series/epochs from Palaeocene to Recent. See Table, TERTIARY.

cairn, carn (Gaelic) Orig. a pile of stones raised as a monument; now a tall pile of stones on a mountain summit in N. England and Scotland. Part of the proper name of many mountains; e.g. Carn Dearg.

calamine An ore of ZINC, commonly known in Britain as *smithsonite* ($ZnCO_3$).

calcareous Containing an appreciable proportion of calcium carbonate ($CaCO_3$); hence c OOZE, c TUFA (or travertine), c rocks, c soils. From Lat., *calx*.

calcicole A plant which requires a soil rich in lime (adj. *calcicolous*); e.g. clematis, spindle-tree, rock-rose, traveller's joy, dogwood, viburnum, box.

calcification A soil-forming process in arid and semi-arid climates, so that calcium carbonate accumulates in the B-HORIZON.

calcimorphic soil A s developed on a calcium-rich parent material; e.g. RENDZINA ss developed on chalk; red and brown Mediterranean ss developed on TERRA ROSSA.

calciphobe, calcifuge A lime-hating plant, restricted to soils with marked acid character; e.g. azalea, rhododendron, *Calluna*, most *Ericaceae*, and other heath-plants. The two terms are virtually syn., though calciphobe is stronger; a calcifuge can survive but not flourish in a chalk-soil, a calciphobe will quickly die.

calcite The crystalline form of calcium carbonate ($CaCO_3$), the main constituent of limestone, and a common 'cement' in other coarse-grained sedimentary rocks. The calcium is derived from weathering of various igneous rocks, carried away in solution as calcium bicarbonate $Ca(HCO_3)_2$, and deposited as c when the bicarbonate decomposes.

C crystals have a hardness of 3 (on MOHS' SCALE), specific gravity 2·7, and usually colourless unless stained by impurities. C is deposited on the beds of streams, and as STALACTITES and STALAGMITES in caves. *Iceland spar* is a pure variety.

calcium carbonate $CaCO_3$, with two crystalline forms, *calcite* and *aragonite*, and occurring in limestone, marble, chalk and coral. It is insoluble in pure water, but dissolves in water containing carbon dioxide (e.g. rainwater) to form readily soluble calcium bicarbonate, $Ca(HCO_3)_2$, present in most surface waters of the earth. When calcium bicarbonate decomposes, calcite is deposited at temperatures below 30°C, aragonite above.

calcrete A SEDIMENTARY deposit derived from coarse fragments of other rocks 'cemented' by calcium carbonate. It is formed by capillary rise from ground water, evaporation at or near the surface, deposition on the floor of now dried lakes, or from algal entrapment of calcium carbonate. It may occur as a massive indurated sheet 0·5 to 2 m thick, in a tightly packed nodular form, or as a crumbly powder. Commonly found in deserts or in areas of markedly seasonal rainfall.

caldera (also **basal wreck**) The large shallow cavity which remains: (i) when a PAROXYSMAL ERUPTION removes the top of a former cone; (ii) as a result of ENGULFMENT. E.g. Askja in Iceland; Crater Lake, Oregon, lying in a c nearly 10 km across. Numerous examples are in Alaska and Aleutian Is., incl. Katmai, formed by an explosion in 1912, when a peak rising to 2285 m (7496 ft) was reduced to a c 4 km across, with a jagged summit-rim reaching only 1370 m (4500 ft). The largest c is said to be Aso in Japan, 27 by 16 km. [*f*]

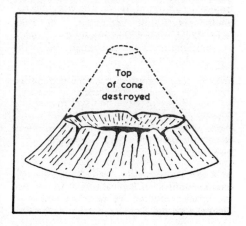

Caledonian First used by E. Suess (1888) to describe ancient mountain ranges which can be traced through W. Scandinavia and Scotland into N.W. Ireland. Sometimes called '*Caledonides*', worn-down relics of once lofty fold mountain systems. Folds of the same age are found in S. Uplands of Scotland, English Lake District and N. Wales; in these cases the trend is more W. to E. C is now mainly used to describe the period of mountain-building (late Silurian-Devonian) which led to the upfolding of these mountains, typically with a N.E. to S.W. trend in N.W. Europe.

calf (Norse *kalv*) (i) An islet adjacent to a larger island; e.g. C. of Man. (ii) A fragment of floating ice; hence CALVING, to form an iceberg.

caliche (Sp.) (i) A crust-like accumulation of impure sodium nitrate ($NaNO_3$) in soils of arid areas, esp. Atacama Desert of Chile and Peru, where it was first exploited in 1825. Its origin is not understood; it may be formed by leaching of bird guano, by bacterial fixation of nitrogen, by leaching from volcanic tuffs, and by drying-up of shallow lakes. In the 19th century it was the world's chief source of fertilizers and nitrogenous compounds; still important in the chemical industry, but nitrogen is now more easily obtained from coal and the atmosphere. (ii) Applied in S.W. USA to a hard surface encrustation of calcium carbonate, as the result of solutions rising to the surface by CAPILLARITY; cf. DURICRUST.

calina (Sp.) A leaden dust-haze during summer in lands bordering the Mediterranean Sea.

calm Force 0 on the BEAUFORT WIND SCALE, with velocity of less than 1 knot. A c may occur in any latitude at any time, partic. under anticyclonic conditions. The chief areas of calms are within 5° latitude N. and S. of the Equator in the DOLDRUMS and in the 'HORSE LATITUDES'.

calorie The amount of heat-energy required to raise 1cc (or 1 g) of water at sea-level pressure from 14·5°C to 15·5°C (technically known as the *15°C calorie* or *gram-colorie*). Now replaced by the SI *joule* (J). 1 calorie = $4.186J = 3.968 \times 10^{-3}$ BTU. 1 langley = 1 c per cm^2.

calving The formation of an ICEBERG through breaking-off of an ice-mass from the tongue of a glacier reaching the sea or the edge of an ICE-BARRIER; e.g. Ross Ice-Barrier, Antarctica.

camber An arching of rock strata caused by some superficial disturbance of the rock itself, rather than by earth-movements, esp. when it overlies plastic clay. Thus a cap-rock may be inclined towards the valleys, as in the Northamptonshire Ironstone fields and in the Cotswolds, where oolites and sandstones are cambered over the Lias. [*f*]

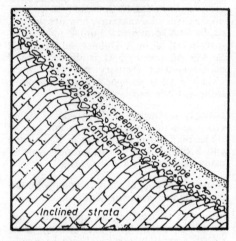

Cambrian The 1st geological period of the Palaeozoic era, and the system of rocks laid down during that time. Name given by A. Sedgwick in 1836 to rocks now divided into C and Ordovician. The C was gen. believed to have lasted from 500 to 410 million years ago, though recent research has dated its start at 600 ± 20 million years. Rocks of this age are found in Gt. Britain in N. Wales (e.g. Harlech Series of grits and slates, Tremadoc Series, Llanberis Slates); S. Wales, Welsh borders and Midlands (quartzites, sandstones, limestones); and N.W. Scotland (white quartzite and limestone). There is usually a distinct break between C rocks and earlier (PRECAMBRIAN) ones.

Campbell-Stokes recorder A device for measuring and recording the duration of bright sunshine over a period of time. A fixed spherical lens burns an image of the sun on to a sensitized recording-card; as the sun moves, so does the position of the image, and a line will be burnt out by continuous sunlight.

campo (Portuguese) (i) SAVANNA in central Brazil, prob. not a natural CLIMAX vegetation, but the result of human interference. It is gen. woodland and parkland, which varies considerably in aspect. It covers an area in Brazil of some 2750 km from N. to S. and 2400 km from W. to E. Usually classified as: (*a*) *c cerrado* (with scattered trees); (*b*) *c limpo* (mostly treeless, with tall coarse grass). (ii) Locally in Argentina, c is used for an intermont depression, usually trough-shaped and defined by faults.

canal Occas. applied to a narrow piece of water connecting 2 larger stretches of sea; e.g. Lynn C. (Alaska). Note *canale*, plur. *canali*, long narrow gulfs along the Adriatic coast of Yugoslavia.

Cancer, Tropic of The N. t, at 23° 32′ N. latitude, where the sun's midday rays are vertical about 21 June; the most N. point of the ECLIPTIC.

cane-break A thick mass of reeds, growing in places up to 10 m high, forming a dense swampy growth along the banks of the Mississippi, Red, Arkansas and other rivers.

cannel coal A dull, non-laminated coal, usually dark grey in colour. It has a high content of ash and volatile matter, is non-caking, and burns easily with a rather smoky yellow flame.

cannon-shot gravel Very coarse gravel deposits, comprising large and almost perfectly rounded stones, formed by Pleistocene glacial streams e.g. in north Norfolk.

canopy An almost continuous stratum of foliage formed by crowns of tall trees, as in Tropical Rain-forest.

canyon (Sp. *cañon*) (i) A deep, steep-sided gorge with a river at the bottom; mainly found in arid or semi-arid areas, where a rapidly eroding river maintains its volume from snow-melt on distant mountains, but where weathering is slight, so maintaining steepness of the walls; e.g. Grand C., Zion C., C. de Chelly in S.W. USA. The c is more pronounced if the land has been slowly uplifted by earth-movements at more or less the same rate as the river has cut down, as in the Colorado Plateau. (ii) A submarine trough, found beneath the sea on the continental shelf and beyond, which has all the appearance of a steep-sided winding valley; e.g. off the mouth of Hudson R. and Zaïre R.; Fosse de Cap Breton off the Biscay coast of France. Its origin is uncertain, possibly the result of: (*a*) faulting; (*b*) erosion by powerful mud-laden submarine turbidity currents or by springs bursting out on the sea-floor; (*c*) former erosion by rivers when the land was higher.

capacity, of a stream Introduced by G. K. Gilbert in 1914 to indicate the total load a s can carry. Ct. COMPETENCE, concerned with the size of the largest particle carried. Both concepts were worked out under laboratory conditions; it is not always easy to apply them to natural ss, and they have met much criticism.

cape A prominent headland or promontory projecting into the sea; C. Cod, Finisterre, Comorin, 'The C.' specif. refers to C. of Good Hope, Republic of S. Africa.

'Cape Doctor' A S.E. wind, blowing chiefly in summer, often strongly, so called because it produces stimulatingly fresh conditions in Cape Town, Republic of S. Africa.

capillarity The property of holding water in soil by surface tension as films around individual particles and in capillary spaces (minute hair-like tubes), above the height at which it could be held by hydrostatic pressure. This water may be absorbed into plant roots, or drawn to the surface through hair-tubes by evaporation. The zone in which c can successfully act against gravity in respect of soil-water is known as the *capillary fringe*.

Capricorn, Tropic of The S. t, at 23° 32′ S. latitude, where the sun's midday rays are vertical about 21 Dec.; the most S. point of the ECLIPTIC.

cap-rock (i) A resistant stratum; it may be responsible for a waterfall (where a river spills over its edge, eroding less resistant rock below; e.g. Niagara), or for a prominent hill, where the capping protects the rock below (see BUTTE). (ii) An impermeable stratum covering an AQUIFER, OIL-DOME, or SALT-DOME. (iii) A mass of barren rock covering an ore-body.
[*f* BUTTE]

capture, river The diversion of headwaters of a river system into the basin of a neighbour with greater erosional activity flowing at a lower level; also known as *beheading*, *river piracy* and *abstraction*. The development of contiguous river systems must lead to one becoming more powerful; gradually it becomes the 'master-stream', more deeply entrenched, and affording a lower BASE-LEVEL for its tributaries. These push back their watersheds by headward erosion and ultimately capture neighbouring streams, thus increasing the volume of the main river and making it still more powerful. At the point of diversion there is usually a marked bend, known as *elbow of c*. Examples are widespread; e.g. the rivers of Northumberland and Yorkshire; the c of the upper Blackwater by the Wey (Surrey); of a former headstream of the Rhine by the Doubs; the many cs in the scarplands of the Paris Basin. 'It is a normal incident in a veritable struggle for existence between rivers, and may occur in very varied circumstances' (S. W. Wooldridge). [*f*]

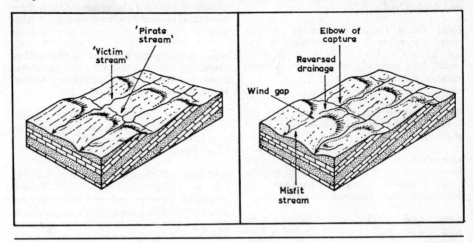

carbonaceous App. to sedimentary rocks containing an appreciable amount of organic material and its derivatives: peat, brown-coal, lignite, coal, c shales. Carbon is widely distributed in the world, occurring in the free state as diamond and graphite, more gen. in combination; e.g. coal (avge carbon content, 80%), petroleum (84%), timber (50%). It forms only 0·2% of the Earth's crust.

carbonation (c-solution) A form of chemical weathering of rocks by rain-water containing carbon dioxide in solution (e.g. weak carbonic acid, H_2CO_3), esp. on limestones and on rocks containing other basic oxides. The limestone is dissolved and removed in solution in the form of calcium bicarbonate. It is commonly referred to simply, though not wholly accurately, as solution. It forms a main agent of sculpture in limestone country, both on the surface and underground. See pl. 47.

carbon cycle The movement of c into the atmosphere as carbon dioxide and its return to the Earth's surface, to be absorbed and stored in vegetation (*photosynthesized*) and other systems. A major contribution is the combustion of fossil fuels, with further small amounts released from the Earth's interior by VULCANICITY, from the soil, and from respiration and decay. It is estimated that the atmosphere contains 2300×10^9 tons of c, incr. by several per cent each decade, which will incr. affect solar radiant energy. The c exchange between ocean and atmosphere is almost in balance (estim. to total 100×10^9 tons annually); a small amount is stored in marine shells.

carbon dating, carbon 14 See RADIOCARBON DATING.

Carboniferous The 5th of the geological periods of the Palaeozoic era, and the system of rocks laid down (C. Limestone, Millstone Grit, Coal Measures). Geologists divide the C into the *Avonian* (Lower C), *Namurian*, *Westphalian* and *Stephanian* (youngest). It occurred from c. 345–280 million years ago. In USA, the C is divided into two: lower or MISSISSIPPIAN, upper or PENNSYLVANIAN, the boundary at c. 325 million years. Lower C formations are mainly of marine origin, the Upper of freshwater or lacustrine origin.

Carboniferous Limestone (i) Stratigraphically the lowest of the formations of the C system, incl. sandstones and shales (e.g. Yoredale Shales), corresponding approx. to the Mississipian in N. America. In UK it is divided into the Lower and Upper *Avonian*, corresponding in W. Europe to the *Tournaisian* and *Viséan*. (ii) Lithologically the most important rock in the Lower C, also known as *Mountain Limestone*, consisting of hard, grey, crystalline, well-jointed limestone, markedly fossiliferous (containing crinoids, corals and brachiopods). It occurs in N.W. Yorkshire, Peak District, Mendips, S. Wales, N. Wales, Forest of Dean, Devonshire, on the S.W. and S. margins of English Lake District, in Midland Valley of Scotland, and over much of Ireland. Many areas form plateaus (e.g. N. and S. Pennines, Mendips), with a marked absence of surface drainage, SWALLOW-HOLES, intricate cave-systems, dry valleys and gorges, and pavements of bare rock with CLINTS and GRIKES. Cf. KARST. See pl. 25, 47, 51, 62.

Carbo-Permian The end of the Carboniferous and beginning of the Permian periods; sometimes used to denote the HERCYNIAN phase of mountain-building.

cardinal points The 4 major points of the compass: N., S., E. and W.

carr A FEN containing, apart from reeds (esp. *Carex paniculata*) and other aquatic plants, shrubs such as alder (which is dominant), sallow, osier and willow. It occurs on waterlogged terrain which is neither too acid nor too poor in mineral elements; e.g. E. Norfolk, parts of Lincolnshire. Cognate with Icelandic *kjörr* and Swedish *kärr*.

carse (Scottish) Fertile alluvial lands along the estuary of a river; e.g. C. of Gowrie.

cascade A stepped series of small waterfalls, often of an artificial ornamental character.

Cascadian In USA mountain-building movements at the end of the Tertiary *c.* 2 million years ago, when the Cascade Mtns were uplifted, accompanied by widespread VULCANICITY, and the FAULT-BLOCK of the Sierra Nevada was bodily uplifted and tilted. Following the 'C Revolution' (*C orogeny* is sometimes applied, though this implies folding which did not occur), there was a period of quiescence in the early Quaternary, followed by renewed volcanic activity when the Cascade peaks were formed (Mt Baker, Rainier, St Helens, Adams, Hood, Jefferson, Washington, Shasta and Lassen Peak).

cassiterite The main ore of TIN as an oxide (SnO_2).

castellanus clouds A mass of cs of turret shape when seen from the side. Formerly known as *castellatus*.

castellated iceberg An i with a towering pinnacled superstructure above the water, in ct. to horizontal or tabular masses.

cataclinal Applied by J. W. Powell (1875) to streams which flow in the direction of the DIP. Now little used (see ANACLINAL and DIACLINAL).

cataclysmal Formerly applied to processes and results of exceptional deluges, earthquakes and catastrophes in gen., which *catastrophists* believed to be responsible for most relief features of the world, in ct. to *uniformitarians* who did not.

cataract Orig. large waterfall; now used for a series of rapids, as on the Nile, with 5 main cs (numbered upstream from the 1st near Aswan to the 5th above the Atbara confluence), and several minor named ones. The Nile has cut its way down vertically through Nubian Sandstone until in places it has reached ancient crystalline rocks. These offer greater resistance to erosion, forming complicated rapids, divided channels and broken water.

catastrophism The concept that the Earth's features are the result of sudden major catastrophic events, not of slow continuous processes operating with gradual inevitability. Ct. UNIFORMITARIANISM. Certain catastrophes may occur (e.g. floods, eruptions, earthquakes), but these are essentially temporary and local.

catchment (i) Area drained by a single river; a natural drainage area which may coincide with a river basin, in which DIVIDES direct water derived from rainfall and PERCOLATION into a river. Where underground flow is involved, the c area may be larger or smaller than that apparent from surface relief. *C Board:* a statutory authority in UK, responsible for drainage and flood control within the drainage basin of a river (e.g. Ouse C. Board), though boundaries of some C Boards extend beyond a single river basin. *Note:* In USA WATERSHED is applied to c. (ii) The intake area of a single AQUIFER.

catena A sequence of different soils which varies with relief and drainage, though normally derived from the same parent-material. Such a sequence may be seen when following a line of profile from a hill-top to a valley-bottom.

'cat's paw' Used by yachtsmen and others for a light breeze or puff of wind affecting only a small area of water.

cauldron subsidence The foundering and collapse of a block of country rock into underlying MAGMA. Glencoe in W. Scotland is an example of a BATHOLITH mass which has collapsed, demarcated by *ring-faults* and RING-DYKES.

causse(s) (Fr.) Sometimes used for limestone country in gen. Specif. the Grands Causses in S.W. of Central Massif of France. From patois word *cau (chaux)*, lime.

cave, cavern (usually syn., though the latter is sometimes regarded as larger) A subterranean chamber, usually natural, with an entrance from the surface. Found partic. at the base of sea-cliffs where wave action has enlarged natural lines of weakness (*sea-cs*) (see *f* STACK) and in limestone country as a result of CARBONATION-SOLUTION, partic. along JOINT-PLANES. Used also as: (i) *caving*, to explore caves; (ii) to collapse (*c-in*); (iii) adjectivally (*c-art, c-man*). The study and exploration of cs is *speleology*. C is occas. used for artificial hollows; e.g. Chislehurst Cs., champagne cs near Reims, cs near Maastricht cut for lime-working, and Tilly Whim Cs. near Swanage, Dorset, produced by quarrying.

cavitation The wearing of rocks by running water, when bubbles of air and water-vapour (formed by rapid increase in the velocity of a stream) collapse, causing minute shock-waves against bed and banks. May occur in high velocity sub-glacial streams, producing 'sand-blasted' potholes and other forms of intensive erosion.

cay (key, kay) A low island of sand and CORAL fragments, built up by waves on a reef-flat, at or just above high tide, drying at low water; e.g. Florida Keys (terminated by Key West), Marquesas Keys, Grand Cs. (Bahamas), West Indies gen., Cayo Grande (off Venezuela). Similar features occur in Indonesia and on the Great Barrier Reef of Australia.

cedar-tree laccolith An intrusion consisting of a series of LACCOLITHS one above another, fed from a single vent or pipe from the MAGMA reservoir.

'ceiling' (i) Some specif. level in the atmosphere, notably the lowest substantial layer of cloud; e.g. 'c zero' means fog at ground level. (ii) The height to which a particular aircraft or balloon can climb. (iii) A physiological limit to which a man can climb without oxygen; this varies with individuals.

Celestial Sphere The imaginary 'bowl' of the heavens, representing a sphere with the Earth at the centre, on the 'inner surface' of which heavenly bodies appear to be placed. Distances are so vast (the nearest star is 4·29 light-years away) that the radius of the C S may be regarded as infinite. The plane of the Earth's Equator produced cuts the C S in the *C Equator*, and the Earth's axis produced meets the C S at *C N. Pole* and *C S. Pole* respectively; ZENITH is a point on the C S directly above the observer. The C S is used as the basis for astronomical and navigational problems, solved by spherical trigonometry. [*f* RIGHT ASCENSION]

cell, atmospheric (i) A large area of predominantly high or low a pressure resulting from the interruption of the planetary pressure system, mainly because of unequal solar heating caused by irregular distribution of continents and oceans. In N. winter the Eurasiatic (Siberian) and American high pressure cs dominate the N. continents, while the Icelandic (N. Atlantic) and Aleutian (N. Pacific) cs are over the oceans. In N. Summer the N. Atlantic and N. Pacific are dominated by the Azorean and Hawaiian 'highs', with 'lows' over N. America and S. Asia. A c is 3-dimensional, and the height to which the air-mass extends is of fundamental importance. (ii) A vertical c in the TROPOPAUSE, involving a *Polar c* (direct thermal drive), a *westerly* (indirect frictional drive), and an easterly (direct thermal drive), from 30°N. and S. to the Equator. See ATMOSPHERIC CIRCULATION, HADLEY C. (iii) A more local 'bubble' in the atmosphere, rising because of convection: a THERMAL or *convection c*.

Celsius scale The internationally accepted name for the CENTIGRADE S of temperature. A. Celsius, a Swedish astronomer (1701–44), developed a thermometer, with the division of the interval between freezing and boiling points of water into 100, in 1742. Actually he used 0 for boiling and 100 for freezing point, but this was reversed in 1743. The name C officially replaced Centigrade in 1948, but the latter is still widely used. For conversion table from °C to °F, see FAHRENHEIT.

'cement' Siliceous, calcareous or ferruginous material, deposited from circulating water, which has converted loose deposits, such as sand and gravel, into a hard compact rock. The nature of the c has a marked influence on the subsequent weathering of the rock. A siliceous c usually produces a hard, resistant rock; e.g. quartzite. Hence *cementation*, a process involved in DIAGENESIS.

Centigrade scale A graduated s of temperature on which 0° represents the melting point of ice, 100° the boiling point of water, at sea level.

$$C = \frac{5}{9}(F-32), \text{ or } \frac{C}{100} = \frac{F-32}{180}.$$

Officially known as CELSIUS since 1948. For a conversion table from °C to °F, see FAHRENHEIT.

centimetre A unit of metric measurement. 1 cm = 0·01 m = 0·3937 in.; 1 in. = 2·54 cm; 1 cm² = 0·155 sq. in.; 1 cm³ = 0·06103 cu. in.; 1000 cm³ = 1 litre = 0·22 UK gallon. A non-preferred SI unit.

central eruption An eruptive form of volcanic activity which proceeds from a single vent (point source) or a group of closely related vents, in ct. to linear or FISSURE E. The product of a c e is a cone.

centripetal drainage A d pattern in which numerous rivers converge on a main stream. E.g. in Katmandu valley of Nepal, streams converge upon the Bagmati R., which drains S. through a gorge cutting across surrounding mountain ranges. In extreme cases a basin of INLAND D is formed.

centrocline See PERICLINE.

centrosphere Syn. with BARYSPHERE.

cerrado (Portuguese) A type of SAVANNA in Brazil, a mixture of low contorted trees, 4 to 7 m high, and tall grass, with closely spaced tangled growth. The *cerradao* has similar species, but trees are taller, 9 to 15 m.

chain Any linear sequence of related physical features; e.g. c of lakes, islands, reefs. Used esp. of mountains with a complex series of more or less parallel or *en échelon* ranges, as distinct from a single range.

chalk Lithologically a soft, amorphous whitish limestone, almost entirely (c. 97%) of calcium carbonate ($CaCO_3$), up to 600 m thick in England. Formerly thought to be wholly an organic deposit, consisting of tests of foraminifera, coccoliths and other marine micro-organisms in a matrix of finely divided calcite; some writers believed it to be a fossil abyssal ooze or calcareous mud, including the foraminifera *Globigerina*. Its origin is still problematical, but it is clear that while some chalk is undoubtedly of organic origin (laid down in shallow water, as indicated by large shells of shallow-water life), with little terrigenous matter (possibly the result of low-lying adjacent land where little erosion was in progress), other c may be due in part to chemical precipitation. Often contains nodules of FLINT, usually in bands. It forms characteristic relief features: rolling hills and undulating plateaus, with open expanses of downland, covered with short turf growing on a thin layer of soil, now mainly cultivated (esp. cereals). Where the strata are tilted, a CUESTA may be formed. Surface drainage is slight, because of numerous close joints, and DRY VALLEYS are common. C occurs widespread in E. and S.E. England, incl. Yorkshire Wolds, Lincoln Wolds, E. Anglian Heights, Chiltern Hills, Salisbury Plain, Dorset Downs, N. and S. Downs. In France c outcrops extensively in the Paris Basin (Picardy, Artois, 'Champagne pouilleuse').

CHANDLER 53

Chalk, The A stratigraphical name, applied to the upper beds of the Upper Cretaceous system (usually considered also to include the Gault and the Upper Greensand). In S. England, this sequence can be recognized:

Upper Chalk	White chalk, with flints Chalk Rock
Middle Chalk	Soft white chalk, with some beds of marl, a few flints Melbourn Rock
Lower Chalk	Grey chalk Chalk-marl

Chalk-marl A stratigraphical horizon near the base of the Chalk, consisting of greyish calcareous material containing up to 30% argillaceous (clay) material; e.g. near Cambridge. See MARL. *Note:* If the rock is referred to, other than the horizon, the term used is chalky-marl.

Chalky Boulder-clay A type of BOULDER-CLAY found especially in East Anglia, greyish in colour, containing fragments of chalk and numerous FLINTS. Formerly classified into Great C B-c (now renamed LOWESTOFT TILL) and Little C B-c (now GIPPING TILL).

chalybeate Water containing hydrated iron compounds, which may appear on the surface as a c spring, of alleged medicinal value; e.g. at Harrogate.

chalybite Natural ferrous carbonate ($FeCO_3$), in its bedded form (SIDERITE) a major source of iron; e.g. in the Coal Measures and Jurassic limestones of England.

champagne (champaign, champain, campagne) (Fr.) An area of open hedgeless plains, as in N. France (ct. BOCAGE). From *campania*, Lat. 'plain'. Used as a proper name for a former French duchy, and now for the PAYS of an open, gently undulating landscape, incl. *C pouilleuse* (chalk-country to the E. of the Ile de France) and *C humide* (Gault Clay farther E. still). Also used for other smaller pays (e.g. *C berrichonne*, *C charentaise*). The name has been transferred to the sparkling wine produced in the neighbourhood of Reims.

chañaral (Sp.) An area of thorny scrub, largely composed of *chañar* bushes, with large thorns; found in N. Argentina and central Chile.

Chandler wobble A w of the Earth in relation to its rotational axis, with an amplitude of 0° 0′ 5″ and a periodicity of about 14 months, first detected in 1891. Its cause is unknown, but it may be the cumulative result of periodic earthquakes.

channel (i) The deepest part of a river-bed, containing its main current, shaped by the force of water flowing in it. It can be measured in terms of: (*a*) depth, from the surface of water to the bed; (*b*) width; (*c*) cross-sectional area; (*d*) wetted perimeter (length of line of contact between water and bed); (*e*) HYDRAULIC RADIUS; (*f*) FORM-RATIO; (*g*) GRADIENT or slope. (ii) A funnel-shaped estuary; e.g. Bristol C. (iii) An irrigation ditch. (iv) A stretch of sea (wider than a strait) between 2 land-areas, linking 2 more extensive seas; e.g. English C., St George's C. (v) The main shipping-lane or fairway, usually dredged, within a wide estuary, or between shoals.

channel flow The RUNOFF of surface water in a more or less narrowly defined trough between banks, rather than spread out laterally over a wide area (SHEETFLOOD or OVERLAND FLOW). Cf. also THROUGHFLOW.

channel storage The hold-up of flood-water within a section of river-c receiving inflow more rapidly than can be passed downstream. The max. storage before flooding occurs is *c capacity*, or BANKFULL stage.

chapada (Portuguese) A tableland in Brazil, formed of an extensive, rel. horizontal, sheet of sedimentary rocks, such as sandstone, lying over crystalline basement rocks of the Brazilian Plateau.

chaparral (Sp.) Evergreen scrub vegetation in California and N.W. Mexico, similar to MAQUIS in S. Europe, the result of long summer drought. It consists of tough, broadleaf, evergreen scrub-oak, interlaced with vines and with scanty grass.

characteristic angles Defined by A. Young as those angles 'which most frequently occur, either on all slopes, under particular conditions of rock or climate, or in a local area'. C as show as modes on a graph of angle frequency distribution; the class with the highest frequency of occurrence is the *primary c a*. Studies in a wide range of environments show that **c as** occur commonly as follows: 33–35°, 25–26°, 5–9°, 1–4°.

Charnian (i) A stratigraphic division of the Precambrian, involving mainly volcanic rocks. (ii) An OROGENY which took place in late Precambrian times, the worn-down remnants seen in the Midlands of England (Charnwood Forest, the Wrekin, Malvern Hills, Caer Caradoc).

chart (i) A map of the sea and coastline for navigators, prepared by the Hydrographic Office of the Admiralty for British use, usually drawn on a MERCATOR PROJECTION, because this shows constant bearings as straight lines. (ii) A map prepared specially for aviation use, emphasizing aeronautical information (landing-fields, pylons, high towers, peaks and other dangers). (iii) A weather map, showing atmospheric pressure, winds and other information; e.g. Daily Weather C.

chart datum The plane from which soundings on a navigational chart are computed. The British Admiralty uses low-water springs; i.e. the worst possible navigational state of depth for a particular chart. A Port Authority (e.g. Southampton Harbour Board, Mersey Docks and Harbour Co.) has its own **c d**, so fixed as to avoid minus quantities in tidal data. **C d** at Southampton was 2·28 m below ORDNANCE DATUM until 1 Jan., 1965, when it was lowered to 2·74 m below O.D.

chatter marks, chattermarks (i) A 'bruising' or 'scarring' (as distinct from scratches), curved or crescentic in pattern, on the rock-floor formerly beneath a glacier. Probably caused by vibratory 'knocking' of loosely embedded boulders in GROUND-MORAINE carried at the base of the glacier. (ii) Applied to marks on wave-worn pebbles.

cheesewring Proper name of a striking granite TOR on Bodmin Moor (Cornwall), with narrow stem and overhanging upper block. Now used by some writers for any similar features, including GARA (plur. *gour*) in hot deserts. Ct. HOODOO.

chemical weathering See CORROSION.

chernozem (Russian) A 'black-earth', of loose crumbly texture, rich in humus and bases, covering large areas of land in middle latitudes where the prevailing natural vegetation was grassland. The A-HORIZON is a black layer, 0·6 to 1 m thick, grading into the B-HORIZON, brown or yellow-brown, in which colloids and bases have accumulated. C extends across central Asia from Manchuria, through S. Siberia and central USSR into the Ukraine, and into Roumania and Hungary. Similar soils are found in central Canada, and in USA from N. Dakota to Texas. Parts of the Argentine Pampas and S.E. Australian 'downs' have similar characteristics. These soils afford a striking example of origin due to climate (warm summers, cold winters, some early summer rainfall but considerable periods of drought). Their texture and richness in plant-foods, together with the wide extent of gently rolling land and climate favouring annual grasses, make them good cereal-growing soils.

chert A layer of irregular CONCRETIONS of an amorphous rock composed of hardened chalcedonic or opaline silica, which splinters easily, with flat fractures (ct. conchoidal of FLINT), occurring in calcareous formations other than chalk; e.g. Portland Stone Cherty Series.

chestnut soil A zonal s, characteristic of areas of sparse dry steppe with 200–250 mm of rain; loose, friable and dark brown in colour. Actually a variety of CHERNOZEM, modified by greater aridity, little leached because of low rainfall. The A-HORIZON is dark brown, becoming paler with depth. It has a wide distribution, covering drier STEPPES to the S. of the chernozem in Russia, extending W. into Roumania and Hungary, in the High Plains of USA, and in the drier parts of the Argentine PAMPA and the S. African VELD.

chevron crevasse A cr near the margins of a glacier which has been rotated or twisted into the pattern of a chevron as a result of glacier motion.

Chezy equation A 'flow equation', designed to relate the velocity of stream flow to characteristics of the channel, in particular hydraulic radius and slope.

$$V = C\sqrt{Rs}$$

in which V is mean velocity, C is the Chezy coefficient (a coefficient of channel resistance), R is hydraulic radius, S is slope. In wide, shallow channels R is approx. equal to channel depth (d); thus v is proportional to the square root of $d \times s$. In simple terms the C e states that, for a channel with a given resistance to flow (constant C), velocity increases as a function of increased hydraulic radius and increased slope.

chili A type of SIROCCO wind, experienced in Tunisia, dry and very warm.

chilling Used esp. in USA in agricultural geography to denote a period of low temperature, though not below freezing, required by specif. crops; e.g. 7°C (45°F) for almonds. This crop actually requires 200–500 *c hours* for the satisfactory development of flower buds.

chimney (i) A steep vertical cleft in a cliff, wider than a crack; e.g. Kern Knotts C on Great Gable, English Lake District. (ii) Sometimes applied to the vent of a volcano. (iii) In USA, a vertical pillar of rock.

china-clay See KAOLIN.

china-stone Granite which has been partially kaolinized and is free from dark minerals. A hard rock which requires grinding for KAOLIN extraction. Also used for other whitish porcelainous rocks; e.g. found in the Ordovician rocks of the Welsh Borderland.

chine A narrow cleft in soft earthy cliffs, through which a stream descends to the sea, esp. in I. of Wight and Hampshire. Usually consists of a steep-sided 'inner' valley, contained within a wider 'outer' valley. Some were probably formed by small tributaries flowing to the former 'Solent River' system, while the inner valley was cut under present conditions. Marine erosion is cutting back the cliffs, and small streams in the inner valley cut down more rapidly; e.g. Branksome C. (near Bournemouth), Blackgang C. (I. of Wight).

'Chinese wall' glacier The edge of an ice-sheet which reaches the sea as a vertical or even overhanging ice-cliff; e.g. along the coast of Greenland.

chinook A dry warm S.W. wind blowing down the E. slopes of the Rockies in Alberta, W. Saskatchewan and Montana, warmed ADIABATICALLY; in spring it causes a swift rise of temperature and rapidly melts snow (Indian name = 'snow-eater'). Cf. FÖHN.

C-horizon In soil science, the layer of rel. little-altered material underlying the soil proper; the parent-material from which the soil has been derived by various soil-forming processes, loosely called 'sub-soil'.

[*f* SOIL PROFILE]

choromorphographic map Type of m which delimits and classifies areas of land (Gk. *khora*) according to surface configuration (Gk. *morphe*). G. M. Lewis's c m of W.-central USA shows 12 types; e.g. 'flat sand plains', 'closely dissected low plains', 'sand hills'.

chott An alt. form of SHOTT.

chrono-isopleth diagram A graph in which hourly values of pressure, temperature, etc. are plotted as abscissae and their times of occurrence in the month as ordinates. Similar values are joined by isopleths.

cinder cone A co around a volcanic vent, composed exclusively of small fragmentary material, sometimes defined as 3–5 mm in diameter. When lava is highly charged with gas, rapid release of pressure causes explosive eruptions, which fragment the lava which solidifies while flying through the air. Ci is a misnomer, orig. it was believed that it was the product of combustion. E.g. Sunset Crater, Arizona, a **co** 300 m high rising from the plateau at 2100 m, with an unbreached crater 120 m deep and about 400 m in diameter. It was the result of an eruption, c. A.D. 1064. Other examples are Cinder Cone (proper name) in Lassen Volcanic National Park (California); Parícutin in Mexico, which first erupted in 1943 and formed a 450 m **co** within a year.

circumference, of Earth Equatorial **c** = 40 076 km (24 902 mi.); Polar **c**, 40 008 km (24 860 mi.).

cirque (Fr.) A steep-walled rock-basin of glacial origin (also *corrie, coire, combre, cwm, kar*). A shallow pre-glacial hollow was progressively enlarged, first by alternate thaw-freeze of a snow-patch within it, which causes rocks to disintegrate (NIVATION). Melt-water helps to move resulting debris. As the snow-patch grows, it develops into a FIRN-field, from which may issue a small C-GLACIER. Freeze-thaw eats into the back-wall of the **c** (BASAL SAPPING), thus maintaining its steepness, and into the floor, thus maintaining and emphasizing its basin shape; it also provides debris which freezes into the base of the ice and acts as an abrasive. The out-moving ice in the **c** pivots about a central point (ROTATIONAL SLIP), which also emphasizes the basin shape. A common feature in glaciated mountain ranges, often containing a small lake. It varies from a tiny rock basin in N. Wales and the English Lake District to the Walcott C. in Antarctica with a BACK-WALL 3000 m high. See pl. 5, 34. [*f*]

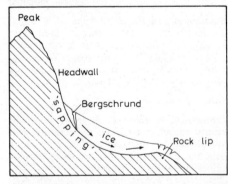

cirque-glacier A short-tongued glacier which barely protrudes from the basin in which the FIRN accumulates; e.g. in the Sierra Nevada.

cirque stairway A succession of CIRQUES one above the other e.g. on Snowdon, where a **c s** comprises a small cirque below the summit, the basin of Llyn Glaslyn, and the larger basin of Llyn Llydaw. The development of **c ss** is possibly related to geological structure, with lines of rock weakness forming the sites of individual cirque depressions. Alt. lower cirque glaciers (and eventually basins) may form from avalanche debris from higher cirque glaciers.

cirrocumulus (*Cc*) A high cloud, usually consisting of ice-crystals, in lines of small globular masses with a rippled appearance, with blue sky between ('mackerel sky'). It usually occurs above 6000 m.

cirrostratus (*Cs*) A uniform milky layer or veil of high sheet-cloud, above 6000 m (20 000 ft), through which the sun may shine with a distinct HALO. It affords indication of the approach of a WARM FRONT. It may thicken and develop into ALTOSTRATUS. See pl. 9.

cirrus (*Ci*) A high-flying (6–12 000 m), delicate, fibrous, wispy cloud, consisting of tiny ice-spicules, so light that it hardly interferes with sunlight or even moonlight. Often a fair-weather cloud, but if it thickens to form CIRROSTRATUS it may be an indication of an approaching DEPRESSION. When **c** is drawn out as 'mare's tails' or 'stringers', it indicates strong winds in the upper atmosphere. Adj. *cirriform*.

clarain structure, in coal Very thin, finely banded structure in **c**, usually parallel to the BEDDING-PLANE. Though lustrous, the bands are less glassy than VITRAIN, and are often embedded in the duller DURAIN. Now obsolescent.

clarke of abundance The % by weight of metallic elements in the avge crustal rock; the highest **c**s are aluminium (8·1%), iron (5·0%), magnesium (2·1%) and titanium (0·44%).

clastic (Gk. *clastos*, broken) App. to rock composed of broken fragments (*clasts*) from other rocks, usually transported from some distance, deposited, and converted by the process of DIAGENESIS into a consolidated coherent rock, hence LITHIFICATION; this takes place by: (i) cementation; (ii) compaction; and (iii) desiccation. They include: (*a*) sedimentary rocks: conglomerate, sandstone, clay, shale; (*b*) pyroclastic rocks; tuff, volcanic ash and agglomerate. A consolidated mass of comminuted shell-fragments, though essentially organic in origin, can also be regarded as a **c** rock.

clatter A Devonshire dialect-word for SCREE or individual boulders.

Plate 9. CIRROSTRATUS CLOUDS, note the FRACTOCUMULUS clouds in the background. (*Royal Meteorological Society*)

clay (i) A fine-textured, plastic, sedimentary rock (*argillaceous*), derived from compaction of mud, consisting mainly of hydrous aluminium silicates, derived from weathering and resultant decomposition of various feldspathic rocks. No structure is developed, and when it dries out c is traversed by irregular cracks. When wet, it forms a virtually IMPERMEABLE rock, since the minute pore spaces between fine particles are filled with water held by surface tension, so sealing the rock against downward passage of water. Many varieties of diverse age are distinguished; e.g. greyish-blue Lias C, dark grey Kimmeridge C, bluish-grey Oxford C, yellow or brownish Wealden C, bluish-grey London C. (ii) In soil science, a soil with individual particles less than 0·002 mm (in USA, 0·005 mm) in size; a c soil has 30% or more of its bulk of c. (iii) A complex group of minerals, composed largely of aluminium and iron silicates. See also LATERITE, RED C, BOULDER-C, KAOLIN.

clayband A layer of clay-ironstone, consisting of ferrous carbonate or *siderite* ($FeCO_3$) mixed with earthy material, found within the Coal Measures. The iron occurs in nodular bands or thin seams, separated by shales and sandstones.

clay-humus complex A c of particles of fine clay and HUMUS, which has the property of attracting and holding ions, partic. *cations* (positively charged) disassociated from salts dissolved in soil moisture. In a balanced productive soil, the **c-h c** thus holds sufficient cations to provide essential nutrients for plants. In an acid soil, hydrogen ions become dominant on the surface of the **c** and the pH VALUE falls.

claypan A stratum of stiff compact clay, forming an impermeable layer below the surface of the soil, causing impeded drainage and waterlogging. Ct. HARDPAN.

clay-slate A category of slate derived from compacted clay (i.e. argillaceous slate), in ct. to slate compacted from volcanic ash (e.g. green slate of Buttermere).

Clay-with-Flints A mass of reddish-brown clay containing flint fragments, lying unevenly on the surface of chalk country, and also in funnel-shaped PIPES which may penetrate to a considerable depth. Its age and origin are problematical and complex. In part it may represent the insoluble residue of the Chalk, and in part it may be derived from formerly overlying Tertiary rocks.

cleavage The natural tendency of a rock, such as slate or compacted volcanic ashes, to split into thin sheets along parallel planes, formed as the result of past metamorphic pressure. It may make any angle with the BEDDING-PLANES (ct. LAMINATION), and occurs particularly in fine-grained rocks. The geological interpretation of the causes of c is a complex problem, however. There are 4 types of c: (i) *fracture*; (ii) slaty; (iii) flow; and (iv) axial plane. *Fracture c* occurs when rocks respond to folding by slipping along shear-planes (see SHEARING), producing numerous small, closely-spaced joints. C is of importance to quarrymen in the extraction and dressing of commercial slates for roofing purposes. *Note*: A crystallographer uses the term to indicate the tendency of a crystal to split along planes determined by its molecular structure. [*f*]

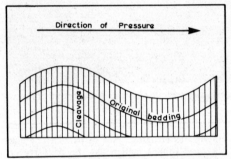

cliff A rock-face along the sea-coast where marine denudation is active and the land rises steeply and appreciably in height inland. The character of a *sea-c* depends on the nature of the rocks, their bedding and jointing, homogeneity or heterogeneity, the presence of bands of weakness such as the SHATTER-BELTS of faults, their gen. resistance both to wave attack and weathering. Cs are formed in many rocks, esp. massive ones; e.g. granite (Cornwall), Old Red Sandstone (Orkneys, W. Scotland, Dyfed), Jurassic limestone (Durlston Head, Dorset), New Red Sandstone (St. Bees Head, Cumbria), Carboniferous Limestone (Dyfed). They also occur in softer rocks, where marine erosion at their base is rapid; e.g. chalk (Dorset, Isle of Wight, Sussex, Kent, E. Yorkshire), Tertiary clays and sandstones (Hants.), Eocene clays and sandstones (Alum Bay, Isle of Wight), glacial clay (Norfolk; Holderness, Yorkshire). (ii) Any high steep rock-face or precipice in the mountains, or rising above a lake shore, (iii) the side of a deeply incised river valley; see MEANDER.

climate The total complex of weather conditions, its avge characteristics and range of variation over an appreciable area of the earth's surface. Usually conditions over many years (e.g. 30 or 35) are taken into consideration. From Gk., meaning 'slope' (possibly the slope of the earth's axis), later considered as a zone lying in a partic. latitude, and then applied to characteristic weather conditions. C is studied in terms of the various elements: TEMPERATURE, incl. RADIATION, ATMOSPHERIC PRESSURE, WIND, HUMIDITY (WATER-VAPOUR, CLOUDS, PRECIPITATION, EVAPORATION). These result from the interplay of factors: latitude, altitude, distribution of land and sea, ocean currents, relief features, and influence of soil and vegetation. The most convenient summary for the geographer is as CLIMATIC REGIONS.

climatic optimum See ATLANTIC STAGE.

climatic region The delimitation of specif. areas in which various repeating combinations of c elements (*c types*) can be distinguished. The earliest were latitudinal temperature zones: *torrid, temperate, frigid*. Successive classifications have been produced, based on more complex data, using seasonal and annual isotherms, significant temperatures (e.g. 0°C), precipitation totals and seasonal distribution. The aim is to arrive at indices which can be plotted on a map and lines bounding rs interpolated. The major problem is that climatic conditions change gradually through transition zones. The most widely used system was devised by W. KÖPPEN. See also FLOHN.

climatograph A circular graph for depicting seasonal temperature conditions. Mean monthly temperatures are plotted from a centre with the aid of a graduated table. The distance from the centre of the circle to 100°F is taken as 10 times the distance from the centre to 0°F. If the latter distance is x, then the difference y for any temperature $t°$ is given by the formula $y = x \frac{(\text{colog } t)}{100}$. If the limiting temperatures of hot, warm, cool and cold seasons are assumed to be 68°, 50°, and 32°F, then the length and nature of such seasons at any place may be read from the graph by noting where the temperature curve cuts the lines representing the limiting temperatures. The slope of the temperature curve correctly represents the degree of change in temperature from month to month. [*f, page 59*]

climatology The scientific study of climate; an expression, description and where possible explanation of: (i) its distribution and regional patterns; and (ii) its contribution to the environment of life. Ct. METEOROLOGY, MICRO-CLIMATOLOGY, LOCAL CLIMATE.

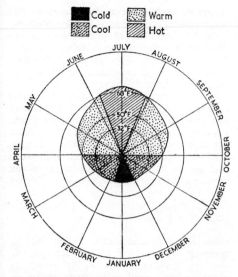

climatomorphology ('climatic geomorphology') The scientific study of the development of land-forms under different climatic conditions, as in tropical, arid, glacial, periglacial and humid-temperate regions; the study of weathering, geomorphological processes, landforms and soils specif. associated with major climatic environments.

climax vegetation, c community When the optimum vegetation of a particular long-term plant SUCCESSION, left undisturbed, has been established in relation to the particular physical environment (climate, relief, soil), a state of equilibrium is attained. Thus the c v of a hot, wet climate within a few degrees of the Equator is 'Tropical Rain-Forest'. *Climatic c v* is now preferred to the obsolete 'Natural Vegetation', while some authorities prefer *potential natural v.*

climograph, climogram A diagram in which data for 2 climatic elements (e.g. WET-BULB and DRY-BULB temperatures; temperature and precipitation) at a particular place are plotted against each other on a graph as abscissae and ordinates resp. The shape and position of the resultant graph provides an index of the gen. climatic character at that place. First used by T. Griffith Taylor to correlate and depict climatic conditions in terms of human physiological comfort; see also HOMOCLIME and HYTHERGRAPH. E. Raisz used *climatogram*. [*f, opposite*]

clinographic curve A graphed **cu** illustrating the slope of an area as it varies with altitude, in practice plotting the avge gradient in degrees between pairs of successive contours.

The *f* indicates a method of finding the mean angle of slope between 2 successive contours, by drawing concentric circles equal in area to that enclosed by each contour. The **c c** is drawn using contour intervals as vertical components, and inserting each section of avge slope between each pair of contours with a protractor. [*f*]

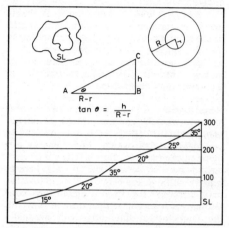

clinometer An instrument for measuring a vertical angle. The *Indian c* consists of a base with a spirit-level, a peep-sight at one end, and a vertical leaf with a thin central slit at the other. Degrees above and below the horizontal (0°) are marked on the leaf. There are other more complex kinds of **c**; e.g. the ABNEY LEVEL. Another type, with a graduated arc and a pendulum, is used for measuring the DIP of strata.

clint A low flat-topped ridge between furrows or fissures (GRIKES) worn along the lines of the joints in the surface of a bare plateau of Carboniferous Limestone, down which rain-water (containing carbon dioxide)

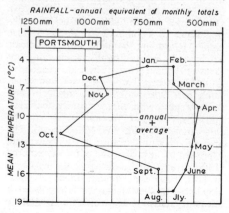

has percolated, enlarging them by CARBONATION-SOLUTION. E.g. Malham Cove, Yorkshire, where cs and grikes are aligned mainly N.W. to S.E. See pl. 47. [*f* GRIKE]

clitter An extensive field of granite boulders, mainly angular, but in some instances rounded, surrounding TORS and occupying hill-slopes on Dartmoor. The c below Belstone Tor exceeds 400 000 m³ in volume. Probably many cs comprise joint-bounded blocks prised from tor margins by frost action during the Pleistocene. The arrangement of c boulders into 'lines' at Great Staple Tor indicates the role of periglacial solifluction in their transport down-slope.

closed system An aspect of GENERAL SYSTEMS THEORY, whereby an assemblage of phenomena exists isolated from the rest by a boundary through which neither energy nor material can pass; ct. OPEN s. The s is characterized by the destruction of any heterogeneity within it, and hence by a trend towards max. ENTROPY. E.g. in an isolated tank containing gas higher in temperature at one end, the temperature differential will gradually decay, with the ultimate establishment of a homogeneous condition. Some authorities claim that in geographical studies it is difficult or impossible to envisage any s that is wholly c.

cloud A mass of tiny visible particles, usually of water (0·02–0·06 mm in diameter), sometimes of ice, which form by condensation on nuclei such as dust and smoke particles, salt, pollen and negative ions. They float in masses at various heights above sea-level, ranging from near the ground (FOG or MIST) to over 13 000 m. They are classified: (i) by height: low cs up to 2000 m, medium cs 2000–7000 m, high cs 7000–13 000 m; (ii) *by form*: feathery or fibrous (CIRRUS), globular or heaped (CUMULUS) and sheet or layer (STRATUS). Other types (see individual names) are distinguished by combinations of the 3 form names, by adding the suffix *alto* to indicate height, and *nimbus* to signify falling rain, thus forming 10 main genera. These are subdivided into 14 *c species* (based on shape and structure), and 9 *c varieties* (arrangement and transparency, with supplementary features), and 9 *accessory c formations*, as in *The International C Atlas* (WMO, Geneva). The amount and nature of the c cover is recorded, expressed in the proportion of the sky covered (10ths or 8ths) and indicated on a map by a proportionally shaded disc. Lines of equal cloudiness (*isonephs*) can be interpolated if adequate statistics are available. Definite c sequences can be observed in connection with the passage of pressure-systems. The cloudiest parts of the world are in mid-latitudes on the W. coasts of continents, and near the Equator. The least cloudy are hot deserts, continental interiors, and Mediterranean lands in summer. See NEPHANALYSIS.

cloudburst A fanciful name for a sudden, concentrated downpour of rain. In USA sometimes defined as a rate of fall of 100 mm per hour; see INTENSITY OF RAINFALL. The area affected is usually quite small, since the c is caused by intense local convectional rising, accompanied by a thunderstorm.

clough, cleugh A steep-sided valley in N. England, esp. Derbyshire.

cluse (Fr.) A steep-sided valley cutting transversely across a limestone ridge, esp. in the Jura Mts and in Fore-Alps of Savoy. Its existence may involve the development of ANTECEDENT drainage; a river, existing before an anticline was upraised, maintained its course by erosion as uplift progressed. The drainage reveals an alternation between longitudinal valleys (*vaux*, sing. *val*) and transverse *cluses*, with sudden changes of direction, elbow-bends, and frequent river-CAPTURES. Note the course of a river such as the Doubs. [*f*]

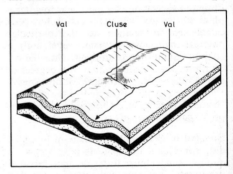

coal Applied to a wide range of combustible deposits derived from accumulated layers of vegetation, classified according to content of carbon and volatile matter, together with colour, hardness and age. The vegetation decomposes under water in the absence of air (ANAEROBIC), with associated pressure and temperature increase resulting from its burial under overlying sediments and from earth-movements. The series comprises: PEAT, BROWN-COAL, LIGNITE, CANNEL COAL, BITUMINOUS COAL, STEAM-COAL and ANTHRACITE. See also CLARAIN, DURAIN, FUSAIN and VITRAIN structures.

Coal Measures The series of coal seams and intervening strata of shale, clay, sandstone and ironstone in the Upper CARBONIFEROUS formation (approx. equiv. to Pennsylvanian in USA). Although **c ms** (l.c.) also occur in Lower Carboniferous of Scotland, these are not incl. in the **C M** series, which refer to the particular stratigraphical division. At the end of the Carboniferous occurred the HERCYNIAN orogeny, which folded, faulted and in places crushed the **C M**. Subsequent denudation removed them from more elevated areas, so they survive in individual basins flanking, sometimes among, uplands, sometimes covered by new deposits (CONCEALED COAL-FIELD).

coast The zone of contact between land and sea; the strip of land bordering the sea, appreciably wider than SHORE. [*f* COASTLINE]

coastal plain (i) Any area of lowland bordering the sea or ocean, sloping gently seaward. It may result from the accumulation of material: alluvium (*deltaic plain*), sand, mud and peat, with or without a rel. fall of sea-level, together with the contribution of dyking and draining by man; e.g. N. Sea coast of Europe from Denmark to Calais; Adriatic coast of N. Italy. (ii) In USA **c p** is restricted to an area of CONTINENTAL SHELF which has emerged as a result of a rel. fall of sea-level; e.g. S.E. USA, extending inland for 150–450 km along the Atlantic and Gulf of Mexico for a distance of 3000 km. Outcropping sedimentary rocks (Cretaceous and Tertiary age) reveal in plan a series of belts parallel to the coast, hence *belted c p*.

coastline Used as a gen. term (the edge of the land as viewed from the sea), but specif. signifies either: (i) the line reached by the highest storm waves; (ii) the high-water mark of medium tides; or (iii) on a steep coast, the base of the cliffs. The Ordnance Survey uses (ii), and on the One-inch Series stipples the area between mean high-water and low-water marks. [*f*]

cobalt Metal obtained from its sulphide and arsenide ore, usually in association with copper and silver ores.

cobble A water-worn stone, larger than a pebble, smaller than a boulder, 60–200 mm in diameter. In USA defined by the WENTWORTH SCALE as between 64–256 mm.

cockpit karst See KEGELKARST.

coesite A dense form of silica first synthesized in the laboratory under very high pressure conditions by L. Coes. Found naturally in rocks beneath Meteor Crater, Arizona, the result of impact METAMORPHISM when a meteorite struck the Earth at a high velocity.

coire (Gaelic) See CIRQUE.

col (Fr.) (i) A conspicuous depression or notch in a ridge or range, providing a pass from one side to the other, commonly resulting from development of back-to-back CIRQUES. Used widely in the French Alps; e.g. **C** du Midi, **C** du Bonhomme, **C** de la

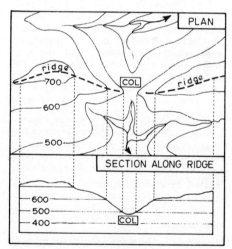

Seigne. Sometimes called *pass* or *saddle*. A **c** may develop in a CUESTA from the beheading of a back-slope valley by scarp-retreat, leaving a **c** in the cuesta at the head of the latter; e.g. Cocking Gap (alt. 100 m) in the W. of the S. Downs near Midhurst, and several distinct **c**s in the S. Downs near South

LAND	COAST
HIGHEST H.W.M. or CLIFFLINE	COAST LINE
	BACKSHORE
AVERAGE H.W.M.	
	FORESHORE
LOWEST L.W.M.	
SEA	OFFSHORE

Harting. [*f*] (ii) ('*atmospheric*'): Used analogously with ref. to weather maps where ridges represent high pressure, valleys low pressure. In a c the weather is extremely variable and presents a difficult problem to the forecaster.

colatitude The complement of the latitude = 90° − latitude. Also known as *Polar distance*.

col channels Glacial meltwater channels cut into previously existing cols by subglacial streams forced across the cols by hydrostatic pressure beneath great thicknesses of ice.

cold desert (i) Gen. term for TUNDRA and POLAR regions, where low temperatures limit or prohibit vegetation (W. Köppen's *ET*, *EF* types). (ii) Sometimes used for areas in continental interiors, poleward of 50°N., and further S. in central Asia where plateaus are shut off by high mountains from maritime influences; defined by A. A. Miller as having one or more months each with a mean temperature below 6°C (43°F) (W. Köppen's *BWk* type); e.g. Luktchun in the Tarim Basin, precipitation 190 mm, mean January temperature −11°C (12°F), mean July temperature 32°C (90°F).

cold front The boundary-zone between a mass of warm air and an advancing, undercutting wedge of cold air which forces the warm air upward; the rear of the warm sector of a DEPRESSION (ct. WARM FRONT). There is an appreciable drop in temperature, extensive CUMULONIMBUS and FRACTO-clouds develop, rain falls in heavy showers (sometimes accompanied by thunder), and the wind freshens from a N. or N.W. direction (in N. hemisphere). An extreme form of a **c f** is a LINE-SQUALL. See ANAFRONT, KATAFRONT.

[*f* DEPRESSION]

'**cold glacier**' In ct. to a 'WARM G', this is a moving ice-mass in which temperatures may be as low as −20° or −30°C throughout the year, with little or no surface melting, as in parts of Greenland and Antarctica. Called by H. W. Ahlman a '*polar g*'.

cold occlusion In an OCCLUSION [*f*] where the overtaking cold air is colder than the airmass in front.

'**Cold Pole**' The point with the lowest mean annual temperature located near Verkhoyansk (at 67° 33′ N., 133° 24′ E.) in N.E. Asia, with mean annual temperature −16·3°C (2·7°F), mean January temperature −50°C (−58°F), mean min. January temperature −64°C (−83°F), lowest recorded temperature −70°C (−94°F), mean annual range 65·5°C (118·6°F). Another **c p** must occur somewhere in Antarctica; the USSR base at Vostok (78° 27′ S., 106° 52′ E.) has recorded −90°C (−130°F).

cold wall A DISCONTINUITY LAYER between water of markedly contrasting temperature; e.g. in N. Atlantic Ocean between Labrador Current and Gulf Stream; in Pacific Ocean between Okhotsk Current and Kuroshio.

cold-water desert Where a hot desert extends to the W. coast of a continent, its climate is affected by equatorward-flowing cold currents and by upwelling cold water ('*cold-water coast*') (W. Köppen's *BWk* climatic type). These **c-w d**s are restricted to narrow strips along the coasts of N.W. Africa, S.W. Africa, N. Chile, N.W. Australia. Cool air moves from over the sea on to the land, producing summer temperatures low for the latitude; e.g. mean July temps. for Walvis Bay, 19°C (66°F), Iquique, 22°C (71°F). The seasonal range is therefore reduced; e.g. Walvis Bay, 5°C (9°F). Fogs may form over the sea and roll inland for a few km; heavy dews are common.

cold wave A sudden inrush of cold Polar air behind a DEPRESSION, causing a marked fall of temperature. In USA, implies a specif. fall below a definite figure in a certain time, depending on season and location.

colloid A substance in a state of extremely fine subdivision with particles from 10^{-3} to 10^{-5} mm in diameter, of both mineral and organic material. It plays a vital yet highly complex role in soil chemistry. Its function is: (i) *physical*: it imparts an adhesive quality to certain constituents, notably clay particles; (ii) *chemical*: it can attract and hold ions of dissolved substances, esp. bases such as calcium. These properties are the result of the electrical forces of different molecules combined along interfaces. Cs carried by rivers are FLOCCULATED where they enter the sea.

colluvium A heterogeneous mixture of loose, incoherent rock fragments, scree and mud, which has moved down to the base of a slope under gravity, the result of MASS-WASTING. In USA also called *slope-wash*.

colony Biologically, a group of closely associated similar organisms; e.g. a **c** of corals. A botanist uses colonization to indicate the gradual spread of a plant species into an area.

columnar structure The result of the cooling of igneous rocks, when internal contraction has set up regular JOINTS vertical to the cooling surface, thus producing columns; e.g. hexagonal columns of basalt in Giant's Causeway (N. Ireland) and Fingal's Cave (Staffa);

Plate 10. The strongly jointed DOLERITE of the Whin Sill, COLUMNAR STRUCTURE exposed by marine erosion at Cullernose Point, Northumberland, England. (*Eric Kay*)

pentagonal columns of phonolite in Devil's Tower (Wyoming), and Devil's Postpile in E. Sierra Nevada (California). See pl. 10.

combe, coombe (i) A CIRQUE in Cumbria; e.g. Birkness C near Buttermere. (ii) A short valley in S. England with a steep head, commonly found in chalk country; e.g. Pebblecombe in N. Downs, E. of Dorking. (iii) A short steep valley running down to the sea; e.g. Wollacombe and Combe Martin, in N. Devon. (iv) A high-lying longitudinal depression along the crest of an ANTICLINE in the folded Jura Mts, cut by an actively eroding stream, with infacing cliffs (*crêts*) of limestone; e.g. C. Berthod, in which rises a headstream of the R. Bienne. See pl. 11.

combe rock See COOMBE DEPOSIT.

comb-ridge A sharp-edged serrated ridge, near-horizontal, formed when two CIRQUES have developed back to back; an ARÊTE in a strict sense.

comet A celestial body moving about the sun in a parabolic orbit, consisting probably of a gaseous nucleus and a long drawn-out tail; e.g. Halley's C (last seen 1910), Kohoutek (appeared Dec. 1973–Jan. 1974).

Plate 11. The Devil's Dyke, a deeply incised COMBE near Brighton, Sussex, England. This dry valley may have been formed by SPRING SAPPING or PERIGLACIAL erosion, or a combination of the two. The flat valley-floor has resulted from the build-up of COMBE DEPOSITS. (*R. J. Small*)

COMFORT

'comfort zone' The range of temperature and RELATIVE HUMIDITY which is physiologically most comfortable to human beings. In England this is around 15°C (60°F), with a relative humidity of 60%. As temperature rises, the rel. humidity should be lower for **c**. A broad **c z** for mid-latitudes is defined by dry-bulb temperatures of 20–25°C, rel. humidity of 25–75%. A *C Chart* has been proposed in USA for offices, factories, schools, etc., derived from air temperature and rel. humidity. An empirical *Discomfort Index* involves air temperature, dew point and rel. humidity, on which 70 is comfortable, at 75 half the people and at 80 all people experience discomfort, while above 85 stress is experienced and offices and factories close. The highest yet obtained on this scale was 92 at Yuma, Arizona.

comminution The reduction of rock material to progressively smaller particles by agencies of weathering and erosion, or by earth movements.

community A group of plants growing in a particular area, usually of distinctive character, and requiring certain physical conditions which satisfy them. The **c** can be of different scales in the plant hierarchy: (*a*) formation (e.g. Temperate Deciduous Forest); (*b*) association (e.g. oak forest).

compaction One of the processes of LITHIFICATION by which unconsolidated materials are consolidated, either by the weight of subsequent overlying deposits (thus eliminating water and closing pores: DIAGENESIS) or by the compressional pressure of earth-movements (METAMORPHISM). This chiefly affects fine-grained deposits, such as silt and clay; these are compacted into mudstone and shale.

compass An instrument used to find direction, consisting of a free-swinging magnetized needle which points to the N. and S. Magnetic Poles along the local line of magnetic force (see MAGNETIC DECLINATION), mounted on a card graduated in degrees, with cardinal and ordinal points marked. More elaborate models have a peep-sight and prism (PRISMATIC C). See also GYRO-C.

competence, of a stream A measure of the ability of a stream to move particles of a certain size, indicated by the weight of the largest fragment that it can transport. As stream velocity increases, the max. particle weight increases, though not in direct ratio. The SIXTH POWER LAW, postulated by W. Hopkins in 1842, suggests that the weight of the largest fragment that can be carried increases with the sixth power of the stream velocity. Thus for each stream velocity there is a corresponding max. weight of particle that can be carried. Likewise, for a given particle size there is a critical water velocity which must be attained before that particle can be picked up. However, once in motion the particle may be transported by a much lower velocity current. The difference between 'pick up' and 'carry' velocities is partic. marked with very small grain sizes; thus it is difficult for slow-moving water to erode a mud bank, though mud particles can be carried by an extremely low velocity current. Another concept postulates that the diameter of a particle which a stream can carry varies with the square of the velocity. Load-carrying ability may also be measured by CAPACITY, which is the total load. A large slowly flowing stream may have a high capacity, but a low **c**; its load consists of a large quantity of fine material in suspension.

stream velocity		diameter of particle	
(km.p.h.)	(m.p.h.)	(mm)	(ins.)
0·4	0·25	0·5	0·02 coarse sand
0·8	0·50	2·0	0·08 coarse grit
1·6	1·00	6·4	0·25 small stones

competent bed A rock-stratum sufficiently strong to bend during folding movements, rather than be distorted by plastic flow and deformation. Ct. INCOMPETENT BED.

complex climatology An analysis of the climate of a place in respect of the frequency of weather types experienced, these defined in terms of the various climatic ELEMENTS.

composite profile A p constructed to represent the surface of any area of relief, as viewed in the horizontal plane of summit-levels from an infinite distance, and including only the highest points of a series of parallel ps. Ct. PROJECTED and SUPERIMPOSED PS.
[*f* PROFILE]

composite volcanic cone A c built up over a long period of time as the result of a number of ERUPTIONS, consisting of layers of ash, cinder and lava fed from the main pipe, which culminates in a crater; e.g. Stromboli and Etna (Italy), Fujiyama (Japan), Mt Hood (Oregon), and most of the world's highest cs. Often known in USA as a *strato-volcano*.

compressed profile A device by which a series of COMPOSITE PS is **c** together, and when viewed at right-angles only features not obscured by higher ones in the foreground are visible.

compressing flow A type of glacier flow, postulated by J. F. Nye, in which a thickening of the glacier accompanies a reduction of surface velocity. C f occurs upstream of valley constrictions, at the base of an ice-fall, and towards the glacier snout. A diagnostic feature of **c f** is the development of 'slip planes' rising up towards the glacier surface; these may be denoted by actual shear planes, as commonly seen at glacier snouts. (Ct. EXTENDING FLOW).

compressional movement, of the Earth's crust A strain developing in a horizontal plane when rocks are exposed to stress. It involves a contraction of the surface rocks, resulting in: (i) FOLDS; (ii) REVERSED FAULTS; (iii) THRUST-FAULTS. It is believed that some RIFT-VALLEYS are caused by **c ms**.

concavity A form of slope, characterized by progressively declining angle towards the slope base. Concave slopes are usually the lowermost elements of composite slope profiles (e.g. the '*basal c*' of convexo-rectilinear-concave profiles). The development of the **c** is often attributed to the dominant activity of surface wash processes (by analogy with the concave curve of water erosion). See also PEDIMENT.

concealed coalfield An area of workable coal in deposits covered by newer rocks.

concordance of summit-levels See ACCORDANT SUMMIT-LEVELS (and *f*).

concordant Parallel to the gen. lines of structure or 'grain' of the country, or to the gen. strata; e.g. a drainage pattern which has developed in a systematic relationship with, and consequent upon, structure. See also C COAST, C INTRUSION. Ct. DISCORDANT.

concordant coast A coastline parallel to the gen. trend-lines of relief; sometimes called *longitudinal* or *Pacific* type. Such a coast tends to be straight and regular, unless a considerable rel. rise of sea-level occurs, when outer ranges become lines of islands and parallel valleys form SOUNDS; e.g. coast of British Columbia; the E. Adriatic coast of Yugoslavia; Cork Harbour in S. Ireland.
[*f, opposite*]

concordant intrusion An **i** of igneous material that lies parallel to the stratification of rocks into which it was intruded; e.g. SILL. Ct. DYKE, a DISCORDANT I.

concretion A more or less rounded mass or nodule of hard material within a bed of different rock. It was probably formed by localized concentration of a cementing material (calcite, dolomite, ferrous oxide, silica) during consolidation of the bed; e.g.

DOGGERS in shales: Coal Measure, Corallian, Kimmeridge Clay, Lias. When the nodules are calcareous they can be used for cement; ironstone nodules form valuable ores. See also GEODE.

condensation The physical process by which vapour passes into the liquid or solid form. It occurs either when air is cooled to its DEW-POINT or when air becomes saturated by evaporation into it. Further cooling will cause excess vapour in the air to be condensed on nuclei as water droplets or (if the dew-point is below 0°C) into the solid form of HOAR-FROST. See also DEW, MIST, FOG, CLOUD, RAINFALL and SNOW.

condensation trail (or **vapour trail**) A line or stream of white cloud-like particles formed behind an aircraft flying in cold, clear, humid air. It results from the **c** of water-vapour derived from combustion of fuel through the aircraft's exhaust, and as a result of reduction of pressure behind the wing-tips. Abbr. *contrail*. Over UK **c ts** rarely form below 8500 m in summer, 6000 m in winter.

conditional instability The state of an airmass with an ENVIRONMENTAL LAPSE-RATE less than the DRY ADIABATIC LAPSE-RATE, but greater than the SATURATED ADIABATIC LAPSE-RATE. A pocket of *unsaturated* air at ground level, if given an upward impulse, will become colder than the surrounding air and sink back. A pocket of *saturated* air will remain relatively warmer, and continue to rise.

conduction, thermal The process of direct heat transfer through matter from a point of high temperature to one of low; ct. CONVECTION, RADIATION.

cone (i) A volcanic peak with a broad base tapering to a summit; see ASH C, CINDER C, COMPOSITE C. (ii) See ALLUVIAL C.

cone of exhaustion A local lowering of the WATER-TABLE around a well, as the result of pumping out water more rapidly than it can percolate laterally through the AQUIFER. In USA a *c of depression*. [*f*]

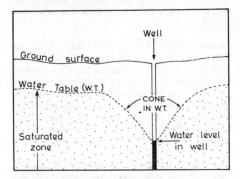

cone sheet A funnel-shaped zone of DYKES or fissures surrounding, in more or less arcuate form, a circular or dome-shaped igneous intrusion which has exerted pressure and so caused fractures within the country-rock. These are inclined inwards towards the top of the intrusion. E.g. in Mull, Skye (Black Cuillins) and Arran. Ct. RING-DYKE.

configuration map A much simplified relief m, on which ridges, crests, spurs and other striking features are represented by bold lines.

confluence (i) The point at which a tributary (*confluent*) joins a main stream. (ii) The body of water so produced, hence a combined flood. *Confluent* is sometimes restricted to a stream joining another of approx. equal size; ct. AFFLUENT.

conformable Where one stratum lies parallel upon another in correct geological sequence as a result of uninterrupted deposition, without any break or interruption by denudation or earth-movements. Noun: *conformity*. Ct. UNCONFORMITY.

conformal projection A class of p (also known as *orthomorphic*) in which shape is maintained over a small area, and at any point the scale is the same in every direction, and angles around every point are correctly represented; e.g. MERCATOR, LAMBERT'S CONFORMAL, STEREOGRAPHIC PS.

congelifluction The flow of earth (see SOLIFLUCTION) under PERIGLACIAL conditions affecting permanently frozen sub-soil. A term introduced by K. Bryan in 1946, along with several others. Each has a precise scientific meaning, which perhaps compensates for their ugliness.

congelifraction Frost-splitting. See pl. 68.

congeliturbation Frost action, involving frost-heaving and churning of the ground, and differential mass-movements such as SOLIFLUCTION, which leads to the disturbance of the soil and sub-soil. Types of PATTERNED GROUND, such as *stone stripes* and *stone polygons*, are produced. See pl. 37.

conglomerate A rock composed of rounded, waterworn pebbles, 'cemented' in a matrix of calcium carbonate, silica or iron oxide; ct. BRECCIA, FAULT-BRECCIA, AGGLOMERATE. Pop. 'puddingstone'. The emphasis is on its rounded constituents, by comparison with angularity of breccia, but there is no precise division.

conic(al) projection One of a group of ps in which part of the globe is projected upon a tangent cone, which is opened up and laid out flat. Most **c** ps have concentric circular parallels, some have straight, others curved, meridians. The cone touches the globe along one or more *standard parallels (s.p.)*.

Radii of Standard Parallels
(Earth's radius $R=1$)
$r = R . \cot latitude$

latitude	
0°	∞
10°	5·671
20°	2·747
30°	1·732
40°	1·192
50°	0·839
60°	0·577
70°	0·364
80°	0·176
90°	0·000

For the ordinary (or *Simple*) **c** p, the selected standard parallel is drawn to scale with the above radius, and divided truly, i.e.
$$1° \text{ of long.} = \frac{2\pi R . \cos lat.}{360}.$$
Draw the central meridian, and divide this truly (i.e. $1° = \frac{2\pi R}{360}$), and draw other parallels concentric to the standard parallel through these divisions. Scale is true along all meridians and on the standard parallel, elsewhere too large. See C WITH TWO S PS, ALBERS' (C

EQUAL AREA WITH TWO S PS), LAMBERT'S CONFORMAL WITH TWO S PS, BONNE, POLYCONIC.
[*f*]

```
θ = latitude = angle
    of standard parallel
R = radius of globe
r = radius of st. parl.
In △ SON, r = R cot. θ
```

Conic(al) Projection with Two Standard Parallels A CONIC(AL) P with 2 standard parallels, spaced their true distance apart. All parallels are concentric circles, meridians are radiating straight lines. Scale is true on each s p, less between them, and progressively larger beyond, so that the overall scale error, as cf. the C P, is reduced. Sometimes (incorrectly) called SECANT C P. The s ps are chosen to fit a partic. country or continent; e.g. for Europe, between 40°N. and 60°N.

coniferous forest A type of f in which trees are mostly evergreen (though not larch), cone-bearing (belonging to the order *Coniferales*), with needle-leaves, shallow root-systems, and soft-wood timber. They can grow under widely differing conditions of climate (from the tropics to the subarctic), of relief (from coastal plains and swamps to high steep mountains), and of soil (from heavy clays to poor sands). Many varieties are tolerant of thin, acid soils and extreme winter cold. They are gen. rapid growing, and so used for AFFORESTATION. The chief types are: (i) HYGROPHYTIC, found in W. and S.E. of N. America, S. Chile, W. Europe, parts of China and Japan, S.E. Australia, New Zealand; these are mostly tall trees, incl. Douglas fir, sequoia, red cedar, Sitka spruce, hemlock, white pine, yellow pine; (ii) MESOPHYTIC, found in high latitudes, known as TAIGA, and at high altitudes, with extensive stands of species of pine, spruce, fir, larch; (iii) XEROPHYTIC, found at high elevation in semi-arid areas, e.g. juniper, pinyon pine (S.W. USA). Conifers grow in many other parts of the world; e.g. in the Mediterranean area (Corsican pine and Aleppo pine), in California (Monterey cypress), on tropical coasts (casuarina).

conjunction The position of 2 heavenly bodies in a straight line on the same side of the Earth (SYZYGY). The Earth, moon and sun in c result in the new moon, when tide-producing forces are at their max., hence SPRING-TIDES. Ct. QUADRATURE, OPPOSITION.

connate water W retained in sedimentary rocks since their formation (sometimes called *fossil w*). Ct. JUVENILE, METEORIC W.

consequent stream A s whose direction of flow is directly related to or c upon the orig. slope of the land; ct. SUBSEQUENT, OBSEQUENT. First used by J. W. Powell in 1875, intended to signify something more than that water on a slope will flow downhill, but in relation to initial fold-structures. Now sometimes a noun, 'a consequent' ('river' understood).
[*f* SUBSEQUENT STREAM]

conservation The preservation from destruction of natural resources (soil, vegetation, animals) by careful control and management, esp. for the benefit of posterity. This is not so much a 'holding-back' as the maintenance of a favourable balance in the use of the environment. Hence c area.

consociation A vegetation unit, dominated by a single species; e.g. beechwood, dominated by the beech tree. Ct. ASSOCIATION.

constant of channel maintenance The area of a drainage basin surface needed to sustain a unit length of stream channel. C of c m is thus the inverse of DRAINAGE DENSITY. In areas of close fluvial dissection the constant is low (e.g. in the Perth Amboy badlands of New Jersey only 2·6 m² of basin area supports each m of channel). By contrast, in S. England 600 m² of basin area support 1 m of channel in chalk country. C of c m varies according to many factors, incl. rock-type, permeability, climate, vegetation and relief, all of which influence the proportion of precipitation which 'runs off'.

constant slope That part of a s profile of accumulation which lies below the FREE FACE [*f*] above, and cts. with the WAXING and WANING SS.

constructive wave One of a series of gentle ws rolling in steadily on to a coast at about 6–8 a minute, which has a powerful push of the SWASH, and because of frictional retardation a less powerful BACKWASH. It therefore tends to move material (esp. shingle) up a beach, so building ridges.

contact metamorphism See THERMAL M.

continent One of the Earth's major constituent land-masses, composed of SIALIC rocks, rising from the ocean floor. Structurally it includes shallowly submerged adjacent areas (CONTINENTAL SHELF) and neighbouring islands; in this sense, the cs occupy about 30% of the Earth's surface. Actual dry land comprises 29·2% of land on Earth's surface; of N. hemisphere = 39·3; of S. hemisphere = 19·1.

Areas (UN figures)

	million km^2	sq. mi.
Africa	30·6	11·8
N. America	17·9	6·9
S. and Central America	24·3	9·4
Asia	45·6	17·6
Europe	9·8	3·8
Australia (with N. Z. and Oceania)	8·5	3·3
Antarctica	11·4	4·4

Europe without the USSR has an area of 5·0 million km² (1·9 million sq. mi.), Asia without the USSR of 28 million km² (10·9 million sq. mi.), the USSR of 22 million km² (8·6 million sq. mi.).
Note: 'The Continent' indicates Europe to people of Gt Britain.

continental air-mass An a-m whose source-region is a high-pressure area over a c interior; usually of low humidity. Denoted by c in the terminology, and may be of either high latitude (Polar), hence Pc, or low latitude (Tropical), hence Tc. In USA and by UK Meteorological Office the letters are reversed cP, cT.

continental climate The climate of a c interior, characterized by seasonal extremes of temperature, with low rainfall occurring chiefly in early summer. This is mainly the result of great distance from the sea, hence *continentality*, for which several indices are available to plot its distribution; e.g. by W. Köppen, G. T. Trewartha and J. A. Shear, and C. Troll, involving mean temperatures of coldest and warmest months, and mean annual temperature.

continental divide A major d which separates the drainage basins of a continent. E.g. in N. America a 'T'-shaped d separates streams flowing W. to the Pacific, E. and S. to the Atlantic, and N. to the Arctic. *Note:* Triple D. Peak, in Glacier National Park, Montana, is at the junction of the 'T'.

continental drift The hypothesis that c masses have changed their relative positions, the result of fragmentation and moving apart of orig. larger masses. Put forward by A. Snider (1858), developed by F. B. Taylor (1908) and esp. by A. Wegener (1915). The earliest theories were based on apparent similarity of coastlines along each side of the Atlantic Ocean, esp. the 'fitting' of S. America into Africa. It was suggested that GONDWANALAND and LAURASIA were major fragments of Pangaea, which broke up, and portions moved apart. The main problem was the nature of the energy required; later, convection currents created by the accumulation of radioactive heat were postulated. There has been a recent revival of interest in the hypothesis, based on the evidence of PALAEOMAGNETISM, widespread fossil discoveries (e.g. *lystrosaurus*), and the concept of an expanding (as distinct from a contracting) Earth. See PLATE TECTONICS.

continental ice-sheet An i-s of c dimensions: e.g. in Antarctica; the Quaternary i-ss which covered the N. parts of Europe and N. America. See pl. 21.

continental island An i which stands close to, and is structurally related to, a continent, rising from the C SHELF, formed by a rel. rise of sea-level; e.g. British Isles, Newfoundland, Sri Lanka. The ct. is with *oceanic i*s, rising from floors of ocean deeps.

continental platform A continent and its surrounding shelf as far as the edge of the C SLOPE; i.e. the true structural continent, of SIALIC material, in ct. to ocean basins.

continental sea A partially enclosed s, lying on or within a continent in the structural sense, linked with the open ocean; e.g. Baltic S., North S., Hudson Bay, Yellow S.

continental shelf The gently sloping (1° or less) margins of a continent, submerged beneath the sea, extending from the coast to where the seaward slope increases markedly. This outer edge has a depth variously ascribed between 120 m (65 fthms) and 370 m (200 fthms); an avge of 130 m (430 ft) has been suggested. It is well developed off W. Europe [320 km W. of Lands' End], 240 km off Florida, 1200 km off the Arctic coast of Siberia, 560 km off Argentina. Around some continents it is narrower or almost absent, esp. along coasts where fold mountains run parallel and close to the ocean (E. Pacific Ocean). Most c ss are portions of the structural continent inundated by a slight rel. rise of sea-level, though some parts may be the result of: (i) marine planation (e.g. Strandflat off N. Norway); (ii) glacial erosion during a period of low sea-level; (iii) the building-up of an offshore terrace or delta by river deposition; and (iv) deposition by ice-sheets (to which the Grand Banks of Newfoundland are due in part). [*f*]

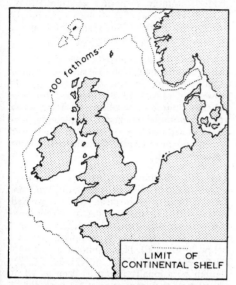

continental slope The marked s from the edge of the C SHELF to the deep-sea or abyssal plain, from about 180–3600 m (100–2000 fthms), in places much further [e.g. to 9000 m (5000 fthms) off the Philippines]. The s is between 2° and 5°. [*f* ABYSSAL]

contorted drift GLACIAL TILL which exhibits foldings, twistings and irregularities, probably due to pressure from an ice-sheet. The contortions are drawn out in the direction of ice-movement; cf. PUSH-MORAINE. The proper name (with capital letters) is used to denote the upper of two TILLS in the cliffs near Cromer, Norfolk, ascribed to the North Sea Glaciation.

contour (-line) A line on a map connecting all points the same distance above (or below) a specif. datum, loosely termed 'sea-level' (see ORDNANCE DATUM). A distinction is drawn between cs based on an instrumental survey, and *formlines* sketched in on maps from gen. observations and from a few located spot-heights. Also called *isohypse*.

contour-interval The vertical distance between 2 successive cs on a map. This is chosen according to the amount of vertical height involved and the scale of the map.

contour-ploughing Ploughing along a slope rather than up and down, to check runoff of rain which might wash away soil; a measure to combat SOIL EROSION.

contrail See CONDENSATION TRAIL.

convection (i) The mass movement of constituent particles within a liquid or gas as a result of different temperatures and therefore different densities within the medium. The movement involves both the medium itself (*c currents*) and the actual heat. In a meteorological context, it involves vertical heat transference within the atmosphere (ct. ADVECTION, horizontal movement). A *c cell* is an updraught of heated air (syn. THERMAL), with a compensatory downward movement of cooler, denser air. (ii) *Forced c* is due to wind turbulence over uneven terrain.

convection rain Rainfall resulting when moist air, having been warmed by CONDUCTION from a heated land surface, expands, rises, and is adiabatically cooled to the DEW-POINT. CUMULUS clouds develop into towering CUMULONIMBUS clouds with an immense vertical range, from which heavy r or hail may fall, accompanied by thunder. C r occurs commonly during the afternoon near the Equator, the result of constant high temperatures and high humidity. With increasing distance from the Equator, the c r becomes assoc. more markedly with summer heating. In maritime temperate climates, most c r is associated with unstable Polar air-masses.

conventional projection A type of MAP P constructed according to mathematical formula, not a PERSPECTIVE P. The formula is selected to preserve some partic. property. See also specif. ps.

conventional sign A standard s used on a map to indicate a partic. feature, wherever the scale of the map is such that this feature cannot be drawn to scale. It may be a letter or a symbol. C ss include boundary lines, which may or may not follow landscape features. A map carries a LEGEND showing the characteristic c ss used.

convergence (i) In *oceans*, sharply defined lines separating converging masses of water, often of differing temperature and salinity; e.g. in the S. Ocean about 50°S., where cold dense Antarctic water meets warmer and more saline water spreading S. (ct. DIVERGENCE, DISCONTINUITY). (ii) In *climatology*, a type of air-flow such that in a given area at a given altitude inflow is greater than outflow, so that air tends to accumulate. If density remains constant, such horizontal c must be accompanied by vertical motion; thus surface c is usually accompanied by an ascending air current. C may be produced either by STREAMLINES approaching each other (*streamline c*), or by a single air current being subject to a progressive reduction of velocity (*isotach c*). Areas with streamline c usually have isotach c as well; e.g. INTERTROPICAL CONVERGENCE ZONE. Pure *velocity c* occurs in the S. Indian Ocean in July. (iii) *C of species* (biological): the increase in degree of similarity between different species which are developing in such a way that their life forms become progressively more alike. This is shown in fossil sequences (*palaeological c*). (iv) The gradual decrease in thickness of a geological formation, so that the upper and lower horizons converge. [*f*]

STREAMLINE CONVERGENCE	ISOTACH CONVERGENCE
constant velocity different directions	20 knots 15 knots 10 knots constant direction changing velocity

convexity A form of slope, characterised by progressively increasing angle towards the slope base. C is usually developed on the upper part of composite slope profiles (e.g. the '*summital c*' of convexo-rectilinear-concave profiles). Many authors have attributed c to the dominant activity of soil creep and other mass movements; other causes include structural controls (e.g. curvilinear sheet jointing giving the convex dome form of tropical INSELBERGS).

coombe See COMBE.

Coombe Deposit A mass of unstratified rubble accumulating in a valley bottom and masking the lower slopes; the product of a past phase of solifluction. *C Rock* is specif. the c deposit of a Chalk landscape, a compacted and hardened chalk mud with sand and flints. A main area of occurrence is on the Sussex coastal plain. See also HEAD. See pl. 11.

cop A small rounded hill in N. and central England; e.g. Mow C., 340 m above sea-level, near Burslem, Staffs.

copper Almost certainly the earliest metal known to Man, since it occurred commonly on the surface as a free (*native*) metal, which could be used for ornaments and vessels, later alloyed with tin to make bronze ('Bronze Age'), and with zinc to make brass. It occurs in nature as a metal and in a group of over 300 different minerals (sulphides, carbonates, oxides), notably c pyrites (c sulphate or chalcopyrite, $CuFeS_2$), which commonly decomposes to form blue (*azurite*) and green (*malachite*) carbonates (Cu_2Co_3–OH_2).

coprolite (i) The fossilized or petrified excrement of creatures, esp. reptiles (saurians), mainly of Jurassic and Cretaceous age. It consists of phosphatic nodules, and forms a valuable source of fertilizer. (ii) Used more gen. for nodules or masses of calcium phosphate in sedimentary rocks, as in the Lower Greensand near Cambridge and near Flamborough Head.

coral (i) A lime-secreting marine polyp, mainly living in colonies in inter-tropical seas. Cs can only grow in clear, well-oxygenated water, with plentiful supplies of microscopic life as food; they cannot live in fresh or silt-laden water, nor if its temperature falls below 20°C (68°F). They cannot live at depths much exceeding 45–55 m (25–30 fthms), though occasionally found as deep as 70 m (40 fthms); they cannot exist for long out of water, and are rarely found above low-tide level. They are confined to seas within 30°N. and 30°S., though exceptionally grow further away (e.g. Bermudas in the path of the warm water of the Gulf Stream), and are absent from W. coasts of continents because of cool currents and up-welling cold deep water. Estimates of coral growth vary from 1–10 cm per annum, up to 25 cm for light feathery extensions. (ii) Also used of the hard calcareous substance (*c rock*), the accumulation of c skeletons. Throughout geological time, cs have been reef-builders; e.g. Wenlock Edge, Shropshire, which is in part a Silurian c reef; Corallian reefs of Jurassic age along the coast of Dorset near Osmington.

coral mud An accumulation of very fine fragments of c found in the bathyal zone of the CONTINENTAL SLOPE around c reefs.

coral reef A reef composed of c limestone, the accumulated skeletons of c polyp colonies. The three main types of r are: (i) FRINGING R; (ii) BARRIER R [*f*]; (iii) ATOLL [*f*].

coral sand Comminuted fragments of c, ground small by waves, forming white s around reefs.

corange line A l on a climatic map joining all places with the same RANGE of temperature between January and July.

cordillera (Sp.) A series of mountain ranges, broadly parallel or closely *en échelon*, belonging to a single OROGENIC belt. Applied broadly to the whole 'Mountain West' of N. America. The Cs de los Andes extend from Venezuela to Cape Horn, and are rarely less than 320 km wide. Also used in S. America for small individual ranges; e.g. C. de Corabaya, C. de Clonche, C. Negra.

core The central mass of the Earth, of about 3476 km (2160 mi.) radius, bounded by the GUTENBERG DISCONTINUITY at 2900 km (1800 mi.) from the surface. It probably consists of a metallic mass of nickel-iron (NIFE), with an overall density of about 12·0. Recent work indicates the presence of an inner core, of 1380–1450 km (860–900 mi.) radius, with a density of about 17·0 at the centre. Temperature in the inner c is estimated to be about 2700 K. See EARTHQUAKE [f].

core sample A s of soil, rock or ice, obtained by driving a hollow tube into the medium concerned, and withdrawing the s intact. Such coring-tubes are used on very long wires to obtain a c s of the material underlying the ocean floor.

corestone See WOOLSACK.

Coriolis Force The effect of the apparent deflecting f produced by rotation of the earth upon a body moving on its surface, deflected to the right in the N. hemisphere, to the left in the S. (see FERREL'S LAW). The C F is proportional to $2vw \sin \phi$, where v=velocity of the object, w=angular velocity of the earth's rotation, and ϕ is the latitude. Called after the French mathematician, G. G. de Coriolis, who discussed its effects in 1835, and the concept was developed by W. Ferrel in 1855.

cornbrash A thin stratum of impure calcium carbonate in the Middle and Upper Jurassic, which weathers to form a stony soil. Its Wiltshire dialect name was 'brash', and as it produced good crops of cereals the name was adopted for the formation by Wm. Smith (c. 1813).

cornice An overhanging edge of compact snow on the lee side of a steep mountain ridge, developed by eddying. Occas. each side of a ridge has a c; e.g. Obergabelhorn in the Pennine Alps, Switzerland.

corniche (Fr.) In a physical sense syn. with CLIFF.

corona Luminous concentric rings around sun or moon, ranging from blue (inner) via green and yellow to red (outer), the result of diffraction of light by water-drops. Ct. HALO. Its angular diameter is much less than that of a halo. A *solar c* around the circumference of the sun is visible at a total eclipse.

corrasion Mechanical erosion, or frictional wearing down, of a rock-surface by material moved under gravity or transported by running water, ice, wind and waves; c is the *process*, ABRASION the *result* on the mass attacked.

corrosion The wearing away of rocks by chemical processes, comprising: (i) SOLUTION (e.g. common salt); (ii) CARBONATION (e.g. limestone); (iii) HYDROLYSIS; (iv) OXIDATION; and (v) HYDRATION. The results are: (*a*) conversion of orig. minerals into secondary weaker minerals more readily removable or soluble; and (*b*) removal of 'cements' in sedimentary rocks, so that the consolidation or adhesion of the particles is weakened, and the rock tends to crumble.

corundum A mineral consisting of aluminium oxide (Al_2O_3), which in its finest form occurs as various gems, notably ruby and sapphire. No. 9 on the MOHS' SCALE of hardness, second only to diamond.

co-seismal line A l connecting points on the Earth's surface at which an earthquake wave has arrived simultaneously. Ct. *isoseismal l*, an indication of earthquake intensity.

cosmic App. to phenomena or features which occur or are situated beyond the Earth's atmosphere; e.g. *c dust*, *c particles*, *c radiation*.

cosmogeny A theory or investigation into the origin of the solar system.

cosmography Used extensively in the past to indicate the description and mapping of the universe, including the Earth; a common title of ancient 'physical geographies'.

cosmology The scientific investigation of the laws of the universe as an ordered entity by astronomers, physicists and mathematicians; e.g. 'the new c' of F. Hoyle.

costa (Sp.) A stretch of coast, e.g. C. Brava; later applied to other resort areas along the coast of Spain; e.g. C. del Sol, Blanca, Dorada.

côte (Fr.) (i) An ESCARPMENT or steep-edge of a hill in France; e.g. C. d'Or. (ii) A section of coast; e.g. C. d'Azur.

coteau (Fr.) Given by early French explorers in N. America to a sharp ridge of hills or a prominent escarpment; e.g. C. des Prairies, Missouri C.

Plate 12. Stair Hole, near Lulworth, Dorset, England. The COVE has resulted from penetration by the waves of the base of a limestone barrier (right) to attack weaker clays (left). Note the small ANTICLINE and SYNCLINE in the exposed strata at the far end of the cove. (*R. J. Small*)

co-tidal line A line on a tidal chart joining points at which high water occurs simultaneously; it radiates from an AMPHIDROMIC POINT, where the water remains at approx. the same level, the height of the tidal rise increasing outwards to the extremities of the c-t l.
[*f* AMPHIDROMIC SYSTEM]

coulée (Fr.) (i) A congealed lava flow, of basalt, rhyolite or obsidian, esp. in USA (no accent); e.g. in the San Francisco Mts, Arizona. (ii) An overflow channel which in the past carried melt-water from an ice-sheet, now dry; e.g. Grand C. in Washington State, the Ice-age channel of the Columbia R. (iii) A solifluction c of material moved by PERI-GLACIAL action, having the appearance of a 'tongue' of debris.

couloir (Fr.) A steep, narrow gully on a mountain-side, esp. in the French Alps; e.g. Whymper C. on the Aiguille Verte, above Chamonix.

countertrades Obsolete. See ANTI-TRADES.

country rock (i) A mass of r traversed and penetrated by later INTRUSIONS of igneous rock. (ii) In USA, gen. bedrock, cfd. unconsolidated material.

cove (i) A small rounded bay, with a narrow entrance; e.g. Lulworth C., Dorset. (ii) A steep-sided rounded hollow among the mountains of the English Lake District; e.g. West C., Green C. and Hind C. on Pillar Mtn, Ennerdale. (iii) Any steeply walled, semicircular opening, esp. at the head of a valley; e.g. Malham C., Yorkshire. See pl. 12.

cover-crop A quick-maturing c grown between main cs as a protective mat over the soil, so as to reduce the danger of SOIL EROSION.

crag (i) A steep, rugged rock projecting from a mountain side, esp. in the English Lake District; e.g. Dow C. (Coniston), Gimmer C. (Langdale). (ii) Compacted shelly sands in E. Anglia, hence proper name of certain geological formations: Coralline C. of Pliocene age, Red, Norwich and Weybourne Cs. of Pleistocene age. Probably from Celtic *cregga*, a shell.

crag-and-tail A mass of rock ('the crag'), lying in the path of an oncoming ice-sheet, so protecting softer rocks in its lee from erosion, and leaving a gently sloping 'tail'; e.g. the hard basalt plug of Edinburgh Castle Rock (the c), with a t of Carboniferous Limestone lying on Old Red Sandstone, sloping E., along which now extends the Royal Mile. Clack-mannan in Scotland has the ruins of a tower on a 58 m high c with the main street extending to the E. down the t. The t may also comprise TILL preserved in the area of stagnation in the lee of the protective mass; e.g. Arthur's Seat, Edinburgh. See STOSS. [*f*]

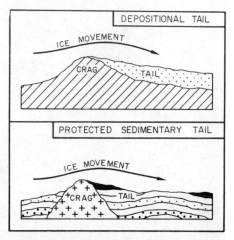

crater (i) The rounded funnel-shaped hollow at the summit of a volcano (cf. CALDERA), the result of either a PAROXYSMAL ERUPTION or ENGULFMENT. (ii) A depression caused by the impact of a meteorite; e.g. Meteor C., Arizona, USA. See pl. 13.

crater-lake A l accumulated in the c of a volcano; e.g. L. Toba in N. Sumatra; Öskjuvatn in Iceland; C. L., Oregon, 600 m deep, the water surface at 1877 m above sea-level, with a perimeter of about 48 km. [*f*]

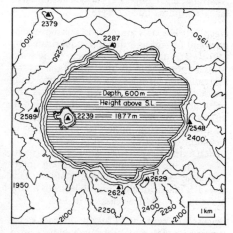

craton, kraton A stable and gen. immobile part of the Earth's crust, usually on a large scale; syn. SHIELD. Hence *cratonic, cratogenic*.

Plate 13. The summit of Surtsey, southern Iceland. Note the two CRATERS, and the small volcanic cone developed more recently in the left-hand crater. (*Eric Kay*)

creek (i) A narrow tidal inlet, espec. on a low-lying coast among mud-banks; e.g. Ashlett C., Southampton Water. (ii) A small stream in USA. (iii) An intermittent stream in Australia; e.g. Coopers C., which (when there is water) drains into L. Eyre.

creep (i) The slow gradual viscous movement downhill of soil and REGOLITH, lubricated by rain-water, under the influence of gravity. May be observed in the landscape as an accumulation of material against stone walls on the up-slope side, leaving bent posts and trees, and in rock outcrops a distinct line of distortion. Basically the result of gravity, c may be caused by numerous minor agencies working together. See CAMBER. (ii) In a cold GLACIER, the slow deformation of ice near its base, the result of pressure causing intermolecular and intergranular motion. This gives it a plastic quality, and so leads to a slow movement of the ice-mass on its bed.

crepuscular rays (i) Clearly defined rs of sunshine which break through chinks in a heavy layer of STRATOCUMULUS clouds toward the Earth's surface, made luminous by water or dust particles in the atmosphere. Pop. called 'Jacob's Ladder'. This is a transferred term, since it lit. means 'twilight'. (ii) Dark and light bands, seen just after the sun has passed below the horizon, and diverging upwards from the sun's position. The light bands are areas still illuminated by sunshine, the dark bands indicate where hills or clouds cut off the sun's rs.

crescentic dune See BARKHAN.

crest (i) In hydrology, the max. stage of a flood-wave which passes down a river; e.g. on the Mississippi. (ii) The highest point of an ocean WAVE; the distance between 2 successive cs is its *wavelength*. (iii) The highest part of a summit ridge.

crêt (Fr.) An in-facing ESCARPMENT in the French Jura, forming a wall of a COMBE worn along the crest of an anticline by river erosion, culminating in a near-vertical limestone cliff (*la corniche calcaire*), below which a gradual slope, covered with scree and in part wooded, descends to the floor of the combe. The highest point of the Jura is a summit on such an escarpment, the C. de la Neige (1723 m, 5653 ft).

Plate 14. Twin CREVASSES produced by tensional forces near the snout of the Mont Miné glacier, Switzerland. Note the thin layer of stones and dirt exposed by ABLATION of the ice surface. (*R. J. Small*)

Cretaceous The 3rd of the geological periods of the Mesozoic era, and the system of rocks laid down during that time, dated c. 136–65 million years ago. The constituent series in England are as follows:

Chalk ⎰ Upper   Pure white limestone
        ⎨ Middle
        ⎱ Lower
Upper Green-
    sand   Sandstone
Gault   Bluish clay
Lower Green-
    sand   Sandstone
Weald Clay   Thick clay
Hastings Sands   Sands and clays

The boundary between the Lower and Upper C is variously interpreted; some geologists place this at the base of Lower Chalk, as the horizon of most marked lithological change, others include Gault and Upper Greensand in the Upper C.

crevasse (Fr.) (i) A deep crack or fissure in a glacier, trending either transversely across it where the slope increases, or longitudinally down it where the glacier spreads out because of the widening of its valley; in either case, differential movement within the ice causes tension and shearing. Intersecting **c**s on a steep slope will produce an ICEFALL; pinnacles of ice isolated by **c**s are SÉRACS. (ii) A break in a levee along a river bank. See pl. 14, 35.

crevasse filling A straight ridge of stratified sand and gravel, formed by the **f** of a **c** in a stagnant ice-sheet which later melted. Similar in form to an ESKER (*sensu lato*), but not winding or branching.

crib (Welsh) A high summit ridge; e.g. C. Goch on the Snowdon 'horse-shoe' of peaks.

critical temperature A t of specif. importance for vegetation; e.g. freezing-point (0°C, 32°F), since many plants are vulnerable to frost, esp. in blossom. Another is 6°C, 43°F, since for most plants active growth cannot take place below this point.

critical tractive force The force needed to initiate movement of particles in a stream channel. The controlling factors are (i) water depth; (ii) water surface slope. These together largely determine stream velocity (see CHEZY EQUATION). F. Hjulstrom has shown that the **c t f** needed for movement of sand grains is less than that needed for gravel or fine clay particles; hence channels developed in sand are very prone to bank erosion.

Cromerian The ante-penultimate interglacial period in Britain, represented typically by lake deposits exposed at the cliff-foot at West Runton, Norfolk (the 'Cromer Forest Beds').

cross-bedding See CURRENT-BEDDING.

cross-cutting relationships, Law of An igneous rock is younger in age than any other rock across which it cuts; one method of establishing the rel. age of rocks.

cross-faulting Two intersecting series of FAULTS.

cross grading A process leading to the development of secondary 'rills' on slopes. R. E. Horton has shown that on a slope where parallel rills are initiated by surface run-off, 'cross flow' will occur under conditions of high discharge when running water overtops the divide between adjacent rill channels. Channels receiving such increments of water develop more rapidly, to take over near-by 'catchments' as the overtopped divides are eroded away. Eventually major rill channels, with extensive side slopes, will be joined by tributary rills crossing the lines of the eroded divides. [*f*]

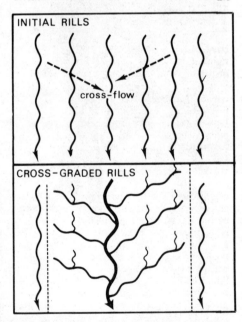

crossover point Points in a meandering channel defining half-meander lengths, and occurring where the channel crosses the 'meander belt axis'. The **c p** thus occurs at a change in direction of the stream channel (from leftwards to rightwards, or vice versa).

cross-profile, of a valley A p drawn transversely across a river valley, roughly at right-angles to the stream on its floor. The form of the **c-p** is determined by: (i) nature of the rocks; (ii) erosive power of the stream; (iii) effects of weathering and rainwash on the valley sides; (iv) the stage which the processes in (ii) and (iii) have attained in their effects on (i). [*f*]

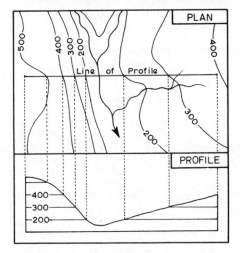

cross-wind A w blowing more or less at right-angles to the course of a moving object or person; esp. important to aircraft.

crumb-structure, of soil The accumulation of fine soil-particles into crumbs, thus coarsening the texture of a clay, and helping to make it more workable. This is assisted by the flocculating effect of lime.

crush-breccia See FAULT-B.

crust, of the Earth The upper granite (SIAL) and intermediate basic (crustal SIMA) layers of the Earth, about 16–48 km in thickness, lying above the MOHOROVIČIĆ discontinuity, and forming the outermost 'shell' (see LITHOSPHERE). Its usage dates from the time when geologists thought of the Earth as a hot liquid body, with an outer cool, solid 'skin'. It is now realized that this is too simple a picture. The analysis of the rocks comprising the c by weight is: (i) *by elements:* oxygen 47%; silicon, 28%; aluminium, 8%; iron, 5%; calcium, 3·5%; sodium, 2·5%; potassium, 2·5%; magnesium, 2·2%; titanium, 0·5%; hydrogen, 0·2%; carbon, 0·2%; phosphorus, 0·1%; sulphur, 0·1%; (ii) *by oxides:* silica, 59·1%; alumina, 15·2%; iron, 6·8%; lime, 5·1%; soda, 3·7%; magnesia, 3·5%; potash, 3·1%; water, 1·3%; titania, 1·0%.
[*f* ISOSTASY]

cryergic App. to physical phenomena resulting from frost action.

cryolaccolith See HYDROLACCOLITH.

cryolite A mineral, Na_3AlF_6, sodium aluminium fluoride, found in PEGMATITE veins in Greenland.

cryology The scientific study of snow and ice; one of the 4 divisions (*sensu lato*), the others being POTAMOLOGY, LIMNOLOGY and HYDROLOGY (*sensu stricto*).

cryopedology Study of structures in the ground resulting from intensive frost action, incl.: (i) the processes which have formed them; (ii) their occurrence; and (iii) civil engineering problems produced in overcoming the difficulties which such structures present.

cryoplanation Land reduction by intensive frost action, incl. SOLIFLUCTION and the work of rivers in transporting material produced by such action.

cryoplankton Microscopic organisms, vegetable and animal, living under conditions of permanent snow and ice. See CRYOVEGETATION.

cryoturbation Frost action which disturbs and modifies superficial layers in the PERIGLACIAL zone. Gen. replaced by CONGELITURBATION, which has a more precise scientific meaning.

cryovegetation Plant communities developing in permanent snow and ice, hence *cryophyte*; e.g. purplish-brown 'bloom' on the glaciers of S. Alaska, denoting the presence of algae.

cryptovolcano An area of volcanic activity, where the explosive energy of gases has shattered the basement rocks to great depths, forming numerous PIPES as a result of FLUIDIZATION, and has ejected large quantities of shattered material (PYROCLASTS). E.g. Swabian Jura (W. Germany), esp. the Ries Basin, where activity is still manifest as hot springs around its margins.

cryptozoic aeon (eon) The span of time between the formation of the Earth and the beginning of the Cambrian period, syn. with Precambrian. Most rocks of this age are IGNEOUS or METAMORPHIC, and the few unaltered SEDIMENTARY rocks contain few traces of primitive life-forms. E.g. algal fossils, trace fossils, leaf-like fossils, primitive worms; gen. these are vegetation, and animal remains appear only at the very end. Lit. 'hidden life'. Ct. PHANEROZOIC.

crystal A solid aggregate of molecules with an orderly atomic arrangement, which may be bounded by symmetrically arranged plane surfaces; e.g. quartz is a 6-sided prism with pyramid terminations, a diamond 8-sided, pyrite cubic.

crystalline rock A r with constituent minerals of c form, developed either through cooling from a molten state (i.e. most igneous rocks, esp. plutonic ones which cool slowly at depth), or as the result of metamorphism (limestone recrystallized to form marble, shales and slates to form schist).

cuesta (Sp.) A ridge with a steep SCARP-SLOPE (*escarpment*) and a gentler BACK-SLOPE. Now used partic. in connection with an asymmetrical ridge resulting from differential denudation of gently inclined strata; a more resistant bed (limestone, chalk) stands out as a low ridge separated by intervening vales worn in clay. In this case the gentle back-slope of the ridge is practically parallel to the dip of the strata. Now more commonly used than *scarp* or *escarpment*, which should refer to only the steep slope, and more succinct than 'scarped ridge'. Adj. *cuestaform*. [*f, opposite*]

cuirass (Fr.) A hardened crust of the REGOLITH, found esp. on level terraces, indurated subaerially through evaporation and the precipitation of metallic oxides, esp. of aluminium and iron, usually under tropical conditions of alt. wet and dry seasons. A c is syn. with FERRICRETE and with LATERITE in the modern restricted sense, though *lateritic c* may be preferable in the more restricted sense (in ct. to *lateritic clay*).

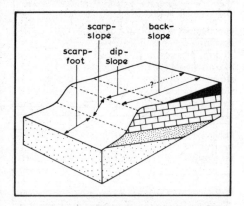

culmination (tectonic) When the strike of a NAPPE (i.e. a direction at right-angles to the folding movements) is traced, it seems to undulate, with higher cs alternating with DEPRESSIONS. The c thus carried the overlying rocks of the nappes upwards, so that they have been more exposed to denudation and therefore largely removed, usually, as a result, exposing crystalline basement rocks in the Alps; e.g. Aiguilles Rouges—Mont Blanc, the Aar—Gotthard massifs. [*f*]

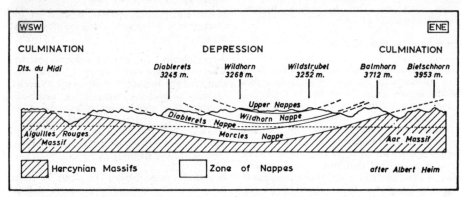

Culm Measures (i) A formation, chiefly of shales, sandstones and thin layers of impure anthracite, found mainly in Devon and E. Cornwall. (ii) Fine dusty coal in S. Wales.

cultural vegetation All types of v altered or modified by man, directly or indirectly, ranging from *slight* (light grazing, selective logging) to *extreme* (wholesale removal of NATURAL v and its replacement by plantations or fields). Natural plant COMMUNITIES have thus been replaced by *substitute communities*. See also SEMI-NATURAL v. Subdivisions incl. MESSICOL v (cultivated *v sensu stricto*), *ruderal* *v* (plants on waste land, building lots, railway sidings and roadsides); and *segetal v* (weed communities among cultivated plants).

cumulative frequency curve A f c in which values are added successively to each other and then converted into percentages. The 50th percentile on such a curve is the median.

cumulo-dome A dome-shaped, apparently craterless, volcano, built up of successive flows of viscous silicic lava; e.g. Grand Sarcoui in Auvergne, France. Syn. with MAMÉLON.

cumulonimbus (*Cb*) A CUMULUS cloud which develops to an immense vertical height 10–11 km, often with the upper part spread out like an anvil; usually associated with thunderstorms and torrential rain or hail. From the side the cloud is dazzling white, from below its base may be almost black. Its upper portion is pop. a 'thunderhead'. See pl. 15.

cumulose deposit A superficial mantle of organic d*s*, such as peat and other swamp materials.

cumulus (*Cu*) A convection cloud which grows vertically from a flat base (the condensation level) into a large white globular or domed summit, sometimes of immense height. 'Fair weather **c**' dies away in the evening as convection currents lessen, others may develop into CUMULONIMBUS. Adj. *cumuliform*. See pl. 16.

cupola (i) A dome-like mass of rock protruding from the 'roof' of a BATHOLITH. (ii) A small isolated mass of intrusive rock lying near, though separate from, a batholith.

current (i) The distinct and defined movement of water in the channel of a river. (ii) The vertical motion of air in an air-mass (CONVECTION **c**). (iii) A permanent or seasonal movement of surface sea-water (DRIFT); e.g. N. Atlantic Drift, Florida C., Benguela C. (iv) See THERMOHALINE C. (v) A *tidal c* through a restricted channel.

current-bedding Thin strata or LAMINAE inclined at varying oblique angles to the gen.

Plate 15. CUMULONIMBUS CLOUD. (*Royal Meteorological Society*)

Plate 16. CUMULUS CLOUDS (viewed from above). Note the irregular rounded summits, indicative of considerable air ascent. (*The Times*)

stratification, esp. in sandstone, resulting from changing currents of water and wind, responsible for the deposition of constituent sand-grains. At intervals a change of direction of the c may truncate the existing beds, and new layers at a different angle may be added later. Syn. with *cross-bedding* and *false-bedding*. [*f*]

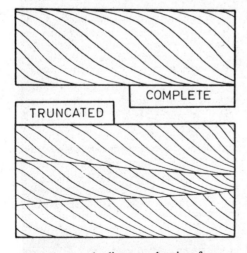

current rose A diagram showing for any point in the sea the % frequency of c flow in each direction by proportional radiating rays.

curve-parallels (i) Superimposed graphs (in series) of related climatic phenomena; e.g. rainfall with rel. humidity, vapour pressure or evaporation. (ii) Graphs used to demonstrate similarities of climatic trends over a period of time at different stations, with a horizontal time scale and a vertical scale in degrees of temperature or mm of rainfall.

cusec A unit of measure of the discharge of a river; i.e. the number of cu ft per second passing a partic. reach. 1 cusec = 538 000 gallons per 24 hours. The number of cusecs is obtained by multiplying rate of flow in ft per second by the cross-section in sq. ft of the river at the point of observation. Now being replaced by m^3 per second ('cumec'). 1 cusec = 0.028 $m^3 s^{-1}$; 1 $m^3 s^{-1}$ = 35·313 cusecs.

cuspate delta A d formed where a river reaches a straight coastline along which wave-action is vigorous, so that material is spread out uniformly on either side of the river mouth; e.g. R. Tiber, Italy.

cuspate foreland A broadly triangular **f** of shingle or sand projecting into the sea, caused by the convergence to an apex of separate SPITS or beach-ridges broadly at right-angles, formed by 2 sets of powerful constructive waves; e.g. Dungeness; Cape Kennedy, Florida; the Darss in E. Germany. [*f*]

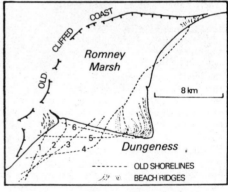

cusum chart A diagram in which cumulative sum techniques are used to analyse a series of slope-angles produced at regular distances along a SLOPE-PROFILE LINE.

cutoff A channel cut by a river as a short-circuit across the neck of an acute MEANDER, leaving an OXBOW. [*f* OXBOW]

cuvette (Fr.) A large-scale basin, non-tectonic in origin, in which deposition of sediment has taken or is taking place; e.g. the Anglo-French C. (S.E. England and N.E. France).

CVP Index An i involving the relationship between climate, vegetation and productivity, with specif. ref. to timber growth. Factors involved include mean temperature of the warmest month, mean annual temperature range, precipitation, EVAPOTRANSPIRATION and length of growing season.

cwm (Welsh) See CIRQUE.

cycle While this would appear strictly to mean a complete round of events or circumstances, its use is well established in geographical literature for a succession of stages or events which occur repeatedly in the same order, notably the passage of energy or material through successive systems. Used partic. of c of EROSION, hence c of NORMAL E, of ARID E, of GLACIAL E, of MARINE E. See also HYDROLOGICAL C, CARBON C, ROCK C, C OF SEDIMENTATION. Used by A. C. Lawson in 1894 and C. W. Hayes in 1899, but chiefly developed by W. M. Davis (from 1889).

cycle of erosion The modification of the physical landscape as a result of the action of natural agencies in an orderly progressive sequence; the full hypothetical c ranges from the uplift of the land into an upland to an ultimate low, almost featureless plain. The concept was first developed and formalized by W. M. Davis (1889). It involves the concept of age: *youth, maturity, old age, rejuvenation*. There has been much recent criticism of these evocative age-terms in descriptive analysis of land-forms, a belief that using the human life-c as a simile can be carried too far. In the long term the time element cannot of course be ignored, and there must be changes between the landscape of the past, the present and the future, though not with rigorous cyclic inevitability. A c implies and involves a repetition of a series of changes which return to a point of origin, as in the case of the HYDROLOGICAL C. Some prefer *sequence* for the time/stage element in landform development. However, in this last context many authorities maintain that the concept of a c of e is logical and exact, since almost all the Earth's surface (at any rate any mountain range folded before the Cainozoic) has been planed at least once and therefore has passed through at least 1 c of e.

cycle of sedimentation The deposition of material in a basin during one complete phase of a marine TRANSGRESSION; i.e. dry land, shallow water, deep water, shallow water, dry land, with their respective associated deposits. See CYCLOTHEM.

cyclic time One of three categories of 'geomorphological time-scale' devised by S. A. Schumm and R. W. Lichty (see GRADED and STEADY TIME). C t corresponds approximately to the length of time required for the completion of erosion cycles (i.e. in the order of tens of millions of years).

cyclogenesis The atmospheric process by which an intense tropical storm develops at a local heat-source ('warm-core') over the ocean; a tremendous vortical disturbance is set up, indicated by a whorl of towering CUMULUS clouds.

cyclone A small tropical low pressure system, with a diameter of 80–400 km, occurring in the Arabian Sea, Bay of Bengal and the Indonesian seas between about 6°N. and 20°N., mainly in April–June and Sept.–Dec. The BAROMETRIC GRADIENT is very steep, falling to 965 mb. (28·5 ins) or lower, sometimes by about 40 mb. in a few hours; lowest recorded 877 mb. in 1958, in Pacific Ocean E. of Guam. The centre ('eye') of the storm is a small area about 20 km across, but round it whirl winds of tremendous force (120–280+ km.p.h.). Torrential rain, assoc. with thunderstorms, occurs. It forms a great menace to shipping. In Dec. 1974, c 'Tracey' destroyed much of Darwin in N. Australia. At one time c was used to denote any small unit-area of low barometric pressure, but in middle latitudes it should be avoided in favour of DEPRESSION, 'low' or 'disturbance', although CYCLONIC RAIN is still used.

cyclonic (frontal) rain Precipitation along the FRONTS and in the warm sector of a DEPRESSION in middle and high latitudes, caused by one air-mass overriding or undercutting another, or by CONVERGENCE. Drizzling rain falls in a broad belt, followed by more concentrated squally rain as the COLD FRONT passes. This rainfall is intensified by the effect of relief, as a depression crosses transversely a coast backed by uplands; e.g. W. Britain.

cyclostrophic wind Air movement on a strongly curved path around a low pressure system where the CORIOLIS FORCE is weak or negligible cfd. centripetal acceleration, thus creating a gradient wind parallel to the isobars. This is partic. evident in a TORNADO and a tropical CYCLONE.

cyclothem A cycle of sedimentation involving a change of environmental conditions, hence the sequential deposition of non-marine sandstone, shale, limestone, the growth of swamp forest (from which coal was derived), followed by a very slow rise of sea-level with deposition of marine limestones and shales.

Cylindrical Equal Area Projection A p with meridians and parallels as straight lines, perpendicular to each other. The Equator is made its true length to scale ($2\pi R$), and divided equally for meridian intersections; e.g. 10° meridian interval $=\frac{2\pi R}{36}$. To make this an equal area **p**, the area between any 2 parallels is made the same as on the globe to scale. Distance of each parallel from the Equator $= R \sin \theta$, where $\theta =$ latitude. The intervals between the parallels decrease from the Equator, so that the increasing E. to W. exaggeration produced by the parallel meridians is balanced by compression in a S. to N. direction by increasing the closeness of the parallels. This **p** is not much used, except in its oblique form.

cylindrical projection A type of **p** in which the globe is regarded as projected upon a cylinder at a tangent to or intersecting its surface. In effect this is opened up and laid flat. The meridians and parallels are straight lines intersecting at right-angles; e.g. MERCATOR, GALL, CYLINDRICAL EQUAL AREA, EQUIRECTANGULAR (or *Simple Cylindrical* or *Plate Carrée*) P.

cymatogeny A term proposed by L. C. King (1961) to describe the large-scale warping of crustal blocks e.g. the Archaean shield or 'basal complex' of Africa. 'By such deformations, in which only vertical displacement is involved, the Earth's surface is thrown into gigantic undulations, sometimes measuring hundreds of miles across and with vertical displacements (of) thousands or even tens of thousands of feet'. A major cymatogenic warping affected E. Africa in late-Tertiary and Pleistocene times, producing broad domes whose crests collapsed to form the western (Albertine) and eastern (Gregory) rift valleys.

dacite A fine-grained EXTRUSIVE igneous rock, of the same composition as quartz-diorite. Mt Lassen (Cascade Mtns, USA) is largely formed of it.

dale A broad open valley, as in the Pennines (e.g. Swaledale, Wensleydale) and English Lake District (e.g. Borrowdale, Langdale).

Dalmatian coast A **c**-line where the trend of the relief is broadly parallel to the **c**, and a rise of sea-level has affected its margins. The outer ranges have become lines of islands, the parallel valleys long inlets; e.g. the type-region of the Yugoslavian coast of the Adriatic Sea (hence 'Dalmatian'). [*f* CONCORDANT COAST]

dam A barrier built across the course of a river, to control and impound the flow of water. Also used of natural features: ice-**d**, lava-**d**, vegetation-**d**, calcareous-**d**, causing the ponding of lakes.

dambo A river valley with a wide marshy floor, developed in seasonally humid tropical plains in central Africa. Stream channels are poorly defined, and, to quote M. F. Thomas, **d**s 'often appear to function as little more than concentrated zones of sheet wash'. The underlying rock is usually deeply weathered and often lateritized.

Daniglacial stage A **s** of the retreat of the Quaternary ice-sheets in N.W. Europe, c. 18–15 000 B.C. The ice left Denmark during this **s**, though Norway and Sweden were still covered.

Darcy's law An equation expressing the rate of flow of ground water through an AQUIFER, or of oil in a stratum, put forward in 1856 by a Frenchman, H. Darcy. $V = P\frac{h}{l}$, where $V =$ velocity, $h =$ HEAD, $l =$ length of distance of flow between 2 points, and $P =$ coefficient of PERMEABILITY of the particular aquifer. A unit of permeability is now exactly defined as a *darcy*, the working unit a *millidarcy* (0·001 darcy).

datum A fact, point or level known or taken as the basis for reasoning, deduction or measurement. Thus **d**-level is the zero from which land altitudes and sea depths are determined; see ORDNANCE D.

day-degree See DEGREE-DAY.

dead cliff A former sea-cliff no longer directly affected by wave erosion, owing to (i) accumulation of beach material, or (ii) a fall in sea-level. **D c**s are degraded by weathering to angles of 30° or less, and become covered in time by vegetation.

dead ground An area invisible to an observer because of the form of the intervening land.

[*f*]

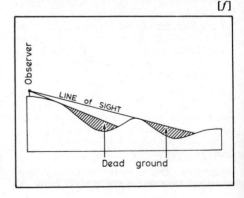

dead ice Stagnant i, usually covered with earthy material and boulders, at the margin of a stationary glacier or ice-sheet. This slowly melts, depositing the material as ABLATION TILL.

dead reckoning Navigation of a ship by careful calculation from its speed and direction, with no external assistance such as radio or astronomical observations.

debris (Fr.) Superficial accumulation of loose material, disintegrated rocks, sand and clay that have been moved from their orig. site and redeposited. Used in English without an accent.

deciduous (forest) The property of annual autumnal leaf-shedding ('the fall') of broad-leaved trees and a few varieties of conifer in middle and high latitudes, as opposed to evergreens. In a Tropical Rain-forest leaf-shed may occur at any time, in a Monsoon Forest during the hot season.

decken structure (Germ., *Deckenstruktur*) A series of RECUMBENT FOLDS or OVERTHRUST masses, lying above each other.

declination (i) Of a *heavenly body*, the angular measurement along a GREAT CIRCLE through both Celestial Poles and the body, measured from the Celestial Equator to the body; see CELESTIAL SPHERE. D is analogous to latitude in a terrestrial sense. The **d** of the sun varies from $23\frac{1}{2}°$N. (about 21 June) to $23\frac{1}{2}°$S. (about 22 Dec.); this can be found in the *Nautical Almanac*. The complement of the **d** $(90°-d)$ is the *Polar Distance*. The **d** of stars varies very little during the year. (ii) *Magnetic*, the angular distance (either E. or W.) between true N. and magnetic N.; see AGONIC [*f*]. D at Hartland in Devon (1967) was 9° 20·3′, decreasing annually by 4·8′.
[*f* RIGHT ASCENSION]

décollement (Fr.) Earth-folding of a superficial nature, so that overlying strata move easily over a basement surface of low friction. E.g. in the Jura Mtns, where basement beds are of Tertiary ROCK-SALT and ANHYDRIDE.

decomposition The break-down of minerals in rocks, and hence of the rocks themselves, by chemical weathering (CORROSION).

deep, ocean An elongated trough or trench below the DEEP-SEA PLAIN, exceeding 5500 m (3000 fthms) in depth, near island arcs or coasts bordered by fold mountain ranges, hence sometimes called *fore-deep*. Most occur in the Pacific; e.g. Mariana Trench off Guam (11 033 m, 36 198 ft, the deepest known sounding), Emden D. off the Philippines (10 794 m, 35 412 ft), Ramapo D. (10 554 m, 34 626 ft), Mansyu D. in the Mariana Trench (9866 m, 32 370 ft). Ds are commonly asymmetrical, with a steeper slope near the land than the open ocean side. [*f* ABYSSAL ZONE]

deepening In a meteorological context, a decrease in atmospheric pressure at the centre of a low pressure system; e.g. 'a **d** disturbance', involving worsening of the weather, increase of wind-speed, and probably an increase in precipitation. The opposite of *filling* (up).

deep focus The SEISMIC F of an earthquake occurring at a depth greater than about 300 km.

deep-sea plain An undulating **p** lying between 3600 m–5500 m (2–3000 fthms), occupying about 2/3rds of the entire ocean floor. Its surface is gen. covered with PELAGIC OOZES. From it rise extensive submarine plateaus, curving ridges, SEAMOUNTS, GUYOTS and volcanic islands.

deep weathering The production of thick REGOLITH (or SAPROLITE) by prolonged chemical weathering. D w is esp. associated with areas of low relief in humid tropical environments, though it is also encountered in temperate regions and deserts (in both of which it is probably a 'relict' feature). Beneath land surfaces of similar age, **d w** reaches on average 30 m in humid tropics, 25 m in the savannas, and 3 m in arid regions. Major controlling factors are climate and vegetation, for chemical reactions are accelerated by high temperatures and abundant water, plus the presence of humic acids derived from decaying vegetation.

deferred junction, of rivers A tributary, unable to join a main **r** because of LEVEES along the latter's course, may flow parallel for some distance before effecting its confluence. Usually occurs on a flood-plain; e.g. the R. Yazoo enters the Mississippi Bottomlands and flows parallel to the main **r** for 280 km before it can effect a j. [*f, page 84*]

defile A narrow steep-sided pass through mountains.

deflation Removal of dry unconsolidated material, esp. dust and sand, from the surface by wind. The finest material is borne high in air and carried for many km; coarser material is swept away in sand-storms; still heavier material moves along the surface in swirling hops. D can take place in any arid or semi-arid area, esp. where the protective mat of vegetation has been removed (e.g. 'Dust

84 DEFLATION

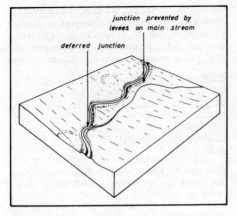

Bowl' of S.W. USA), or along a sand-dune coast (e.g. Landes of S.W. France). See BLOW-OUT. [*f*]

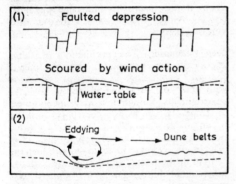

deforestation The complete felling and clearance of a forest.

deformation A change in shape, volume or structure of a mass of rock by folding, faulting, compression, shearing or solid flow, the result of STRESS. The **d** may be: (i) *elastic* (the rock will return to its orig. form); (ii) *plastic* (the rock will 'flow' and the change is permanent); or (iii) *rupture* (causing jointing and faulting).

deglaciation The withdrawal of an ice-sheet *in the past* from an area of land; e.g. from N. Germany and Denmark in the latter stages of the Quaternary Glaciation.

deglacierization The gradual withdrawal of a glacier from a land-mass *at the present time*. Partic. relevant to those areas of the Earth where the present glaciers are shrinking.

degradation (i) The gen. lowering of the land-surface by physical processes (esp. by rivers), and the removal of material to be deposited elsewhere. (ii) Used in the more limited sense of vertical erosion by rivers in order to maintain a GRADED state (ct. AGGRADATION). See also DENUDATION.

degree (i) A unit of temperature on any thermometric scale (FAHRENHEIT, CENTIGRADE or CELSIUS, RÉAUMUR and ABSOLUTE or KELVIN). For conversion table from Fahrenheit to Centigrade see entry under the former. (ii) A unit of angular measurement = 1/360th of a circle. (iii) The unit of angular measurement of LATITUDE and LONGITUDE. 1° lat. at Equator = 110·569 km (68·704 mi.); 1° lat. at 45° = 111·132 km (69·504 mi.); 1° long. at Equator = 111·322 km (69·172 mi.); 1° long. at 45° = 78·850 km (48·995 mi.) (based on Clarke Ellipsoid) (see EARTH).

degree-day A measure of the departure of the mean daily temperature from a given temperature. The values above or below significant CRITICAL TEMPERATURES (e.g. 0°C, 6°C, 18°C) are added together for the period under consideration, and expressed as a single ACCUMULATED TEMPERATURE.

dejection cone See ALLUVIAL CONE.

dell A small, shallow dry valley, developed in periglacial environments and partially occupied by frost rubble. In cross-profile a **d** is gently U-shaped, and sometimes asymmetrical. The main cause is stream erosion, in association with intense frost weathering and solifluction which causes valley widening and slope decline.

delta A tract of alluvium formed at the mouth of a river where the deposition of some of its load exceeds its rate of removal, crossed by the divergent channels (DISTRIBUTARIES) of the river. Orig. named after the Nile D., whose outline broadly resembles the Gk. letter Δ, though not all **d**s are of that shape (ct. BIRD'S FOOT D). Not only is a river current checked where it enters the sea, causing deposition of sediment, but fine particles of clay coagulate and settle rapidly when they mix with salt water (FLOCCULATION). The deposition of TOPSET, FORESET and BOTTOMSET [*f*] BEDS follows a regular sequence. Chief types of **d**: (i) ARCUATE [*f*]; (ii) BIRD'S FOOT (or birdfoot) [*f*]; (iii) CUSPATE. See also LACUSTRINE D.
[*f, page 85*]

démoiselle (Fr.) An earth-pillar capped by a boulder which has protected underlying material from weathering. Partic. common in TILL and volcanic BRECCIA; e.g. in the Chamonix valley, French Alps.

dendritic drainage A tree-like pattern of converging tributaries upon a main river

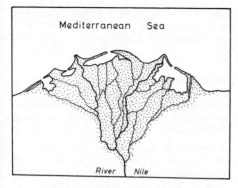

(Gk. *dendron*, a tree), usually where the rocks do not vary over its basin, or where there is no structural control; i.e. a type of INSEQUENT D.
[*f*]

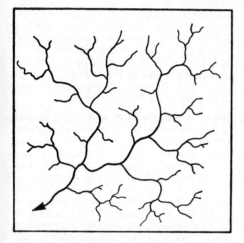

dendrochronology The working-out of past climatic chronology, using the evidence of tree-rings, since each reflects in its width the temperature and precipitation of the year in which formed. E.g. in S.W. USA, where a master-index has been constructed from the Douglas fir, the *Sequoia gigantea* and the *bristlecone pine* (the oldest of which is 3600 years): this is of great value to archaeologists dating pre-historic Indian cultures, and studying the climatic and hydrological changes of the last 3000–4000 years.

dene, den, dean (i) A steep-sided wooded valley; e.g. Lockeridge D., well-known for its SARSENS, and the Dean near Aldbourne, both in Wiltshire. (ii) An area of sand-hills near the coast; e.g. the Denes near Yarmouth.

density The concentration of matter, expressed as mass per unit volume, in g cm^{-3}.

The unit of **d** employed is that of water; at 0°C, 1 cm of water weighs 0·999878 g which is near enough to 1·0. **D** is an *absolute* quantity, while SPECIFIC GRAVITY (numerically the same) is a *relative* quantity. Warm water is less dense; e.g. **d** at 15°C = 0·999154 g cm^{-3}. **D** of ice at 0°C = 0·91752; of charcoal = 0·34; of granite, 2·70; of lead, 11·36; of platinum, 21·53. **D** of SIAL = 2·65–2·70; of SIMA = 2·9–3·0; of olivine, 3·27–3·37; of the Earth as a whole = 5·527; of the core of the Earth = about 12·0; of air at 0°C and 766 mm barometric pressure = 0·001293. The **d** of sea-water depends on both salinity and temperature. For a particular salinity, **d** varies inversely with the temperature. The mean **d** of surface water for the whole ocean = 1·0252; from 3600 m (2000 fthms) down it remains constant at 1·0280. At 15·5°C, fresh water has a **d** of 1000; water with salinity of 30‰ = 1·0220; of 40‰, 1·0300 g cm^{-3}. Muddy water has a greater **d** than clear. The **d** of a gas is proportional to its molecular weight. If the **d** of air is taken as 1·0, that of hydrogen = 0·0693, and for all other gases **d** = 0·0693 × molecular weight. *Conversion:* 1 lb. per cu. ft = 16·019 kg m^{-3}; 1 kg m^{-3} = 0·062 lb. per cu. ft; 1 g m^{-3} = 0·036 lb. per cu. in.

density current An ocean **c** caused by differences in **d** of the water (the result of varying salinity and temperature); e.g. salty dense water from the Mediterranean Sea flows W. along the bottom of the Strait of Gibraltar, while less saline, less dense water flows E. on the surface. It is estimated that the entire water in the Mediterranean is thereby changed every 75 years. Dense cold water from the Arctic Ocean sinks and flows S. under the less dense warm water of the Atlantic Ocean.

denudation The operation of all natural agencies by which the Earth's surface undergoes destruction, wastage and loss through WEATHERING, MASS-MOVEMENT, EROSION and TRANSPORT. (Sometimes **d** is used syn. with erosion, but the latter excludes weathering.) Orig. the 'laying bare' of rocks by removal of material covering them. Now commonly used as syn. with DEGRADATION, though W. M. Davis distinguished between **d** as the *early* very active processes in the cycle of landform development, and degradation as the *latter* more leisurely processes. Others regard **d** as the actual *processes*, degradation as the collective *result* on the landscape.

denudation chronology The study in geomorphology of the sequence of events leading to the formation of the present landscape.

deposition The laying-down of material transported by running water, wind, ice, the tides and currents in the sea; the complement of DENUDATION in the whole sum of processes of physiographic change. **D** includes not only mechanical laying-down of material, but also chemical precipitation, the formation of a crust or layer by evaporation, and the growth, accumulation and decay of living organisms (e.g. coral, humus, peat).

depression (i) (*physiographic*) A hollow in the earth's surface; more specif. a low-lying area wholly surrounded by higher ground, with no outlet for surface drainage; i.e. a basin of INTERNAL (or inland) DRAINAGE; e.g. Lop Nor **d** in central Asia. (ii) (*tectonic*) In a **d**, NAPPE structures are carried downwards and therefore largely preserved in the present mountain ranges; see CULMINATION [*f*]. (iii) (*atmospheric*) A low-pressure system in mid and high latitudes, formerly called a CYCLONE, now often a 'low' or a 'disturbance'. It is 'shallow' or 'deep' according to the number of surrounding isobars and the difference in pressure between the inner and outer ones. Used more commonly in UK than elsewhere. [*f*]

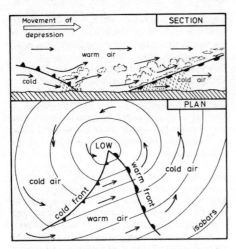

depression storage The proportion of rainfall temporarily retained by irregularities of the ground surface (i.e. as small 'puddles'). OVERLAND FLOW can occur only when depression storage is exceeded.

depth hoar The surface area of a snowfield is cooled at night, because of RADIATION, while lower layers of snow retain rel. warmer air within the pores. This air rises through the snow layers, and becomes cool and supersaturated; its moisture is deposited directly on to snow crystals which grow as prisms and cups, with weak points of contact. The result is the creation of a layer of fragile ice, which may form a 'bridge' across a CREVASSE, dangerous for a mountaineer to attempt to cross.

deranged drainage In drift-covered areas, the irregular deposition of material may produce a disordered **d** pattern, with a confused intermingling of isles, marshes and streams; e.g. in Finland; the Canadian Shield.

desalinization, desalination The removal of salt from soil or water. (i) After reclamation of an area formerly covered by sea, the marine clays are salt impregnated; this is removed by repeated deep ploughing, exposure to weathering, continual pumping away of rain-water, application of gypsum (which makes a highly soluble compound, calcium chloride), and cultivation of salt-tolerant and absorbing plants; e.g. the sea-polders of the Netherlands. (ii) In an area of constant irrigation in a hot climate where evaporation is rapid, the saline content of the soil steadily increases; this is avoided (e.g. in Imperial Valley, California) by flushing fresh water at intervals through pipe-drains. (iii) The production of fresh from salt water; e.g. in Kuwait, Malta.

desert An area of land with scanty rainfall and therefore little vegetation and (under natural conditions) limited human use. The main categories of **d** are: (i) hot trade-wind **d**s, areas of high atmospheric pressure, with rainfall less than 250 mm, high summer temperature (W. Köppen's *BWh* group); e.g. the Sahara; (ii) coastal **d**s, on the W. margins of continents, in latitudes 15°–30°, with cold offshore currents, and summer temperatures (about 18°C) low for latitude (*BWk*); e.g. Atacama, Kalahari; (iii) mid-latitude **d**s of continental interiors, with high summer and low winter temperatures (*BWk*, *BSk*); e.g. Gobi D.; (iv) ice and snow **d**s of Polar lands (*EF*); e.g. Greenland, Antarctica. See also ARID.

desert pavement An area of REG, or pebbly **d**, where the abrasive action of wind carrying a load of quartz-sand has ground and polished the upper surfaces of the pebbles, leaving a closely packed layer; sometimes called a *d mosaic*. The pebbles are commonly cemented by the precipitation of salts drawn to the surface in solution by capillarity, and left by evaporation.

desert varnish The hard, glazed surface of a rock in a **d**, the result of a film of iron or manganese oxide, deposited by evaporation from strong solutions brought to the surface by capillarity.

desiccation (i) Gen., drying up. Partic. implies progressive increase in aridity, usually the long-term result of a climatic change towards a decreasing precipitation, e.g. **d** of central Asia, but sometimes the result of deforestation, over-cropping, or a failure to continue irrigation (e.g. due to war or pestilence). (ii) The loss of moisture from pore-spaces in a soil or a sediment, usually causing cracks.

desiccation breccia A mass of angular material, the result of drying-out and cracking into irregular sun-baked polygons of formerly wet clay or mud; these broken slabs were later washed away by flood waters, redeposited and compacted.

desilication The removal of silica (SiO_2) from soil in areas of heavy rainfall; thus soils of tropical rain-forests are gen. deficient in silica, as well as in bases.

destructive wave One of a series of storm-ws, following each other in rapid succession, with an almost vertical plunge of water on the beach as each w breaks, and with a more powerful BACKWASH than SWASH. Thus they 'comb down' the beach, and move material seaward.

detritus Fragments of disintegrated rock-material that have been moved from their orig. site; DEBRIS is now more commonly used.

Deuterozoic An obsolete term for the Newer or Younger PALAEOZOIC; i.e. Devonian, Carboniferous, Permian, as ct. with the also obsolete PROTOZOIC.

Devensian The final glacial period in Britain, equivalent to the WEICHSEL period in N.W. Europe. Represented typically by glacial tills of the Cheshire Plain and the Hunstanton Brown Boulder Clay of Norfolk.

Devonian The 4th of the geological periods of the Palaeozoic era, dated c. 395–345 million years ago, and the system of rocks laid down during that time. In the British Isles these rocks occur in Cornwall; Devon (whence the name); S. and central Wales; the S. Uplands, the Midland Valley and N.E. of Scotland; the Orkneys; and S.W. Ireland. In Devon and Cornwall the system consists of sandstones, grits, slates and limestones, for the most part laid down in the sea. Elsewhere it comprises red and brown sandstones, conglomerates, marls and limestone, probably laid down in lakes; these rocks are known as Old Red Sandstone. The Caledonian orogeny, which began in the Silurian, continued well into the **D**. It was a period of vulcanicity, with intrusions of granite in the Grampians, the S. Uplands (Criffel), the Cheviots, and the English Lake District (Shap, Eskdale).

dew The condensation of water-droplets on the surfaces of plants and other ground objects, the result of cooling by nocturnal radiation to a temperature below the D-POINT of the layer of air immediately resting on the earth's surface. **D** forms partic. under conditions of a clear, still atmosphere; it requires high humidity at the surface, and favourable radiating surfaces surrounded by air; e.g. blades of grass. The **d** of spring and early summer is mainly derived from water-vapour in the lowest layers of the atmosphere, while in autumn it comes from the ground itself, both directly and transpired by plants (GUTTATION D). *Dewfall* involves the downward turbulent transfer of water vapour on to the cooled surface. **D** is small in UK, resulting in an annual avge of only 2·5 to 5·0 mm.

dew-mound Used in drier parts of Israel, Libya and other desert lands, esp. in the Middle East. A mound of earth is covered with flat stones on which **d** condenses, trickling between them into the earth, thus kept moist. Olives, oranges and other fruit-trees grow out of the mounds.

dew-point A critical temperature at which air, being cooled, becomes saturated with water-vapour (i.e. the RELATIVE HUMIDITY = 100) and below which condensation of excess vapour causes the formation of minute drops of water, provided that nuclei for condensation are present. **D-p** can be determined using a HYGROMETER, with tables, or by a special d-p hygrometer, using a polished metal surface which can be cooled until a film of moisture appears at a temperature which is recorded.

dew-pond A shallow artificial hollow containing water, usually lined with puddled clay and straw, found commonly in the chalklands of S. England. Most water is derived from rainfall, some by condensation of sea-mists from the Channel, but only a small amount actually from DEW.

D-horizon The fresh unaltered rock beneath a SOIL PROFILE.

diabase The American term for dolerite in gen., but used specif. in Britain for dolerite which has been much altered by decomposition of feldspars and mafic minerals.

diabatic A thermodynamic process involving actual loss or gain of heat from a system such

88 DIABATIC

as an airmass, in ct. to ADIABATIC. It involves evaporation, condensation, conduction, radiation and mixing by turbulence. *Non-a* is sometimes preferred.

diachronous Applied to sedimentary strata which appear to be continuous beds, but are actually TIME-TRANSGRESSIVE; i.e. the same lithological facies can be found in different places and are of different ages. This may make difficult the definition of geological boundaries, as in the CAINOZOIC, giving rise to stimulating controversy. Means lit. 'across time'.

diaclinal Applied by J. W. Powell in 1875 to streams and valleys which have a direction at right-angles to the strike of the rocks, crossing the axis of a folded structure.

diagenesis The process of compaction and cementation of sediments leading to LITHIFICATION.

diagonal scale A s in which unit-lengths shown can be sub-divided with precision. In the [*f*] below, the upper arrow indicates a length of 22 mi., 4 furlongs, the lower of 37 mi., 7 furlongs. [*f*]

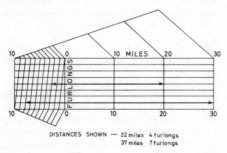

DISTANCES SHOWN — 22 miles 4 furlongs
37 miles 7 furlongs

diamond A crystalline form of carbon, formed under conditions of extreme heat and pressure, the hardest known mineral, with the highest number (10) in MOHS' (HARDNESS) SCALE. Usually found embedded in 'pipes' of igneous rock, or may be washed out and redeposited in 'placers'.

diapir The upward piercing of an anticlinal fold, forming cracks and fissures, by either (i) igneous material; (ii) masses of salt forced up through cracks from a SALT-DOME (e.g. S.W. France); or (iii) mud (hence MUD-VOLCANO).

diastrophism, diastrophic Forces which have disturbed or deformed the Earth's crust, including folding, faulting, uplift and depression, though not vulcanicity. The forces are classified as: (i) EPEIROGENIC; (ii) OROGENIC. Ct. TECTONIC.

diatomaceous earth A friable deposit consisting largely of aquatic vegetable organisms (*diatoms*) which secrete an external casing of silica; on death these accumulate on the floors of lakes and the sea. They are so minute that 6–10 million form 1 cm^3. It is found in Kentmere in Westmorland, in N.E. Skye, Japan, and in vast beds (over 300 m thick) of Miocene age in California. The hydrous earth is dried to form *diatomite*, of chalk-like appearance.

diatom ooze A siliceous o consisting of skeletons of microscopic plants (*diatoms*) flourishing in the cold waters of oceans, deposited on the ABYSSAL ZONE of their floors. Found in a continuous band round the S. Ocean in latitudes 50°–60°S., and in the N. Pacific Ocean.

diatreme A vent drilled through the crust of the earth as a result of explosive volcanic activity. Through it lava and volcanic debris reach the surface. It may choke with intrusive rocks, BRECCIA, PYROCLASTS, and may form a 'diamond pipe'.

die-back A diseased condition of plants, often applied gen. to the dying-off of large tracts of similar species at the same time, sometimes for no apparent reason; e.g. *Spartina Townsendii* on the marshes around the coast of S. England, partic. in the Lymington and Beaulieu estuaries, where both '*pan d-b*' and '*channel d-b*' can be seen.

differential ablation The differential melting of a glacier surface to give varied relief forms, such as PERCHED BLOCKS, DIRT CONES, moraine ridges etc. Factors influencing **d a** are (i) ice colour—white bubbly ice, of high ALBEDO, melts less rapidly than darker bubble-free ice; (ii) till cover—a layer of debris in excess of 1 cm protects the ice from solar radiation. Observations on the Tsidjiore Nouve glacier, Switzerland, show that ablation of ice covered by 6 cm debris is only 75% of that of near-by bare ice.

differential denudation The results of weathering and erosion on an area where the rocks display very varied resistance; different rock-types are worn down to differing degrees, sometimes resulting in striking relief-forms. See pl. 17. [*f, page 89*]

diffluence, difluence The lateral branching of a glacier, so that part flows away from the main ice-stream. This usually results from some down-valley blocking, either by a narrowing of the valley profile or at the junction of a tributary glacier. The ice in the main glacier will build up, and when thick

Plate 17. DIFFERENTIAL DENUDATION, weathering and erosion of strongly folded sedimentary rocks, Jordan. Note the many HOGBACKS, with intervening STRIKE VALLEYS, formed by the hard and soft strata. (*Aerofilms*)

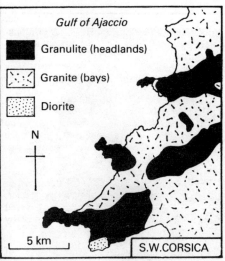

enough it may flow over a col in the valley side into a neighbouring valley. This ice-flow may cut down the col so much that the pre-glacial watershed is breached and in post-glacial times the river occupying the valley may take a new course through the overflow-channel. E.g. in Scotland, between the Cairngorms and Grampians, the upper Feshie flows N.W. through a channel across a pre-glacial watershed, the result of past **d**. In the Karakoram Mtns the N. Rimu glacier sends off a lobe of ice N.E., its melt-water flowing to the Yarkand R., and on into the Lop Nor depression; the other lobe joins the main Rimu glacier, hence to the Shayok R. and the Indus. See TRANSFLUENCE.

diffraction In meteorology, the process of radiation spreading akin to scattering, caused by the bending of light rays by an obstacle.

diffusion In meteorology the apparently random mixing of air bodies, either by *molecular d* (which is a slow process of little rel. importance), or by *eddy d* (the result of turbulent motion). (Also used in respect of liquids and light.)

dike See DYKE. The form dike is usual in USA.

dilatation See PRESSURE RELEASE.

dilatation joint A sheet-plane **j** in rock, possibly produced by PRESSURE RELEASE; usually curvilinear and broadly parallel to the surface of a mass of rock, resulting in SHEETING and a tendency towards the formation of a DOME.

diluvium, diluvial An almost obsolete term, orig. applied to materials thought to be deposited by the Genesis Flood, then to glacial drift. It is still sometimes used to denote the older Quaternary deposits, in ct. to the younger ALLUVIUM.

diorite A coarse-grained intrusive igneous rock, with much plagioclase feldspar, the rest various ferro-magnesian silicates; it is INTERMEDIATE in composition. It occurs in N. Wales, in the Malvern Hills, and near Nuneaton (Warwickshire) where it is quarried.

dip The max. angle of slope of a stratum of sedimentary rock at a given point. The angle of inclination is given in degrees from the horizontal, not the surface slope of the land, and direction is expressed as a compass bearing. The direction of **d** is at right-angles to the STRIKE. Ct. TRUE D and APPARENT D.
[*f* STRIKE]

dip-fault A FAULT with its STRIKE approx. at right-angles to the strike of the BEDDING-PLANES, hence virtually parallel to the DIP.

dip-slope A slope whose surface inclination is in the same direction and of the same amount as the **d** of the underlying strata. There is rarely exact parallelism between the land-surface and the **d** of the rock; this is often noticeable on a CUESTA, and some prefer the use of the term BACK-S to **d**-s where the angles involved are partic. divergent. See pl. 25. [*f* CUESTA]

dip-stream A s flowing in the direction of the DIP of the strata.

dirt band A b of debris within the ice of a glacier, originating between the annual accumulation-layers of fresh FIRN. Its exposure at the surface of the glacier may form one type of OGIVE.

dirt cone A conical or elongated mound of debris, usually up to 1–2 m in height, standing above a glacier surface. D cs invariably comprise an ice-core, covered by a silty or gritty layer of debris 1–2 cm in thickness. D cs develop owing to DIFFERENTIAL ABLATION, caused when the ice is protected by a small patch of fine detritus. This may accumulate in any depression in the ice surface (such as a CREVASSE, MOULIN or stream channel). The subsequent development of the cone, over a period of only a few weeks, involves a form of INVERSION OF RELIEF. See pl. 18.

disappearing stream A surface s which vanishes underground, usually down a SWALLOW-HOLE when it passes on to limestone or other partic. pervious rock; the water works its way down through joints, enlarging

Plate 18. A series of small DIRT-CONES, about a metre high, on the surface of the Tsidjiore Nouve glacier, Switzerland. (*R. J. Small*)

them by both carbonation-solution and erosion, towards the base of the limestone, where it emerges as a RESURGENCE; e.g. R. Jonte in the Grands Causses, Central Massif of France, which disappears underground near Meyrueis and reappears at Les Douzes.
[ƒ SWALLOW-HOLE]

discharge, of a river The quantity of water passing down a stream, depending on its volume and velocity, expressed in CUSECS, or in USA in *c.f.s.* (the number of cu ft per second passing a specific section), or in $m^3 s^{-1}$. 1 cusec = 0.078658 $m^3 s^{-1}$; 1 UK gallon per hour = 0.004546 $m^3 h^{-1}$; 1 $m^3 h^{-1}$ = 219.969 UK gallons per hour. River **d** is measured at a gauging-station, using a current-meter and a gauge for ascertaining the depth of water. See RATING CURVE. Mean **d** of major rivers near their mouths (in 1000 $m^3 s^{-1}$): Amazon, 180; Zaïre (Congo), 39; Yangtse, 22; Mississippi, 18; Yenesei, 17; Irrawaddy, 14; Brahmaputra, 12; Ganges, 12; Mekong, 11; Nile, 3.

disconformity A type of UNCONFORMITY involving a non-sequence of beds but not a contrast in their dip and strike. Sedimentary rocks were subject to prolonged denudation, which removed certain strata, and then material was deposited on them to form new strata parallel to the underlying ones, without any folding or other earth-movements involving angular change either of the under- or over-lying rocks. A **d** may be very difficult to recognize, usually requiring the evidence of distinctive fossils. E.g. the section revealed in the walls of the Grand Canyon, S.W. USA, where: (i) the Redwall Limestone of Devonian age was deposited directly on the Muav Limestone of Cambrian age, representing an interruption of deposition of about 150 million years; and (ii) the 'Great Disconformity', where Cambrian rocks rest on the eroded surface of the Precambrian basement rocks.
[ƒ UNCONFORMITY]

discontinuity (i) A plane or thin layer of separation in the oceans between masses of water of contrasting temperature and salinity; e.g. in the N.W. Atlantic between the Labrador Current and the Gulf Stream Drift. (ii) A FRONT or frontal zone between AIR-MASSES of contrasted temperature and humidity; e.g. Pacific and Atlantic Polar Fronts, Mediterranean Front, Arctic Front. (iii) A sudden change of character in the structure of the Earth at great depths; the two main **d**s are the MOHOROVIČIĆ and the GUTENBERG. (iv) In recent geomorphological studies, a marked change of slope is referred to as a **d**, most commonly the result of an outcrop of a resistant band of rock, or of the interaction of different slope processes above and below the **d**, or of different erosional histories of various parts of the slope PROFILE.
[ƒ ISOSTASY]

discordance (i) A DISCORDANT INTRUSION. (ii) An ANGULAR UNCONFORMITY; i.e., where the BEDDING-PLANES of adjacent strata are not parallel.

discordant (adj.) Cutting across the gen. lines of the structure or 'grain' of the country. A drainage pattern which has not developed in a systematic relationship with, and is not consequent upon, the present structure. See also D COAST, D DRAINAGE, D INTRUSION. Ct. CONCORDANT.

discordant coast A coastline trending transversely across the 'grain' of ridges and valleys (a transverse or 'Atlantic' coastline); e.g. S.W. Ireland, N.W. Spain. See pl. 19.
[ƒ RIA]

discordant drainage A river which is discordant with the present geological structures over which it flows, the result of SUPERIMPOSITION, ANTECEDENCE, CAPTURE or GLACIAL DIVERSION. [ƒ ACCORDANT DRAINAGE]

discordant intrusion An intrusion of igneous rock that cuts across the gen. stratification of the rocks into which it is intruded; e.g. DYKE. Ct. SILL, which is an accordant **i**.

discordant junction Where a tributary stream abruptly joins a main river flowing at a markedly different level, as from a HANGING VALLEY. Ct. ACCORDANT JUNCTION.

dislocation Syn. with FAULT, where strata are displaced rel. to each other.

dismembered drainage A complex **d** system that has been broken up by the submergence of its lower parts, thus creating a series of streams draining individually to the sea; e.g. Rs. Frome, Stour, Avon, Test, Itchen, Hamble, Meon, which may once have been tributaries of the so-named 'Solent R.', now reach the English Channel, Solent and Southampton Water independently.

dispersion diagram A **d** used to indicate the distribution of any quantity for any unit of time over some period. In a rainfall **d d** a vertical column is used, with a scale of mm, in which one dot is placed to indicate each year's or month's total precipitation for a specif. place. From it the median and quartile values can be established.

Plate 19. Truncation of small east-west HERCYNIAN fold-structures of a DISCORDANT COAST near Bude, Cornwall, England. Note the effects of differential wave attack, with hard strata forming PROMONTORIES and rock-ribs. (*Aerofilms*)

dissection In geomorphology, the cutting-up of a land-surface by eroding streams, specif. of an uplifted PENEPLAIN. Hence *dissected plateau.*

distributary An individual channel formed by the splitting of a river, as in a DELTA, which does not rejoin the main stream, but reaches the sea independently; e.g. Grande Rhône and Petite Rhône; Rosetta and Damietta branches of the Nile.

disturbance In meteorology, a DEPRESSION or 'low' of no particular intensity.

disturbance line An intense linear pressure system which travels W. across W. Africa, esp. in spring and autumn, and brings heavy falls of rain.

diurnal range The difference between the max. and min. value within a period of 24 hours, esp. of temperature; e.g. there is a high **d r** in a hot desert, where clear skies allow great INSOLATION during day and rapid terrestrial RADIATION by night, with a fall of 14–17°C (25–30°F) in the 2 hours succeeding sunset. Total **d** rs of 33–40°C (60–70°F) are commonly recorded. Azizia in Tripoli has the highest ever recorded **d r**, between 52°C and −3°C (126°F and 26°F).

diurnal tide In some sea areas, (e.g. Gulf of Mexico, around the Philippine Is., off the coast of Alaska, and off parts of the coast of China), only one high and one low **t** occur in each 24 hours. In these areas, because of the shape of the water-body, the *diurnal* component of the **t**-producing forces is dominant. [*f* TIDE]

divagation The lateral shifting of a river's course as a result of extensive deposition in its bed, esp. in conjunction with the development of MEANDERS.

divergence (i) In the oceans, a line or zone from which surface water moves away under the influence of wind-drift, thus causing upwelling of deep water. (ii) In climatology, a type of air-flow such that in a certain area at a given altitude the outflow is greater than the inflow, so that the mass of air contained tends to decrease. If the density remains constant, such horizontal **d** must be accompanied by some compensating inflow from a descending air current. Like CONVERGENCE, **d** takes two forms: *streamline* **d** and *isotach* **d**. **D** is considered as a positive quantity, convergence as a negative form of **d**. An example of **d** is in the area of descending and out-blowing air in the HORSE LATITUDES. (iii) See GLACIAL D. (iv) In biology, **d** *of species*: the evolution of orig. similar species in such a way that their life forms become progressively less similar. (v) The distance between a line representing the spheroidal surface of the Earth from any point, and the tangent (light ray) originating at the same point. This **d** is decreased by ¼th by the effects of diffraction on the light ray. An empirical formula, where h is the **d** in m and K is the distance in km from the observer, $K = 3 \cdot 80\sqrt{h}$, and $h = 0 \cdot 069\ K^2$. At a distance of 10 km, **d** is 6·92 m. In UK measure, $h = 0 \cdot 574\ K^2$, and $K = 1 \cdot 317\sqrt{h}$. At a distance of 10 mi. **d** = 57·4 ft. [*f*]

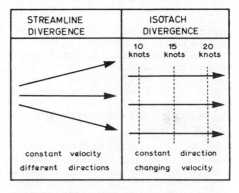

divide A ridge or area of high ground between river basins; the highest line of an INTERFLUVE. Syn. with English use of WATERSHED. Used specif. in N. America as the *Continental Divide*, a line of separation between Pacific and Atlantic drainage.

doab A low-lying, alluvium-covered area between two converging rivers, specif. in the Indo-Gangetic plain of India and Pakistan.

Doctor The HARMATTAN wind in West Africa.

dod, dodd A rounded hill or summit in N.W. England and S. Scotland; e.g. Dodd on S. slopes of Skiddaw, Starling Dodd on N. side of Ennerdale, Great Dod in N. part of the Helvellyn range.

'dog-days' A pop. name, esp. in USA, for the hottest time of summer, assoc. with the star Sirius ('the Dog Star').

dogger A large nodule or CONCRETION found in sedimentary rocks; e.g. ironstone nodules in Jurassic rocks of Yorkshire; siliceous **d**s in Corallian beds of Dorset, and in Wenlock Shales of Silurian age in Shropshire.

dog-tooth erosion Used mainly in USA for the acute **e** of the crest of an ANTICLINE, producing a pattern of sharp crests separated by torrent-worn gorges.

doldrums A belt of light indeterminate winds in equatorial latitudes notably over the E. part of the oceans, though displaced a few degrees of latitude with the seasons, lagging behind the overhead sun. The belt is characterized by high temperatures and humidity, and forms part of the INTERTROPICAL CONVERGENCE ZONE.

dolerite A basic igneous rock of HYPABYSSAL occurrence in minor intrusions, as SILLS and DYKES. Usually dark coloured and fine or medium textured, and sometimes solidifies in columns; e.g. Gt. Whin Sill of N. England, where the rock is known as 'whinstone' by quarrymen. Called *diabase* in USA. See pl. 10.

doline (Fr), alt. **dolina** (It., Slavonic) A shallow funnel- or saucer-shaped depression in limestone country (KARST). It commonly leads down into a vertical shaft. Recent international usage defines it as large enough to contain cultivable soil on its floor. A *collapse*-d is a depression on the surface caused by sub-surface solution-enlargement, with subsequent collapse.

dolomite (i) A yellowish or brownish mineral consisting of calcium magnesium carbonate, $CaMg(CO_3)_2$, with up to about 44% of $MgCO_3$. Named after an 18th-century French geologist, D. G. Dolomieu. Its origin is varied; in some cases it is precipitated directly from sea-water, possibly under warm shallow conditions; in others 'dolomitization' of limestone has taken place, i.e. post-depositional replacement has occurred of some of the calcium by magnesium. D rock is sometimes called *magnesian limestone*, occurring in the Permian; it forms cliffs along the Durham coast, and a distinctive escarpment, over 180 m high, overlooking the Durham coalfield to the W. (ii) the **D**s, the proper name of ranges of pinnacled mountains in N.E. Italy.

dome (i) Any **d**-shaped landform; e.g. English Lake District. (ii) A broad, short ANTICLINE or PERICLINE; e.g. Kingsclere Pericline in N. Hampshire; Weald. (iii) An underground **d**-shaped structure containing salt, with oil and gas; in places great plugs of salt were forced upwards through sedimentary strata, with a cap of limestone, gypsum and anhydrite; e.g. SALT-DS of Texas and Louisiana. (iv) An *oil-d* in gently flexed sedimentary strata; e.g. Rock Springs D. and Teapot D. in Wyoming; Dominguez Hills near Los Angeles, California. (v) An igneous intrusion of **d** shape, including LACCOLITH (Navajo Mtn, Utah) and BATHOLITH (the Black Hills of Dakota). (vi) A rounded granite outcrop; e.g. **d**s of Yosemite (California); Stone Mtn on the Piedmont upland, Georgia; *bâlons* in the Vosges, France. (vii) A rounded snow-peak; e.g. Dôme de Gouter (Mont Blanc), Dôme de Neige (Dauphiné), in the French Alps. (viii) An acid volcano; e.g. Puy de Dôme, Auvergne, France. Hence *domite*, a fine-grained igneous rock largely forming the Puy de Dôme. *Note:* A **d** may be either structural or physiographic, or both. Thus the Weald is a dome structure, not a physiographic dome, while the Black Hills of S. Dakota are both.

dome-on-dome inselberg A composite inselberg, comprising a large lower dome (usually of granite or gneiss), surmounted by a smaller upper dome or collection of weathered boulders. J. C. Pugh suggests that d-o-d is develop in two cycles of erosion, involving both deep weathering and exhumation of a solid core and 'scarp retreat' (as in the PEDIPLANATION CYCLE). [*f*]

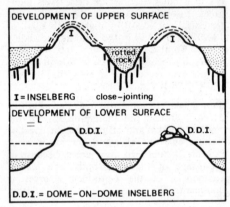

dominant The characteristic species in a partic. plant COMMUNITY, contributing most to the gen. 'plant landscape', and influencing which other plants will grow there. In an oak-wood, the **d** is the oak; in a heathland, it is ling (*Calluna*). The **d** tends to be the largest or strongest plant in the community.

dominant wave The largest ws along a section of coast; i.e. those capable of most powerful work.

dominant wind The most *significant* **w** at a given place, not necessarily the one which blows most frequently (i.e. PREVAILING). Thus a shingle-beach may be aligned at right-angles to the **d w**, not necessarily to the prevailing **w**.

donga (Bantu) (i) A steep-sided gully in S. Africa, occupied temporarily by flood waters. (ii) A gully produced by soil erosion. See pl. 20.

Plate 20. A DONGA or WADI-like valley, cut by seasonal streams into weak volcanic TUFFS and ALLUVIUM deposited in old PLEISTOCENE lakes near Lake Magadi, southern Kenya. (*R. J. Small*)

dore An opening in a ridge between masses of rocks; e.g. D. Head, above Mosedale, English Lake District. In Gaelic, *dorus*; e.g. An Dorus, a gap in the Black Cuillin ridge, I. of Skye.

dormant volcano A v which has not erupted in historic times, but not thought to be extinct; e.g. most vs in the Andes and Cascade Mtns.

double surface of levelling A term proposed by J. Budel (in Germ. Doppelten Einebnungsflächen) to describe the major denudational process of the seasonally humid tropics. On the upper (ground) surface, transportation of weathered debris is effected by wet-season streams and surface wash; the lower 'surface' (lying up to 30 m or more beneath the ground surface) is represented by the BASAL SURFACE OF WEATHERING, produced by protracted DEEP WEATHERING. In the right circumstances, surface stripping can expose the basal surface; where the relief of the latter is strong, inselberg landscapes are formed.

double tide A tidal régime in which the t rises to a max., falls slightly and then rises to a 2nd max.; e.g. in Southampton Water, along the Solent and as far as Bournemouth, and on the French side of the Channel near Honfleur. Alt., a **d** low **t** may occur, as at Portland, Dorset (GULDER). The **d t** is attributed to the effects of: (*a*) configuration of the coast; (*b*) shallow water, causing the deformation of a PROGRESSIVE WAVE and the superimposition of a quarter-diurnal t upon a semi-diurnal t. If the harmonic curves of a semi-diurnal and quarter-diurnal t are combined, they will produce a **d** high **t** if the initial phase difference between the constituents is near 180°. If the initial phase difference is near zero a **d** low **t** results. [*f* TIDE]

doup A rounded hollow in N. England, esp. in English Lake District; e.g. Great **D**. beside the Pillar Rock, Ennerdale.

down (land) (i) gently undulating upland, usually of chalk; e.g. N., S., Dorset, Hampshire Ds. (ii) Temperate grasslands in Australia (e.g. Darling Ds) and New Zealand. (iii) The proper name for part of the N. Sea near Goodwin Sands ('The Downs').

downthrow The vertical change of level of the strata in a fault; the strata are lowered on the **d** side. [*f* FAULT]

down-warping A gentle downward deformation of the crust, without any distinct folding or faulting. This can be caused by the weight of a continental ice-sheet; e.g. in the Gulf of Bothnia, where the 'recoil' resulting from the relief of weight by ice-melting has averaged

11·18 mm per year for the last 150 years. Similar **d-w** and recoil have taken place in the Great Lakes area of USA and Canada. **D-w** may also be caused by widespread sedimentation; e.g. beneath the Mississippi delta. Several extensive shallow lakes are the result of **d-w** or 'sagging'; e.g. L. Victoria (E. Africa); L. Eyre (Australia); L. Chad (W. Africa). **D-w** contributed to the creation of GEOSYNCLINES.

down-wearing   Denudational processes resulting in: (i) the decline of a slope, the creation of a slope of progressively lower angle as the PENEPLAIN stage is reached; ct. BACK-WEARING; (ii) on a near-level surface its lowering as a result of deep weathering and transportation, esp. in a humid tropical climate.

draa   Large sand-waves in the desert, upon which dune-patterns may be superimposed.

drag fold   (i) A minor **f** formed subsidiary to a main **f**, or along the sides of a FAULT where the vertical displacement has resulted in flexures and puckers in the rocks. (ii) Used in geology with specif. ref. to COMPETENT and INCOMPETENT strata, the **d f** being produced in the latter by the movement of adjacent more rigid competent beds.

drain   An artificial channel for carrying off excess water; e.g. in English Fen District (Hundred Foot D, also called New Bedford R., N.W. of Ely); in the Somerset Levels (North D., near Wedmore; King's Sedgemoor D.).

drainage   (i) The act of removing water from a previously marshy area; e.g. **d** of the Fens. (ii) The discharge of water through a system of natural streams; hence **d** *basin*, a unit-area drained by a single river system; **d** *pattern, system* or *network*, the actual arrangement of the main river and its tributaries; **d** *density*, the ratio of the total length of all streams in a single system to the area drained by that system. See specif. drainage types.

drawn   One of a maze of channels or ditches, used to drain water-meadows in S. England.

Dreikanter (Germ.)   A stone scoured on 3 sides or facets by the sandblast effect of wind (i.e. a VENTIFACT) in a desert. Ct. EINKANTER.

drift   (i) All materials (boulders, sand, clay, gravel) derived from glacial erosion, and deposited either directly from the ice or by melt-water; i.e. incl. glacial and fluvio-glacial deposits. (ii) Used in a wider sense by the British Geological Survey to distinguish all superficial deposits from solid bed-rock. Maps are published in 'drift' and 'solid' editions. (iii) A horizontal passage in a mine, following a vein of metal ore or a seam of coal; ct. CROSS-CUT, which intersects it. (iv) The slow movement of surface ocean water under the influence of prevailing winds; e.g. N. Atlantic D., moving at about 5 nautical mi. (9 km) per day. (v) Any surface movement of loose material by the wind, with accumulations; e.g. *snow-d, sand-d*.

drift-ice   Loose, detachable pieces of floating ice, separated by open water, moving with the wind or a current, through which a ship can readily penetrate.

dripstone   A calcite formation (TUFA) formed by dripping water in limestone country, esp. in caves; see STALACTITE, STALAGMITE.

drizzle   Fine continuous rainfall, in which raindrops are very small, usually defined as less than 0·5 mm in diameter. Esp. associated with a WARM FRONT.

drought   A continuous period of dry weather, specif. defined as ABSOLUTE DROUGHT, PARTIAL DROUGHT or DRY SPELL (now obsolete by UK Met. Office).

drowned valley   A **v** filled with water by a positive change of sea-level, the result of either subsidence of land or rising of sea. See FIORD, RIA, DALMATIAN COAST.

drumlin (Irish)   A smooth elongated hummock of glacial till, with a long axis parallel to the direction of the moving ice-sheet responsible for its deposition; it varies from a small mound to a hillock 2 km long and 90 m high. They commonly occur in swarms *en échelon*, sometimes called 'basket of eggs' relief. The ice deposited each mass of clay, possibly from part of the ice-base locally more loaded with material, because friction between the clay and the underlying floor was greater than that between the clay and the overlying ice. The shape was then streamlined by subsequent ice-movement. In gen., where ice moved most rapidly and exerted most pressure, the **d** was most elongated. E.g. in N. England (Aire Gap, coastal plain of Morecambe Bay, Solway Plain); Midland Valley of Scotland; N. Ireland. See also ROCK-DRUMLIN. See pl. 21.   [*f, page 97*]

dry adiabatic lapse-rate   A l-r of temperature with height, occurring when a 'parcel' of unsaturated air rises through the atmosphere in equilibrium, so expanding and cooling dynamically. The **d a l-r** is 1°C in 100 m (5·4°F in 1000 ft) of ascent, and is a physical constant. See ADIABATIC.

Plate 21. A large field of DRUMLINS, deposited by a CONTINENTAL ICE-SHEET near Cape Krusenstern, Northwest Territories, Canada. (*Canadian Government*)

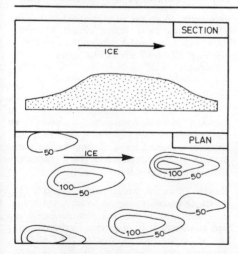

dry-bulb thermometer A mercury t, used in conjunction with a WET-BULB T to obtain RELATIVE HUMIDITY.

dry delta See ALLUVIAL CONE, ALLUVIAL FAN.

dry-gap See WIND-GAP.

dry spell Any period of drought; sometimes defined specif. as 15 consecutive days, none of which has more than 1·0 mm of precipitation (British definition, now obsolete.). In USA there is no measurable precipitation for 14 days.

dry valley A v, notably in chalk and limestone, which contains no permanent stream, though one may break out following heavy rainfall (see BOURNE). In chalk country, the pattern of **d** vs is like that of a normal river system. Many dissect chalk CUESTAS, have steep walls and abrupt heads, and sometimes follow zig-zag courses; e.g. Rake Bottom, near Butser Hill, Hants. There are various theories of origin: (i) a gradual lowering of the WATER-TABLE following a gen. reduction of precipitation; (ii) the cutting back of an escarpment, with resultant lowering of the SPRING-LINE; (iii) surface erosion under PERIGLACIAL conditions, when the chalk was frozen and therefore impermeable; (iv) powerful SPRING-SAPPING, cutting back along lines of weakness such as jointing, which helps to account for the frequent angularity of the course and the steep valley-head. Some **d** vs are caused by river CAPTURE, some by glacial derangement. **D** vs in limestone may also be due to: (i) a former surface stream disappearing down a joint; (ii) collapse of an underground cavern; (iii) lowered water-table.

Dumpy level An instrument for LEVELLING in survey work, consisting of a short telescope fixed to a base, with an attached spirit-level.

dune A low ridge or hillock of drifted sand, mainly moved by wind, occurring: (i) in deserts; e.g. ERG of the Sahara and KOUM of Turkestan; and (ii) along low-lying coasts, above high-tide level; e.g. Culbin Sands along the Moray Firth; Formby ds in Lancashire; Landes in S.W. France; along the coast of N.W. Europe from the E. Baltic to the Belgo-French frontier. Factors in **d** formation include: (*a*) load of sand; (*b*) strength and direction of the wind; (*c*) nature of the surface over which the sand is moved (deep sand or bare rock); (*d*) presence of an obstacle as a nucleus; (*e*) presence of vegetation, wild or deliberately planted (acacia, eucalypts in deserts, marram grass and pines on coast); (*f*) presence of ground-water reaching the surface. See also BARKHAN, SEIF-D. Elaborate classifications of **d**s include HEAD-D, TAIL-D, ADVANCED-D, LATERAL D, WAKE-D, STAR-D, SWORD-D. [*f*]

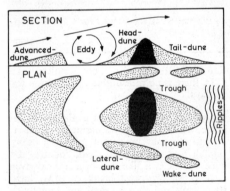

dunite An ultrabasic rock consisting mainly of *olivine* (magnesium iron silicate, Mg_2SiO_4 with Fe_2SiO_4).

durain structure, in coal In banded coal, layers of dull grey to brown material, hard and tough in quality. Now obsolescent.

duricrust A concentration in the upper part of the soil of various materials (aluminous, calcareous, siliceous, ferruginous), drawn up in strong solutions by capillarity to form a hard surface layer (hence resp. BAUXITE or ALCRETE, CALCRETE, SILCRETE, FERRICRETE). See also LATERITE, CUIRASS. It is usually found in semi-arid areas with a brief rainy season and a long hot dry season. At some depth it forms DURIPAN. See pl. 59.

duripan A HARDPAN produced through cementation by siliceous minerals.

dust Small (less than 0·6 mm in diameter) particles of comminuted matter, fine enough to be carried in suspension by the wind. *Cosmic* **d** is of meteoric origin, *volcanic* **d** the product of an eruption, *hygroscopic* **d** (i.e. with a marked affinity for water) acts as nuclei for the condensation of water-vapour. An avge 1 km^3 of air contains 600 tonnes of **d**. Hence D-STORM. **D** may cause HAZE and vivid SUNSET COLOURS.

dust bowl A semi-arid area, from which surface soil is being or has been removed by the wind, esp. where vegetation has been destroyed by injudicious cultivation or overgrazing. Orig. a proper name for parts of S.W. and W. USA, esp. in Kansas; now used of any area where similar processes are in operation. Once the orig. protective cover of grassland had been removed by ploughing for cereal cultivation, a period of drought exposed the loose surface to the wind. One category of SOIL EROSION.

dust-devil A short-lived swirling wind round a small low pressure nucleus, the result of intense local surface heating and convection. It whips up dust to form a rapidly moving pillar; e.g. in the Sahara, Kalahari, central and W. Australia, parts of US Mid-West.

dust-storm A storm in a semi-arid area, in which winds carry dense clouds of dust, sometimes to a great height, often obscuring visibility. Often of whirlwind form, or may comprise a 'wall' of **d** up to 3000 m high. Termed *haboob* in Sudan. In a **d-s**, a km^3 of air contains about 2300 tonnes of **d**. Ct. SANDSTORM.

dust-well A small hollow on the surface of a glacier, formed by a patch of **d** absorbing the sun's rays, causing a rise of temperature, and melting the surrounding ice. The **d** gradually sinks down into the ice.

Dwyka Tillite A widespread **t** found in S. Africa, in places resting on an ice-worn rock-pavement, and containing striated boulders and ERRATICS. It is of Upper Carboniferous age, and is regarded as evidence of a Carboniferous-Permian glaciation. Similar **t**s have been found in Orissa in India (*Talchir* T), Australia, Argentina and Brazil. It is thought that there were at least 3 distinct advances of the Carboniferous-Permian glaciation.

dyke (dike) (i) A mass of intrusive rock which cuts discordantly (i.e. transgressively) across bedding-planes of the country rock. Sometimes **d**s occur in parallel or radial 'swarms', as in N.W. Scotland, Mull and Arran. When affected by denudation, a **d** will either stand up as a ridge (where the **d**-rock is more resistant than its surroundings), or be worn away to a ditchlike depression. (ii) A drainage ditch or watercourse; e.g. Friar D., Vale of Pickering, Yorkshire. (iii) An artificial embankment to protect low-lying land from flooding; e.g. around parts of English Fen District, Netherlands [*f* POLDER], W. Denmark. (iv) An embankment in the flood-plain of a river parallel to its course; e.g. Mississippi, lower Rhine. (v) A man-made defensive earthwork; e.g. Offa's **D**., built in the 8th century from the estuary of the Dee to that of the Wye.

[*f*]

Harder than surrounding rocks

Softer than surrounding rocks

dyke-spring A spring thrown out along the line of a DYKE formed of an impermeable rock such as basalt or dolerite, which penetrates a permeable sedimentary rock or a pervious igneous one; e.g. in Black Cuillins, (I. of Skye) made of well-jointed GABBRO, with several high-level springs along lines of dykes.

[*f*]

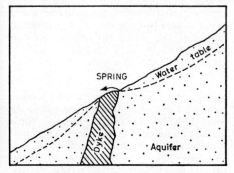

dyke swarm A collection of **d**'s often parallel or radial to each other, near and around a PLUTONIC intrusion; e.g. Is. of Mull and Arran, in W. Scotland. See CONE-SHEET, RING-DYKE.

Dymaxion Projection A mathematically derived **p** formulated and developed by R. Buckminster Fuller, the name obtained from a contraction of *dynamic, maximum* and *ion*. It claims to have numerous advantages, the chief being the provision of a world map of negligible dimensional distortion. The **p** is based on an all GREAT CIRCLE grid, so producing a basic pattern of equal-sided great circle triangles, with matching edges on which scale is true. All spherical great circle co-ordinates are transferred to a plane surface by means of the mathematically complex Fuller Transformation; this adjustment is carried out by interior contraction of the data, rather than by exterior stretching used by most other **p**s. The triangles can be reassembled in a continuous 'flat skin' or 'world mosaic' in any selected way, so as to bring focus to bear on any of the dynamic interrelationships of the world's surface. The radii of the **p** are always perpendicular to the transforming surfaces, which facilitates the accurate transference of astronomical data to the plane surface. Thus the **p** is highly suitable for assembling a comprehensive mosaic for aerial photographs, and also for automatically plotting and guiding missiles and aircraft on a 'world-around' scale. One application of this **p** (the *D Airocean World Fuller Projective-Transformation*) was published in 1954.

dynamic climatology C approached from the viewpoint of its relationship to the gen. circulation of the atmosphere, energy processes, atmospheric dynamics and thermodynamics. Ct. SYNOPTIC C.

dynamic equilibrium The concept that conditions of energy balance (**e**, steady state) can develop in the physical landscape; e.g. between the rate of rock weathering and rate of removal of weathered products on a slope. The existence of such an **e** may be indicated by uniform slope angles within a small region of uniform relief, rock-type, climate and vegetation. The **e** is not maintained for ever, but may be disturbed, e.g. by climatic change affecting rates of weathering and removal. However, it is believed that in time **e** will be restored by an adjustment of slope angle to meet the requirement of the new situation. Thus the **e** is **d** (changing). Again, a beach can be explained in terms of currently acting processes, such as wave, wind and tidal conditions, in a changing condition of balance, rather than in terms of a cyclic evolution from a hypothetical initial form. Acceptance of **d e** theory implies the abandonment of landscape analysis in terms of structure, process and stage, and its replacement by explanation of landforms simply as the outcome of an interaction between structure and process. Obviously in the long term the time element cannot be ignored; there must be changes between the landscape of the past, the present and the future, though not with a rigorous cyclic inevitability, interrupted in the short term by conditions of a quasi steady state. Cf. STAGE.

dynamic lapse-rate A r of ADIABATIC temperature change applying to rising 'parcels of air'. Ct. STATIC LAPSE-RATE.

dynamic metamorphism Now strictly defined as a category of m involving only localized stresses which breaks up rocks. On a large scale, this contributes to REGIONAL M.

dynamic rejuvenation R of a river caused by a change of BASE-LEVEL, due to a rel. rise or fall of sea-level, or from an alteration of the land surface through TECTONIC causes (folding, faulting, tilting, uplift). Ct. STATIC REJUVENATION.

dyne An absolute unit of force, producing an acceleration of 1 cm s^{-1} per second when acting upon a mass of 1 gr; i.e. 1 g cm s^{-2} in the CGS system (now obsolete). At sea-level and at 45°N. and 45°S., a mass of 1 g experiences a gravitational force of 980·616 ds 1 mb. = 1000 d per cm^2 = 100 NEWTONS per m^2.

eagre See BORE.

Earth (i) The planet inhabited by Man, with its orbit around the sun between those of Venus and Mars. Its shape is an oblate spheroid (i.e. a flattened sphere), with its equatorial diameter 12 755 km (7926 mi.), polar diameter 12 714 km (7900 mi.), with an oblateness of 1/293 or 0·33%. Mean radius = 6371·229 km. The International ELLIPSOID OF REFERENCE (1924) has an ellipticity of 1/297. The Airy spheroid (1870), used by the British Ordnance Survey, has 1/299. The Clarke spheroid (1866), used for N. America, has 1/295. The Clarke spheroid (1880), used by the British Admiralty, has 1/293·5. Equatorial circumference, 40 076 km (24 902 mi.), polar circumference, 40 008 km (24 860 mi.). Area = 510·1 × 10^6 km^2 (197·0 × 10^6 sq. mi.). Mean density = 5·517; mass = 5·975 × 10^{21} tonnes (5·882 × 10^{21} tons). Volume, 1083 × 10^9 km^3 (259·9 × 10^9 cu. mi.). Recent research, based on study of the gravitational modification of paths of satellites, indicates that the E is slightly pear-shaped, that the North Pole is about 45 m farther from the equatorial plane than is the South Pole, if both were projected on to a theoretical sea-level plane. (ii) The solid material on the surface of this planet, in ct. to water. (iii) Loose, disintegrated surface material, in ct. to solid rock, sometimes even applied to soil; e.g. 'the good earth'. (iv) A proper name for certain amorphous fine-grained materials; e.g. Fuller's E.

earth-flow A type of MASS-MOVEMENT of surface material down a slope under the influence of gravity, with lubrication from rain-water. A mass of saturated material pulls away from the slope as a SLIP, leaving a low arcuate cliff, and moves down as a plastic mass, forming bulging lobes or tongues, and lower down becoming a MUD-FLOW. Shallow e-fs are esp. common over impermeable rocks. Even when turf-covered, the lower slope may bulge out, with the sods wrinkling and cracking. A special form of e-f is SOLIFLUCTION, under PERIGLACIAL conditions. [*f*]

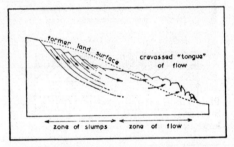

earth-movement A m caused by internal forces (compressive, tensional, uplifting, depressing, folding, faulting), on both a major and minor scale, and both rapid (earthquakes) and slow, which affect the CRUST of the E; known collectively as DIASTROPHISM.

earth pillar A mass of soft earthy material, capped and protected by a rock, the result of rapid subaerial denudation, esp. in a semi-arid BADLAND area (see DÉMOISELLE). It is common in glacial till, where the mixture of fine material and boulders provides a suitable medium; e.g. the Chamonix valley, French Alps; near Bolzano, Italian Tyrol. See pl. 22.

earthquake A rapid and perceptible tremor, movement and adjustment of and within the rocks of the Earth's crust, causing the propagation of a series of elastic shock-waves outwards in all directions. These are classified as: (i) *primary* (longitudinal, push, or '*P*' waves); (ii) *secondary* (transverse, SHAKE or '*S*' waves); (iii) *surface* ('*L*' waves). The most severe es are associated with distinct fault-lines; e.g. the San Francisco e (1906) along the San Andreas Fault. Es were measured on

Plate 22. A group of EARTH PILLARS, fashioned from glacial morainic deposits at Euseigne, Switzerland. Note the manner in which the pinnacles of weaker clay and sand are capped by large protective boulders. (*Eric Kay*)

the Rossi-Forel Scale, replaced first by the Mercalli and in 1935 by the RICHTER scales. [*f, opposite*]

Earth science A currently pop. term for a wide discipline involving geology, geography, geophysics, geodesy, climatology, oceanography and biology.

earth-slide A type of MASS-MOVEMENT where a large block of e moves downhill *en masse*, but retains its character without breaking up.

easting Distance E. from the origin of a GRID, providing the first half of a standard grid-reference.

ebb channel The route by which a tidal current flows seawards after high tide. This may differ from the flood c, esp. among complex offshore banks, causing progressive changes in the form of the interplay of ebb and flood tidal currents on the Goodwin Sands, which tend to rotate anticlockwise, the N. end moving W., the S. end moving E.

ebb tide The recession or outward movement of water, after high t, as the level falls

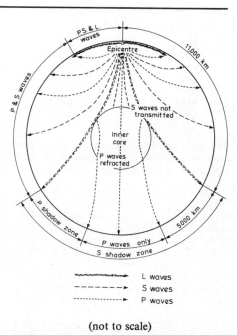

(not to scale)

towards low **t**; the reverse of FLOOD T. In estuaries, harbours and other inlets, the ebbing of the **t** may cause a powerful outward-flowing tidal stream or current, followed by a period of slack, and then by the rise of the next flood **t**. Where the range of the **t** is large, as in shallow marginal seas and estuaries (e.g. 13 m at Avonmouth, 21 m in the Bay of Fundy), the amount of water transferred is large, and the ebb **t** is powerful.

echo-sounder A surveying instrument used to determine depths of oceans, whereby sonic or ultra-sonic vibrations are transmitted down through the water to the ocean floor, to return as an echo, electrically recorded as an *echogram*. For very accurate work the ship is anchored, but for normal purposes a continuous profile is obtained while under way. It is also used, mounted on a sledge, for measuring the thickness of ice, as in Greenland and Antarctica; and for measuring the depth of different densities of soil or rock.

Eckert Projections A series of 6 **p**s, used for world maps, broadly similar in appearance to MOLLWEIDE, with each pole represented by a line half the length of the Equator, rather than by a point, the parallels straight lines. In the Fourth E P meridians are ellipses passing through the Equator, which is divided truly, and parallels are spaced so as to make the **p** equal area, as follows:

Distance of parallels from Equator
0°	0·0
10°	0·155
20°	0·308
30°	0·454
40°	0·592
50°	0·718
60°	0·827
70°	0·915
80°	0·976
90°	1·0

The shape of the continents is quite good, much better than on Mollweide. In the Sixth E P, the meridians are *sine*-curves.

eclipse *Of the moon* (*lunar*), caused by the interposition of the earth when it is in line between the sun and the moon (OPPOSITION); *of the sun* (*solar*), caused by the interposition of the moon when it is in line between the sun and the Earth (CONJUNCTION). The **e** may be *total* or *partial*. A *lunar* **e** can occur at the time of full moon, a solar **e** at new moon, though only on rare occasions since the moon is rel. small and the plane of its orbit is tilted at 5° 09′ to that of the ECLIPTIC. *Solar* **e**s are more frequent though much more rarely seen at any one place, since a total solar **e** can be seen only along a belt of about 145 km width, the result of the narrow cone of lunar shadow. A total lunar **e** may last for 2 hours, a solar **e** only a few minutes.

ecliptic The apparent path of the sun through the sky during a complete revolution in a year (about $365\frac{1}{4}$ days); the plane of this path (the plane of the **e**) is tilted at an angle (which varies slightly) of 66° 73′ 50·5″ to the earth's axis; i.e. an obliquity of 23° 26′ 49·5″ from the plane of the Earth's Equator. It has been calculated that this obliquity varies in about 40 000 years between 22° and 24·5°. The **e** is divided into 12 sections, each denoted by a Sign of the ZODIAC. [*f* AXIS, OF EARTH]

eclogite A gabbroid rock, composed chiefly of pyroxene and garnet, which has been subject to intense METAMORPHISM, with attractive green and red colouring. Found among the Lewisian rocks in N.W. Scotland.

ecoclimate The study of climate in relation to plant and animal life. Cf. ECOSYSTEM.

ecological energetics The study of complex energy flow throughout an ECOSYSTEM: some radiant energy (combined with carbon, water, nutrients) is converted into chemical energy to be stored in plant tissue (*photosynthesis*), some is directly reflected, some is used to sustain plant life processes. Hence the output of plant food, which may provide energy input to another ecosystem; carbon dioxide is returned to the atmosphere; nutrients (such as organic nitrogen) are recycled between organisms and the environment. See FOOD CHAIN, along which there is a flow or transfer of energy, though with progressive loss.

ecology The science of the mutual relationship of organisms to their environment. Hence *ecological region*. *Human* or *cultural* **e** stresses the relationship of 'people and place' (as defined by geographers), of 'people and people' (as defined by sociologists). Without any qualification, the term **e** is botanical in concept. Note also the *Journal of Animal Ecology*.

ecosphere By astronomers and others, that zone concentric to the sun, roughly between Venus and Mars, where life as we know it is possible. Beyond it temperatures are too low, within it too high.

ecosystem An organic COMMUNITY (or *biotic complex*) of plants and animals (BIOME), viewed within its physical environment or habitat; 'a segment of nature', the result of soil, climate, vegetation and animals. Often used in ECOLOGY for the 'physical background'. The study of an e provides a methodological basis for complex synthesis between organisms and their environment.

ecotone A transition zone where plant COMMUNITIES naturally merge into one another, rather than change abruptly. This may be an area in which the two communities actually compete, thereby giving an impression of gradual mergence.

ecotope In one use, syn. with ECOSYSTEM; in another, the area actually occupied by an ecosystem.

edaphic App. to soil features which affect the growth of plants and other organisms: variations in texture, acidity and alkalinity, presence of minerals, and water content. Also used gen. as relating to, or app. to, the soil. Gk. *edaphos*, basis or floor.

eddy A movement of a fluid substance, partic. air and water, within a large one. Thus winds are thought of as having partic. characteristics such as force and direction, the net result of the movement of es. The scale of es may vary considerably, from DEPRESSIONS and ANTICYCLONES within the gen. circulation of the atmosphere to current es in a small air-stream, channelled by the Earth's rotation into a predominantly E.–W. direction.

edge Used in place-names for arêtes, ridges, mountain crests and plateau margins; e.g. Swirral E. on Helvellyn; Cross Fell E. in the N. Pennines; the E., the N. margin of Kinder Scout near Edale; Wenlock E., Shropshire. See pl. 23.

edge-line, of relief The description of a sudden sharp change or break of slope on a relief map by a heavy line; known in Germany as *Kantographie*, after *Kanten* (edges), and in USA as *kantography*.

Eemian The last interglacial period in N.W. Europe (see IPSWICHIAN).

effectiveness of precipitation, effective p (i) The actual total p, minus the max. possible evaporation. A high evaporation rate obviously reduces the value of the p for agriculture and water supply. E.g. in N. Sri Lanka the mean annual rainfall is 1270 mm, which would be heavy in middle latitudes, yet known as the 'Dry Zone', with much semi-arid scrub-jungle, because of the high evaporation rate. (ii) In hydrology, that portion of the p which flows away in a stream channel. (iii) In irrigation, that portion of the p which passes into the soil and is available for cultivated plants.

Plate 23. Curbar EDGE, a bold outcrop of hand Millstone Grit, southern Pennines, England. Note the effects of BLOCK DISINTEGRATION, the result of FROST WEDGING in past cold periods. (*R. J. Small*)

effective temperature See COMFORT ZONE, WIND CHILL.

efficiency, of a stream An indication of the load-carrying ability of a stream, the joint result of its COMPETENCE and CAPACITY. G. K. Gilbert suggested (1914) a formula, where e = capacity in grams per second/discharge in CUSECS × percentage slope of channel bed.

effluent A stream issuing from a lake.

egre See BORE.

Einkanter (Germ.) A stone on the surface of the desert upon which the scouring effect of wind-blown sand has cut a single facet; this implies a steady, unchanging wind-direction. Ct. DREIKANTER.

elastic rebound When rocks are subject to stress, the pressure or strain accumulates until they reach breaking point, when they tear apart, to snap back into a position of little or no strain. This is a major cause of FAULTS and EARTHQUAKES. Strain accumulated along the San Andreas Fault for a century before the sudden release of e energy resulted in the San Francisco earthquake of 1906.

E-layer The HEAVISIDE-KENNELLY LAYER.

Elbe glaciation The 1st glacial period in the N. European Plain, by those authorities who support a 4-fold glaciation in that area and ascribe certain morainic remnants near the Elbe valley to it; i.e. they equate the E g with the Günz in the sequence of Alpine glaciations. Others deny that the Scandinavian ice-sheet advanced so far S., and believe that the 1st glaciation to affect the N. European Plain was the Elster; i.e. equivalent to the 2nd or Mindel Alpine glaciation.

elbow of capture See CAPTURE, RIVER.

element (i) A combination of protons, neutrons and electrons; each is identified by the number of protons (atomic number) in its nucleus. 103 es are known, 92 occurring in nature, 11 made in the laboratory. They range from hydrogen (atomic number 1) to lawrencium (103). See also ISOTOPE. (ii) Each of the physical constituents which make up the sum total of climate: temperature, pressure, wind, humidity and precipitation; hence 'the elements'.

elevation (i) The height or altitude above some partic. level; e.g. above sea-level, or specif. above some DATUM. (ii) The vertical angle between the horizontal and some higher point, as in surveying or astronomical observations.

ellipsoid of reference The figure of the EARTH expressed as the ratio between the difference between the equatorial and polar radii, and the equatorial radius. If equatorial radius $=a$, and polar radius $=b$, then ellipticity $=\frac{a-b}{a}$. If the International E of R (1924) is used, ellipticity $=1/297$. For alt. values, see EARTH.

Elster glaciation A main stage in the g of the N. European Plain, correlated with the Mindel (Alpine) and Kansan (American) glacial periods. The moraines of the E g are deeply weathered, and form part of the Older Drift in Europe.

eluviation The removal of material, esp. bases, in solution (*chemical* e or LEACHING), or of finer fractions in suspension (*mechanical* e), from the upper to middle and lower layers of the soil profile by downward percolation of water, or horizontally through the same layer. The upper horizon from which material has been eluviated is the *eluvial* (A_2 or E) *horizon;* ct. ILLUVIATION. *Note:* In USA chemical solution is known as *leaching*, and e is confined to mechanical removal of fine particles or colloids. [*f* SOIL PROFILE]

elvan A vein of intrusive igneous rock, usually quartz-porphyry, found in Cornwall and Devon, penetrating the granitic rocks. 'Blue e' is sometimes applied by quarrymen to DOLERITE.

embayment (i) An open rounded bay in the coastline. (ii) In a structural sense, a large-scale 'sag' near the border of a continental mass, where a considerable thickness of sediment may be deposited. (iii) Also in a structural sense, an area of sedimentary rocks projecting into a mass of crystalline rocks; cf. Germ. use of '*Bucht*'; e.g. Kölnische Bucht.

emergence A rise of level of the land relative to the sea, so that areas formerly under water become dry land; specif. emerged coasts, characterized by RAISED BEACHES, and some coastal plains; e.g. S.E. USA.

emplacement The creation or formation in position of igneous rock masses, either by displacement, replacement or METAMORPHISM of the country rock; e.g. e of granite rocks of the Ben Nevis massif during early Devonian times, the result of down-faulting of a central core among older metamorphic rocks, and its infilling by successive upwellings of MAGMA.

endogenetic (endogenic, endogenous) 'Internal' forces of contraction, expansion, uplift, depression, distortion, disruption and outpouring which have contributed to the present landforms. Ct. EXOGENETIC.

endorheic drainage A centripetal drainage pattern, characteristic of the arid margins of the tropics, giving rise to large-scale alluvial accumulations and sometimes PLAYA lakes.

Endrumpf (Germ.) The end-product, the final land-surface of denudation, as envisaged by W. Penck. Though he actually equated it with the PENEPLAIN of W. M. Davis, it may be likened rather to L. C. King's PEDIPLAIN.

energy The power of doing work possessed by a body or system of bodies. This includes: (i) *chemical* (e stored in the molecules of compounds); (ii) *elastic* (e assoc. with a condition of mechanical strain); (iii) *geothermal* (e derived from the Earth's interior); (iv) POTENTIAL; (v) *heat* (the internal random motion of molecules); (VI) KINETIC; (vii) *nuclear* (e contained in the nucleus of an atom); (viii) *solar radiant* (e transferred from the sun by electromagnetic radiation). Partic. important in physical geography is the e assoc. with various phase changes of water.

energy balance See HEAT B.

energy grade line An expression for the rate of KINETIC ENERGY dissipation along a stream channel. Where the profile is steep and velocity high, the e g l is correspondingly steep. The e g l is, however, also affected by channel ROUGHNESS and channel plan. In a meandering stream the e g l may be smooth owing to (i) high energy dissipation where the stream negotiates bends, and (ii) equally high energy dissipation where the stream crosses shallow bars at CROSSOVER POINTS.

englacial Contained within the mass of a glacier or ice-sheet (debris, boulder, moraine, stream). Ct. SUB-GLACIAL. See pl. 45.

englacial river A melt-water stream flowing through a tunnel within a glacier or ice-sheet.

engulfment The inward collapse of a volcanic cone, the result of molten lava being drawn off under the surface of the earth or through a fissure in the flanks of a volcano. This is one way in which a CALDERA may be formed. E.g. Crater L., Oregon, formed from a peak (now denoted by the name Mt Mazama), partially destroyed by an eruption and accompanying e 6500 years ago. It was once believed that this peak 'blew its top' by a PAROXYSMAL ERUPTION, but now thought that it collapsed inwards; of an estim. 130 km³ of material which disappeared from the probable cone, only 14 km³ can be accounted for as debris; the rest must have collapsed within the underlying magma.

entrenched meander See INTRENCHED M.

entropy (i) In GENERAL SYSTEMS THEORY the amount of free energy in a system. A definition with a negative quality, because maxi. e relates to the min. amount of free energy in a system, whereas systems possessing a great deal of energy have min. e. A decreasing amount of free energy in a system (i.e. the trend toward max. e) signifies the progressive destruction of the heterogeneity of the system, the levelling of differences that formerly existed. Thus W. M. Davis's cyclical concept of landscape development relied on PENE-PLANATION, the progressive levelling of differences within the system, and thus involved a trend toward max. e. (ii) In geology, applied to sediments in respect of the uniformity or otherwise of constituent grades or particle size; mixed sediments have low e, uniform sediments high e.

environment The whole sum of the surrounding external conditions within which an organism, a community or an object exists. Often used in a limited way in geography; e.g. *natural* e, meaning either the non-cultural and non-social e, or the landscape before Man came. The *geographical* e means factors of the e whose relationships are considered in terms of spatial location. The *physical* e includes all phenomena apart from Man and the things he creates, while the *non-human* e includes everything not in a social system, whether made by Man or not. These are all slightly different, and unqualified use of the term can be misleading. In recent years Environmental Studies have developed as a strong cross- and inter-disciplinary approach.

environmental geoscience The study of the interaction between Man and the Earth's natural systems.

environmental lapse-rate The actual rate of decrease of temperature with incr. altitude at a given place at a given moment, with a mean of about 0·6°C per 100 m (1°F for 300 ft or 3·56°F for 1000 ft). See PROCESS L-R.

Eocene (i) In UK the 1st (lowest) of the geological periods of the Cainozoic era, lasting from about 65–38 million years ago, and the system of rocks laid down during that time. In England it is represented in the London and Hampshire Basins, chiefly comprising marine, deltaic and estuarine sands, clays and loams.

Hampshire Basin	London Basin
6 Barton Beds	—
5 Bracklesham Beds	Middle and Upper Bagshot Sands
4 Lower Bagshot Sands	Lower Bagshot Sands
3 London Clay	London Clay
2 Reading Beds	Woolwich and Reading Beds
1 Absent	Thanet Sand

(ii) In USA and incr. in UK, the E is regarded as an epoch within the Tertiary period and the Cainozoic era (see Table, TERTIARY) (i.e. in a lower grade in the geological hierarchy than in Britain), preceded by the Palaeocene and succeeded by the Oligocene. Some American authorities, however, still dispense with the Palaeocene, and regard the E as the oldest Tertiary epoch and series of rocks.

Eogene Syn., esp. in USA, with PALAEOGENE (PALEOGENE).

eohypse A 'restored contour' of a former land-surface (e.g. a dissected plateau), inserted by plotting surviving portions of land and by EXTRAPOLATION of the original contours.

eolith A flint found in E. Anglia (specif. within the CRAG formations of the Tertiary), claimed by some authorities to have been chipped and shaped by very early Man; i.e. one of the earliest known artifacts. Others believe that an e is a naturally formed flint, though possibly used by early Man without being fashioned or modified by him.

eon, alt. **aeon** Used gen. of an immeasurably long period of time, more specif. as 2 divisions of all earth-time: (i) CRYPTOZOIC E; (ii) PHANEROZOIC E.

Eozoic (i) By some authorities applied to the earliest era, corresponding to the Precambrian group of rocks; Gk., 'dawn of life'. In the sequence E, Palaeozoic, Mesozoic and Cainozoic. (ii) Others narrow its application to the earliest era of Precambrian time, followed by Archaeozoic and Proterozoic.

epeiric sea A rel. shallow s on the continental shelf, the result of EPEIROGENIC movement; syn. with *epicontinental*; e.g. North S., Irish S., Baltic S., Hudson Bay.

epeirogenic, epeirogenetic (Gk. *epeiros*, a continent) Large-scale continent-building forces, hence applied to all *en masse* vertical or radial crustal movements, both uplift and depression. Strata are not thereby folded or crumpled, though they may be slightly tilted or warped. Ct. OROGENIC. E.g. the e sinking of the N. part of N. America, hence formation of the Arctic islands and Hudson Bay

epicentre A point on the surface of the earth vertically above the SEISMIC FOCUS (ORIGIN) of an earthquake shock. [*f* EARTHQUAKE]

epicontinental sea See EPEIRIC S.

epigene Geological processes taking place at the surface of the earth, and the rocks formed there. Hence *epigenic, epigenesis, epigenetic*.

epigenetic drainage See SUPERIMPOSED DRAINAGE.

epigenetic ice A lense of underground i formed in earth material; if on a massive scale, known as *segregated i*.

epigenetic mineral An ore m formed later than the rock which contains it.

epiphyte A plant growing on another, not parasitic, but using it for support; e.g. lichens, mosses, orchids.

epoch The 3rd category of the sub-divisions of geological time, in the hierarchy *era, period, e, age*. The rocks deposited during an e are called a *series;* e.g. Coal Measures, Bunter, Lias Clay. 42 es are distinguished in Gt. Britain.

equal-area projection A type of MAP P in which areas are accurately shown, though at the expense of shape and direction, resulting in considerable distortion; e.g. MOLLWEIDE, ALBERS', BONNE, AZIMUTHAL EQUAL AREA, CYLINDRICAL EQUAL AREA. Also referred to as *equivalent* or *authalic p*s.

Equation of Time The difference between MEAN SOLAR T. and APPARENT (*local*) T. (i.e. as shown on a sundial) at noon. On about 24 Dec., 15 April, 14 June and 1 Sept. these ts coincide. The max. positive value (sun 'fast') is 14 min. 25 sec. on 11 Feb., and the max. negative value (sun 'slow') is 16 min. 22 sec. on 2 Nov. The value for every day is given in *Nautical Almanac;* see ANALEMMA [*f*]. The variation in interval between successive meridian passages of the sun occurs because: (i) the Earth's orbit round the sun is not circular; (ii) the Earth's velocity in its orbit is not constant; (iii) the ECLIPTIC is inclined to the Equator.

Equator (terrestrial) The parallel of latitude 0°, midway between the Poles in a plane at right-angles to the Earth's axis, length 40 076 km (24 901·92 mi.); a GREAT CIRCLE. See also CELESTIAL E.

equatorial climate A climatic type occurring between about 10°N. and 10°S. near sea-level, with constant high temperatures and humidity, and with about 12 hours day and night. There is little seasonal change, since climate is dominated throughout the year by a near overhead midday sun, converging tropical maritime air-masses associated with atmospheric disturbances, and heavy CONVECTIONAL rainfall. Under the Köppen system, this is type *Af*, with *Am* where monsoonal influences are experienced (as in parts of Indonesia and the coast of W. Africa). See INTERTROPICAL CONVERGENCE ZONE.

equatorial current A surface movement of oceanic water in e latitudes: (i) towards S.W. or W. in the N. hemisphere (the N. E C); (ii) towards N.W. or W. in the S. hemisphere (the S. E C); and (iii) an E.-moving countercurrent between the two in the DOLDRUMS. The N. and S. E Cs flow at about 28 km (15 nautical mi.) per day.

equatorial rain-forest The luxuriant forest found within about 7° latitude N. and S. of the Equator, with constant rainfall, high temperatures, and no seasonal periodicity. It is found mainly in the Amazon Basin, Zaïre

(Congo) Basin, and parts of Indonesia (esp. Borneo and New Guinea). Characterized by a profusion and variety of species, and a distinctive 'layered' arrangement, with a 'canopy' or upper layer of foliage. As there is virtually no seasonal climatic change, the gen. aspect of monotonous uniform greenness varies little. The most valuable trees, though scattered, are hardwoods; e.g. rosewood, mahogany, greenheart, ebony, ironwood.

equatorial trough A shallow t of low pressure lying on or near the Equator, the convergence zone of air moving from the subtropical high pressure zones, broadly corresponding to the belt of the DOLDRUMS over the oceans, though not present in the N. hemisphere summer over the continents. Like the Doldrums, it is displaced a few degrees of latitude N. and S. with the seasons, lagging behind the overhead sun. An area of high temperature and humidity with stagnant or sluggish air near sea-level. With incr. altitude, air-flow velocity increases. The **t** corresponds to the INTERTROPICAL CONVERGENCE ZONE.

equatorial westerlies A zone of semi-permanent W. winds near the Equator (in the INTERTROPICAL CONVERGENCE ZONE) between the converging TRADE WIND systems, which are deflected on crossing the Equator and acquire a W. component. These ws are best defined in the E. Indian Ocean, over Indonesia and the E. Pacific Ocean.

equidistance, equidistant projection A **p**, or its quality, of possessing a property which maintains the scale either radically from the point of zero distortion (e.g. AZIMUTHAL or ZENITHAL E P), or at right-angles to the line of zero distortion (e.g. CYLINDRICAL E P).

equifinality The state of a system of inter-related objects which start from a variety of initial conditions, and suffer a series of changes so that they end in the same result. Applied partic. to landforms: e.g., an amphitheatre-like DRY VALLEY (which may be the result of a CIRQUE-GLACIER, ROTATIONAL MASS-MOVEMENT, SPRING SAPPING, etc.), or a TOR (which may be the product of sub-surface chemical weathering or frost disintegration).

equilibrium A state of balance, when various forces have created a state or form which will not be altered with passage of time unless controlling factors change; otherwise, if a temporary tendency towards change is introduced, it is countered by an opposing force which tends to restore the system to its former state; e.g. .isostatic e, the shore-profile of e, the river long-profile of **e**, stable and unstable

e of a column of air in the atmosphere. See DYNAMIC E.

equilibrium line The line dividing the 'accumulation area' of a glacier from the 'ablation zone'. The **e l** is not exactly coincident with the FIRN LINE, as on many glaciers freezing of surface meltwater below the level of the firn line builds up the ice surface. The position of the **e l** varies from year to year; it is lower than average after a season of heavy winter snow and/or a cool summer ablation period.

equinoctial The Celestial Equator.

equinoctial gale A period of strong winds which may be experienced about each EQUINOX. A period of storm in mid-September does frequently occur in Britain. G. Manley found in the records for 52 years (1889–1940) such stormy periods in 31 years, with a frequency peak about 20 Sept.; on 16–17 Sept., 1935, winds up to 158 km h^{-1}, were experienced in Cornwall. The last week in March is also notably stormy. No physical cause for this has been discovered, but it is a well-established sailors' tradition, possibly related to the frequency of Caribbean hurricanes in September.

equinox One of the 2 points of intersection of the sun's apparent path during the year with the plane of the terrestrial Equator; hence the time when the sun is directly overhead at noon at the Equator, about 21 March (the *spring* or *vernal e*) and 22 Sept. (the *autumnal e*). At this time, day and night are approx. equal throughout the world, since the circle of illumination by the sun of half the globe passes through the Poles, hence each parallel is half in light, half in darkness.

equipluve A line on a map joining places with the same PLUVIOMETRIC COEFFICIENT.

Equirectangular Projection A very simple **p** with a network of horizontal parallels and vertical meridians. A standard parallel near the centre of the area to be shown is divided truly $= \dfrac{2\pi R \cdot \cos lat.}{meridian\ interval}$. On a world map the Equator ($2\pi R$) can be used as the standard parallel. Meridians are drawn vertically through divisions of the standard parallel, themselves divided true to scale, and other parallels are drawn as horizontal lines. If the Equator is the standard parallel, the graticule will consist of squares; any other standard parallel will produce rectangles with the N. to S. dimension longer. The **p** is neither equal area nor conformal. It is used for large-scale city and estate maps for geographic reference. The US Air Force have used since 1951 a world map on the E P, divided into 15° squares, starting at an origin at 180° longitude and 90°S. latitude in the S.W. corner (GEOREF system).

equivalence The quality of a PROJECTION in which the product of the scale ratios *a* and *b*, at right-angles to each other, is everywhere the same; i.e. areas of any size are represented in correct proportion to each other.

equivariable An isopleth interpolated to join places with equal COEFFICIENTS OF VARIABILITY: e.g. with a similar deviation from their avge. climatic conditions, usually expressed as a percentage.

era A major division of geological time and chrono-stratigraphic units; Palaeozoic, Mesozoic, Cainozoic and Quaternary. (In recent international usage, incr. in UK, the Quaternary is demoted to the status of a period, thus leaving 3 PHANEROZOIC es.). Precambrian times are sometimes: (i) included in a single era, Eozoic; (ii) divided into 3 es: Eozoic (oldest), Archaeozoic and Proterozoic. An e in time corresponds to a *group* of rocks.

erg (Arabic) The sandy desert of the Sahara, with areas of dunes, sand-sheets and undulating 'sand-seas'.

erosion Processes of earth-sculpture by agents (running water, ice, wind, waves) that involve transport of material, not including static WEATHERING nor MASS-MOVEMENT through gravity. Includes *physical e* (CORRASION) and *chemical e* (SOLUTION, CARBONATION). Commonly (but incorrectly) used syn. with DENUDATION.

erosion surface Commonly used in the sense of a level **s** formed by **e**, rather than by deposition. The term can be misleading, since most of the Earth's land-**s** is in some sense an **e s**, whereas the term is usually intended to refer only to near-horizontal portions. In this context PLANATION surface is preferable. E *platform* is used for a **s** of limited extent.

erratic, erratic block A fragment of rock carried by a glacier or ice-sheet and deposited some distance from the outcrop from which it was derived (see also PERCHED B); e.g. boulders of Shap granite found in the Scarborough area, Ribble valley and near Wolverhampton; riebeckite-microgranite from Ailsa Craig (in the Firth of Clyde) found in Merseyside, Anglesey, the I. of Man, S. Wales, S.E. Ireland; dark Silurian boulders lying on Carboniferous Limestone in the Pennines (the Norber Stone); boulders of red jasper, found in Boone County, Kentucky, at a distance of nearly 1000 km from the nearest solid rock of this type, N. of L. Superior. The Madison Boulder in New Hampshire, USA, moved 3·2 km by the ice-sheet, weighs 4662 tons.

eruption The process by which solid, liquid or gaseous materials are extruded or emitted on to the Earth's surface as a result of volcanic activity. The **e** may be: (i) *central*, from a single vent or a group of closely related vents; (ii) *linear* or FISSURE [*f*], when lava wells up along a line of crustal weakness. Materials erupted include gaseous compounds of sulphur, hydrogen and carbon dioxide; steam and water-vapour; SCORIA, PUMICE, CINDERS, DUST, ASH; acid and basic LAVA (hence *eruptive* rocks). Es range from quiet outflows and outwelling, to explosive and paroxysmal activity. An **e** may be small-scale, *periodic* or *intermittent*; see GEYSER. See pl. 24.

escarpment (i) The steep side of a CUESTA, equivalent (but preferable) to SCARP; more

Plate 24. ERUPTION of the central-vent volcano, Popocatepetl, Mexico, emitting gaseous compounds, steam and scoria; note also the lava flow descending the western slopes of the cone. (*Popperphoto*)

Plate 25. The ESCARPMENT face of the CARBONIFEROUS LIMESTONE at Eglwyseg, North Wales. Note the STRATIFICATION of the limestone, the occurrence of TALUS slopes at the base of the scarp-face. (*Aerofilms*)

loosely the whole cuesta. (ii) Any steep slope interrupting the gen. continuity of the landscape; e.g. a line of sea-cliffs; a 'step' or edge where a stratum of hard limestone, or an igneous SILL, appears on the surface. See pl. 25.

esker A narrow winding ridge consisting of stratified deposits of coarse sand and gravel. Common in Finland, the former East Prussia and Sweden, where they wind across country among lakes and marshes, and also in parts of N. England and Scotland. Eskers over 150 km long can be traced in Maine. One theory postulates that they are the 'casts' of stream courses in tunnels within an icesheet, which survived when the ice finally melted. Possibly they represent a rapidly receding delta formed at the edge of an icesheet or glacier by an englacial or subglacial stream. Owing to the enclosed nature of the stream, hydrostatic pressure was considerable, causing rapid flow and a heavy load. See also BEADED E. Es seem to occur longitudinal or parallel to the direction of ice-flow and at right-angles to the ice-front. Ct. KAME. See pl. 26.

[*f, page 110*]

espiñal (S. Am.) An area of thorny scrub.

estuary The tidal mouth of a river, where the channel broadens out in a V-shape, within which the tide flows and ebbs; e.g. Thames, Severn. Hence *estuarine*. Most es are the result of a rel. rise of sea-level. Some are the scene of extensive deposition (e.g. Dee in Cheshire), other may be kept rel. clear by scouring effects of tidal streams (e.g. bottle-neck of lower Mersey).

étang (Fr.) A shallow lake, esp. lying among sand-dunes or beach-gravels; e.g. E. de Berre, N.W. of Marseilles; E. de Biscarosse along the Landes coast of France.

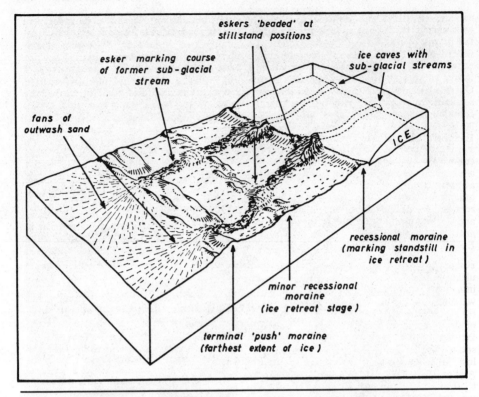

Plate 26. An ESKER system in Lake Rörströmsjon, Sweden. (*Erling Lindstrom*)

etchplain A land surface produced by denudation under savanna conditions of seasonal aridity. The rock is deeply etched and weathered; the surface layer is then stripped off by rainwash and episodic streams, and hence down-wearing predominates and maintains a broadly horizontal plain. Ct. PEDIPLAIN, PENEPLAIN.

etesian winds, etesians Strong N.E. to N.W. ws blowing during summer in the E. Mediterranean Sea, the result of a steep pressure gradient towards the thermal low pressure over the Sahara. They are at their strongest (20–30, sometimes 45 knots) in the late afternoon when convection is at its max., but weaken and die away towards evening. They frequently cause rough seas, and over land may bring clouds of dust, sometimes with fog in coastal areas.

étroit (Fr.) A narrow part of a stream or stream valley e.g. the R. Tarn, in the Grands Causses of S. France, where it passes through deep gorges cut into massive dolomitic limestone.

eugeosyncline A GEOSYNCLINE assoc. with VULCANICITY during its infilling by sedimentation. Ct. MIOGEOSYNCLINE.

eustasy, eustatism, *adj.* **eustatic** A worldwide change of sea-level, indicating an actual fall or rise of the sea (e.g. by abstraction of water to form ice-sheets or the return of water after their melting), but not by a local movement of the land. At their max., the Quaternary ice-sheets must have lowered sea-level by 90 m, and their remnants still contain enough water to raise it a further 30 m; this is known as *glacial-e.* Created by E. Suess (1888).

eutrophic Applied to an environment (e.g. a lake) rich in plant nutrients, and hence with a great abundance of plant and animal organisms. Nutrients are derived from fertilizers and lime applied to surrounding farms, and from animal and human waste from nearby settlements. E.g. Esthwaite Water, English Lake District. Ct. OLIGOTROPHIC.

evaporation The physical process of molecular transfer by which a liquid is changed into a gas. In climatology, the rate of e of water is a function of: (i) VAPOUR-PRESSURE; (ii) air temperature; (iii) wind; (iv) the nature of the surface: ct. bare soil and rock, where the rate of e is high, with a fine protective surface tilth or plant-cover, where the rate is low. High e rates occur in hot deserts; e.g. Atbara (Sudan) has a potential e rate of 6250 mm (246 in.), Helwan (Egypt) of 2390 mm (94 in.). Rates are lowest in equatorial climates (500–750 mm, 20–30 in.) because of high humidity, and in cool midclimates (London, 330 mm, 13 in.) because of rel. low temperatures.

evaporimeter A device for measuring the amount and rate of evaporation. In the *Piche e,* distilled water in a tube is allowed to evaporate from a piece of porous paper, and the loss in a certain time is measured on a graduated scale along the tube. A more common method in UK is to expose water in a large shallow tank (the Symons Tank), and measure the falling level at intervals. Syn. with *atmometer* in USA.

evaporite A SEDIMENTARY rock composed of minerals precipitated from a solution, and dried out by evaporation; e.g. GYPSUM, ROCK-SALT. This occurred during major periods of oceanic transgression, as the Permo-Triassic deposits in UK. A shrinking inland sea produces sheets of es; e.g. Gt Salt Lake, Utah; Dead Sea. The greatest deposition is in areas of high evaporation from inland seas of high salinity, esp. where a minor embayment is subject to constant evaporation and replenishment; e.g. Gulf of Kara-Bogaz, Caspian Sea.

evapotranspiration The loss of moisture from the terrain by direct evaporation plus transpiration from vegetation; e.g. a single maize plant in USA evapotranspired 54 gallons between 5 May and 8 Sept.; an acre of maize lost 324 000 gallons in that time, equivalent to 280 mm of rainfall. At Slaidburn, Yorks., the Fylde Water Board found that in a spruce plantation, with 984 mm rainfall, only 255 mm penetrated the soil; i.e. there was a loss through e of 1 million gallons a day on 1500 acres. (i) *Potential e,* a theoretical max. loss assuming a continuous supply of water to the surface and vegetation (e.g. by irrigation); (ii) *actual e,* the observed or true amount of e, which lessens as soil-moisture diminishes or if no rainfall or irrigation water is received. Actual e will equal potential e in areas of high rainfall and low evaporation. If evaporation exceeds precipitation, the area is subject to desiccation. The avge annual e in S. England is 500–550 mm.

everglade(s) A marshy area, with tall grass, canes and some trees. Specif. (proper name) in Florida and along Gulf coast of S. USA.

evorsion Erosion by eddies in the rock-bed of a stream, forming POT-HOLES.

exaration Glacial erosion carried out by ice alone, i.e. PLUCKING, not by ice laden with debris. Ct. ABRASION.

exfoliation Strains cause a 'shell', 'scale' or concentric sheet to pull away and split or peel off an exposed rock surface, a process known as e. This may be partic. effective on the face of a crag which receives and loses the sun's heat rapidly, esp. where the rock is composed of different minerals, with varying coefficients of expansion, resulting in complex strains. The process of e probably requires the presence of moisture to be effective; some authorities now insist that e involves chemical change. It seems that a massive rock with curvilinear joints is most affected. The term *spalling* as used in USA is limited to a superficial 'scaling'. What was regarded as large-scale e is more probably SHEETING. See pl. 7, 27.

exhumation In geomorphology, the uncovering by denudation of a former surface or feature buried by later deposition, hence 'exhumed relief', 'e surface', 'e landscape'; sometimes the result is a 'fossil erosion surface'; e.g. e of Charnwood Forest (Leicestershire), formerly buried under Triassic rocks; of part of the chalk dip-slopes of London Basin; of parts of Central Massif of France.

112 EXOGENETIC

Plate 27. EXFOLIATION in a sheet of granite, approximately 1 m thick, peeling from the upper surface of Sobi Hill, a prominent INSELBERG near Ilorin, western Nigeria. (*R. J. Small*)

exogenetic (exogenic, exogenous) The external forces of denudation (weathering, mass-wasting, erosion, transport, deposition). These combine with internal (ENDOGENETIC) forces to produce land-forms.

exosphere A zone in the IONOSPHERE above about 700 km, characterized by the presence of neutral atomic oxygen, ionized oxygen and hydrogen atoms. This is taken to be the outer fringe of the Earth's atmosphere, since there begins the escape of these neutral particles into space.

exotic A plant or animal introduced from some other area where it flourished naturally; e.g. rhododendron into England from the Himalayas; citrus fruit into Mediterranean lands from China; rabbit into Australia.

exotic river A r deriving most of its volume from headstreams in another region, as from snow-melt or heavy rainfall on distant mountains; e.g. Colorado R. in S.W. USA; Blue and White Niles; Tigris and Euphrates.

expanded-foot glacier A g which issues beyond the mouth of a valley to expand into a broad lobe of ice on an adjacent plain; e.g. Skeidarajökull on the edge of Vatnajökull, in Iceland. An e-f g is a small version of a PIEDMONT G.

exposure (i) The position of any place relative to sunshine, prevailing winds, oceanic influences; cf. ASPECT. (ii) The situation of a meteorological instrument, which should be standardized to give observational uniformity.

(iii) In a geological context, a place where rocks are naturally or artificially exposed to view. A rock covered with soil derived from its own disintegration would be mapped as an OUTCROP, not as an e.

extending flow A type of glacier flow, postulated by J. F. Nye, in which an attenuation of the glacier accompanies an increase of surface velocity. E f occurs where glaciers pass over large bedrock irregularities, and to the rear of glacier 'surges'. Diagnostic features are large transverse CREVASSES (ct. the shear planes developed by COMPRESSING FLOW) and small step-faults.

extinct volcano A v formed in long past geological time, in an area where now no sign of activity. The orig. form may be largely destroyed by denudation; e.g. Arthur's Seat and Castle Rock, Edinburgh; Mont Dore, Central Massif of France.

extrapolation Extending the values of a series of variables on either side of some known values. In a cartographic or diagrammatic sense, implies the reconstruction of past or future patterns or trends by extending and developing present patterns or trends as mapped or graphed. E.g. extension and projection of a graph of population trends; reconstruction of former contours (EOHYPSES) from surviving fragments of a dissected land-surface; reconstruction of the former profile of a river valley before rejuvenation by e from the KNICKPOINT. It can also be based upon finding a mathematical expression to fit a segment of river profile, and then e of the curve downstream on that basis.

extreme climate A c with a considerable difference or *range* between the temperatures of the warmest and coolest months. This occurs mainly within continental interiors, far from moderating marine influences; e.g. Cold Continental C in central Asia (Verkhoyansk, Jan. mean, $-51°C$ ($-59°F$), July mean $15\cdot5°C$ ($60°F$), with an absolute minimum of $-69\cdot5°C$ ($-93°F$); this is W. Köppen's *Dwd* type.

extremes, of temperature The highest and lowest shade ts recorded during a day (*diurnal e*), month, year or any period of available records at a meteorological station. It may be used of the mean highest and mean lowest t over a period of time. E.g. Ajaccio (Corsica), in Dec., extreme max.$=20°C$ ($68°F$), extreme min.$=-2°C$ ($28°F$). *Note:* the mean monthly t for Ajaccio in Dec., the figure usually quoted in climatic tables, is $9°C$ ($48°F$). The world's extreme max. $t=57\cdot8°C$ ($136°F$) (Azizia, Tripoli), extreme min.$=-88\cdot3°C$ ($-127°F$) (in Vostok in Antarctica in 1960). UKs extreme max.$=38\cdot1°C$ ($100\cdot5°F$) (Tonbridge, Kent), extreme min.$=-27\cdot2°C$ ($-17°F$) (Braemar, Scotland).

extrusion The emission of solid, liquid or gaseous materials on to the earth's surface during an eruption; hence EXTRUSIVE ROCK. Syn. in this sense with VOLCANIC. Ct. INTRUSION.

extrusion flow, of ice Movement caused by great thickness and resultant pressure of ice within an ice-sheet. This causes an *en masse* subsidence within the ice, which towards the margins becomes a more horizontal and outward movement, probably facilitated by PLASTIC DEFORMATION in the basal layers. So in the Pleistocene an ice-sheet gradually crept S. over extensive lowlands (N.W. Europe, N. America), despite the gentle gradient. It was formerly believed that valley glaciers also experienced e f, but this is not now generally accepted.

extrusive rock An igneous r formed by pouring out of lava on the surface, where it solidified; syn. with volcanic r, as opposed to PLUTONIC or INTRUSIVE. Cooling is rapid, resulting in small-crystalled or even glassy rs; e.g. basalt, obsidian, rhyolite, andesite.

exudation basin A depression in the surface of the ice at the head of a glacier where it leaves an ice-sheet; e.g. in Greenland.

'eye' The central area of a HURRICANE or other tropical storm, a small region 20–60 km across, of calms or light variable winds, surrounded by the storm-area. Atmospheric pressure at the centre may be as low as 965 mb.

eyot See AIT.

fabric Used to indicate the physical composition or makeup, both textural and structural, of some compound; e.g. TILL f, an assemblage of clay, stones and sand, which may be statistically analysed in various ways, incl. the orientation and dip of the stones within its matrix. A specialized application is PETROFABRIC ANALYSIS.

facet Used in land classification to denote groupings of similar land ELEMENTS (e.g. the grouping of individual alluvial fans at the base of an escarpment).

facies (i) Appearance, nature or character, applied esp. to descriptions of the composite character of a rock. (ii) A certain classificatory implication, distinguishing one stratigraphic body from another, with ref. both to lithological character and fossil content. (iii) Applied to rocks of a certain age where lateral changes of lithology imply changing conditions of formation, with accompanying differences in the fossil content.

factor Used in 2 senses: (i) a gen. cause or control contributing to a partic. result, effect or condition; e.g. in climate, fs include latitude, altitude, distribution of land and sea, ocean currents, presence of lakes, influence of relief barriers. (ii) In the analysis of situations involving variables, implies one category or 'family' of variables; *f analysis* is an important aspect of geographical research, enabling data to be mathematically analysed, by means of a computer, on a very large scale.

Fahrenheit A graduated scale of temperature, introduced about 1724 by Gabriel Fahrenheit, on which $0°$ represents the melting-point of ice in a mixture of sal-ammoniac and water, $32°$ represents the melting-point of ice in water, and $96°$ was taken as the temperature of a healthy person (subsequently found to be $98\cdot4°$). He did not use the boiling-point of water, but found by experiment that on his scale it equalled $212°$. Fahrenheit is remembered also for his improvement of the barometer and thermometer.

Conversion: $°F = \frac{9}{5}(C° + 32°)$
$°C = \frac{5}{9}(F° - 32°)$

114 FAHRENHEIT

°F	°C	°F	°C
100	37·7	46·6	8
98·6	37	45	7·2
96·8	36	44·6	7
95	35	42·8	6
93·2	34	41	5
91·4	33	40	4·4
90	32·2	39·2	4
89·6	32	37·4	3
87·8	31	35·6	2
86	30	35	1·6
85	29·4	33·8	1
84·2	29	33	0·5
82·4	28	32	0
80·6	27	31	−0·5
80	26·6	30·2	−1
78·8	26	30	−1·1
77	25	28·4	−2
75·2	24	26·6	−3
75	23·8	25	−3·8
73·4	23	24·8	−4
71·6	22	23	−5
70	21·1	21·2	−6
69·8	21	20	−6·6
68	20	19·4	−7
66·2	19	17·6	−8
65	18·3	15·8	−9
64·4	18	15	−9·4
62·6	17	14	−10
60·8	16	12·2	−11
60	15·5	10·4	−12
59	15	10	−12·2
57·2	14	8·6	−13
55·4	13	6·8	−14
55	12·7	5	−15
53·6	12	3·2	−16
51·8	11	1·4	−17
50	10	1	−17·2
48·2	9	0	−17·7

Note: −40° is common to both scales.

fairway The main navigable channel of a river or estuary, usually buoyed and lighted, leading up to a port or harbour.

falaise (Fr.) A low cliff, not necessarily marine; e.g. F. de l'Ile de France.

fall (i) Used in USA for autumn. (ii) See WATERFALL.

fall-line A line or narrow belt joining the points where a series of near-parallel rivers descends by falls or rapids from a plateau edge on to a lowland; e.g. the F-l in S.E. USA between the ancient rocks of the Piedmont Plateau and the newer rocks of the Atlantic coastal plain. [*f, opposite*]

false-bedding See CURRENT-B.

false cirrus Thick, greyish CIRRUS cloud, often assoc. with the upper part of a thundercloud, and gen. a sign of bad weather.

false drumlin A mass of rock over which a thin cover of DRIFT has been laid by an ice-sheet, giving the superficial appearance of a true D. [*f* DRUMLIN]

false origin A point of o in a GRID system, from which the position of any place is uniquely defined in terms of its coordinates. The o is transferred from the intersection of the projection axes to the f position to avoid negative quantities. The f o of UK National Grid lies S.W. of the Scilly Is., transferred 400 km W. and 100 km N. from the intersection of the central meridian 2°W. and standard parallel 49°N. The f o of American military maps on the UTM GRID for each of the 60 zones is a point 500 km W. of the central meridian on the Equator for the N. hemisphere, 10 000 km S. of this point for the S. hemisphere.

fan See ALLUVIAL F.

fan-folding An ANTICLINORIUM with the axial planes of the folds converging towards a centre.

fanglomerate Material deposited in an ALLUVIAL FAN, consisting of heterogeneous fragments of all sizes, which have been compacted and/or cemented into solid rock.

fast ice I which covers the frozen surface of the sea, but remains f to the actual coast.

fathom A nautical measurement of depth = 6 ft = 1·829 m; 100 fthms = 1 cable; 10 cables = 1 nautical mile.

fault A surface of fracture or rupture of strata, involving permanent dislocation and displacement within the Earth's crust, as a result of the accumulation of strain; see ELASTIC REBOUND, Hence faulting. See also

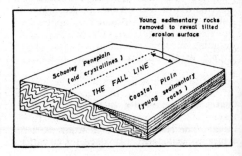

DIP-, NORMAL, OBLIQUE, REVERSED, STRIKE-, TEAR and THRUST F, HADE, HEAVE, THROW, FOOT-WALL, HANGING-WALL. [*f*]

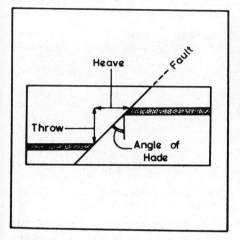

fault-block A section of the Earth's crust sharply defined by **f**s; it may stand up prominently (BLOCK MOUNTAIN, HORST), be tilted (TILT-BLOCK), or be denuded to the level of the surrounding country (faulted inlier). Ct. GRABEN.

fault-breccia Rock occurring in the SHATTER-BELT of a fault, consisting of crushed angular fragments. [*f* HANGING WALL]

fault-line scarp Where faulting brings rocks of varying resistance into close juxtaposition, differential denudation may wear away less resistant rock on one side of the fault, forming a cliff or scarp along its line. Ct. FAULT-SCARP. The resultant higher ground may be on either the upthrow or downthrow side. E.g. a **f-l s** has developed along the Mid-Craven F. in Yorkshire. To the N. is a plateau of Carboniferous Limestone, which has been down-faulted to the S., so that less resistant Bowland Shales have been brought into juxtaposition along the **f**. Denudation has caused the line of the **f** to stand out as a **f-l s**, in places 90 m high, including Malham Cove and a line of 'scars'. The E. face of the Sierra Nevada, California, is a major **f-l s**, as is the E. face of the Grand Tetons, Wyoming. As denudation progresses, the **f-l s** may develop into an OBSEQUENT F-L S (facing the opposite direction), then into a RESEQUENT F-L S (facing the original direction).
[*f*, *opposite*]

fault-scarp A visible steep edge or slope of a recent fault, due to earth-movements. It is an initial land-form, present only in the earliest stages of denudation succeeding the earth-movement which caused it; e.g. 2 **f**-ss were produced in Montana and Wyoming by the Madison earthquake of 1959, forming long 'steps' across country about 4–6 m high. Most **f**-ss are soon obliterated by weathering and erosion, or later develop into FAULT-LINE SCARPS. See pl. 28. [*f*]

fault-spring A SPRING thrown out along a fault, where a permeable bed, such as sandstone, is brought up against an impermeable rock, such as shale.

fauna The animal life of any region, geological formation or period; hence *faunal province*, an assemblage of **f**, both currently (e.g. Australasian) or fossilized.

feather-edge The fine edge of a bed of rock, usually sedimentary (though it includes also igneous intrusions), where it thins or PINCHES-OUT and disappears. It can be seen on the surface of a DIP-SLOPE, as in the N. Downs, where an Eocene **f-e** occurs, from beneath which the Chalk emerges; and at the base of the Coal Measures in N. Staffordshire ('F-e Coal').

feedback Defined by R. J. Chorley and B. A. Kennedy as 'the property of a system ... such that, when change is introduced via one of the system variables, its transmission through the structure leads the effect of the change back to the initial variable, to give a circularity of action'. The most common type is *negative f*, in which the effect is to *counteract* the initial

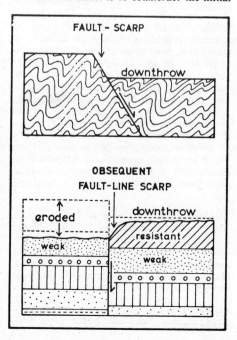

116 FEEDBACK

Plate 28. A low FAULT SCARP (approximately 100 m high) produced by late-PLEISTOCENE faulting of LAVA on the floor of the RIFT VALLEY, southern Kenya. (R. J. Small)

change e.g. a sudden influx of load will cause stream overloading; deposition of part of the load will steepen the channel and increase stream energy, so that effective transport of the larger load is possible. The less common *positive f* has a *snowballing* effect, increasing the impact of the initial change e.g. removal of vegetation will cause run-off and soil erosion; the more permeable A horizon will be removed to expose less permeable B and C horizons, thus further increasing run-off and erosion.

feldspar, felspar A complicated group of minerals consisting mainly of silicates of aluminium, with those of potassium, sodium or calcium. *Alkali* includes orthoclase (mainly potassium), soda-orthoclase (containing some sodium), and anorthoclase (mainly sodium). *Soda-lime f* (with silicates of sodium and calcium) is known as plagioclase-f, a series varying in composition from albite (sodium) to anorthite (calcium); the opalescent variety of albite is moonstone. Plagioclase-fs form important constituents of such sialic rocks as diorite and andesite, and occur also in a calcic form in such basic rocks as gabbro and basalt.

fell (Norwegian) An open hill- or mountainside in N. England, esp. English Lake District, hence *f-pasture*, and '*the f*' as the upper part of a sheep-farm. Also used for culminating summits; e.g. Scafell, Bowfell (English Lake District); Crossfell (N. Pennines).

Felsenmeer (Germ.) A continuous spread of large angular boulders, hence *block-field*, *block-spread*, *boulder-field*, formed *in situ* by acute frost action on well jointed rocks, covering the surface of a high plateau or flat-topped mountain. Lit. a 'rock-sea'; e.g summits of the Scafell range in English Lake District, and of Glyder Fach and Glyder Fawr in N. Wales.

felsic Used esp. in USA of minerals consisting mainly of quartz and the feldspars, light in colour and of low density. Ct. MAFIC.

fen A water-logged area, with pond-weeds, reeds, rushes and alders, in which peat is accumulating, but in which the ground-water is alkaline (usually with calcium carbonate) or neutral; e.g. F. District of Lincolnshire and Norfolk; Lancashire Mosses; Somerset Levels; '*laagveen*' of the Netherlands. When drained, it affords good humus- and mineral-rich soil, of excellent texture, dark in colour. Ct. BOG, MARSH.

fenland An area of FEN. Proper name for the area in E. England around the Wash.

Fenster (Germ.), **fenêtre** (Fr.), *lit.* a window An opening worn through the upper strata by denudation, where rocks have been overfolded, thus exposing underlying younger rocks in its floor. Should not be used simply for the exposure of older underlying rocks by denudation, as along the crest of an anticline; see INLIER. E.g. Fenêtre de Theux in the Ardennes, Belgium, is a small massif mainly of Devonian rocks, outlined by perimeter faults. It is a piece of a NAPPE, revealed by the removal of overthrust Cambrian-Silurian rock by denudation. In the E. Alps, denudation has cut through the Austride nappes, exposing underlying nappes in 3 places: Hohe Tauern, Semmering and Lower Engadine districts, the classic 'windows of the Alps'. [*f*]

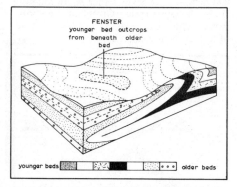

feral relief Used by C. A. Cotton to describe a type of relief in New Zealand in which the main valley sides are dissected by many insequent streams (ct. the undissected slopes of humid temperate landscapes in Great Britain). F r is related to the occurrence of winter max. rainfall which, in the absence of evapotranspiration, leads to rapid run-off and intense dissection.

Ferrel's Law As the result of the Earth's rotation, a body moving on its surface is subject to an apparent deflection to the right in the N. hemisphere, to the left in the S. (the CORIOLIS FORCE). This apparent force is zero at the Equator, but incr. progressively N. and S. Its effects are shown most specif. in air and water. Postulated by W. Ferrel, an American scientist (1856).

ferric In chemical terms, a trivalent (or tervalent) iron compound. The two most economically valuable are *haematite* (red f oxide, Fe_2O_3) and *limonite* (hydrated f oxide, $2Fe_2O_3.3H_2O$).

ferricrete A DURICRUST in which iron oxides form the cementing agent; sometimes used as syn. with LATERITE (*sensu stricto*).

ferromagnesian mineral A m characterized by an abundance of iron and magnesium, of which the chief are augite, biotite, hornblende and olivine. Gen. these ms are dark in colour and of high density. Syn. with *mafic*. Ct. FELSIC.

ferrous App. to iron compounds not saturated with oxygen; in chemical terms, a divalent (or bivalent) compound of iron. Usually formed by the reduction of ferric compounds; e.g. *siderite* (f carbonate, $FeCO_3$), ore found in the Jurassic oolites of England.

ferruginous Rocks containing some iron, often reddish in colour.

fetch The distance of open water over which a wind-blown sea-wave has travelled, or over which a wind blows. This helps to determine the height and energy of a wave, and hence its erosive and depositive effects on the coast. Many coastal features of deposition seem to be orientated to face the direction of max. f; e.g. CUSPATE FORELAND [*f*], such as Dungeness. Wave attack is emphasized and concentrated on headlands exposed to a long f.

fiard (fjard) (Swedish, *fjärd*) A coastal indentation resulting from a rise of sea-level on the margins of a glaciated rocky lowland or PENEPLAIN, with lower shores and broader profiles than a FIORD, usually with a threshold and deeper than a RIA, and with numerous fringing islands; e.g. the coast of S.E. Sweden. This English usage has taken on a more specialized meaning not apparent in Sweden, where the term means simply large open areas of water surrounded by islands, as in this area near Stockholm. [*f*]

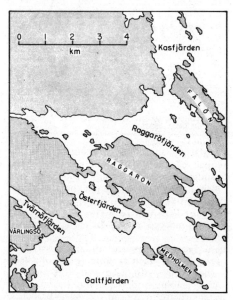

field capacity The state of the soil when all 'gravity water' has been drained, usually over a period of several days or even weeks after cessation of rainfall. The remaining 'capillary water', held to individual soil particles by surface tension, is sufficient to provide the needs of growing plants. However, in a major drought this will be reduced by evapotranspiration well below f c, and plant wilting will occur.

FILLING

filling An increase in the atmospheric pressure near the centre of a low pressure system (DEPRESSION), hence 'filling-up', or dying away, as opposed to '*deepening*'.

finger lake A long, narrow l occupying a U-shaped glacial trough; e.g. in English Lake District; N. Wales; Highlands of Scotland (*lochs*); and N. Italy (Maggiore, Como, upper part of Garda). Most are the result partly of glacial erosion of the containing valley in solid rock, partly of the deposition of a morainic dam across the lower end of the valley. The F Ls of USA occur in New York State, near Syracuse; they now drain N. to L. Ontario, though their preglacial valleys drained S. to the Susquehanna R. [*f*]

Sogne F. is 160 km long and mostly less than 5 km in width, with walls sloping at 28°–40° from the 1500 m plateau-surface to over 900 m below sea-level. Other fs are found in W. Scotland (sea-lochs), Greenland, Labrador, British Columbia, Alaska, S. Chile, and W. of S. I. of New Zealand. [*f*]

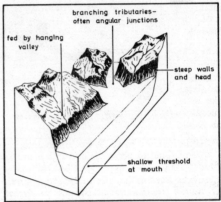

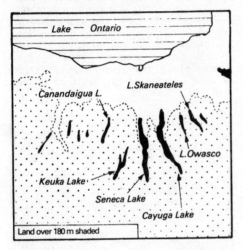

fire-clay A fossil clay capable of withstanding great heat, used for making refractory bricks for the linings of furnaces and for glazed drain-pipes. It occurs commonly as an underclay (SEAT-EARTH) beneath seams of coal in the Lower Coal Measures, and may be the bed in which roots of the 'coal forest' grew. See also GANISTER.

fire-damp An explosive mixture of air and methane (CH_4), which may occur in coal-mines.

firn (*Germ.*) Syn. with *névé*, but many glaciologists prefer **f**. Others define **f** as compacted snow, *névé* as the area of **f**. Lit. 'last year's' or 'old' snow. Where snow is able to accumulate in a basin or hollow, from direct snowfall and from avalanches down surrounding slopes, it is compressed by the weight of the addition of successive layers, and is gradually changed into a more compact form. Air is retained between the particles, forming a mass of whitish granular ice. During surface melting on summer days, water percolates into the **f** and then re-freezes at night to form a more compact mass. SUBLIMATION assists, whereby molecules of water vapour escape from snow-flakes and re-attach themselves, so that granules of crystals become progressively more tightly packed. Its density is greater than 0·4, less than 0·82 (Some American glaciologists give a precise density of 0·55.) Some degree of stratification can be seen, representing each year's contribution of snow. From **f** are

Finiglacial stage The 3rd main stage in the retreat of the continental ice-sheet from Scandinavia, from about 8000–6500 B.C. During this stage, the ice left the area which is now Finland. The land was gradually uplifted by ISOSTATIC recovery, and the Baltic Sea was reduced to land-locked L. Ancylus.

fiord, fjord (Norwegian) A long narrow arm of the sea, the result of submergence of a deep glacial valley, with steep walls (the angle of which continues beneath the water), deep water virtually to its head, rectilinear branches, and a bar or threshold near its mouth (usually of solid rock, with sometimes a cover of glacial debris). The fs of Norway are deeply cut into the high plateau of Scandinavia. During the Ice Age, glaciers eroded along lines of least resistance, such as a pre-glacial river valley, a major fault-line, or a line of weak unresistant rocks (e.g. Hardanger F., which lies along a syncline of schist between two masses of hard crystalline rock). The

derived: *firnification*, the creation of **f**; *f-line* or *f-limit*, the highest level attained by the snow-line during the year or alt. the edge of the snow-cover at the end of summer; *f-field*, the mass of accumulated compact snow; *f-basin* in which it accumulates.

firth A Scottish word for some water-areas, notably: (i) the lower part of an estuary (e.g. F. of Forth, Solway F.); and (ii) a strait (e.g. Pentland F.).

fissile The quality of a rock being easily split, where BEDDING-PLANES, JOINT-PLANES, LAMINAE or CLEAVAGE are well developed and defined; e.g. slate (a property of which a quarryman takes advantage), shale, flags, schist and oolitic limestones (as in the Cotswolds, where used for roofing and locally called 'slates', e.g. Stonesfield Slate, a sandy limestone of the Great Oolite Series). F bedding has laminae less than 2 mm thick.

fissure eruption A linear volcanic e, in which lava, gen. basic and of low viscosity, wells up to the surface along a line of crustal weakness, usually without any explosive activity. The gen. result is an extensive basalt plateau. In the Laki (Iceland) **f e** of 1783, an outpouring of lava occurred along a **f** 30 km long, and also a string of ASH CONES (or 'eruptive conelets') was built up along its line. [*f*]

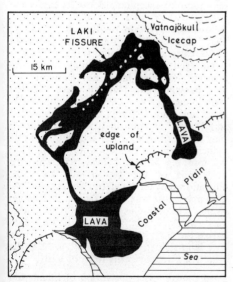

fjard See FIARD.

fjeld, fjell (Norwegian), **fjäll** (Swedish) Cognate to *fell* in N. England. A plateau surface, often rocky, bleak and monotonous, above the tree-line, covered with snow in winter. Poor pasture occurs on lower parts, and lichens and mosses on higher, with occas. patches of dwarf birch, willow, alder, juniper and berried plants of a TUNDRA aspect. Used by Lapps in the N., and for summer grazing of cattle and goats on the lower areas further S. E.g. Dovre F. lying at 900–1200 m.

fjord (Norwegian) See FIORD.

flagstone, flag (Old Norse, *flaga*, slab) A micaceous sandstone or sandy limestone which splits along the BEDDING-PLANES, and is hard enough to be used for paving and building. E.g. Old Red Sandstone (though grey in colour) of Caithness (N. Scotland); Carboniferous sandstones of Yorkshire; Silurian flags of S. Cumbria.

Flandrian Transgression A gen. rise of sea-level from c. 8000 B.C., when it was approx. 55 m below present. The rise covered the basin of the N. Sea lowlands, and formed the N. Sea and English Channel. The Strait of Dover was breached c. 5000 B.C. By c. 3000 B.C., sea-level stood about 6 m lower than at present.

flash (i) A sudden rise of water in a stream. (ii) A small lake in a hollow caused by subsidence due to underground mineral working, esp. salt; e.g. Bottom F., Top F. near Winsford, Cheshire.

flash-flood A sudden but short-lived torrent in a usually dry valley, notably in a semi-arid area after a rare, brief, but intense rainstorm, carrying an immense load of solid matter, the product of desert weathering. The stream turns into a mud-flow and soon comes to rest. The flow may be torrential for a short time, and there are records of people swept away and drowned. E.g. Arizona, Utah, S. California. It may also be caused by the collapse of a dam; e.g. Fréjus in France (1959); or of an ice- or log-jam, as in N. Canada.

flat (i) A low-lying area of marsh or swamp in a river valley; e.g. Altcar Fs., Lancs. (ii) A mud-bank exposed at low tide (*mud-f*). (iii) In geomorphology, any nearly level area within a region of marked relief.

flatiron A triangular mass of rock outcropping on the end of a spur where the strata are inclined steeply. The term is used esp. in USA; e.g. the Flat-Irons, great slabs of coarse sandstone and conglomerate, the Fountain formation of Pennsylvanian (Upper Carboniferous) age, outcropping along the Front Range of the Rockies, and overlooking

Boulder, Colorado. The smooth steep E. faces are actually tilted bedding planes. [*f*]

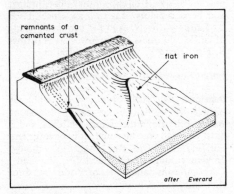

flatland ratio map A m constructed by gridding a contour m, calculating the percentage of each square occupied by land below any selected critical slope, plotting these values, and inserting isopleths.

fleet (i) A small lagoon, usually of salt or brackish water, separated from open sea by a long sand- or shingle-bank parallel to the coast; e.g. the Fleet behind Chesil Bank, Dorset. (ii) A small creek or inlet, the name preserved in the case of Fleet St., London called after the Fleet R.

flexure Gentle bending of strata as a result of tensional stress.

flint A black or dark grey mass or nodule of a dense, fine-grained silica, usually occurring in bands along bedding- or joint-planes in chalk. They appear prominently in the Upper Chalk, but rarely found in the Lower Chalk. Probably small siliceous sponge-spicules, orig. scattered throughout the Chalk, were carried away in solution by percolating water, and were redeposited where some silica already existed; i.e. at the bedding-planes, so that large amounts of silica were concentrated there. Thus the f appears to grow as in the manner of a concretion. The silica in f is very pure. A nodule is hard and tough, but under impact breaks with a conchoidal fracture, providing a sharp edge. F was the chief raw material of tools and weapons of the Stone Ages. Cf. EOLITH. F also is found as detrital pebbles in Tertiary deposits.

flocculation, *vb*. to **flocculate** (i) The process by which soils and soil-colloids coagulate or aggregate into 'crumbs', coarsening the texture of a fine clay soil; stimulated by liming. (ii) The deposition of clay particles in salt water.

floe A thin, detached, floating horizontal sheet of sea-ice. See pl. 29.

Flohn's climate classification A c, produced in 1950, based on global wind belts and precipitation features, producing 7 major categories; thus zone 1 is 'Equatorial westerly belt: continuously humid'; zone 2 is 'Tropical Zone, winter Trades: summer rainfall'. Temperature as such is not used as a specif. criterion, though is obviously implied by terms 'tropical', 'boreal', etc.

flood The inundation of any area not normally covered with water, through a temporary rise in level of a river, lake or sea. (i) A *river* floods when its channel is inadequate to accommodate discharge from its CATCHMENT; i.e. rises from BANKFULL to f-stage, so spreading over its f-plain. This may occur after exceptional rainfall or snow-melt; e.g. on R. Ohio, Jan., 1963. On some rivers, flooding is part of the seasonal régime; e.g. R. Nile, on which irrigation depends. If a river flows above its f-plain as a result of sedimentation in its bed, so that it is contained between natural or artificial embankments (LEVEES), the danger of widespread flooding if the levees are breached is very great; e.g. lower Mississippi; R. Po (esp. in the winter of 1951-2 and the autumn of 1966); Hwang-Ho ('China's Sorrow'). Concentrated rainfall may produce exceptional local fs; e.g. in the valleys of the E. and W. Lyn R., N. Devon, following heavy rain on Exmoor, 18 Aug., 1952; discharge of water here exceeded 18 000 CUSECS for a few hours. (ii) A *lake* may be enlarged by exceptional rainfall, so flooding adjacent low-lying land; e.g. Derwentwater and Bassenthwaite (English Lake District) were joined in Aug., 1938, flooding the interlacustrine land; L. Eyre (Australia) expanded to 8000 km^2 of water in 1950-1. (iii) The *sea* may flood low-lying coastlands if a high tide coincides with a STORM-SURGE, esp. if dikes collapse; e.g. 31 Jan.-1 Feb., 1953, when the coastlands of the Netherlands and S.E. England were inundated. (iv) The *collapse of a dam* may cause disastrous fs; e.g. Dolgarrog, Wales (1925); Fréjus in France (1959); near Los Angeles (Dec., 1963). A natural dam across the Gros Ventre R., Wyoming, formed by a land-slide in 1925, collapsed in 1926; the resultant f destroyed the town of Kelly.

flood control The prevention of inundation of land by flooding, including: (i) reduction of concentrated runoff by maintaining a vegetation cover; e.g. Tennessee Valley, USA; (ii) creation of storage basins to hold water temporarily; e.g. in Iraq; (iii) creation of barrages to hold water until it can be

Plate 29. A series of tabular masses of ice FLOES, resulting from summer melting of sea-ice in the Angmagasalik Fjord, western Greenland. (*Eric Kay*)

passed down-stream; e.g. Ohio R.; (iv) dredging and straightening of channels to facilitate outfall, and the creation of more direct outfalls to the sea; e.g. R. Ouse in E. Anglia.

flood-plain The floor of a valley over which a river may spread in time of flood, depositing ALLUVIUM. The **f-p** is usually bounded by low BLUFFS well back from the channel; an area of gentle or even imperceptible gradients, across which wanders the alluvium-laden river, beyond whose channel lie marshes, OXBOWS and stagnant creeks, with man-made embankments set well back. The alluvium may be deposited as only a thin veneer, or in great thickness; in the lower Nile valley no boring has ever reached the rock-floor. The **f-ps** of Nile, Tigris-Euphrates, Indus, Ganges are intensively cultivated. [*f* MEANDER-BELT]

flood-stage The state of river flow beyond BANKFULL stage; i.e. water overflows beyond the banks.

flood tide (i) The rising **t** across mud-flats, sand-flats and beach between low-water and high-water marks on a gently shelving coast; this may be very rapid; e.g. in the Solway Firth, Morecambe Bay and Dee estuary. (ii) The inflowing tidal stream up an estuary or other inlet, flowing strongly for about 3–4 hours until a period of slack occurs at high **t**; ships 'come up with the **t**.' There are anomalies; see YOUNG FLOOD STAND. The **f t** in an estuary into which a river flows with a powerful current may cause a BORE.

flora (i) The plant-life of any region, formation or period. (ii) A work listing plant-life by species, or describing the plant-life of a partic. area. Comprises (i) *macroflora* (trees, shrubs, herbs); (ii) *microflora* (fungi, bacteria).

flow The movement of a fluid. (i) The **f** of water or lava, under the influence of gravity. (ii) The movement of air (wind), broadly from areas of high to low pressure, modified by effects of the CORIOLIS FORCE and friction with the earth's surface. (iii) *Plastic f* of solid rocks under the influence of stress, without any fracture or rupture. (iv) A petrological result of the flow of a magma as it cools, resulting in parallel orientation of crystals, hence textural or mineralogical striping; e.g. banded rhyolite in the Ordovician Borrowdale Volcanic Series. (v) The movement of glacier-ice; see GLACIER FLOW.

flow duration curve A graph plot depicting the percentage of time various discharge rates in rivers are equalled or exceeded e.g. the **f d c** for the Bighorn R. at Thermopolis, Wyoming, shows that for 34% of the time discharge lies between 28–56 m^3 s^{-1} (1000–2000 cu secs), whereas for 6% it lies between 56–84 m^3 s^{-1} (2000–3000 cu secs).

flow till Surface debris released by the top melting of stagnant ice which—owing to lubrication by water—moves rapidly down the ice surface. Till-structures (e.g. layering) derived from the parent ice are thus destroyed. Ct. 'melt-out tills', where the debris collects in a stable position thus preserving inherited till-structures.

fluidization Igneous activity when hot gases pass through fine-grained material, causing it to flow, and resulting in the creation and enlargement of cracks in rocks by physical and chemical means. Probably one way in which PIPES are formed, as circular enlargements of cracks or faults.

flume (i) An artificial stream channel constructed for industrial purposes: to provide water for power, to float logs and for water supply. (ii) In parts of USA a narrow ravine or gorge.

fluorine dating Vertebrate remains (bones and teeth) absorb **f** at a constant rate from percolating ground-water; thus those of the same age in the same horizon should have the same % of **f**, which would confirm their contemporaneity. Since the amount of **f** varies greatly from place to place, this method is useful only to corroborate that all remains in one site are of identical age. No absolute **d**, or correlation of deposits at different sites, are possible.

fluorspar, fluorite Calcium fluoride (CaF$_2$), a glassy translucent mineral, usually in the form of cubic crystals, occurring in clusters or aggregates, commonly as a GANGUE-filling or 'veinstone' in a vein of a metallic mineral, notably lead and zinc. It has a wide range of colour, from white to deep purple; a massive dark blue variety, mined at Castleton in Derbyshire, is 'Blue John'. It occurs commonly in limestone, esp. Derbyshire, Illinois and Kentucky in USA.

flush (i) A sudden rush of water down a stream. (ii) An area of minerally enriched soil; see FLUSHED SOIL.

flushed soil The ss on lower hill-slopes, areas of accumulation which are base-rich because of downhill movement to them of products of leaching from higher up. F ss are classified as: (i) *damp f*, around spring-sources and rivulets, where water with mineral salts in solution enriches a small area, often indicated by a patch of bright green vegetation; (ii) *dry f*, freshly weathered rock particles at the foot of scree, stone-slide, gully or crag, where new rock is constantly exposed to weathering, yielding a continuous supply of bases.

fluvial, fluviatile App. to a river. There is no real difference between them, but the practice is to use fluvial in respect of river flow and erosive activity, and fluviatile for results of river action (*f deposits*), and for river life (*f fauna* and *flora*).

fluvioglacial App. to effects of melt-water streams issuing from a glacier-snout or ice-sheet margin. In USA *glaciofluvial* is preferred, since logically *glac-* precedes *fluv-*. (i) *F erosion*: (*a*) cutting of a channel or trough by melt-water along the edge of an ice-sheet (e.g. URSTROMTAL); (*b*) cutting of an overflow channel (SPILLWAY) across a preglacial watershed. (ii) *F deposition*: laying-down of an OUTWASH plain of gravels, sands and clays, in that order from the edge of the glacier; laying down of fine materials in a PROGLACIAL LAKE, forming lacustrine sediments (e.g. 'L. Agassiz' in N. America); deposition of VARVES.

Flysch Coarse sandstones, calcareous shales, conglomerate, marls and clays, occurring on the borders of the Alpine ranges. Of early Tertiary (possible also late Cretaceous) age, and composed of materials eroded from the rising fold-ranges before the fold mountain-building max. in Miocene times; probably deposited under shallow-water marine conditions. The ct. is with MOLASSE, composed of materials worn away and redeposited *after* the orogenic max.

focus See SEISMIC FOCUS.

Foehn See FÖHN.

fog Obscurity of ground-layers of the atmosphere, the result of condensation of water-droplets, together with particles of smoke and dust held in suspension. Under the International Meteorological Code, defined as a visibility of less than 1 km. In UK a thick **f** has a visibility of less than 180 m. See also ADVECTION F, ARCTIC SMOKE, FRONTAL F, RADIATION F, SMOG, STEAM F. Ct. MIST, HAZE.

fog-bow A type of rainbow of about 40° radius, formed with the sun behind the observer, and with an area of f in front. The droplets are so small that break-up of light by refraction and reflection into colours of the spectrum is prevented, and the bow is thus mainly colourless, giving a white effect, though sometimes the outer margin may have a slight reddish tint, the inner bluish.

fog-drip Precipitation of water droplets from a bank of wet f, esp. along a COLD WATER DESERT coast, where humidity is high; e.g. California, N. Chile and Peru, S.W. Africa. The amount of moisture may nurture vegetation; cf. DEW-MOUND.

Föhn, Foehn (Germ.) A warm, very dry wind, descending esp. the N. slopes of the Alps. It blows when a depression lies N. of the Alps. Moist air is drawn from over the Mediterranean Sea, and rises over the mountains, cooling at the SATURATED ADIABATIC LAPSE-RATE. Air-mass turbulence on the crest and leeward side causes eddying and descent of the air, warming at the DRY ADIABATIC LAPSE-RATE. Temperatures can rise by 10°C or more in a few hours; a rise of 17°C in 3 minutes has been recorded. In spring snow is rapidly melted, so clearing pastures. Sporadic periods of f early in the year can do harm, causing avalanches and premature budding of trees and plants. *F-effect* is used of any similar wind; e.g. SAMUN (Persia), NOR'WESTER (New Zealand), BERG (S. Africa), SANTA ANA (California), CHINOOK (E. Rockies of N. America), ZONDA (Argentine).

Föhrde (Germ.) (*plur*. **Föhrden**) A long straight-sided inlet in a glacial clay coastline, the result of submergence by a rise of sea-level of a narrow valley. These valleys were probably eroded by rivers flowing in tunnels under the ice-sheet of Quaternary times. Confusingly known as *fjords* in E. Denmark. E.g. Flensburger F., Kieler F. [*f, opposite*]

fold, folding The bending or crumpling of strata as a result of compressive forces in the Earth's crust, usually along well-marked zones which indicate lines of weakness. F ranges from a gentle flexure (*inflexion*), through simple upfolds (ANTICLINES) and downfolds (SYNCLINES), to ASYMMETRICAL FS, OVERFOLDS, RECUMBENT FS and NAPPES. See also ANTICLINORIUM, SYNCLINORIUM, PERICLINE, OROGENY. [*f* ANTICLINE]

foliation (Lat., *folia*, leaves) (i) A wavy laminated or banded fabric in such rocks as schist and gneiss, the result of METAMORPHISM re-crystallizing and segregating different minerals into parallel layers. (ii) A wavy structure in bands of ice in deeper parts of a glacier.

food chain A series of organisms with inter-related feeding habits, each serving as food for the next in the c; a series of interwoven f cs is a *food-web*. E.g. plants: herbivores: carnivores: higher carnivores: scavengers: parasites. The c starts with *autotrophs* (organisms which can produce matter directly from sunlight and inorganic materials) and leads to *heterotrophs* (animals which feed directly on organic matter).

foot A measure of length = 12 ins, $\frac{1}{3}$ yd, 0·30479 m; 1 cu. ft = 1728 cu ins, 28316·1 cm^3, 6·23 gallons of water; 1 ft-pound weight = a unit of energy, the work done in lifting 1 lb. mass through 1 ft vertically; 1 sq. ft = 144 sq. ins = 0·093 m^2.

foothills A transitional line of hills, lying between and more or less parallel to a main range of mountains and a plain, the result either of the intermediate zone of uplift or of active denudation; e.g. fs of the Rockies, Himalayas.

foot wall (i) The lower side of a FAULT; i.e. beneath the fault. (ii) Used by miners to denote solid rock beneath a vein, lode or other ore-body. [*f* HANGING WALL]

force (Old Norse, *fors*) A waterfall in English Lake District and Pennines; e.g. Scale F., Buttermere; High F., Teesdale.

ford The shallow part of a river which can easily be crossed; hence a common place-name for a riverside settlement; e.g. Oxford. Note stages of development (f, then bridge) implied by Fordingbridge (Hampshire).

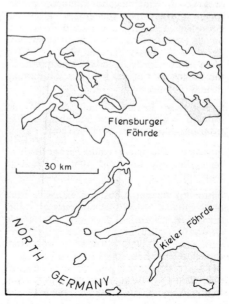

forecast Adopted since 1860 for the weather anticipated for the future, to avoid such terms as 'prophecy'. In 1963 the UK Meteorological Office introd. 'long range fs' for a month ahead, following a long period of 'daily fs'. A f is a probabilistic statement issued with some degree of confidence on the basis of available evidence. In Europe the f tends to be a definite prediction; in USA the probability is stated; e.g. a 50% probability of rain during the next 24 hours.

foredeep A deep elongated trough in the ocean floor, lying near and parallel to an island arc or mountainous land areas, esp. in the Pacific Ocean. See DEEP.

foredune On the sea-shore, a DUNE, or a line of dunes, nearest the sea.

foreland (i) A low promontory projecting into the sea; e.g. N. Foreland, S. Foreland in Kent. See also CUSPATE F. (ii) An ancient continental mass of great stability bordering a GEOSYNCLINE [*f*]. Orogenic forces were directed towards the f and folded ranges were squeezed on to it during mountain-building periods; e.g. Hercynian F. of Europe, during the Alpine orogeny. Earlier authorities thought that the motive force came from one side of a geosyncline only (*backland* or *hinterland*), directed on to the other side (the f). This has been largely replaced by the idea of two fs moving together, as in the Himalayas. But in the Alps it is clear that the force was mainly directed to the N., and the Hercynian continent acted as a f. (iii) *Glacial f:* an area of lowland, lying beyond mountain ranges which during the Quaternary glaciation nourished glaciers; these moved down and out as PIEDMONT GLACIERS on to the f; e.g. Bavarian F., Swiss Plateau, N. Part of the Plain of Lombardy in Italy, Lannemezan on the French side of the Pyrenees, Carpathian F. in S. Poland.

foreset beds Inclined b built outward and forward in a DELTA, each one above and in front of the previous. [*f* BOTTOMSET BEDS]

foreshore The area extending from the lowest spring tide low-water line to the avge high-water line. [*f* COASTLINE]

forest (i) A continuous and extensive tract of woodland, usually of commercial value. (ii) In UK an area of former f, now largely cleared for agriculture and settlement, perhaps with small surviving portions of woodland, though the district retains the f name; e.g. F. of Bere (Hampshire), Ashdown F. (Kent) Sherwood F. (Nottinghamshire). (iii) A royal hunting ground, outside the common law and subject to f law; e.g. New F. (iv) A waste or uncultivated area of heath or moorland, used for hunting and stalking (as in Scotland); e.g. Mamore F. (Invernessshire), F. of Atholl (Perthshire).

forked lightning A type of cloud-to-ground electrical discharge, with distinctive down-pointing and branching flashes.

formation The 4th category of the geological division of the hierarchy of rocks, corresponding to the time interval of an age. It consists of a stratum, or a series of strata, with some distinct and well-defined lithological properties. It can be gen., e.g. coal-seam; or specif., with its own name, e.g. Nothe Grit, Nothe Clay, Bencliff Grit and Osmington Oolite are fs within the Corallian Series within the Jurassic system within the Mesozoic group.

form-line A contour drawn in by eye, usually with the help of spot-heights, but not instrumentally surveyed.

form-ratio A simple relationship between the depth and width of a river, expressed as a fraction (1/100) or ratio (1:100).

Fortin barometer A pattern of mercury b, though the KEW pattern is more gen. used. The main feature of the F b is that the level of the mercury in the cistern is brought to zero on the scale by a screw adjustment before a reading is taken.

fosse (i) Lat. *fossa*, a ditch or trench around an ancient earthwork, forming a line of defence. It survives in place-names; e.g. F. Way. (ii) In French usage, a linear deep or submarine trench; e.g. F. de Cap Breton in Bay of Biscay [*f* SUBMARINE CANYON]. (iii) In USA, a depression between the side of a valley-glacier and the containing-wall of a valley, along the line of which a KAME-TERRACE (*f*) may develop. (iv) In N. England a waterfall, usually spelt 'foss'; e.g. Janet's Foss, Malham, Yorkshire; prob. a corruption of FORCE.

fossil The hard part of an organism or its exact replacement by mineral matter (silica, calcite, limonite, pyrite), or its impression, preserved in sedimentary rocks. Some fs may have pyritized shells; e.g. ammonites in the Jurassic. Fs include fauna, flora, foot-marks of animals, and impressions of soft-bodied animals (*organic trace*); e.g. worms. PALAEONTOLOGY is the study of fs, since 'they stamp the sediments with a characteristic mark' (F. Hodson).

fossil erosion surface A s which has been worn down, buried by subsequent deposition, and later 'exhumed' by removal of the 'cover' of newer deposits by denudation; e.g. sub-Eocene s in S.E. England. Occas. small patches of the 'cover' may survive.

fossil soil See FIRE-CLAY, GANNISTER, SEAT-EARTH.

fossil water See CONNATE W.

Foucault's pendulum A p consisting of a heavy metal ball suspended from a wire, set swinging in a certain direction. Its path appears to move gradually round to the right, and in due course it arrives back at its original direction, the result of the Earth's rotation. A p located at the Pole would show one complete rotation in 24 hours (i.e. 15° in 1 hour), in latitude 50° it takes about 31 hours, at the Equator there is no turning at all. The amount of turning indicated by the p varies with the *sine* of the latitude; i.e. number of degrees of turning per hour = 15 × sin θ, where θ is the latitude. Originally devised by L. Foucault (1851).

fractional crystallization The separation of constituents of MAGMA as it cools into successive minerals. Gen. the order is the inverse of the melting point, one of incr. silicity: first accessory minerals, then ferromagnesian silicates, feldspars and finally quartz. The process is complex, differing with the chemical nature of the magma and whether solidifying at depth or on the surface. Some magmas solidify in 2 distinct phases.

fractus A cloud species used as a prefix to other c-names (as *fracto-nimbus, fracto-stratus, fractocumulus*) to indicate tattered, shredded, ragged cs, an indication of high winds and stormy conditions in the upper atmosphere, which may possibly affect the weather near sea-level. See pl. 9.

fracture A clean-cut break in a rock stratum subjected to strain resulting from either tension or compressions. Cf. FAULT.

frazil Needle-like ice-crystals which develop as a spongy mass in super-cooled water in motion, so that sheet-ice cannot be formed.

fragmental rock A r made from recognizable fragments of others, compacted or cemented; syn. with CLASTIC R; e.g. conglomerate, sandstone, clay. See also PYROCLAST.

free face A rock wall too steep for weathered material to rest, which falls to the bottom, where it forms SCREE. The f f is one element distinguished by W. Penck and A. Wood as part of the hillside slope profile. See also CONSTANT SLOPE, WAXING SLOPE, WANING SLOPE. See pl. 68. [*f*]

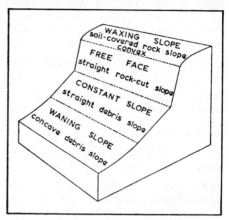

freestone A fine-grained, even-textured rock that can be sawn freely in any direction into blocks, occurring in thick beds; ct. thin beds of FLAGSTONES. Includes several Jurassic limestones (Bath, Ham, Portland Stone); and various sandstones (e.g. New Red in the Penrith valley and near Liverpool, where they are used for the Anglican cathedral; Fell Sandstone in the Lower Carboniferous of Northumberland). Some types, esp. limestone, can be cut easily, but harden on exposure; e.g. limestones of Malta.

freeze-thaw A very active form of weathering under PERIGLACIAL conditions, in mountains, and in middle latitudes in winter, wherever temperatures fluctuate above and below 0°C. The action of frost breaks up rock, while melt-water removes fragments. Results include the production of hollows by NIVATION, gen. break-up of the surface, sorting of coarse and fine material, formation of PATTERNED GROUND, and movement of material by SOLIFLUCTION.

freezing front The downward or lateral limit of freezing, as it affects a soil or debris layer in a periglacial environment. Migration of the f f is important geomorphologically, as laboratory experiments show that finer material 'moves ahead' of the migrating f f, while coarser material is 'entrapped'. Thus debris sorting occurs both vertically and laterally assisting the formation of PATTERNED GROUND.

freezing-point The temperature at which a liquid or gas changes to the solid state. Water freezes at 0°C (32°F). Ct. mercury at −39°C, nitrogen at −210°C.

126 FREQUENCY

frequency curve A graph showing the number of occurrences of values (e.g. temperature, summit-levels); the horizontal axis indicates the range in size of the variable, the vertical axis either the actual or percentage frequency for each value of the variable. Values may be grouped into classes, forming a HISTOGRAM.

fresh breeze Force 5 in the BEAUFORT WIND SCALE.

freshet A sudden surge of flood-water down a small stream, the result of a heavy rainstorm in the upper basin, or snow-melt following a rapid rise of temperature.

fret A sea-mist that rolls up suddenly over a shore, but does not extend far inland.

friable (i) An easily crumbled rock; e.g. soft, poorly cemented sandstone. (ii) Loamy soils with a well-marked crumb-structure.

friagem A cold strong wind experienced on the CAMPOS of Brazil and in E. Bolivia, the result of an anticyclone, with air movement from the S. It may cause a distinct cold spell during the dry season in winter.

friction cracks Used by S. E. Harris for various rock fractures caused by ice moving over bedrock. Common forms are 'crescentic gouges', 'crescentic fractures' and 'lunate fractures' (small depressions resembling inverted BARKHANS, with the 'horns' pointing in the direction of glacier flow).

frigid climate One of 3 climatic types or zones postulated by classical and later scholars, the others being *torrid* and *temperate*. Currently applied in a gen. sense to an area with a snow cover for much of the year and with a permanently frozen subsoil (PERMAFROST), or to the polar-arctic group of climates (W. Köppen's *ET* and *EF* types).

fringing reef An uneven platform of CORAL, fringing and attached to the coast, with a shallow narrow lagoon between it and the mainland, or with no lagoon at all, and with its seaward edge sloping steeply into deep water. See pl. 30. [*f*]

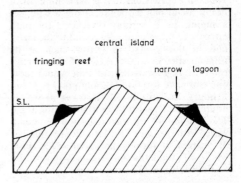

front The boundary surface, interface or transition zone which separates two AIRMASSES of markedly different temperature and humidity. This may occur: (i) on a large scale, the fs between major air-masses; e.g. *Pacific Polar F, Atlantic Polar F, Mediterranean F, Arctic F*; (ii) on a smaller scale in a local DEPRESSION: WARM F, COLD F, OCCLUDED F.
[*f* DEPRESSION]

frontal fog Short-lived f, really thick, fine drizzle, sometimes assoc. with the passage of the WARM FRONT of a DEPRESSION. The warm

Plate 30. A small coral FRINGING REEF, exposed at low tide and separated from the coast by a shallow lagoon, south of Mombasa, Kenya. (*R. J. Small*)

rain falls into cold underlying air near the ground, saturating the air to form fog.

frontal rain See CYCLONE R.

frontogenesis The physical processes in the atmosphere by which a FRONT is developed or intensified, part of the formation of an individual DEPRESSION. Its appreciation demands the study of conditions in the upper atmosphere, partic. the development of CELLS, and of JET STREAMS which may separate them.

frontolysis The dissipation of a frontal zone or FRONT.

frost (i) The condition of the atmosphere when it is at or below the freezing-point of water. Usually descriptive adjectives are applied to the degree of **f**: *light, heavy, hard, sharp, killing, black*. (ii) An agent of weathering which freezes water contained in cracks and fissures in rocks and soil. As the volume of water when frozen increases by 10%, it exerts great pressure and rock tends to shatter. Esp. potent in high mountains. (iii) Minute ice-crystals formed by freezing of dew, fog or water vapour, sometimes called *white f*. See GLAZED F, HOAR-F, RIME. Many self-evident words are used with **f** as a prefix: *f shattering, f cracking, f splitting, f stirring, f weathering, f wedging, air-f*, GROUND-F. See pl. 23, 42.

frost boundary A **b** indicating the occurrence and incidence of **f**. It may indicate the limit of areas: (i) which have never experienced **f**; (ii) with mean min. temperatures above 0°C; (iii) with a growing season with a spec. number of **f**-free days; e.g. cotton needs 200; (iv) with a lowest mean monthly temperature above 0°C; (v) with no month with a mean temperature above 0°C; i.e. the boundary of a polar climate of perpetual **f** (W. Köppen's *EF* type of climate).

frost heaving The lifting of the soil surface by **f** action within it; gen. the result of formation of 'ice-lenses' which expand within the soil layer, forcing it upwards into a *f* mound. Syn. with *congeliturbation*.

frost 'pocket', hollow A low-lying area into which cold dense air drains by gravity, as radiation of heat into space on a clear night causes rapid cooling of air on upper hill-slopes. The '**p**' may thus contain air below freezing-point, while temperature on higher slopes is still above; see INVERSION OF TEMPERATURE. Fruit-growers seek to avoid such sites, since **f** may cause serious losses should it occur during blossom time, and the chance of this is much increased in a **f** '**p**'.

frost thrusting The *lateral* pressures exerted by the freezing of ground water in periglacial environments (ct. FROST HEAVING, causing upward pressure).

full A ridge of sand or shingle, roughly parallel or at a slight angle to the shore. Formed by constructive waves just in front of the line of their break-point, aligned at right-angles to their direction of approach. Fs are separated by long shallow depressions known as SWALES, *lows, slashes, furrows* or *runnels*. During a storm, with strong onshore winds and destructive waves, the fs are combed down and destroyed. Found on the Lancashire coast esp. at Blackpool and Formby (sand); on the Dorset coast and at Dungeness (shingle).

fumarole (It.) A small hole or vent from which issue steam, hydrochloric acid, sulphur dioxide and ammonium chloride, usually in the form of powerful jets. Ct. SOLFATARA. E.g. Valley of 10 000 Smokes, Alaska; Bumpass Hell on the flanks of Mount Lassen in N.E. California.

fundamental complex An obsolescent term for the great 'shield' areas of Precambrian rocks on the Earth's crust, largely replaced by BASAL C.

funnel cloud A whirling cone of dark grey **c** which projects downwards from the low-lying base of a CUMULONIMBUS **c**, gradually elongating until it touches the surface of the sea; it may become part of a WATERSPOUT. It is often associated with a TORNADO. Its technical name is *tuba*.

fusain In banded coal, layers of 'dirty' material mainly of flakes of charcoal-like substance, extremely friable and dusty, with a high ash content. Obsolescent.

gabbro A dark-coloured plutonic rock, basic in composition, low in silica, with basic plagioclase and sometimes olivine, forming a sharp-textured, coarse-grained crystalline mass. The Black Cuillins of Skye and mtns of Rhum are of **g**; several large masses in N.E. Scotland (Insch Mass, 180 km² in extent, Huntly Mass), and N. Guernsey, Channel Is. G commonly occurs as LOPOLITHS and RING-DYKES.

gale (i) In gen, parlance, a strong wind. Specif. on the BEAUFORT SCALE: (*a*) *moderate g* (Force 7); (*b*) *fresh g* (Force 8); (*c*) *strong g* (Force 9); (*d*) *whole g* (Force 10). Winds of Force 11 and 12 are pop. referred to as **g**s, but on the 'Beaufort Scale' are 'storm' and 'hurricane'.

galena The chief ore of LEAD, lead sulphide (PbS), soft and heavy, with a dull lustre, commonly occurring in cubic crystals, and found usually in Carboniferous Limestone.

gallery forest A dense tangle of trees fringing the banks of a river in open SAVANNA country. The vegetation meets overhead, giving a tunnel-like appearance, hence Sp. *galeria*, used in S. America, from which 'gallery' is corrupted.

gallon Measure of volume. 1 g of distilled water at 15°C weighs 10 lb., occupying 277·3 cu. ins. or 0·16 cu. ft; 1 g = 4545·96 cm^3, 160 liquid oz, 8 pints; 1 g (British) = 1·2009 gs (USA) = 4·55 litres.

Gall's Stereographic Projection A CYLINDRICAL P in which the cylinder intersects the globe at 45°N. and S.; along these parallels scale is correct, between them too small, and poleward of them too large. Parallels are parallel straight lines, their separation obtained: (i) graphically by projecting stereographically (i.e. from the opp. side of the equatorial diameter) on to the edge of the cylinder, intersecting at 45°N. and S.; (ii) trigonometrically from the formula 1·7071 R. tan $\frac{1}{2}\theta$, where θ is the latitude, which gives the distance from the Equator. Meridians are vertical lines, drawn through the Equator, which is divided truly; e.g. 10° meridian interval

$$= \frac{2\pi R . \cos 45°}{36}$$

The p is neither equal area nor orthomorphic, but exaggerates shape and area in high latitudes less than the MERCATOR P. [*f*]

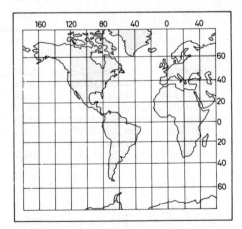

gangue Usually worthless mineral matter, known as *veinstone*, in a vein, commonly quartz, calcite, fluorspar, barytes, etc., within which lie metallic ores. Some g minerals may be useful; e.g. fluorite, barytes.

ganister, gannister A rock with a high silica content, found as a sandy SEAT-EARTH in the Lower Coal Measures, and used for heat-resistant refractory bricks for furnace-linings.

gap A break in a ridge (see WATER-GAP, WIND-GAP). In the Appalachians in USA, crossed by rivers flowing to the Atlantic, such gs are frequent; e.g. Cumberland G. in Kentucky, a defile 500 m deep, followed by the Wilderness Road, a trail blazed by Daniel Boone in 1775.

gara (*pl.* **gour**) (Arabic) A mushroom-shaped rock in desert lands, resulting from undercutting by wind abrasion caused by sand particles carried near the ground. Partic. well marked if in a bedded mass of rock a soft stratum near ground-level underlies a more resistant one. Ct. HOODOO.

garigue, garrigue (Fr.) A stunted evergreen xerophytic scrub, found on limestone in drier areas of Mediterranean climate, incl. stunted evergreen oak, thorny aromatic shrubs, prostrate prickly plants and tuberous perennials, separated by bare rock. This Provençal term has also been applied to uncultivated land with calcareous soil on which such scrub grows, as in S. France, Corsica, Sardinia and Malta.

gas, natural Gaseous hydrocarbons in the form of ethane and methane, found in the Earth's crust, frequently associated with petroleum.

gash breccia A type of Breccia found specif. in the Carboniferous Limestone of S.W. Wales. It consists of angular masses of limestone and dolomite, embedded in an interstitial matrix of calcite, crushed stalagmite and red clay. Some b is in small pockets, other forms are in large masses, forming extensive cliff faces. Its age is attributed variously to Triassic and mid-Tertiary times. It may be the result of solution by underground water, infilling of caverns by collapse of walls and roof, or by tectonic action along shatter belts and fault intersections.

gat (i) Strait or channel between offshore islands or shoals, notably between the Frisian Is (*zeegaten*), through which flow powerful tidal streams; e.g. Texel Zeegat. (ii) An opening in cliffs leading inland from the sea. In Kentish place-names, this usually becomes 'gate'; e.g. Margate, Ramsgate.

gate (i) A valley through a low range of hills (Anglo-Saxon *geat*, an opening), hence

Reigate, Rogate. Sometimes modified to *yate*; e.g. Markyate, where A5 enters the Chilterns. (ii) An opening, wider than a gap, between hilly areas; e.g. Midland G. between the S. end of the Pennines and the Wrekin; G. of Carcassonne (*f* below). (iii) A restricted section in a river valley; e.g. Iron G. on R. Danube. (iv) An entrance to a bay or harbour between promontories; e.g. Golden G., San Francisco. [*f*]

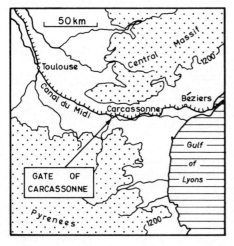

Gauss Conformal Projection The TRANSVERSE MERCATOR P.

Gaussian Curve A mathematically derived frequency distribution which possesses perfect symmetry about the central value.

geanticline, geoanticline An upfold on an Earth scale. GEOSYNCLINES do not gen. have gs as complementary features, and the term is often used for upfolds developing within a geosyncline, as the latter is laterally compressed.

geest (Germ.) An area of coarse sand and gravel, mostly under sparse HEATH vegetation, occurring mainly in N. Germany (where the word occurs frequently in regional names) and in adjacent parts of the Netherlands, Denmark and Poland. The sands and gravels are of fluvioglacial origin, laid down as the 'Older Drift' cover by streams issuing from continental ice-sheets, and later cut into individual areas, partly by fluvioglacial channels, partly by modern river valleys; e.g. Lüneburg Heath between Weser and Elbe valleys, G. of Oldenburg and Hanover between Ems and Weser valleys.

gendarme (Fr.) A sharp rock pinnacle projecting from a ridge, orig. in French Alps, now used widely by mountaineers; e.g. Grand G. on the Grépon (Mont Blanc massif).

genera, cloud The 10 major c types, according to *International Cloud Atlas*. See CLOUD and individual types.

general systems theory A concept developed as a framework for modern science, orig. with partic. ref. to biological sciences. It represents an effort to generalize ideas, processes and functional relationships in any partic. s of interrelated objects or ideas, to work with analogies from one discipline to another, to produce a scientific doctrine of 'wholeness'. Inevitably the approach stresses relationship between form and process, as well as the multivariate character of phenomena, and is incr. utilized by geographers, who analyze ss on 4 levels of abstraction: (i) *morphologic ss* (relations between individual components built up by statistical association; e.g. drainage basin morphometry); (ii) *cascading ss* (relations between components involving transfer of mass or energy, both inputs and outputs; e.g. hydrological cycle); (iii) *process-response ss* (combination of (i) and (ii); e.g. atmospheric circulation); (iv) *control ss* (modification of (iii) by human intervention, governing inputs and outputs or levels of components, so affecting the operation of the s; e.g. a multi-purpose water authority). A certain critical reaction to g s t can be discerned; e.g. 'g s t seems to be an irrelevant distraction' (M. Chisholm). See CLOSED S, OPEN SYSTEM.

generic Phenomena closely linked and similar in type. Thus 'Mediterranean climate' is a g concept summarizing certain climatic features, which may be used to describe climates elsewhere which are broadly similar and belong to the same type. Hence *g classification* based on morphological elements. Cf. GENETIC.

generic region A r distinguished by criteria of a given type, found as a repeating pattern; e.g. chalk cuesta, horst, area of savanna, area of Mediterranean climate. Ct. SPECIFIC R, which is unique, precisely located and usually has a proper name.

genetic description, of landforms. An analysis of landforms on the basis of their origin, in terms of processes which are sculpturing them; the systematic study of landforms.

geo, occasionally **gio** (Norse *gya*, a creek) A long narrow steep-sided inlet, running inland from the edge of a cliff. Worn by marine erosion at the base of the cliff along a line of weakness, such as a joint-plane (as in Old Red

Plate 31. A GEO, narrow inlet (the Devil's Frying Pan, Lizard Peninsula, Cornwall, England) resulting from selective marine erosion. At an early stage the sea formed a deep cave, the roof of which subsequently collapsed (except at one point where a natural bridge remains). (*Aerofilms*)

Sandstone). The cave is driven inland, and its roof is cut away along the same plane by waves surging into it, together with compressional effects on the air within. Ultimately the roof collapses. E.g. in Orkneys, Caithness, S. Skye and Soay, all in Old Red Sandstone; Huntsman's Leap, S.W. Wales, in Carboniferous Limestone. See pl. 31.

geochronology The dating of past events in the earth's history, as indicated in the record of the rocks; the 'science of earth time'. It includes *relative* chronology, in order of occurrence or formation, and *absolute* chronology, involving dating in years, using the measurement of RADIOACTIVE DECAY (uranium/lead, rubidium/strontium, potassium/argon), and RADIOCARBON DATING. Other methods incl. counting of LAMINAE and VARVES, DENDROCHRONOLOGY, TEPHRONOLOGY, CORE SAMPLING, PALAEOMAGNETISM, FLUORINE-PHOSPHATE RATIO. Partic. attention is paid to precise dating of events of the Quaternary by geologists, archaeologists and botanists. The measurement of geological time is sometimes called *geochronometry*.

geode A hollow, near-spherical nodule between 2·5 cm and 25 cm or more in diameter, commonly found in limestone. Its interior is lined with crystals; occasionally a loose piece of material inside may make an intriguing rattle.

geodesic line See ORTHODROME.

geodesy The branch of mathematics dealing with the shape and size of the Earth. From this can be obtained data to enable the exact fixing of control points for TRIANGULATION and LEVELLING of a high degree of accuracy, the basis of a major topographical survey; hence *geodetic*, the application of **g** to surveying. Standard figures for the Earth have been produced by mathematicians and surveyors, notably Everest (1830), Bessel (1841), Airy (1849), Clarke (1858, 1866, 1880), Hayford (1909–10). (See EARTH). Geodetic work involves a combination of precise survey and exact determinations of gravitational force.

geodimeter A surveying instrument which estimates distances between 2 points by measuring the time interval between a light signal sent from one station and its return via a reflector from the other. Cf. TELLUROMETER.

geodynamics The study of deformation forces or processes within the earth.

geographical mile Syn. with NAUTICAL MILE.

geographics A perspective model of the terrain printed by a computer from grid-based source-data, using a viewpoint selected by the observer. A programme is devised to produce either a perspective drawing akin to a block-diagram (using a digital pen-plotter), or a half-tone image in which shadow areas according to a selected light source are delineated by screen ruling, so that each depicted point has a grey tone-value according to the intensity of its illumination. The clarity and effectiveness of the latter may be incr. by the superimposition of an automatic contour-plot.

geoid The terrestrial spheroid, an 'earth-shaped body', regarded either: (i) as a mean sea-level surface continued through the continents; or (ii) as an undulating surface determined in gravitational terms (*geopotential*), rather higher than the surface of the spheroid under the continents, rather lower under the oceans.

geological time The chronology of Earth history, organized in a hierarchy of time intervals (in descending scale): *era, period, epoch, age, moment*, corresponding in terms of rocks to *group, system, series, stage, zone*. See GEOCHRONOLOGY.

geology The history, composition, structure and processes of the Earth. Its accepted sub-divisions are: (i) *Stratigraphy*; (ii) *Palaeontology*; (iii) *Mineralogy*; (iv) *Petrology*; (v) *Physical G*; (vi) *Structural G*; (vii) *Geophysics*; (viii) *Engineering G*; (ix) *Economic G*. An alt. sub-division is (*a*) *physical g*; (*b*) *historical g*. Physical g is itself sometimes divided into *structural g* and *dynamic g*. *Geomorphology* is regarded as a branch of g in USA, more usually as a branch of geography in UK.

geomagnetism The Earth's magnetic field and the study thereof.

geometronics Used in USA for aspects of cartographical work involving *geo* (the Earth) and *metron* (measurement), utilizing electronic and other modern techniques.

geomorphology The scientific interpretation of the origin and development of landforms of the Earth; the modern development of *physiography*.

geomorphological map A map depicting the landforms of an era, constructed to show the *origin* or *history* of these landforms. G ms are necessarily selective, depicting such features as EROSION SURFACES (incl. age and mode of formation), significant deposits etc. Ct. MORPHOLOGICAL MAPS.

geophysics The scientific study of the physics of the Earth's crust and its interior, involving consideration of earthquake waves, both natural and those made by deliberate explosions (SEISMOLOGY), magnetism, gravitational fields and electrical conductivity, using precise quantitative methods. Applied g is mainly concerned with techniques involved in the discovery and location of structures assoc. with economic minerals (petroleum, iron ore, radio-active ores) by means of geophysical surveys. A recent tendency, esp. in USA, is to widen the term to include the physics of the Earth's environment also, incl. meteorology, astrophysics, etc.; this does not meet universal acceptance.

Georef system A world reference s for location of points on the Earth's surface, used by US Air Force for the direction of long-range aircraft and missiles, and other strategic purposes, introduced in 1951. The world-map is covered with a rectangular graticule, with 15° meridian intervals, each band lettered A to Z (omitting I and O), starting at 180° and working E., and with 15° parallel intervals, lettered A to M (omitting I), starting from the S. Pole. The origin of the map ref. is the bottom left corner, i.e. intersection of 180° and the line representing the S. Pole. Thus 288 segments are located by two letters each; UK is in sq. KM. Each segment is sub-divided into degrees (225 quadrangles), then minutes, then 10ths and 100ths of minutes (not seconds).

geostrophic flow The concept of a wind blowing parallel to the isobars as a result of balanced forces exerted by the pressure gradient in one direction and by CORIOLIS deflection in the opposite direction. Winds approaching pure g f are found in the upper atmosphere, but nearer to the surface of the earth friction causes them to blow at an oblique angle to the isobars towards low pressure. See FERREL'S LAW. [*f*]

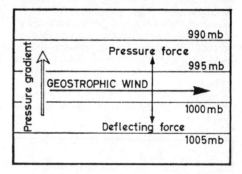

geosyncline A linear depression or a downfold in the crust on an Earth scale. A g probably developed as a slow continuous down-warping or subsidence; the floor gradually sank over a long period of geological time at much the same rate as thicknesses of sediment, worn away from land-masses on either side, were deposited within it. Some maintain that this deposition was an actual cause of the sinking of its floor through its own weight, others that it was a consequence of the existence of a convenient repository.

[*f* MEDIAN MASS]

geothermal gradient The incr. in temperature in the Earth's crust with incr. depth. Various figures are quoted; evidence indicates that the g g is not constant, but steepens with incr. depth. Estimates vary from 1°C in 28 m (1°F in 51·1 ft) to 1°C in 40 m (1°F in 72·9 ft); an avge for the SIAL layer could be 1°C per

28·6 m (1°F in 52·1 ft). The temperature of 986°C (1806·8°F) has been given for the base of the SIALIC rocks.

Gestalt concept (Germ.) Used in synoptic climatology to denote a complex of climatic elements occurring in a well-known and recognizable pattern; e.g. a COLD FRONT and its weather pattern.

geyser (From Icelandic *geysir*, lit. a 'gusher' or 'roarer') An intermittent fountain of hot water ejected from a hole in the Earth's crust with considerable force, accompanied by steam. Some gs erupt at regular intervals, others more irregularly and spasmodically. Superheating in the long, bending pipe of a g causes build-up of water pressure at a temperature above 100°C, until suddenly part is converted into superheated steam, and the water in the upper part of the pipe is violently emitted. Cooler water flows into the pipe, and the heat incr. begins again. A cone of *sinter*, deposited from the hot water, accumulates around the vent. E.g. Iceland, Yellowstone National Park (Wyoming, USA), N. Island of New Zealand. 'Old Faithful', in Yellowstone, has an average eruption interval of about 65 mins., ranging between as little as 33 mins. and as much as 95 mins.; it throws 45–90 000 litres of near-boiling water, accompanied by steam, 37–55 m into the air, lasting from 2 to 5 mins. Between 1870 and the end of 1973, Old Faithful erupted approx. 830 000 times. [*f*]

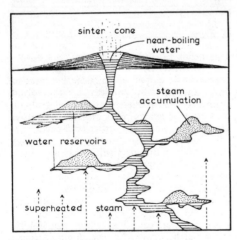

geyserite A siliceous deposit (*sinter*) formed around thermal springs and geysers; e.g. a low cone has been built up around the vent of Old Faithful, Yellowstone National Park. Ct. TRAVERTINE, which is calcareous.

ghat (Hindi) A mountain pass in India; e.g. Thalgat and Bhorghat behind Bombay. Probably the term was transferred by Europeans to the mountains through which the passes led, hence W. Ghats and, by analogy, E. Ghats.

gibber, g plain Used in Australia for an area of arid gravel-desert, sometimes produced by the dissection and break-up of a duricrust sheet (e.g. of SILCRETE).

gibli, ghibli (Arabic) An extreme form of SIROCCO in Tunisia; sometimes so intense that it feels like the hot blast from an opened oven-door.

gill, ghyll (Norse) (latter obsolescent) (i) A swift-flowing mountain torrent in English Lake District (e.g. Piers G., Lingmell G., both in Wasdale), and in Yorkshire (e.g. Tennant G., Craven). (ii) A fanciful name for a wooded ravine.

Gipfelflur (Germ.) The surface of uniformity of summit levels, independent of structure and rock type, found in the Alps ('*die G der Alpen*') and in the Rockies, but not thought to imply the remnant of a PENEPLAIN.

Gipping Till An extensive glacial deposit in the W. parts of E. Anglia. G t overlies the LOWESTOFT TILL, contains much chalky debris and igneous ERRATICS from N. England, and was probably deposited during the WOLSTONIAN glacial.

glacial App. to a glacier, ice-sheet or ice-age, its effects and results; hence g epoch, advance, cycle, drift, erosion, deposition, lake. See under specif. terms.

glacial breaching See DIFFLUENCE.

glacial control theory A t, postulated by R. A. Daly, explaining the occurrence of CORAL at depths below its normal habitat by a fall in sea-level occasioned by the withdrawal of water contained in the ice-caps of Quaternary times. This meant: (i) water in tropical latitudes must have been so much cooler that all existing living coral was destroyed; and (ii) preglacial reefs and other islands were planed down by marine erosion to the sea-level of the time. These platforms provided bases for the upward growth of coral, as the temperature of the sea again incr. and the declining ice-sheets returned their melt-water to the oceans, thus causing a gradual rise of sea-level with which growing coral could keep pace. For recent research concepts, see ATOLL.

glacial divergence The interruption of a drainage pattern by the advance of a glacier or ice-sheet; e.g. R. Derwent (Yorkshire) once flowed E. to reach N. Sea, but when N. Sea ice-sheet created a barrier, the river was forced to turn S. into the Vale of Pickering, a course which it has maintained in post-g times.
[*f* OVERFLOW CHANNEL]

glacial drainage channel A stream-cut c assoc. with glaciation and melt-water. Formerly suggested that such cs were mainly OVERFLOW CS from glacially dammed lakes, but recent research has indicated that streams flowing under glaciers may give rise to cs on the bed which are revealed on ice retreat.

glacial lake A small sheet of water ponded in an angle between a valley-wall and the edge of a glacier; e.g. Märjelen See on the margin of the Aletsch Glacier; many ls around the edge of Vatnajökull ice-sheet in Iceland (e.g. Vatnsdalur); Tulsequah L., Alaska. [*f*]

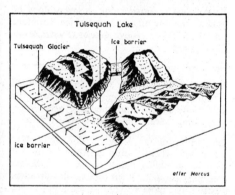

glacial outburst A catastrophic flood resulting from the sudden release of meltwater contained within, or dammed up by, glaciers. G os are frequent during periods of glacial recession (e.g. in the Alps since 1850) when meltwater is abundant, e.g. the g o in the Ferpècle valley, Switzerland, in August 1943 when a large englacial lake in the Ferpècle glacier was drained in a few hours.

glacial stairway The irregular long-profile of glaciated valleys, formed by successive rock basins and rock steps. Many g ss are structurally determined. The basins result from abrasion and plucking of weaker rock; the steps are developed in harder, more massive outcrops. However, other explanations include Garwood's 'Protectionist Theory', which infers that the steps represent the former margins of protective ice, beyond which the valley was lowered by fluvial erosion.
[*f, opposite*]

glaciation (i) The covering or occupation in the past of an area by an ice-sheet or glacier. (ii) The period of time during which this occurred; hence Quaternary g. (iii) Associated processes and results of all glacial activity.

glacier (Fr.) A mass of ice of limited width (ct. ICE-SHEET), moving outward from an area of accumulation. Sometimes called mountain-g, VALLEY-G, or alpine-g. It moves continuously from higher to lower ground, and is enclosed within distinct valley walls. Gs vary in length from tongues which barely protrude from a FIRN-basin, to the Petermann G. in Greenland (200 km), and many in the Antarctic over 160 km long. The longest in Europe is the Aletsch (16 km) in Switzerland. The world's longest is the Lambert G. in the Australian Antarctic Territory (400 km), discovered in 1957. A g may be *active*, i.e. rapidly moving and transporting large amounts of material, or *passive*, with low rates of accumulation and ABLATION, and a slight gradient, hence slow-moving and virtually stationary, transporting little or no material. See CIRQUE-G, EXPANDED-FOOT G, GLACIERET, HANGING G, PIEDMONT G, TIDAL G, WALL-SIDED G. There are many self-evident compound words and phrases. See also 'WARM GLACIER' and 'COLD GLACIER'.

glacier band A banded structure of some kind which may be seen in a g, either in plan on its surface, or in section in a CREVASSE; it may be a dirt b, or a b of ice of contrasting colour and texture. See OGIVE.

glacieret A small glacier, esp. in USA; e.g. in the Sierra Nevada of California and the Olympic Mtns of N.W. Washington State.

glacier flow A not very satisfactory term to cover the complex physical forces involved in the outward movement of ice from its FIRN-field of origin. This may be either: (i) EXTRUSION F; or (ii) gravity f, involving:

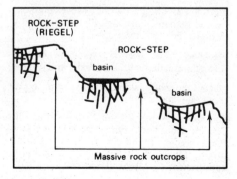

(a) REGELATION; (b) INTERGRANULAR TRANSLATION; (c) PLASTIC DEFORMATION; and (d) LAMINAR F. The avge rate of movement of an Alpine g is about 0·3 m per day; in Greenland 30 m per day has been measured, and 18 m per day is common. The Black Rapids G. in Alaska attained 76 m per day for a short time in 1937. The record measured movement for 1 year was 1710 m for the Storström G. in Greenland. The sides of a g move less rapidly then the centre, sometimes at only half the rate, as the result of friction.

glacier karst The irregular relief sometimes formed by DIFFERENTIAL ABLATION of stagnant ice masses e.g. on the lower Malaspina glacier. An uneven cover of till results in varied ablation rates, and the formation of numerous depressions and 'sink holes'; these may collapse to expose subglacial streams or become flooded by water to give lakes.

glacierization The gradual spread of glaciers or ice-sheets over an area during the present time; ct. DEGLACIERIZATION.

glacier milk A white turbid stream of meltwater laden with pulverized rock ('rock-flour'), issuing from a glacier SNOUT.

glacier mill See MOULIN.

'glacier mouse' A small rounded stone, almost covered with moss, found on Icelandic glaciers. The stones either lie on superficial morainic material, or have rolled off on to adjacent ice.

glacier table A block of stone on a g, protecting an underlying pedestal of ice from melting by the sun's rays. When the pedestal exceeds approx 1 m in height, it is exposed to solar radiation, melts and the g t collapses. See pl. 32.

glacio-eustatism A change of sea-level resulting from the advance or shrinkage of ice-sheets, hence withdrawal from or return to of water in the oceans. See EUSTASY.

glaciofluvial, alt. **glacifluvial** More commonly, and perhaps more logically, used in USA in preference to FLUVIOGLACIAL.

glaciology The scientific study of ice, its form, nature, distribution, activity and results.

glacis A gently sloping bank, esp. on a mountain-side.

glauconite A greenish mineral, hydrous silicate of iron and potassium (Gk. *glaukos*, bluish-green), found in rocks of marine origin, to which it may give a greenish appearance; e.g. Upper Greensand and Chalk

Plate 32. A GLACIER TABLE, a large block of gneiss, standing on a pedestal of ice 1 m high, on the Tsidjiore Nouve glacier, Switzerland. (*R. J. Small*)

Marl (where g grains seem to be casts of foraminifera). At present g is forming in submarine banks, and therefore sandstones containing it are probably of marine origin.

glazed frost, glaze A layer or coating of clear ice formed when: (i) rain falls on to a surface (e.g. road) which has a temperature below freezing ('*black ice*'); (ii) supercooled droplets freeze on impact with telegraph wires, branches, leading edges of aircraft wings; (iii) renewed f occurs after a partial thaw.

glei (gley) soil A s HORIZON where intermittent waterlogging, the result of poor or impeded drainage, reduces oxidation or causes deoxidation of ferric compounds, so that resulting ferrous compounds have a bluish-grey mottled appearance, with sticky, clayey, compact and usually structureless texture. Hence *gleiization*, used esp. in USA for the development of g characters. Occurs in bog-ss and meadow-ss (*humic-g*).

glen A long steep-sided flat-bottomed valley, esp. in Scotland, narrower than a STRATH; e.g. G. More, G. Brittle, G. Roy. Commonly used in compound place-names; e.g. Glencoe, Gleneagles, Glenfinnan.

glint-line A marked edge at the margins of much denuded rocks; used specif. for the boundary between an ancient SHIELD and younger rocks; e.g. in Scandinavia and along the edge of the Canadian Shield. From Norwegian *glint*, a boundary. [*f*]

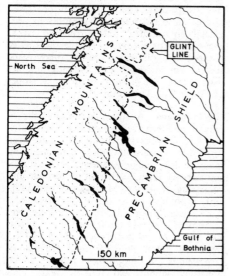

glint-line lake A l formed along a g-l. In Scandinavia a series of long, narrow ls is strung out along the valley of each stream flowing more or less parallel to the Baltic Sea. The steep edge of the Baltic ice-sheet at one stage of the Quaternary Glaciation lay to the E., forming a dam, ponding water between this and the main watershed to the W. Some melt-water escaped W., eroding deep overflow channels. When the ice withdrew, the rivers resumed their preglacial courses to the E. and the overdeepened valleys remained l-filled. Morainic blocking at their E. ends contributed. E.g. Store Lule L., 185 km long.

[*f* GLINT-LINE]

globe (i) A commonly used name for the Earth, as 'sailing round the **g**'. (ii) A spherical model on which is shown the pattern of continents and oceans. The earliest **g** to which ref. is made was by Crates (*c*. 160 B.C.), which has not survived. The earliest existing **g** was by Martin Behaim (1492), made in Nürnberg, which does not show the New World. A modern **g** is made with a shell of several alt. media (metal, card, plastic, glass, rubber).

Globigerina ooze A calcareous deposit on the deep-sea floor, though not in the great ocean depths, consisting of remains of minute one-celled foraminifera, *Globigerina* the most common. Very widely spread in the N. and S. Atlantic, Indian and the S. (but not N.) Pacific Oceans. Its rate of deposition is very slow, taking about 3250 years for a thickness of 25 mm.

Globular Projection A **p**, neither equal-area nor conformal, but with small distortion, commonly used in pairs in atlases, showing each hemisphere. Draw a circle of radius $r = \sqrt{2} \cdot R$, for $\pi r^2 = 2\pi R^2$ (half the area of a sphere), where R = radius of the Earth to scale; put in the Equator and central meridian for the hemisphere; divide these and the circumference equally; and draw parallels and meridians as arcs of circles.

glory See BROCKENSPECTRE.

gloup, gloap See BLOW-HOLE.

gneiss A coarse-grained crystalline rock, gen. of foliated texture, and of streaked, wavy or banded appearance. Formed by DYNAMIC METAMORPHISM of granite and other igneous rocks ('*orthogneiss*'), or of sedimentary rocks ('*paragneiss*') that have been penetrated by *magmatic intrusion* ('*injection g*'). Covers a wide range of gneissic rocks derived from granite, syenite, diorite and gabbro. Note '*augen-g*', which contains porphyritic crystals surrounded by mica or hornblende in an elliptical pattern, giving the appearance of an eye. Used sometimes in a stratigraphical sense, as Lewisian G. of Precambrian age.

Gnomonic Projection An AZIMUTHAL or ZENITHAL P, constructed by projecting from the centre of the globe on a tangent plane. The Polar case is easiest to construct [*f* below]. Lines are drawn from the centre of a circle of radius R (= earth, to scale) to a tangent plane touching at the Pole. The distances of these intersections from the Poles are the radii of the respective parallels, which are concentric circles, their distances apart incr. outwards. Alternatively, the radius of each parallel = $R \cdot \cot \theta$, where θ is the latitude. Meridians are radiating straight lines from the centre. Scale greatly increases from the centre and it cannot be practicably used more than 45° from the Pole. In the equatorial case, all parallels other than the Equator are hyperbolas. In the oblique case, parallels in higher latitudes than the parallel of tangency are ellipses, the parallel of tangency is a hyperbola, and the parallels between the parallel of tangency and the Equator are also hyperbolas. In all 3 cases, GREAT CIRCLES are straight lines, and all straight lines are great circles. But a RHUMB-LINE is curved (except for directions from the Pole in the Polar case). A navigator lays out his route as a straight line on the **G P**, then transfers it to a MERCATOR P. The US Hydrographic Office publishes charts for the world's oceans on the **G P**. They are also used for radio and seismic work (as waves travel virtually in great circle directions) and for star-maps. [*f*]

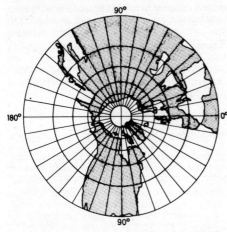

gold A precious metal known since earliest times, mentioned in *Genesis*, and regarded throughout history as a symbol of wealth, as well as being used as ornaments because of its untarnishable quality. Orig. wholly derived from alluvial gravels (PLACER deposits), the **g** originating in a VEIN from which it was removed by weathering and river-action.

Goletz terraces Terraces resulting from selective freeze-thaw and solifluction processes, and cut into hill-sides and summits in periglacial environments. Also referred to as ALTIPLATION or CRYOPLANATION terraces. **G ts**, covered by solifluction debris, may reach several km in length and breadth, but are usually smaller e.g. the **G ts** on the S. slopes of Coxe's Tor, Dartmoor.

Gondwanaland A single land-mass, the S. part of the Precambrian PANGAEA, from which the S. continents are thought to have been formed in Palaeozoic times. Evidence from many fields (geology, PALAEOBOTANY, PALAEOMAGNETISM) is indicative of a former unit consisting of Africa, Madagascar and India and probably parts of S. America and Australasia. Their suggested break-up and movement is part of the theory of CONTINENTAL DRIFT. A recent discovery about 640 km from the South Pole of fossil remains of *lystrosaurus* (a hippo-like reptile which flourished about 200 million years ago, sometimes known as the 'type-fossil' of the Lower Triassic, and long known in Africa), seems to confirm the existence of **G**.

Goode's Interrupted Homolosine Projection An equal-area **p**, based on the MOLLWEIDE and SINUSOIDAL PS, using the Sinusoidal from the Equator to 40°N. and 40°S., Mollweide in higher latitudes. The oceans are 'interrupted' to allow continents to be recentred on several meridians to attain good overall shape. Used widely for maps of economic distributions. [*f*]

---

gore A triangular piece of paper or thin card on which is printed a section of a map of the world, bounded by meridians and tapering to the Poles. Twelve or 24 **gs** for each hemisphere can be glued on to a sphere to fit reasonably well, and so make a complete printed globe.

Plate 33. A deeply cut river course at the Dukan GORGE, north-east Iraq. (*Aerofilms*)

gorge A deep, steep-sided, rocky river valley; e.g. Aar G. in Switzerland; Rhine G. in W. Germany. See pl. 33.

gossan A Cornish miner's name for the ferruginous infilling of a near-surface portion of a mineral vein containing sulphides, from which copper, sulphur, etc. have been leached, moving down in percolating solutions and leaving a mass of hydrated iron oxide. Sometimes called 'iron hat'.

Götiglacial One stage of the retreat of the Quaternary glaciation in Scandinavia, lasting from *c.* 15 000–8000 B.C. The ice-sheet left S. Sweden, and a rise of sea-level resulted in an enlarged Baltic Sea, known as the Yoldia Sea.

gouffre (Fr.) A large vertical shaft in limestone e.g. the Gouffre de Padirac, Causse de Gramat, S. France. (syn with AVEN).

gouge (i) A fault-plane clay or other fine material, in ct. to FAULT-BRECCIA. (ii) The soft earthy material alongside the containing wall of a mineral vein, which facilitates mining of the vein itself, after the miner has 'gouged' it out.

Graben (Germ.) A narrow trough, let down between parallel faults, with throws in opposite directions. It is often regarded as syn. with RIFT-VALLEY, but strictly a **g** is a structural feature, and is not necessarily a valley.

grad A measurement of latitude and longitude found on some old maps, esp. of France and Turkey, using 1g = 1/400th of 360°, rather than a degree.

gradation (i) Syn. with DENUDATION, esp. in USA (see DEGRADATION). (ii) The process of bringing the land-surface to a uniform GRADE.

grade A concept of EQUILIBRIUM applied esp. to stream long-profiles and to ground-slopes. A graded stream is one whose flow is such that at a given point there is a balance between material eroded and deposited. A change in volume, velocity and load, tending to increase either erosion or deposition, will produce a change in the gradient of the long-profile in such a direction that the balance is restored. G may be achieved equally well by changes in factors other than long-profile; e.g. channel cross-section and roughness, and the nature of the load. Thus a smooth profile is not essential for the production of **g**, and this state may be reached at any stage during a stream's development. G is a condition achieved only as the avge state over a period of time; at any given time a graded stretch of river may not represent an exact balance between erosion and deposition. Similar arguments are applied to the production and removal of waste on slopes.

[*f* REJUVENATION]

graded time One of three categories of 'geomorphological time-scale' devised by S. A. Schumm and R. W. Lichty (see CYCLIC and STEADY TIME). G t refers to the scale of time required, say, by a stream in the development of a graded profile.

gradient (i) The steepness of a slope, expressed either as a proportion between its vertical interval (*VI*) (reduced to unity) and its HORIZONTAL EQUIVALENT (*HE*) (e.g. 1 in 20), or as an angular measurement from the horizontal. The latter can be computed by expressing **g** as a fraction, reducing it to a decimal, and then looking up the angle corresponding to this computed tangent in a table. To give an approx. conversion (reasonably correct to about 7°) from the **g** as a fraction to an angle, multiply by 60. Thus,

a slope of 1° = 1 in 60 (actually 57·14)
a slope of 2° = 1 in 30 (actually 28·65)
a slope of 3° = 1 in 20 (actually 19·08)

[*f* SLOPE-LENGTH]

G can be expressed as a percentage; e.g. a 5% slope is 1 in 20 (about 3°). (ii) The degree of variation in various phenomena: temperature, atmospheric pressure (*barometric g*), density, velocity.

gradient profile A p along a road, railway or river, showing distances, slopes of different character and degrees of slope, with names *en route* and possibly geological outcrops.

gradient wind The w resulting from the balance between the horizontal atmospheric pressure force (CORIOLIS), and the centrifugal effect due to curvature of the ISOBARS. This balance is modified by the effect of friction near the surface.

grain (i) The gen. trend of structure, relief and physical features; e.g. the '**g**' of N. Scotland is N.E. to S.W. Ct. transverse g of a RIA coastline with longitudinal g of a DALMATIAN COAST. (ii) A small mineral particle, as sand-g. (iii) As an adjective, the coarseness or fineness of a rock, as 'fine-grained' clay, 'coarse-grained' sandstone.

gramme (gram) 1 g *mass* = 15·432 grains; 1000 g = 1 kilogramme = 2·20462 lb; 1 g = 1 cm^3 of water at 4°C; 1 oz. = 28·35 g.

granite A coarse-grained PLUTONIC rock formed either by slow cooling of a large intrusion of MAGMA (see BATHOLITH), through GRANITIZATION by transformation of pre-existing country rock by magmatic emanations, or by some other process of alteration not yet understood, a form of METASOMATISM. Its chemical composition incl. free quartz (20–40%), alkali feldspar, mica, hornblende and accessory minerals. It occurs as large masses in uplands, exposed by denudation; e.g. Dartmoor, Bodmin Moor and Land's End, Shap, Cheviots, Cairngorms, Wicklow Hills, Mourne Mtns, Mont Blanc massif, Brittany and Limousin in France, the mountains in Rocky Mtn National Park, Sierra Nevada. Used in a wide and gen. sense for any coarse-crystalled, light-coloured igneous rock, or as cfd. BASALTIC and ULTRAMAFIC rocks. Mica may be replaced by hornblende, hence a *hornblende-g*. When the dominant mafic minerals are micas, not hornblende, *biotite-g* or *muscovite-g* is formed. Granodiorite, granulite and GRANOPHYRE are broadly included in g *sensu lato*. See also MICROGRANITE.

granitization A concept that pre-existing rocks may be transformed into GRANITE, or into a granitic magma, by METASOMATISM. This may account for the emplacement of some BATHOLITHS.

granophyre A fine-grained granite or quartz-porphyry with a micrographic structure (i.e. quartz and feldspar are closely interpenetrated on a microscopic scale). G represents a stage in the GRANITIZATION of some other rock. E.g. Ennerdale G., in English Lake District.

granular disintegration The breaking down or crumbling of porous rocks into a g mass, as a result of freezing following absorption of water into the pore-spaces.

granule A fragment of rock with diameter 2–4 mm.

graphicacy The art of depicting spatial relationships in 2, sometimes 3, dimensions, by means of maps and diagrams (CARTOGRAPHY), skilfully chosen photographs, and graphs. '... Graphicacy spills over into literacy on the one hand and numeracy on the other without being more than marginally absorbed in either....' (W. G. V. Balchin and A. Coleman).

graphite A soft, black opaque form of carbon, occurring in veins or lenticular masses, and found in Korea, USSR, Malagasy, Sri Lanka and Mexico. Mixed with fine clay, it is used for 'lead' pencils.

grassland, natural A category of vegetation, occurring in areas which experience a season of prolonged drought, but have some rainfall (though gen. inadequate for tree growth) coinciding with the period of growth. The major types are: (i) *tropical g* (SAVANNA); (ii) *mid-latitude g* (STEPPE, PUSZTA, PRAIRIE, PAMPAS, VELD, DOWNS); (iii) *mountain g*, an altitudinal zone above the tree-line. In Britain g occurs widely, replacing what once must have been forest. *Gen. character:* (i) *permanent g*, orig. sown by Man, but has developed like a natural plant community, yet subject to grazing; (ii) *short-ley g*, sown by Man, and remaining down for a few years within the arable rotation; (iii) *upland g*, such as down- and fell-grazing; all may be improved by liming, fertilizing and drainage. *Type of grass:* (i) *turf* grasses; (ii) *meadow* grasses; (iii) *tussock* grasses. *Variety of grass:* (i) *neutral g*, the permanent grass of the lowlands in fields, with perennial ryegrass (*Lolium perenne*), meadow grass (*Poa pratensis*), timothy (*Phleum pratense*) and white clover (*Trifolium repens*); (ii) *basic g*, on chalk and limestone, with sheep's fescue (*Festuca ovina*) and red fescue (*Festuca rubra*); (iii) *acid g*, on base-deficient soils on siliceous rocks of N.W. Britain, with common bent (*Agrostis tenuis*) and sheep's fescue; (iv) *moor g*, on poorly drained peaty soils, with white bent (*Nardus* *stricta*) and wavy hair grass (*Deschampsia flexuosa*); (v) *grass heath*, with sheep's fescue and wavy hair-grass.

graticule A net of parallels and meridians on a specific MAP PROJECTION.

graupel A form of soft HAIL, or pellets formed of opaque ice-particles, usually falling in showers.

gravel An assemblage of water-worn stones, hence usually rounded, ranging in diameter from 2 to 50 mm, of fine, medium and coarse grades. Some authorities limit g to sizes of 2–10 mm, referring to the larger 10–50 mm stones as pebbles. One definition in the USA uses a diameter of from 4·76 mm to 76 mm, others only up to 64 mm. There is in fact no real agreement. The Wentworth Scale of particle size in the USA excludes the term (a pebble is 4–64 mm).

gravel train A VALLEY TRAIN composed mainly of g.

gravity, gravitation(al) Expressed as Newton's Law (1686): each body in the universe attracts every other body with a force directly proportional to the product of their masses and inversely proportional to the square of the distance between them, reckoned from their centres of mass along a line joining these centres. G is used in compound words where downhill movement is involved; e.g. *g flow* (of a glacier); *g movement* of material on a slope, esp. when lubricated by water (see MASS-MOVEMENT); *g water* held in soil-pores which will drain away downward if and when free drainage conditions develop; *g gliding* of strata which may produce DECOLLEMENT; *g sliding* or downhill SHEARING; *g wind* (see KATABATIC WIND).

gravity anomaly In geophysics, the difference between a computed and observed terrestrial *g* value; a significant anomaly is evident where a mountain range produces less disturbance of gravity than calculated. This must be compensated for by deficiency of mass in depth; i.e. implying mountain 'roots' of less dense rocks; see ISOSTASY. An excess observed g is *positive*, a deficiency is *negative*.

gravity collapse structure The c and down-sliding of strata on either side of an ANTICLINE, partic. when erosion has taken place, and where resistant strata are separated by INCOMPETENT BEDS. E.g. the mountains of S.W. Iran.

gravity slope That portion of a hillside s which is rel. steep; used esp. of a receding ESCARPMENT, usually steeper than 22°, by H. A. Meyerhoff, in considering W. Penck's concepts of S RETREAT. Such a s commonly lies at the angle of rest of material eroded from it, and may be equated with the CONSTANT S [*f*] of A. Wood.

great circle A c on the Earth's surface whose plane passes through its centre; the shortest distance between any 2 points on the surface is an arc of a **g c**. An infinite number of **g c**s can be drawn on a sphere, but only 1 **g c** can be drawn through any 2 points unless they are ANTIPODAL, in which case an infinite number is possible. Every complete **g c** bisects any other at 2 antipodal points. On GNOMONIC PROJECTION all **g c**s are straight lines. These properties are only approx. correct, since the earth is not a perfect sphere.   [*f*]

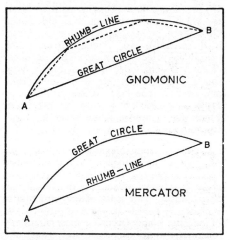

great circle route A long-distance air-route which follows approx. a **g c** e.g. London–Los Angeles, Amsterdam–Vancouver (both over Greenland), Amsterdam–Tokyo (over Greenland and Alaska), the so-called 'polar routes'. In practice, a **g c r** is broken down into a series of RHUMB-LINES to avoid continuous navigational changes of direction.

Great Interglacial The longest of the i periods in the Quaternary glaciation, the MINDEL-RISS, lasting (on the SHORT TIME-SCALE) for 190–240 000 years.   [*f* MINDEL]

'greenhouse effect' A pop. term for the important fact that short-wave RADIATION can pass easily through the atmosphere to the surface of the earth, while a proportion of the resultant heat is retained in the atmosphere, since outgoing long-wave radiation (of about 8·5–11·0 microns wavelength) cannot penetrate the atmosphere as easily, esp. when there is a cloud cover. Thus hard frosts occur during clear nights, when outgoing radiation

Green Mud A fine-grained m on the continental slope, at depths of 180–3300 m (100–2000 fthms), esp. off W. coasts of Africa and N. and S. America, deriving its colour from glauconite, and incl. a considerable proportion of calcium carbonate.

Greenwich Mean Time M SOLAR T at Greenwich Observatory (zero meridian), used as standard t for the British Isles and for parts of western Europe. Standard m ts for almost all other countries and zones are exact hours and half-hours fast or slow on Greenwich. See ZONE STANDARD T.

Greenwich meridian The Standard or Prime m which passes through the old Royal Observatory at Greenwich (a brass line is inset in the ground); =0° longitude, from which meridional values for almost all the world are calculated. Longitude is expressed E. or W. of this to 180°. Other ms which have been employed include Ferro (Canaries), used by Ptolemy, c. A.D. 150 (17° 14' W. of Greenwich); the Panthéon Observatory in Paris; Philadelphia.

gregale A strong blustery N.E. wind, accompanied by showers, in the central Mediterranean Sea. It is sometimes called the *greco*. It is partic. frequent in the Tyrrhenian Sea, and around the coasts of Sicily and Malta; it caused St Paul to be wrecked on the coast of the latter island. Occurring most commonly in winter, it is associated with the passage of a depression to the S.

grey box approach A type of 'systems approach', in which only a *partial* view of the system is obtained. As R. J. Chorley and B. A. Kennedy state, 'interest is centred on a limited number of sub-systems, the internal operations of which are not considered' (ct. BLACK BOX and WHITE BOX APPROACHES).

grèzes litées (Fr.) 'Bedded slope deposits', comprising alternating layers of angular stones and fines, formed in periglacial environments. Frost action provides the coarser debris, which may be solifluscted downslope and periodically eluviated by sheet-flow or rill-wash from melting snowdrifts. Fossil g l is well developed in the limestone of eastern France (e.g. the Jurassic limestones of Lorraine).

grey-brown podzol A type of soil transitional between ps and BROWN FOREST SOIL, more acid than the latter because of leaching under humid conditions, but with less pronounced leaching than the former, and with greater organic content. Found widely in UK, W. Europe and N.E. USA; it carries good grassland and is used for mixed farming.

grey earth A group of soils (*serozem*) in arid areas of mid-temperate latitudes, where low rainfall results in sparse vegetation and small organic content, but little leaching. Calcium carbonate is common, esp. in the upper layer where it may form a lime-crust. The soils are highly deficient in nitrogen, and have a low colloidal content. Found in Great Basin of Utah, USA; S. Argentine; N. Libya; the area between the Aral and Caspian Seas; parts of Mongolia; N.W. Pakistan; S.W. and S. Australia. They are of little value for agriculture unless they are irrigated, adequately drained to flush off salt and so prevent further salt accumulation, and heavily fertilized with organic materials and nitrogenous artificials.

greywacke, graywacke, grauwacke (Germ.) A rather old-fashioned term applied to certain dark, strongly cemented, coarse sandstones or gritstones, containing large angular particles, hard and resistant. Commonly of Lower Palaeozoic age.

greywether Syn. with SARSEN. Such blocks were orig. given this name because of their similarity to grazing sheep, as they occur scattered over grassy downlands; partic. common in Wiltshire.

grid (i) A network of squares covering a map-series, formed by lines drawn parallel and at right-angles to a central axis, and numbered E. and N. of an origin, from which the position of any point can be stated uniquely in terms of its easting and northing. UK uses the NATIONAL G. Each American state has a g (or more than one if its longitudinal extent exceeds 150 mi.), based on either TRANSVERSE MERCATOR or LAMBERT'S CONFORMAL projection, in which a g of 1000-ft squares is drawn, enabling lists of 'state-g co-ordinates' to be compiled. This is used by the US Land Office, hence known as the USLO g; see also LAND SURVEY system. The US military g system, which is metric, is based on 60 zones, each of 6° longitude, drawn on TRANSVERSE MERCATOR PROJECTION; hence known as the UTM g. See also UPS g, for polar regions within 80°N. and 80°S., drawn on a Polar STEREOGRAPHIC projection. (ii) A uniform pattern (usually of squares, equilateral triangles or hexagons) which covers a surface on which data have to be mapped, in order to carry out a spatial analysis of these data (e.g. slopes or altimetric frequencies), or to compute values at the node of each grid-unit for the interpolation of ISOPLETHS. [*f*]

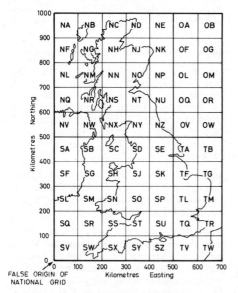

FALSE ORIGIN OF NATIONAL GRID

grid-north The meridional axis of a GRID system. In the UK NATIONAL GRID, based on the TRANSVERSE MERCATOR PROJECTION, **g-n** corresponds to meridian 2°W. Sheet-lines of all Ordnance Survey sheets are g-lines. At 2°W. **g-n** and TRUE NORTH coincide, but there is increasing divergence to E. and W., and the relation of **g-n** and True N. is stated for each corner of each sheet. Thus the most W. sheet of the One-inch and 1:50 000 O.S. series (Land's End) has a 2° 54′ E. divergence at its W. margin, the most E. in East Anglia has 2° 58′ W. divergence at its E. margin.

grike, gryke A deep groove crossing a limestone 'pavement', bounded by ridges (CLINT). The result of solution by CARBONATION through acidulated rain water, concentrated along well-marked joints; e.g. in N. Pennines, esp. near Malham in W. Yorkshire. See pl. 47.
[*f, opposite*]

grit, gritstone A coarse sandstone, usually massive, with grains of uneven size, probably a compacted marine deltaic deposit. Used as a proper name for Millstone Grit, the middle series in the Carboniferous system. The rock forms extensive moorland plateaus in the central Pennines and cappings on mountains rising from the limestone plateaus; e.g. Ingleborough, Pen-y-Ghent and Whernside. Sometimes (incorrectly) used for certain limestones; e.g. Calcareous G., Pea G.

GROUND 141

grivation The angle between GRID-NORTH and MAGNETIC N. on any gridded map.

groove A large-scale striation, produced by the impact of large boulders trapped in the basal layers of glaciers. Gs range in depth from less than one m to 30 m (observed in the Mackenzie valley, Canada). Where **gs** are curving or run transversely to the direction of ice movement, formation by sub-glacial streams or water-soaked till of high mobility is probable.

Grosswetterlage (Germ.) A large-scale circulation pattern of the atmosphere, within which the 'STEERING' over the region remains basically unchanged during a period of time.

grotto A picturesque or poetic term for a natural or artificial cave.

ground fog A low-lying RADIATION F, found partic. in hollows and depressions; by definition, it does not extend vertically to the base of any clouds.

ground-frost A temperature below freezing recorded on grass or ground surface, though air temperature may remain above freezing-point; this was formerly defined as −1°C (30·2°F), and likely to be harmful to tender vegetation. This definition was used by the UK Meteorological Office until 1960, since when it has simply signified a **g** temperature below 0°C.

ground ice (i) **I** formed on or attached to the bed of a sheet of water. (ii) A mass of clear **i** found in frozen ground, esp. under conditions of PERMAFROST. Cf. PIPKRAKE.

ground-moraine Debris carried at the base of a glacier or ice-sheet, deposited as a horizontal sheet of Till when this melts rapidly. See ABLATION TILL, LODGEMENT TILL.

ground swell Waves passing into shallow water, with a resultant increase in wave-height.

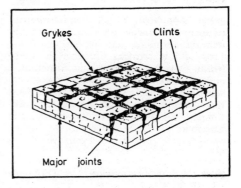

ground water The body of w derived from percolation, contained in the soil, subsoil and underlying rocks above an impermeable layer. (*Note:* this excludes subterranean rivers.) Syn. with *phreatic w*. See also CONNATE, JUVENILE, METEORIC and VADOSE W.

group The rocks formed during the major time-division of an era; i.e. PRECAMBRIAN, PALAEOZOIC, MESOZOIC, CAINOZOIC.

growan The coarse-grained product of granitic decay, called on Dartmoor by this local name.

growing season That part of the year with temperatures high enough to allow plant growth, gen. thought of in terms of cultivated crops. In UK about 6°C. The **g s** is defined as the number of days between the last 'killing frost' of spring and the first of autumn; e.g. 200 days for cotton. The **g s** has an important effect on agricultural systems and patterns; e.g. whether spring or winter wheat is grown; barley can be grown further N. than wheat because of its short **g s**.

groyne A timber, concrete or iron framework running out to sea to arrest LONGSHORE DRIFT, so maintaining material on the beach. On S. coast of England, gs cause an accumulation of material on their W. sides, since it moves E. under the influence of prevailing W. and S.W. winds.

grus Partially decomposed granite.

G-scale A s of geographical measurement based on the Earth's surface area and derived by successive subdivisions of this area in terms of the power of 10. The G-values rise with decr. size; thus the Earth's surface has a G-value of 0 (in areal terms $5 \cdot 101 \times 10^8$ km^2, $1 \cdot 968 \times 10^8$ sq. mi.), the USSR of 1·82, Yorkshire 4·51, Trafalgar Square approx, 10·0.

gryke See GRIKE.

guano Accumulation of bird-droppings, forming a valuable source of phosphatic fertilizer; e.g. along coasts of Peru and Chile; coast of S.W. Africa; Christmas I. in the Indian Ocean; Navassa I. and Sombrero I. in West Indies. G in Peru and Chile is granular, light-coloured and porous, that in the West Indies has been leached and is hard and compact. On Christmas I. some coral reef limestones have been converted into calcium phosphate by percolation from overlying g beds.

gulch A deep rocky ravine in W. USA.

gulder A double low tide, experienced near Portland, Dorset. See DOUBLE TIDE.

gulf A large inlet of the sea, usually more enclosed and more deeply indented than a bay; e.g. G. of Mexico, G. of Carpentaria.

gully A well-defined waterworn channel.

gully erosion The effects of a sudden rainstorm with a localized concentrated run off, thus creating a deep gash in the land-surface, esp. when this consists of soft material; one important aspect of SOIL EROSION. Also *gullying*. Seen well in the Badlands of S. Dakota, where gs have been eroded in Oligocene shales and limestones. See pl. 39.

gumbo A type of sticky fine-grained alkaline clay soil in the midwestern states of USA and in Canada, when saturated with water. It dries out like a brick.

Günz The earliest of a series of 4 periods of fluvioglacial deposition distinguished in the Bavarian Foreland by A. Penck and E. Brückner (1909) as an individual phase of distinct glacial advance during the Pleistocene. They correlated this with a distinct gravel deposit, Older Deckenschotter. Now applied gen. to an early major glacial period, though it is realized that the picture of glacials and inter-glacials is much more complicated and researchers postulate at least one earlier glacial (?*Donau.*); see also MINDEL, RISS, WÜRM. The G is equivalent in age to the NEBRASKAN in N. America, and probably to the ELBE of N. Europe. [*f* MINDEL]

gust A temporary, short-lived incr. in wind speed, followed by a lull. In US terminology wind speed must attain 16 knots.

gustiness factor In meteorology, a measure of the intensity of gusts, obtained by dividing the range of wind speeds between gusts and intervening calm or light wind periods, by mean wind speed.

gut A narrow channel opening into the sea or a large estuary, esp. in E. USA.

Gutenberg Channel At a depth between about 100–200 km below the upper surface of the MANTLE lies a layer of less rigid and more plastic material, in which the speed of earthquakes drops (P-waves from 8·2 to 7·85 km s^{-1}, S-waves from 4·6 to 4·4). The existence of this C was theoretically predicted by B. Gutenberg and confirmed by detailed seismographic evidence in the Chilean earthquake of 1960. The G C is thought to coincide largely with the ASTHENOSPHERE, above which lies the LITHOSPHERE.

Gutenberg Discontinuity The d at a depth of about 2900 km from the surface of the earth, between the MANTLE and the CORE, called after B. Gutenberg who discovered it in 1914. At this depth S-waves of earthquakes disappear, while P-waves travel on at a reduced speed (above the D 16·6 km s^{-1}, below 8·1 km); i.e. it is likely that the G discontinuity marks a change from a solid to a liquid medium, though of much greater density and under enormous pressure. [f EARTHQUAKE]

guttation dew D formed on vegetation, not by condensation from air, but from water transpired from the plants themselves.

guyot A flat-topped mountain, rising from the floor of the Pacific Ocean to within 0·8 km of the surface; a few have been discovered in the Atlantic Ocean. It differs from a SEAMOUNT, which has a pointed summit. There are estim. to be 10 000 gs and seamounts in the Pacific (H. W. Menard), some rising 3 km above its floor. They prob. orig. as volcanoes, their summits worn down by marine planation, and were later covered with water either by rise of sea-level, or by subsidence of their foundations. Many gs are old, as shallow-water Cretaceous and Miocene material has been dredged from some summits.

gypsey A BOURNE.

gypsum Natural calcium sulphate (CaSO$_4$. 2H$_2$O), one of the class of SEDIMENTARY rocks known as EVAPORITES. In a fully crystalline form it is *selenite*. *Alabaster* is very fine-grained variety. Without its water of crystallization it is *anhydrite*. G occurs in Permian and Triassic beds of N. England.

gyre A closed circulatory system or 'cell' occurring in major ocean basins between about 20°N. to 30°N. and 20°S. to 30°S. Its movement is generated by: (i) convection flow of warm surface water poleward; (ii) deflective effect of the Earth's rotation; (iii) effects of prevailing winds. The N. Atlantic g involves the Gulf Stream (N.), N. Atlantic Drift (E. to N.E.), Canaries Current (S. to S.W.), and N. Equatorial Current (W.). Other ocean basins have similar gs, except for the landlocked Indian Ocean, where the triangular peninsula and the MONSOON change of wind direction cause a double g moving in seasonally opposite directions; another double g occurs in the S. Pacific. Sometimes called *gyral* in USA.

gyro-compass A c which does not make use of the Earth's magnetism, consisting of a rotating wheel with a constant direction of axis and plane of rotation.

haar A cold mist experienced in spring and early summer, esp. in E. Scotland.

habitat Used syn. with ENVIRONMENT, esp. in ecological context, as an area in which the requirements of a specif. animal or plant are met.

hachure A line on a relief-map drawn down a slope, thicker and closer together where the gradient is steepest. It enables minor but significant details to be brought out, which would be lost within the contour-interval on a contour-map, and can show striking relief in a dramatic manner. It lacks specif. information about altitude. [f]

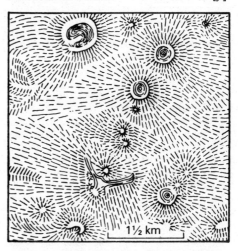

hadal Sometimes applied to the deepest parts of the ocean below about 5500–6100 m, i.e. below the ABYSSAL ZONE. Cf. DEEP, ocean.

hade (i) Angle of inclination of a FAULT-plane from the vertical. (ii) In mining, the angle of a VEIN or LODE, measured from the vertical. [f FAULT]

Hadley cell A feature of the mean atmospheric circulation, comprising a thermally driven circulation c extending from the Equator to about latitude 30° N. and S., with rising air at the Equator, poleward high-altitude flow, descending flow at about 30° N. and S. and surface equatorward flow. The concept was put forward in 1686 by Edmund Halley, improved and modified in 1735 by G. Hadley in his explanation of the Trade Winds.
[f ATMOSPHERIC CIRCULATION]

haematite, hematite An ore of IRON, Fe$_2$O$_3$, grey, black or reddish in colour. Worked for many years in W. Cumbria and Furness, averaging a content of iron 40–62%, silica 5–15%, and very low phosphorus; it occurs as veins, flats and vertical 'sops'; about 150 million tons have been mined to date. Sometimes it occurs in large kidney-shaped nodules, hence 'kidney ore'. The largest deposits of **h** are in the 'iron ranges' of the L. Superior district. In its cubic form, *martite*, **h** is the main ore at Krivoi Rog, Ukraine.

Haff (Germ.) A shallow fresh-water or brackish coastal lagoon, formed by growth of a sand-spit (NEHRUNG) across the mouth of a river, sometimes enlarged by submergence of adjacent lowlands by the sea. The classic examples are on the Baltic coast of E. Germany, Poland and USSR; e.g. Kurisches H., Frisches H. [*f* NEHRUNG]

hag, hagg A steep-sided mass or bank of peat, found on a moorland where erosion is now active; e.g. on gritstone Pennines, such as Kinder Scout plateau.

hail, hailstone An ice-pellet with a diameter of 5 mm or more, which falls from a CUMULO-NIMBUS cloud, at the passage of a COLD FRONT, and in summer or in hot climates (India, S. USA) after exceptional heating, convectional overturning, and rapidly ascending air-currents. Each **h** forms around an 'embryo', a frozen droplet in a cloud. At first it is swept upwards by updraught, until it attains a dimension which enables it to fall through the updraught, hence out of the cloud. This accounts for ordinary **h**s, but the growth of very large **h**s is still only partially understood. In section many show concentric layers of ice, alt. clear and 'milky'. The 'clear' layers are caused by freezing of water under conditions of 'dry-growth', i.e. with low humidity in that part of the cloud, while opaque layers are produced by 'wet-growth', i.e. in part of the cloud with high humidity so that small droplets freeze on impact. The largest **h** recorded in UK fell near Horsham on 5 Sept., 1958; in section it resembled a half grapefruit and weighed 0·14 kg. One weighing 0·7 kg, with a circumference of 430 mm, was recorded in USA in 1928. One weighing 3·4 kg has been reported in India, and one of 4·6 kg in China. They do damage to orchards, crops and glass-houses, and in USA and India men and animals have been killed.

Haldenhang (Germ.) Introduced by W. Penck for a section of slope at the foot of a rock-face, often beneath a TALUS layer, and less steep than the slope above.

half-dome A convex, bare rock outcrop, on a large escarpment face developed partic. in granitic rocks in tropical environment. H-ds represent 'cores' of massive rock exposed by general scarp retreat. Sometimes they are 'massive convex rock spheroids' (M. F. Thomas); in other instances they are divided by joints, and are degenerating into KOPJES.

half-life That period of time in which half the atoms of a radioactive mineral disintegrate and change into the ISOTOPE of another element. The **h-l** of uranium^{235} is 0.713×10^9 years, of uranium^{238} 4.50×10^9 years, of potassium^{40} 1.19×10^9, of rubidium^{87} 47.0×10^9, of thorium^{232} 13.9×10^9, of carbon^{14} 5570 years.

halite Rock-salt, or sodium chloride, an EVAPORITE, which forms thick layers in association with such sedimentary rocks as sandstone and shale.

halo A ring or rings of light, concentric to sun or moon, when the sky is thinly cloud-veiled, the result of refraction of light by water-drops or ice-crystals; the ring may be white, or tinted red on the inner side, or range from red to blue (outer). Ct. CORONA. The most common **h** has a radius of 22°. Ct. CORONA, produced by *diffraction*.

haloclasty The rapid disintegration of rocks, specif. in a hot desert, as a result of weathering effects of water, incl. HYDRATION, salt crystallization and rapid cooling due to surface wetting. Ct. THERMOCLASTY.

halomorphic A soil-type which is predominantly saline. See SOLONCHAK, SOLONETZ.

halophyte A plant growing in salt-impregnated soil on the shore of an estuary or in a salt-marsh, or able to survive in the presence of salt-laden spray. The category incl. marsh samphire (*Salicornia herbacea*), sea manna-grass (*Glyceria maritima*), sea aster (*Aster tripolium*), perennial rice-grass (*Spartina townsendii*). Sand-dune plants are not included.

hamada (Arabic) A rock desert, consisting of a bare wind-scoured pavement, diversified by relict masses (HOODOOS, YARDANGS, ZEUGEN), and with little sand. H plains cover large areas in the Sahara, W. Australia and Gobi Desert in Mongolia.

Hammer Projection A variation, by E. Hammer on the ZENITHAL EQUAL AREA (LAMBERT) P, made by doubling the horizontal distances along each parallel from the central meridian. This transforms the circular shape of Lambert into an ellipse, similar in appearance to MOLLWEIDE P, but with all parallels curved except the Equator, which is a straight line. Sometimes called Aitoff-Hammer, but should be credited to Hammer alone; the AITOFF is quite distinct. [*f*, *page 145*]

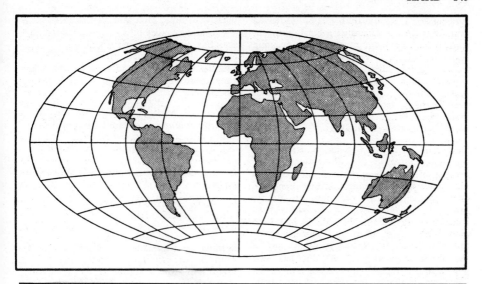

hanger (i) A wood (usually beech) on steep slopes of chalk country; e.g. Selborne H. (ii) The steep slope itself.

hanging glacier A short g protruding from a basin or shelf high on a mountain-side, from which masses periodically break off as ice-avalanches.

hanging valley A high-lying tributary v, which leads with a marked steepening of slope into a main v. Commonly occurs in a glaciated area where the main v has been over-deepened by a glacier; e.g. side-vs of the upper Rhône; Lauterbrunnen (both in Switzerland); Yosemite valley (California). As a result, a stream flowing down a **h v** falls as a cascade into the main valley; e.g. Staubbach (Lauterbrunnen); Bridal Veil Falls (Yosemite). Other factors may cause minor valleys to join main ones in this discordant manner. River erosion in the main v may result in its overdeepening. Sea-cliffs may contain **h** vs cut off by marine erosion; e.g. along the Devonshire coast near Hartland, where Litter Water has a 23 m fall from cliff to shore.
[*f* ALP]

hanging wall The upper face of a FAULT or VEIN. [*f, opposite*]

hardness scale The MOHS' s of hardness of minerals, in which 10 selected minerals have been arranged in an ascending scale: 1. talc; 2. gypsum; 3. calcite; 4. fluorite; 5. apatite; 6. orthoclase; 7. quartz; 8. topaz; 9. corundum; 10. diamond. All others can be ascribed accordingly; e.g. 2·5 galena; 5·5 uraninite. This **s**, while progressive, is not uniform; thus diamond (no. 10) = 10 × corundum (no. 9), but the latter is only 10% harder than topaz (no. 8).

hardpan A thin hard stratum within or beneath the surface soil, consisting of: (i) sand grains or gravel cemented by ferric salts deposited by percolating solutions (*ironpan*); (ii) compact redeposited humus compounds (*moorpan*); (iii) washed-down or synthesized clay (*claypan*); (iv) a layer of calcium carbonate (*limepan*); and (v) siliceous minerals (DURIPAN).

hard water W which does not form a lather with soap, because of the presence in solution of calcium, magnesium and iron compounds. It may be: (i) *temporarily* **h**, containing soluble bicarbonates; (ii) *permanently* **h**, containing sulphates. Water flowing from limestone uplands is **h**, and cannot be used in the textile industry.

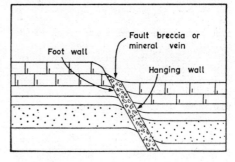

hardwoods Trees and timber of broad-leaved deciduous forests, affording a hard, compact wood. They are gen. much slower growing than softwoods. Temperate varieties incl. walnut, maple, cherry, poplar and oak; in Mediterranean climates are evergreen oaks, in Australia the eucalyptus; monsoon climates produce teak; tropical forests ebony, rosewood, greenheart, mahogany. Most hs are heavy, but not all; e.g. balsa.

harmattan A dry N.E. wind blowing from the Sahara toward the W. African coast; because of its dryness and rel. coolness it is reputed to bring a health contrast (hence its name 'Doctor') to prevailing humidity, though in fact outbreaks of diseases such as spinal meningitis coincide with the h. Inland the wind is dust-laden and unpleasant. It blows much farther S. and more strongly in the N. winter (to about 5°N.); dust has been deposited on ships in the Gulf of Guinea. In July (when planetary pressure- and wind-belts move N.), it blows only as far as about 18°N.

hatchet-planimeter An instrument for measuring an area on a map; the point of a tracer-bar is carefully traced round its perimeter.

haugh Esp. in Scotland and N. England low-lying land along the flood-plain of a river.

Hawaiian eruption A volcanic e in which great quantities of basic lava flow quietly out, with no explosive activity, hardening to form a large, low-angled SHIELD VOLCANO, as in Hawaii.

Hawaiian 'high' One of the atmospheric high-pressure cells of the HORSE LATITUDES in the N. Pacific. Much more markedly developed in summer than in winter.

hazards, natural Dangers due to n causes which may drastically affect the landscape and therefore man—earthquake, drought, flood, avalanche, frost, coastal erosion, volcanic eruption, tropical storm, atmospheric pollution. UNESCO is sponsoring research on a global basis into these hs, together with the human response and national policies.

haze Obscurity of the lower layers of the atmosphere as a result of condensation of water droplets on dust, salt or smoke particles, with a visibility of more than 1 km but less than 2 km. Also frequently used for a gen. obscurity which results from something other than moisture in the atmosphere; e.g. 'heat h', 'dry h.'

head (i) A body of water at some altitude above an outfall. (ii) (with cap.) A compacted mass of rubble, sand and clay of Pleistocene age, found in valley bottoms, lower slopes and sea-cliffs, formed under periglacial conditions as a product of SOLIFLUCTION ('the H'). (iii) The leading wave of a BORE on a river. (vi) A cape, more usually headland; e.g. St. Bees H., Beachy H.

headcut A vertical drop in the bed of a stream channel, normally affected by up-valley migration. Many small gullies originate as a series of hs, each formed at a point of slope increase or line of seepage. The individual hs grow upslope to give eventually a unified gully system.

head-dune A sand-dune accumulated in the 'dead-air' space on the windward side of some obstacle. [f DUNE]

heading A horizontal tunnel driven into an AQUIFER to tap ground-water penetrating fissures, in order to supply wells or reservoirs; e.g. a h in chalk near Frinton 3 km long supplies water to Eastbourne; Brighton has 5 wells into which lead 8 km of hs.

headland A promontory with a steep cliff-face projecting into the sea.

headwall The back of a CIRQUE, rising steeply from its floor, separated from the FIRN-field by a RANDKLUFT. The steepness of the h is incr. by freeze-thaw, the result of water making its way down the randkluft into crevices in the rock-wall, hence *h recession*.

headward erosion The cutting back up-stream of a valley above its original source by rainwash, gullying and SPRING-SAPPING. The source of the stream gradually recedes and ultimately may notch the ridge which forms the orig. watershed; this may lead to river CAPTURE.

headwater The upper part of a river system, used more commonly in the pl., denoting the upper basin and source-streams of a river.

heat balance The equilibrium condition which exists in avge terms between RADIATION received from the sun (INSOLATION) by the Earth and the atmosphere, and that re-radiated or reflected (see ALBEDO). This is a concept of overall balance but varies with latitude and the seasons. Gen. lands between 35°N. and S. and the Equator receive more radiation than they lose, while lands poleward of 35°N. and S. receive less than they lose. This is a *vertical h b*, but also a *horizontal h b* occurs, as when heat is transferred from low to high latitudes by air-masses and ocean currents; this is the 'driving force' of atmospheric circulation.

heat gradient The rise of temperature with increasing depth in the Earth's crust, on avge about 1°C per 30–35 m.

heath, heathland (i) Plants of genus *Calluna* (ling or heath), not botanically heather (*Erica*). (ii) The plant formation dominated by *Calluna*, which incl. *Erica*, bilberry (*Vaccinium myrtillus*), dwarf gorse, broom, whin, wavy hair grass (*Deschampsia flexuosa*), and lichens such as *Cladonia*. (iii) A landscape (*heathland*) of uncultivated, gravelly or sandy soils, leached and acid in character, with a shrubby vegetation of species listed in (ii). Silver birch and dwarf oak occur sporadically. E.g. Dorset hs; Breckland of Norfolk–Suffolk; centre of the Weald. Upland hs or 'moors' of N. England and Scotland can be incl.; e.g. N. Yorks Moors; grouse-moors of the Scottish Highlands.

'heat-island' The area of a city above which the air temperature is slightly higher than surrounding areas, the result of the dissemination of h from the concentration of buildings.

heat-wave A spell of several days of unusually hot weather, not specif. defined, but gen. over about 27°C. (80°F) in UK.

heave, of fault The forward horizontal displacement of strata in an inclined normal fault, expressed as the distance apart in m of the ends of the disrupted strata. [*f* FAULT]

Heaviside-Kennelly Layer A section of the IONOSPHERE at about 100–120 km (the *E-layer*). The ionosphere is characterized by high ion density, but variation of electron density allows it to be subdivided into separate ls. (There is a certain doubt as to whether these zones are in fact continuous ls or not.) It reflects long radio-waves back to earth, which enables them to be received at a distance, rather than disappear into space. It allows the penetration of short radio-waves, which continue to the APPLETON L, 300 km up.

hectare (*ha*) An acceptable though non-SI metric unit of areal measurement.

1 ha = 2·47106 acres
1 acre = 0·4047 ha
1 ha = 10 000 sq m
100 ha = 1 km²

hekistotherm A plant such as reindeer moss or lichen, which can exist where mean temperature of the warmest month is under 10°C (50°F).

Helada, Tierra The highest altitudinal climatic zone in Mexico, including loftiest peaks and permanently frozen or snow-covered areas.

helicoidal flow A kind of 'cork-screw' motion in the current of a river within a bend, so tending to move material from its concave bank to its convex bank emphasizing deposition on the inside and erosion on the outside of a MEANDER.

helictite A formation of calcium carbonate in a cave, of varied shape; it may be thin as a thread, in spirals or loops, or festooned with tentacles. Formed by deposition of calcite from solutions percolating down fine fissures on to ceilings of caves in limestone country, usually where a strong current of air is present.

heliotropism The turning of plants towards the light.

Helm wind An E. or N.E. wind blowing strongly and gustily down the Crossfell escarpment in N. Pennines into the Eden valley; with it is assoc. the *helm-cloud* or Helm, resting above the ridge (a type of BANNER-CLOUD), and the *helm-bar*, a parallel line of cloud a few km to the W. These clouds are mainly the result of LEE-WAVES, eddies and turbulence, together with an INVERSION-layer at about 1800 m.   [*f* LENTICULAR CLOUD]

hemera A palaeontological term for the time-period of max. development of an organism. The corresponding stratigraphical division is an *epibole*.

hemipelagic deposits Material deposited near the shore in shallow water and bathyal zones.

hemisphere Half a sphere. The globe is divided into N. and S. hs by the Equator. There is also a 'land' and a 'water' h. 'W. H' is sometimes used for the Americas.

Hercynian Orig. used in Germany for any mountains produced by folding and faulting, of various ages; later limited to late Palaeozoic (Upper Carboniferous-Permian) orogeny. This is given various names; sometimes the whole mountain system in central Europe is named H (after Lat. name *Hercynia Silva*, of Harz Mtns), sometimes called the *Altaides*. VARISCAN (coined by E. Suess) is used by some as syn. with H, by others as E. representatives of it. The mtns of Brittany and S.W. Britain are of the same age, known specif. as ARMORICAN. In Europe the worn-down, tilted blocks, the eroded remnants of H folds, survive as massifs: S.W. Ireland, Brittany, Central Massif of France, Vosges, Black Forest, Ardennes, Middle Rhine Highlands, Bohemian 'Diamond', Upper Silesia, and Donetz (a syncline). Remains of fold ranges of broadly similar age include Urals, Tien Shan and Nan Shan in C. Asia, Appalachians of N. America, and foothills of the Andes. See pl. 19.

heterogeneous A specif. meaning applied to a gen. word, implying rocks of varying kind or nature in close juxtaposition, or to a rock composed of different materials, and therefore tends to weather unevenly; ct. HOMOGENEOUS.

heterosphere The outer atmosphere above the HOMOSPHERE, on the basis of its gaseous constituents. It consists of a series of distinctive layers: (i) molecular nitrogen at 90–200 km; (ii) oxygen at 200–1100 km; (iii) helium at 100–3500 km; and (iv) hydrogen from 3500 km upward. These layers are separated by transitional boundary zones. The density of the gases is extremely low.

hiatus A gap in the geological succession, the result either of strata not being deposited during a period of earth-movement or denudation (e.g. absent Miocene in Britain), or as the result of their removal by denudation. See UNCONFORMITY.

'high' In meteorology an area of high atmospheric pressure, surrounded by closed isobars; an ANTICYCLONE.

high cohesion slope Used by A. N. Strahler to describe steep slopes (usually in excess of 40°) formed in highly cohesive rock (e.g. granite, certain types of clay). The angle of slope is so steep that weathered products are quickly removed by wash to expose bare rock (ct. REPOSE SLOPES); and the rock itself is so cohesive as to resist physical collapse.

highland High uplands, used (in pl.) specif. as a proper name; e.g. Hs. of Scotland.

high seas The open ss beyond territorial waters, used both expressively and as a technical term in mercantile law.

high water The highest point reached by the sea in any one tidal oscillation. *HWMMT*, High Water Mark Medium Tides, is shown on UK OS maps as High Water Mark.

hill Uplands of less elevation than mountains in the same area; there is no specif. height-definition, because this is a matter of rel. scale. The S. Downs, Chilterns and Cotswolds are hs. Black Hs of Dakota rise to 2207 m, Nilgiri Hs in S. India to 2694 m.

hill-shading A method of relief depiction on maps, by assuming an oblique light from the N.W., hence slopes facing E. and S. are shaded; the steeper the slopes the darker. Ridge-crests, plateaus, valley-bottoms and plains are left unshaded, giving the impression of a relief model. Printed topographical maps make effective use of **h-s** applied in a subdued stipple in a neutral tone in conjunction with contours; e.g. Ordnance Survey 'Tourist' edition, 1 in and 1:50 000 maps.

hindcast In studying a sequence of events, it is possible to extrapolate either forwards (FORECAST) or backwards (**h**). The latter enables a probable sequence of events and processes to be more accurately worked out, using recorded events and processes in conjunction with those obtained by theoretical h. This may be based both on actual laws and on tendencies and probabilities.

hinge The line along a fold where the curvature of the strata is at its max. A *h-fault* is a fracture along which the vertical displacement incr. gradually from zero; it gen. passes into a FLEXURE. A *h-line* is a boundary between stable and moving sections of the Earth's crust.

histogram A graphical representation of a frequency distribution of a variable, such as seasonal frequencies of rainfall. Amounts are plotted as abscissae, scale of frequencies as ordinates.

[*f* ALTIMETRIC FREQUENCY GRAPH]

historical geology The time-record of changes in or on the Earth. One of the 2 major divisions of g, the other PHYSICAL G.

Hjulstrom curve A graph plot depicting the influence of stream velocity (in cm s^{-1}) and particle size (in mm) on the 'entrainment' of debris by streams. The curve for 'erosion velocity' (the minimum velocity needed for movement of particles of a given size) shows surprising features e.g. the relatively high velocity required for movement of ultra-fine and fine particles (in the 0·001–0·01 mm range), and the relatively low velocity required for movement of sand particles (in the range 0·1–1·0 mm).

hoar-frost The deposition of ice-spicules directly from water-vapour on the surface of plants and objects on the ground, cooled by nocturnal radiation to a temperature below DEW-POINT, itself below freezing-point. Sometimes *white-dew* in USA. See also DEPTH HOAR.

hoe A promontory or ridge projecting into the sea or into a lowland. Used commonly as a place-name suffix, hence Ivinghoe, Plymouth H.

hogback, hog's back A CUESTA in which the dip of the strata is great, so that both front- and back-slopes are steep (rather than one steep and one gentle slope), forming a narrow crested hill-ridge; e.g. Hs. B., W. of Guildford; in Colorado, hs of Dakota Sandstone (Lower Cretaceous) parallel to Front Ranges of the Rockies. See pl. 17. [*f*]

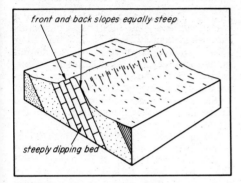

hog's-back cliff A sea-cliff with a long, often well-vegetated slope leading down from the cliff summit to the beach. H-b cs may be caused by (i) past degradation of a vertical cliff by periglacial processes at a time of low sea-level; (ii) seaward dipping strata; (iii) formation initially as a valley-side slope, which has become exposed at the coast by cliff recession; (iv) protection of the cliff base from wave undermining by extensive beach accumulation.

holistic App. to the view that in nature functional organisms are produced from individual structures which act as complete 'wholes'. Geography is said to have this **h**. approach, whereby phenomena are viewed not as individual entites but as interrelated complexes. Hence *holism*, the doctrine itself.

holm, holme (i) A small islet in an estuary or river; e.g. Steep H., Flat H. in the Bristol Channel. (ii) An area of level land liable to flood, along the banks of a river, esp. in N. England and S. Scotland, a common valley place-name; e.g. Milnholm, Murtholm, Broomholm, in the valley of R. Esk, Dumfriesshire; and Atholme, Lincs.

Holocene (i) The most recent (*lit.* 'wholly recent life') of the geological periods (the younger part of the Quaternary era), and the rocks laid down during that time; e.g. alluvium, peat. It occupies the time since the termination of the Ice Age; the Upper H is the last 5000 years. Many American authorities do not use the term, simply speaking of *Recent*, or of *Postglacial*. (ii) By international practice, and incr. in UK, the latest series/epoch in the stratigraphical and chronological column, i.e. all post-PLEISTOCENE time, the younger part of the Quaternary system/period. Syn. with RECENT. See table, TERTIARY.

holokarst An area of KARST, its features developed to the fullest extent; used by J. Cvijić, though by few other authorities.

Holstein The penultimate interglacial period in N.W. Europe (see HOXNIAN).

holt A small wood, a wooded knoll; hence a widely used rural place-name.

homoeostasis The ability of natural systems to undergo self-regulation, usually by the operation of negative FEEDBACK mechanisms (sometimes referred to as *dynamic h*).

homoclimes Places with broadly similar CLIMOGRAPHS [*f*].

homocline, homoclinical A succession of strata dipping gently and continuously in one direction. Used commonly in USA, but UNICLINE, uniclinal, is gen. preferred in UK. In USA used specif. for ridges and valleys formed by denudation from one flank of a folded structure, as in the Appalachians; hence *homoclinal ridge*, *h valley*. Ct. MONOCLINE.

homogeneous Of the same kind, character or nature, applied specif. to rocks, in ct. to HETEROGENEOUS.

homolographic Sometimes used to denote an EQUAL-AREA projection, specif. MOLLWEIDE P.

Homolosine Projection See GOODE'S INTERRUPTED HOMOSINE P.

homoseismal line, homoseism A line on a map linking places affected simultaneously by an earthquake shock.

homosphere The Earth's atmosphere to an altitude of 80–90 km, in which the chemical composition and mean molecular weight of dry air is largely uniform. This incl. TROPOSPHERE, STRATOSPHERE and MESOSPHERE. It is overlain by the HETEROSPHERE.

honeycomb weathering In some types of rock along the shore, the w and erosion of small water-filled pools may produce distinctive features. The pools become enlarged by chemical and physical w, leaving only sharp interconnecting ridges; the deeply pitted rock surface resembles that of a **h**.

hoodoo A fanciful name in USA for a grotesque rock pinnacle or pedestal (esp. of

Plate 34. The Matterhorn, Pennine Alps, Switzerland. Note the HORN, the ARETE leading down from the summit and separating CIRQUE basins to right and left. (*Aerofilms*)

sandstone), the result of weathering in a semi-arid region; sometimes (wrongly) ascribed to scouring and undercutting by wind-transported sand grains. The infrequent rains run down the pinnacle, which stays wet near its base for a longer period, causing more concentrated chemical and physical changes, loosening of grains and hence active undercutting. Cf. YARDANG, ZEUGE.

horizon (i) The *visible* (local, apparent, natural, visible, geographical or sensible) **h** is the boundary of the Earth's surface as viewed from one point, where Earth (or sea) and sky appear to meet. A near-by eminence which interrupts the view is not part of the **h**. Distance of **h** from an observer at height $x = 2xa$, where a = Earth's radius. For a height of 30 m, **h** is theoretically at 20 km, actually at 24 km because of refraction. For a height of 120 m, **h** is at 40 km; 300 m, at 60 km. (ii) In geodetic surveying, a great circle on the celestial sphere whose plane is at right-angles to a line from the zenith to the point of observation; i.e. *true, astronomical,* or *celestial h.* (iii) In geology, the plane of a stratified surface, or a bed with a partic. series of fossils. (iv) Each main layer or zone within a SOIL PROFILE [*f*]; see A-, B-, C-, D-, E- H.

horizontal equivalent The distance between 2 points on the land-surface, projected on to a **h** plane. [*f* SLOPE LENGTH]

Horn (Germ.) A pyramid peak, formed when several CIRQUES develop back to back, leaving a central mass with prominent faces and ridges. Used commonly in the Alps as a suffix; e.g. Wetterhorn, Matterhorn. See pl. 5, 34. [*f* PYRAMIDAL PEAK]

hornblende A rock-forming mineral of the AMPHIBOLE group, occurring as a dark-coloured crystalline mass in igneous and metamorphic rocks. Consists mainly of calcium, iron and magnesium silicate.

Horse Latitudes The sub-tropical high-pressure belts, about 30°N. to 35°N. and 30°S. to 35°S. (though interrupted by the distribution of land and sea); zones of calms and descending air, from which air-masses move poleward and equatorward. These belts may in part be the result of a gen. movement of air in the upper part of the TROPOSPHERE from the Equator, coming under the influence of CORIOLIS deflection so that an accumulation of air occurs in these latitudes. Possibly an equatorward movement of air in the upper troposphere from high latitudes tends to descend in the H L, so incr. the air accumulation. Said to be so called from throwing overboard of horses in transport from Europe to America if the ship's passage were delayed by calms. [*f* ATMOSPHERIC CIRCULATION]

horst (Germ.) A block, usually with a level summit-plane, sharply defined by faults, and left upstanding by differential movement, either by sinking of the crust on either side of a pair of faults, or by bodily uplift of a mass between these faults; e.g. Vosges, Black Forest, Harz, Sinai, Korea, Morvan. The **h** may be denuded so that although the structural pattern remains, the upstanding relief form may disappear. Cf. RIFT-VALLEY and GRABEN. [*f*]

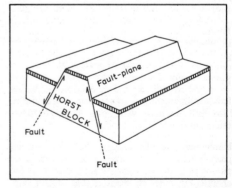

Horton's Laws A series of empirical relationships between stream order and such factors as the no. of streams and area of the drainage basin. The no. of streams of a given order declines sharply with incr. order, and the total length of a drainage net incr. regularly with order. In most drainage basins, the no. of streams in a given order is approx. 3 times greater than in the next higher order; e.g. a basin with 50 1st order streams will have approx. 15 2nd order streams, 5 3rd order streams and 1 or 2 4th order streams. See STREAM ORDER.

hot spring (thermal spring) In some areas assoc. with past or present vulcanicity, hot water (21°C to near boiling-point) flows out of the ground continuously, in contrast to forceful periodic emission of a GEYSER; e.g. Yellowstone National Park, Wyoming; Lassen Park, California; Iceland (where water is piped and used for central heating and swimming-baths); N. I., New Zealand. H ss occur in some non-volcanic areas, e.g. Bath in England, Spa in Belgium.

'hot tower' Pop. name for the central vertical current of heated unstable air in a CUMULONIMBUS cloud within an intense tropical low pressure system.

hour angle Of a heavenly body, the **a** at the pole between the meridian of the observer and the DECLINATION CIRCLE through the heavenly body, one **a** in the spherical triangle. Usually expressed on the scale 24 hours = 360°.

how A low hill, notably in Cumbria, from Old Norse *haugr*; e.g. Pickett How in Loweswater valley. Widely used in W. Lake District as a suffix in farm names. *Note:* Torpenhow, a village in Cumbria, consists of 3 syllables, each meaning hill in a different language or dialect.

howe A depression, or 'hollow place', esp. in E. Scotland; e.g. H. of Fife, H. of the Merse.

Hoxnian The penultimate interglacial period in Britain, equivalent to the HOLSTEIN interglacial of N.W. Europe, and represented typically by old lake deposits at Hoxne, Suffolk.

hum A residual hill-mass, shaped roughly like a haystack, in a limestone area, esp. in the KARST.

humic acid A complex organic **a** formed by partial decay of vegetation through which passes percolating water. This may assist in chemical weathering of rock, esp. of feldspar, and also contributes to the formation of certain soil-types.

humidity The condition of the atmosphere with ref. to its water-vapour content. See ABSOLUTE H, RELATIVE H, SPECIFIC H, MIXING RATIO, SATURATION DEFICIT.

hummock A low rounded hillock of earth, rock or ice.

humus The remains of plants (and to a less extent of animals) in the soil, decomposed into a darkish amorphous mass through the action of bacteria and other organisms. Essentially the organic constituent of soil. See MODER, MULL, MOR.

hurricane (i) An intense tropical storm in the W. Indies and the Gulf of Mexico occurring in July-Oct., accompanied by winds of terrific force (160 km h^{-1} plus), torrential rain and thunderstorms. Hs originate in mid- or W. Atlantic in latitudes 5°N. to 20°N. Their tracks are gen. W.N.W. over the Caribbean Sea and Florida, turning in a N.E. direction from about 30°N. along the Atlantic coast of USA. Their effects are sometimes felt as far as Maine and even the Maritime Provinces. Indicated by a female codename in alphabetical order as they occur. For details of tropical storm gen., see CYCLONE. (ii) Force 12 (121 km, 75 m.p.h. plus) on BEAUFORT WIND SCALE.

hurst Used commonly in placenames (esp. in the Weald of S.E. England) to denote a wood, thicket, knoll, or occas. a sand-bank in a river.

hydration A process of weathering, whereby minerals take up water and expand, causing stresses within rock; e.g. anhydrite becomes gypsum ($CaSO_4.2H_2O$). This can cause SLAKING in a clay-rich sedimentary rock.

hydraulic force The eroding **f** of water on rocks through its sheer power; e.g. a river surging into cracks, sweeping against banks on the outside of bends, with turbulence and eddying. Similarly, waves can exert a **h f** as they pound the base of cliffs, esp. where air is compressed in cracks and fissures. No load of material is involved, or the erosive process becomes ABRASION.

hydraulic geometry Study of the changing geometry of stream channels in relation to variations in discharge. A cross-section of a stream must have the capability to transfer varying amounts of water and sediment; in order to achieve this capability adjustments must be made in channel width, depth, slope and roughness, which in turn determine stream velocity. The relationships between these variables and discharge are expressed by a series of formulae (see AT-A-STATION ANALYSES).

hydraulic gradient The slope of the WATER-TABLE, expressed as the ratio between the HEAD (h) and the actual length of flow (l). Thus if $h = 6$ m and $l = 60$ m, $hg = 10\%$ or $0\cdot1$.

hydraulic radius The ratio between the area of cross-section of a river channel and the length of its wetted perimeter, used by hydrologists and engineers.

hydraulic tidal current A c which forms to compensate for difference in height when high tide occurs at different times at either end of a long narrow stretch of water; e.g. Menai Straits, Pentland Firth.

hydrocarbon A compound of hydrogen and carbon: liquid, solid or gaseous. Many are familiar in commercial use; e.g. coal-gas, petroleum.

hydrogenic deposits A chemically derived sedimentary rock precipitated directly from water, esp. of calcium ions; e.g. calcite. Ct. EVAPORITE.

hydrogen-ion concentration A measure of the acidity of a solution, since the properties of acids are the result of the presence of hydrogen ions, expressed in terms of the negative index of the logarithm of the concentration. In soil science, given as pH VALUE.

hydrograph A graph representing stream DISCHARGE measured at a given point as a function of time.

hydrography The description, surveying and charting of oceans, seas and coastlines, together with the study of tides, currents and winds, mainly and essentially from the point of view of navigation; e.g. Hydrographer to the Admiralty, Hydrographic Office. Ct. HYDROLOGY.

hydrolaccolith A 'blister', approx. 6 m high raised by hydrostatic pressure in TUNDRA or PERIGLACIAL areas, when autumn freezing may trap a layer of water between the frozen surface and underlying PERMAFROST. This may later collapse, leaving a hollow which may become a small pond. Ct. PINGO.

hydrologic(al) cycle The endless interchange of water between the sea, air and land: EVAPORATION from the oceans, movement of WATER-VAPOUR, CONDENSATION, PRECIPITATION, then some surface RUNOFF to the oceans, movement by GROUNDWATER, some EVAPORATION, and some EVAPOTRANSPIRATION. 'All the rivers run into the sea; yet the sea is not full; unto the place from whence the rivers come, thither they return again.' *Ecclesiastes*, i, 7. If 100 units represent avge annual global precipitation (actually about 860 mm), then 77 units fall over the ocean, 23 over land. 84 units enter the atmosphere from the ocean by evaporation, most to be stored there, with 7 units transferred horizontally to atmosphere over the land; of precipitation over the land, 16 units are re-evaporated or transpired, 7 units runoff to the ocean, though part is temporarily stored as ground water and in rivers, marshes and lakes. Nearly 98% of all free water on the globe is held in the oceans ($1\cdot3 \times 10^{18}$ m^3).

hydrology The scientific study of water, esp. (by usage) inland water, both surface and underground, including its properties, phenomena, distribution, movement and utilization. In its widest sense, also includes POTAMOLOGY, LIMNOLOGY and CRYOLOGY. Ct. HYDROGRAPHY.

hydrolysis A form of chemical weathering, a reaction involving water, strictly one in which a salt combines with water to form an acid and a base. E.g. the breakdown of feldspar, whereby colloidal silica is removed in solution and clays are formed.

hydrometeor An ensemble of liquid in the atmosphere: RAIN, DRIZZLE, SNOW, SLEET.

hydromorphic A soil-type syn. with GLEI.

hydrophyte A plant which lives in water or saturated soil, such as floating and submerged acquatic plants.

hydrosphere The waters of the surface of the Earth, complementing the ATMOSPHERE and the LITHOSPHERE, though the h and the atmosphere overlap by way of the HYDROLOGICAL CYCLE.

hydrostatic equation In meteorology, the basic relation in the atmosphere between density, pressure, gravity and altitude. This e can be solved to produce a *barometer-height formula*, relating barometric pressure to temperature and altitude.

hydrothermal App. to geological processes involving heated or superheated water, which can: (i) make strong solutions of silicates, carbonates, etc.; (ii) produce alteration processes in minerals, e.g. kaolinization, change of olivine into SERPENTINE; and (iii) cause deposition of minerals in veins (cf. PNEUMATOLYSIS) and around mineral springs and GEYSERS; e.g. SINTER, TRAVERTINE.

hyetograph (i) A type of self-recording RAIN-GAUGE, with a float attached to an inked pen which indicates the amount of water in the gauge; sometimes called a 'tipping-bucket' gauge. (ii) A columnar diagram which shows max., min. and mean rainfall for each month, also standard deviation and probable deviation from the accepted mean. A chart record of the rate of rainfall is a *hyetogram*.

hygrograph An instrument which makes a continuous record of changes of the RELATIVE HUMIDITY of the atmosphere by transmitting the record from a hair HYGROMETER to an inked pen on a rotating drum.

hygrometer An instrument which measures the RELATIVE HUMIDITY of the air. Several types are available, incl. various PSYCHROMETERS. Other types use a human hair or a lithium-chloride strip. The former lengthens and shortens, and the resistance of the latter varies, these changes being amplified and registered.

hygrophyte A plant which lives in an environment with plentiful water supply, as in the Tropical Rain-forests, usually with large broad leaves. Hence adj. *hygrophilous*, *hygrophytic*.

hygroscopic App. to a substance with a marked affinity for water; e.g. particles of salt in the atmosphere on which condensation may take place.

hypabyssal rock An intrusion of igneous **r** along cracks and lines of weakness, solidified as a DYKE or a SILL; intermediate in physical form between deep-seated PLUTONIC **r** and extrusive VOLCANIC **r**. Some types include well-formed crystals of different minerals embedded in a ground-mass of microcrystalline glassy material; e.g. porphyries.

hypogene Rocks formed within the Earth, including plutonic and metamorphic, excluding sedimentary and those formed by the solidification of volcanic material on the surface. Sometimes used as an adj. (*hypogenic*), in ct. to external or *epigenic* or supergenic features, but has not been gen. accepted.

hypsithermal Used in USA to denote the interval of mild climate which followed the last major phase of the Quaternary glaciation. Its temporal span has been defined by RADIOCARBON DATING as between 9000 and 3000 B.P., though this varies with latitude (i.e. *time transgressive*). In W. Europe the **h** period corresponds broadly to the BOREAL and ATLANTIC climatic phases.

hypsographic (or hypsometric) curve A graph used to indicate the proportion of the area of the surface of an island, continent or globe at various elevations above, or depths below, a given datum (usually sea-level). The vertical axis indicates height, horizontal axis areas of land. A percentage **h c** uses percentages of the total area for the horizontal axis, instead of absolute areas. [*f*]

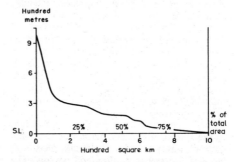

hypsometer An instrument which, with the aid of tables, enables altitude to be determined by measuring the temperature at which water boils at that height, since this varies with pressure of the atmosphere. Boiling-point of water at 1013 mb. is 100°C, at 909 mb. 97°C, at 728 mb. 90°C.

hythergraph A climatic diagram plotting temperature against humidity or precipitation. It summarizes broad climatic differences in relation to human activity, esp. settlement. See CLIMOGRAPH.

ice The solid form of water, formed by: (i) freezing; (ii) compaction of snow; and (iii) condensation (SUBLIMATION) of water-vapour directly into crystals. Density 0·9166; i.e. less than water, and thus ice floats.

ice age A geological period of widespread glacial activity, when **i**-sheets covered large parts of the continents. In pop. usage, the 'I A' or 'Great I A' signifies the latest (Pleistocene or Quaternary) glaciation. There is considerable argument as to its date of origin, with a disputed range of more than a million years. The 'LONG TIME-SCALE' puts this at about 1·8–2·0 million years ago, the 'SHORT TIME-SCALE' at *c.* 600 000 years ago. At least 3 others have occurred, Huronian in Canada, the widespread Precambrian and early Cambrian, and late Carboniferous which mainly affected the S. continents. The glaciation of the Pleistocene took place in distinct glacial phases, separated by interglacial periods of warmer conditions. The elucidation of these stages attracts much research. As these ice-sheets at their max. covered about 47 million km², cfd. the present 15 million km², their impress on the landscape through erosion and deposition was considerable.

[*f* MINDEL]

ice-barrier The edge of the Antarctic ice-sheet; e.g. the Ross I.-B.

iceberg A large mass of ice which has broken off the tongue of a TIDAL GLACIER or the edge of an ice-barrier, then floated away under the influence of currents and winds. The ratio of the volume of submerged ice to that above water depends on the rel. density of ice and sea-water. It was thought that this was about 6:1, some as much as 9:1, but recent opinion believes that it is more like 4 or 3:1. The N. hemisphere is are *castellated*, towering high above the water; 10–15 000 each year are formed, mostly from Greenland, carried S. by the E. Greenland and Labrador Currents. They are rarely found in the Pacific Ocean, as the Bering Strait is shallow and narrow. The S. hemisphere is are mainly *tabular*, large floating islands of ice up to 80–100 km in length. They break off from the edge of the Antarctic ice-barrier, and float N. as far as 60°S. in the Pacific. Is move at about 6 km per day, depending on currents and strength and direction of the wind.

ice blink The glare from the underside of a cloud-layer, produced by reflection from an i surface below, as in an ice-sheet or pack-ice. This may produce eye irritation and even snow-blindness.

ice-cap A permanent mass of ice covering plateaus and high-latitude islands, smaller than an ice-sheet; e.g. Spitsbergen; Novaya Zemlya; Franz Josef Land; part of Iceland (where 37 named examples have been distinguished); Jostedalsbre on the Norwegian plateau.

ice-contact slope A steep-edged fluvio-glacial deposit, believed to be formed against an ice margin and subsequently only slightly modified by slumping of the deposit. The risers of kame terraces may sometimes be interpreted as **i-c ss**.

ice-cored moraine Debris covered ice-ridges and hummocks, usually marginal to existing glaciers; the ice-cores are detected by geophysical soundings. The ice-cores develop (i) as former parts of the glacier protected by an increasing thickness of ablation till (see DIFFERENTIAL ABLATION); (ii) as large snow-banks, collected against the glacier snout and buried by sliding debris. **I-c ms** are often abnormally large in relation to adjacent glaciers. Moraine ridges, both lateral and medial, on existing glaciers are also predominantly ice-cored, and may be classified as **i-c ms**.

ice-dam A dam on a river caused by blocks of **i**, which may cause widespread flooding in spring and early summer, as along the Canadian and Siberian rivers; also called an *ice-jam*. During the Quaternary glaciation, **i-d**s ponded up water to form PROGLACIAL LAKES. Some present lakes result from **i**-damming; e.g. Märjelen See, Switzerland.

ice-edge The boundary between open water and a mass of floating sea-**i**.

icefall A confused labyrinth of deep clefts and **i**-pinnacles (SÉRACS), formed by the intersection of CREVASSES where a glacier steepens; e.g. glaciers moving down from the flanks of Mont Blanc to the Chamonix valley, descending over 1800 m in only 3 km. The huge Khumbu **i** was a major obstacle in the approach to the S. Col of Everest. See pl. 35.

ice-field Strictly a large continuous area of PACK-I or sea-**i**, by both USA and British definition. Gen. used more widely; e.g. Columbia **I-f** in Jasper National Park, Alberta, Canada, 340 km^2 in area.

ice-floe See FLOE.

ice-fog A fog consisting of minute **i**-crystals suspended in the air, in conditions of calm air and low temperature. If the sun is shining above the fog-layer, the effect is dazzling, and without dark goggles the risk of snow-blindness is great.

ice-front A cliff of **i**, the seaward face of a floating mass of **i**, such as an **i**-shelf or tidal glacier.

ice-jam (i) A mass of broken **i**-fragments, esp. during spring-melting, jammed in a narrow channel, causing flooding. (ii) A mound of broken **i** piled up along the shore of a lake, notably Great Lakes of N. America.

Icelandic 'low' The mean sub-polar atmospheric l pressure area in N. Atlantic Ocean between Iceland and Greenland, most marked in winter. It is not a very intense stationary '**l**', but an area of rapidly moving individual 'lows', interrupted by occas. periods of higher pressure. Cf. ALEUTIAN LOW.

ice-rind A thin crust of sea-**i**, formed by the freezing of snow-sludge on the surface of a calm sea.

ice segregation The formation within the soil or regolith of bands and layers of clear ice, resulting in uplift of soil material (FROST HEAVING). The ice results not simply from the freezing of interstitial soil water, but is fed by upward movement from ground water.

Plate 35. The Pigne d'Arolla ICE-FALL, Switzerland. The ice-fall drops about 600 m from the snow and ice-fields marginal to the rounded peak of the Pigne (top left of photograph). The surface of the ice-fall is broken by many CREVASSES and SÉRACS. (*R. J. Small*)

Some calibres of regolith (silts and to a lesser extent clays) are very susceptible to the formation of **i ss**.

ice-sheet Any large continuous area of land-i, incl. i-caps (e.g. Greenland I-cap); the *I Glossary* also includes floating sea-i, such as i-shelves. It is preferable to restrict i-cap to smaller masses, and use **i-s** for: (i) 2 very large existing masses: Antarctica (13 million km^2) and Greenland (1·6 million km^2); and (ii) the i-ss of the Quaternary glaciation, covering at their max. large areas of N. America and N.W. Europe.

ice-shelf A large floating i-sheet attached to the coastline; e.g. Ross I. S.

ice streaming The occurrence of 'lines' of high velocity within slow moving or virtually stagnant ice-sheets. D. E. Sugden inferred **i s** from the morphology of large glacial troughs in the Cairngorms, formed when the massif was covered by an ice-sheet which 'protected' many summit tors and plateau surfaces. **I s** has been observed in the Greenland and Antarctica ice-sheets.

iceway A linear hollow eroded by the movement of lobes from the edge of a lowland ice-sheet, esp. where the ice was compressed between higher land. In the Mersey-Wirral area of N.W. England, 4 parallel **is**, eroded by lobes from the Irish Sea ice-sheet, are now occupied by the Dee estuary, the mid-Wirral depression, the Mersey estuary and the valley of the Ditton Brook in S. Lancashire.

ice-wedge A w-shaped mass of **i** in the ground, tapering downwards. If ground freezes under PERIGLACIAL conditions to low temperatures (*c.* −21°C), espec. where covered with a thick mantle of loose material, cracks will develop. In summer these fill with meltwater, which freezes to form an **i-w**. Each winter the cracks and **ws** may enlarge and ultimately they may penetrate to a depth of 10 m. Repeated melting and re-freezing not only causes an incr. in size, but helps to shatter surrounding material. Sites of former **ws** can be recognized by curving infilling masses of material outlining their shapes. **I ws** commonly form polygon patterns, hence **i w** *polygons*. [*f, opposite*]

ichor Derived from the ethereal fluid believed to fill the veins of the Gk. gods, and used to denote energetic magmatic fluid which penetrates and circulates through existing rocks, causing *granitization*. See also MIGMATITE and GRANITE.

iconic model A **m** in which some aspect of the physical world is presented, showing the same properties, though reduced to scale; these **ms** may be *static* or *working*. An **i m** cts. with an *analogue* or *simulation m*, in which actual properties are represented by different, though analogous, properties.

icing The accumulation of a thickness of clear ice on exposed objects, caused by freezing of SUPER-COOLED droplets in a cloud, as on leading edges of the wings of aircraft, or by rain falling on an aircraft flying in sub-freezing temperatures, though not in a cloud; ice can also form on branches, telegraph wires (which may be brought down by the weight), on the surface of a road, and on railway lines and points; the last is espec. difficult if **i** occurs on the conductor rail of an electrified line. Cf. GLAZED FROST, RIME.

igneous rock A **r** formed by the solidification of molten MAGMA. Its character depends on: (i) *chemical composition* of the magma, whether: (*a*) silica-rich (granite, rhyolite, obsidian); (*b*) BASIC (gabbro, dolerite, basalt); or (*c*) INTERMEDIATE (diorite, andesite); (ii) *mode of cooling*, whether: (*a*) at depth within the crust, slowly and large-crystalled, hence intrusive or PLUTONIC (granite, diorite, gabbro, peridotite); (*b*) on the surface, rapidly, and fine-crystalled or glassy, hence extrusive or VOLCANIC (rhyolite, obsidian, andesite, basalt); or (*c*) intermediately, hence HYPABYSSAL (granophyre, porphyries, dolerite). Some **i rs** may be formed from consolidated fragments of pre-existing ones (BRECCIA, TUFF).

IGY The most recent International Geophysical Year, from 1 July 1957 to 3 Dec. 1958.

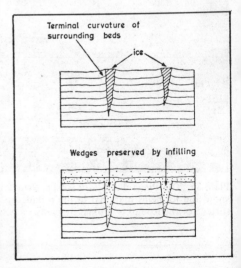

Illinoian In America a major glacial advance, corresponding to the RISS in the Alps. Its deposits consist of clay-till, widely spread over Illinois, with extensions into Indiana, Wisconsin and E. Iowa. Much is covered with LOESS. In N.W. Illinois and S. Wisconsin the drift is stony, with areas of KAME-MORAINE, and much eroded remnants of TERMINAL MORAINE.

illumination, circle of, The GREAT CIRCLE which separates light and darkness, hence day and night, on the surface of the globe.

illuminated relief The representation of r on a map to give a 3-dimensional effect, with either oblique or vertical apparent lighting.

illuviation The redeposition at a lower depth (usually in the B-HORIZON), sometimes forming a HARDPAN, of material removed from the upper soil horizon by ELUVIATION, including colloids, salts and mineral particles.
[*f* SOIL PROFILE]

imbricate(d) structure (i) Wedges of rock piled up on one another as a result of an extreme form of OVERTHRUST; rocks are thrust over in 'slices', without folding, each separated by a THRUST-PLANE, with which each 'slice' makes an oblique angle. E.g. N.W. Highlands of Scotland, where Cambrian quartzites and limestones are sliced between the underlying 'Sole' thrust-plane and overlying Glencoul thrust-plane. Can be seen well near Loch Glencoul, in Sutherland. The 'slices' are inclined to the S.E., the direction from which pressure came. I s is known in Germ. as *Schuppenstruktur*. [*f*]

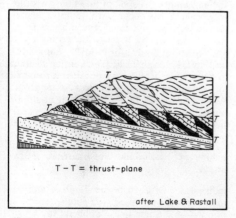

(ii) In conglomerates and pebble beds, where a strong current in the river responsible for deposition caused the long axes of the pebbles to 'lean' downstream, to be cemented in that position.

immature soil See AZONAL S.

immigrant Plants and animals introd. into a totally new environment; see EXOTIC.

impermeable The quality of not allowing the passage or transmission of water through a rock, because it is *non-porous* (with 2 exceptions), or IMPERVIOUS, or both. The exceptions are clay (which is porous, but becomes i because of its fine pores, which when wet are filled with water held by surface tension, so sealing the rock against passage of water), and unjointed chalk.

impervious App. to a rock which is not pervious; i.e. it does not allow the passage of water because it has no joint-planes, cracks or fissures; e.g. slate, shale, gabbro, massive granite.

in-and-out channel A small and discontinous channel or scar cut into a hill-side by melt-water flowing around an ice projection. Formerly escarpment coombes in the Chilterns were interpreted as i cs, but these are now regarded as due to SPRING SAPPING or NIVEO-FLUVIAL processes.

inch (i) (from Gaelic *innis*). A small rocky island; e.g. I. Marnock, in Sound of Bute, W. Scotland. (ii) A flat area of alluvium in a valley, which at times may be flooded; e.g. Inches of Perth in the Tay valley.

incised meander The down-cutting of a river due to REJUVENATION which maintains the pattern of a m at a progressively lower level below the gen. surface; e.g. rivers Wye, Dee, Meuse. A distinction is made between IN-TRENCHED and INGROWN MS though most i rivers show both at different stages of their courses. See pl. 36. [*f*]

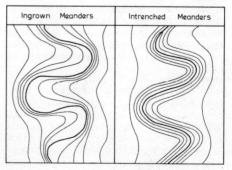

inclination-dip The angle of d of a VEIN, FAULT or STRATUM, measured from the horizontal.

incompetent bed A rock-stratum so weak that it is complexly distorted by plastic flow

158 INCOMPETENT

Plate 36. The INCISED MEANDER of the Afon Rheidol, Cardiganshire, Wales. (*Eric Kay*)

and deformation, rather than by bending, during folding movements. Such a **b** may thicken at the HINGE of a fold, and thin out in its limb; e.g. Upper Jurassic rocks along the Dorset coast, folded to form N. limb of the Ringstead Anticline. A common result in an **i b** is the development of slaty CLEAVAGE. Ct. COMPETENT BED.

inconsequent drainage A **d** pattern which has not developed in a systematic relationship with, and is not consequent upon, the present structure; it is either ANTECEDENT or SUPERIMPOSED, and the river is DISCORDANT; ct. INSEQUENT, where the drainage pattern appears to be independent of surface features.

[*f* ACCORDANT DRAINAGE]

indeterminacy The principle that since landforms are the product of a number of interrelated factors, which adjust to each other in a complex fashion, the manner of this mutual adjustment can vary considerably (i.e. there is no simple 'deterministic' or 'cause-effect' relationship). To take an hypothetical example, an increase in stream discharge can, in 9 cases out of 10, be accommodated by an increase in channel width; in the 10th case it may cause bed scour, and thus increased depth. In the words of Leopold, Wolman and Miller 'the number of cases... in the landscape is so large that one may expect to see a spectrum of results..., a statistical distribution of results. The central tendency in this distribution may ... be possible of forecast, but it may not be possible to specify in any individual case the manner in which the adjustment took place.'

index fossil A **f** characteristic of a special zone in the rocks, named after it.

Indian summer A spell of fine sunny weather which seems to occur with some regularity during October and early November in USA, said to be used by Indians for storing crops and making ready for winter. Sometimes used in UK for an earlier spell of fine weather in September, which may occur as a result of an extension of the Azores anticyclone. According to C. E. P. Brooks, it could be recognized in 43 out of 52 years, the peak being 10 Sept., but there is no real statistical evidence.

indifferent equilibrium See NEUTRAL INSTABILITY.

indigenous Orig. in a specif. place and remaining *in situ*, used of plants, animals, population. Sometimes applied to rocks, minerals and ores orig. *in situ*, not transported from elsewhere.

induration Hardening, used of rocks so transformed by heat, pressure or cementation, and of soil HORIZONS hardened by chemical action to form a HARDPAN. Ct. DIAGENESIS.

inface The steep face of a CUESTA, used esp. in USA.

infiltration The movement of water into the soil from the surface; the max. is the *i rate*, which depends on compactness or friability of surface, vegetation layer, POROSITY, PERMEABILITY and existing moisture content. A dry porous soil can absorb water at an initial rate of 1000 mm per hour, slowing as the upper layers become saturated. Ultimately it reaches FIELD CAPACITY. A vegetation layer of open texture absorbs the initial impact of rain and allows steady **i**.

infiltration capacity The *constant rate* at which water can percolate into a soil. When rain occurs, at first the entry of water into the soil will be rapid. However, as voids are filled infiltration will slow down quite rapidly until **i c** is attained. When rain falls at a rate in excess of **i c**, OVERLAND FLOW is generated.

influent stream A river whose channel lies above the water-table; e.g. a **s** flowing in a chalk valley, which only maintains its flow because its bed is lined with fine silt, forming an impermeable layer. Some rivers may lose part or all of their water down crevices or joints in limestone and chalk; e.g. R. Mole near Dorking; Mymmshall Brook in Hertfordshire.

ingrown meander An INCISED M where valley-sides are asymmetrical, the result of considerable lateral erosion, so that one side presents steep, even undercut banks, while the other is gentle; e.g. lower Seine, which flows 90 m below the flanking chalk plateaus; lower Ribble, Lancashire. Ct. INTRENCHED M.

[*f* INCISED MEANDER]

initial landform A **l** where the orig. features, formed when uplifted by various TECTONIC forces, have been only slightly modified, as opposed to SEQUENTIAL LS.

inland basin See BASIN.

inland drainage See INTERNAL D.

inland sea An isolated sheet of water of large extent, with no link with the open **s**; e.g. Caspian, Aral, Dead S.

inlet A narrow opening of the sea into the land, of a lake into its shores.

inlier An exposed mass of rocks wholly surrounded by younger ones, commonly occurring when the crest of an anticline or an elongated dome is removed by erosion; e.g. a small **i** of Jurassic rocks in the centre of the Weald; an **i** of Silurian rocks is at Woolhope, Hereford and Worcester county. Ct. OUTLIER, FENSTER. [*f*]

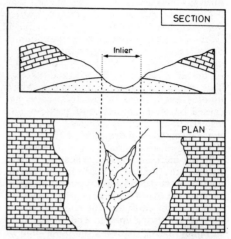

inner lead A stretch of calm, sheltered water between the coast of a mainland and a string of parallel offshore islands; e.g. skerry-guard of Norway; along the coast of British Columbia (the line of Georgia, Queen Charlotte, Hecate, Clarence, Chatham Straits).

inselberg (Germ.) A prominent residual rock-mass, isolated by circumdenudation, rising from a distinctive near-level erosion surface. It may have developed from the extension of PEDIMENTS into upland areas, in time reduced to PEDIPLAINS. The mountain-front on each side recedes through SLOPE RETREAT, ultimately reducing the intervening upland to an **i** (see ZONAL INSELBERG). Some writers suggest that certain **is** result from the exposure and reshaping of 'cores' of resistant rock revealed by removal of a deep layer of weathered REGOLITH; these are sometimes known as BORNHARDTS or SHIELD INSELBERGS. An **i** is a characteristic landform of semi-arid or SAVANNA landscapes in a late stage, though they occur in areas with different contemporary climates. See pl. 7, 27, 38.

insequent drainage A **d** pattern which has developed on the present surface, but which appears to reflect chance development, and is not consequent upon it nor controlled by features of the surface; e.g. DENDRITIC DRAINAGE.

insolation The energy emitted by the sun which reaches the surface of the earth. The sun, a mass of intensely hot gases, with a temperature at the surface estim. to be 6000 K, and at the centre 45 000 000 K, pours out electromagnetic energy in the form of waves: very short wave-length X-rays, gamma rays and ultraviolet rays (wavelength 0·01–0·4 microns); visible light-rays (wavelength 0·4–0·7 microns); and short-wave infra-red waves (wavelength less than 4·0 microns). The Earth receives only about one 2-thousand-millionth of the total **i** poured out by the sun, but this is vital; the amount received at the outer limit of the atmosphere is the SOLAR CONSTANT. This enters the Earth's atmosphere, which absorbs part; some is lost by 'scattering' caused by air molecules, dust particles and water-vapour; part is reflected back into space by clouds and dust. About 47% reaches the earth, of which nearly a ¼ is immediately reflected back into space, depending on the nature of the surface (see ALBEDO). The remainder is converted into long-wave infra-red radiation (greater than 4·0 microns), which heats the surface of the earth and (by conduction) the layer of air resting upon it. The total amount of radiant energy reaching the outer limit of the atmosphere depends on (i) the SOLAR CONSTANT; (ii) the duration of daylight; (iii) the sun's altitude; (iv) the actual distance between Earth and sun (less in January than July, so that energy input in the former is 7% greater). Over the year locations above the Equator receive about 2·5 times as much energy as over the Poles, though at the latter the energy input varies from 1000 langleys in 24 hours (summer) to nil (winter), while the equatorial input is much more uniform throughout (about 800 langleys per 24 hours).

instability, atmospheric The physical condition whereby a parcel of air is warmer, and therefore less dense, than air above, causing bodily rising and expansion. Vertical movement and adiabatic cooling of moist air, the result of **i**, is intimately associated with atmospheric disturbances, and the most potent cause of precipitation.

insular App. to an island; hence **i** location, I CLIMATE.

insular climate A climatic régime experienced by islands and coastal areas, characterized by equable conditions, with low seasonal range of temperature; e.g. Ocean I., 1°S. in the W. Pacific, has an annual range of only 0·01°C, and Jaluit, in the Marshall Is. (6°N.), of 0·3°C. Even in middle latitudes, such **i** cs have rel. small ranges; e.g. Scilly I., Jan. 7°C (44°F), July 16°C (61°F), (range 9°C, 17°F), and frost is almost unknown.

intensify An incr. in the amount of some physical quantity. In meteorology, used of the gen. strength of air-flow around a low pressure system; i.e. a depression intensifies.

intensity of rainfall The rate at which rain falls, ranging from fine drizzle to torrential downpour. The hourly **i of r** is obtained from $\frac{\text{total rainfall}}{\text{no. of hours of rain}}$. The less specif. daily **i of r** is $\frac{\text{total rainfall}}{\text{no. of RAINDAYS}}$. Hourly **i of r** at Boston, USA, is 9·1 mm; at Cherrapunji in N.E. India is 106·0 mm. Réunion on 24 Sept. 1958 experienced a total of 1168 mm in 24 hours.

interception The capture of rain-drops by plant cover, which prevents direct contact with the soil. If rain is prolonged, the retaining capacity of leaves will be exceeded and water will drip to the ground (*throughfall*); some will trickle along branches and down stems or trunk (*stemflow*). Some is retained on the leaves and is evaporated.

interface (i) A zone of sharp change between 2 different substances, or between 2 different characteristics of the same substance; e.g. in a crystal. (ii) The vital line of contact between the atmosphere, biosphere and lithosphere, involving exchanges of matter and energy. At this **i** is situated the BIOSPHERE.

interflow Syn. with THROUGHFLOW.

interfluve The area of land between 2 rivers in the same drainage system. See pl. 59.

interglacial period A **p** of time between 2 advances of continental ice-sheets during the Quaternary glaciation; commonly 'interglacial' alone is used as a noun. Between these advances climate was milder and ice-sheets shrank; from evidence of plant remains (rhododendron, box) some **i** ps were warmer than at present. An early **i p** (Günz-Mindel in the Alps) lasted about 60–80 000 years, the Great I. (Mindel-Riss) about 190–240 000 years, and the Riss-Würm about 60 000 years, on the 'SHORT TIME-SCALE'; see PLEISTOCENE. Corresponding periods in N. America are known as the *Aftonian*, *Yarmouthian* and *Sangamonian* resp., in UK as *Cromerian*, *Hoxnian* and *Ipswichian* resp. Recent research has deduced at least 6 stages prior to the Cromerian, 3 predominantly warm (possibly is), 3 cold (glacials). [*f* MINDEL]

intergranular translation The movement of ice-grains within a mass of glacier-ice, so that they slide over each other (as in a mass of lead shot) and thus contribute to overall glacier movement; one contributory category to the gravity flow of a glacier.

intergranular void The pore spaces in rocks of open texture, coarse-grained constituents and loose cementation (rock of *primary permeability*); e.g. sandstone, oolitic limestone.

interior drainage See INTERNAL D.

interlocking spur A projecting s in a 'young' river valley, where a stream tends to follow a winding course round obstacles, each bend separated by an **i** s; viewed upstream, these ss 'interlock', 'overlap' or 'interdigitate' with each other. [*f, opposite*]

intermediate layer Formerly given to, and still occas. used for, the SIMA **l** in the earth's crust.

intermediate rocks Igneous rs classified according to chemical composition as having less than 10% free quartz, but with some feldspar (either plagioclase, alkali or both); e.g. trachyte, andesite (EXTRUSIVE) diorite, syenite (INTRUSIVE).

intermittent spring A s whose discharge varies intermittently, the result of fluctuations in precipitation and therefore of height of the WATER-TABLE. Sometimes the result of a

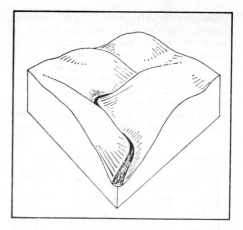

intermittent saturation, zone of A layer, just below the surface soil, which may contain GROUND-WATER after long-continued rain (VADOSE WATER), but dries out after a short period of drought.

SIPHON in a cave-system, which opens out to the surface at the source of the spring; e.g. Ebbing and Flowing Well, Settle, Yorkshire. [*f*]

intermont, intermontane. Between the mountains, as of a high plateau between ranges; e.g. Bolivian and Peruvian Plateaus between E. and W. ranges of the Andes; Great Basin of Utah between the Sierra Nevada and the Wasatch; Tarim Basin between the Altyn Tagh and Tien Shan.

internal drainage A d system whose waters do not reach the ocean. Also known as *inland* or *aretic d*; e.g. R. Jordan and Dead Sea; R. Volga and Caspian Sea; Amu and Syr Darya and Aral Sea; Tarim R. in Lop Nor; many temporary streams converging on L. Eyre, L. Chad.

International Date-Line A line following approx. 180° meridian (with deviations to avoid land areas), where, to compensate for the accumulated time-change of 1 hour in each 15° longitude time-zone, a ship or aircraft moving W. omits a day, while one travelling E. repeats a day. Thus in an aircraft flying W. (e.g. San Francisco to Tokyo) local time at 1100 Monday becomes 1100 Tuesday; when flying E. 1100 Monday becomes 1100 Sunday.
[*f*]

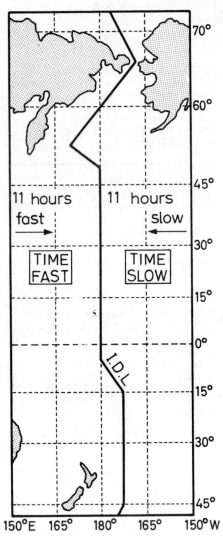

International Million Map A m series on scale 1:1 million (15·782 miles to 1 in.), proposed in 1891 by A. Penck, the principle accepted in 1909. Each sheet covers 4° of latitude and 6° of longitude (except in high latitudes, where it is 4° × 12°), reckoned from Greenwich meridian and the Equator, and is known by an index letter (alphabetically from the Equator) and a number reckoned E. from Greenwich, prefixed by N. or S. according to hemisphere. Drawn on a modified POLYCONIC PROJECTION. The headquarters of the IM Bureau were established at Southampton. Progress was slow and sporadic; by 1931 only 89 sheets had been produced, though others were published by private bodies (e.g. American Geographical Society were responsible for 16 maps of S. America), and the US Army Map Service and the British GSGS published sheets, esp. during 1939–45. After post-war discussions, the functions of the Central Bureau were transferred to the UN Cartographic Office, which gives a summary of progress in its annual report. Modifications have been made to specifications of the **m**.

interpluvial periods Stages in low latitudes which correspond to, though not necessarily synchronous with, interglacial periods of higher latitudes. The correspondence between *pluvials* and *glacials* may be only approx.

interruption of, interrupted, map projection A **p** can be drawn with several standard meridians instead of one; each is centered over a continent, with true scale along it, and a section of the **p** in a lobate form is plotted. Gaps are left in the oceans between each re-centred section to accommodate errors. This reduces overall distortion of shape, esp. towards the margins, and linear scale discrepancy, while retaining equal area property. I **ps** are used for world distribution maps. E.g. MOLLWEIDE, SANSON-FLAMSTEED, GOODE'S I HOMOLOSINE PS, Several **i ps** have been designed for *The Times Atlas*; e.g. *Bartholomew's Regional P* (conformal and near equal area, to show continental relationships), and *Lotus P* (to show oceans, currents, etc.).
[*f, page 163*]

interstadial Lit. between stages, partic. with respect to periods between 2 glacial stages or 2 phases of glacial retreat; commonly used as a noun (an '**i**'). Of a more minor character, or less marked, or of briefer duration, than an INTERGLACIAL. Thus an **i** oscillation occurred in each earlier glaciation of Quaternary times and at least 2 in the WÜRM.

Intertropical Convergence Zone A broad trough of low pressure, more or less along the axis of the Equator, moving slightly N. and S. with the seasons, towards which the Trade Winds blow, i.e. towards which Tropical Maritime (*Tm*) air-masses converge. Usually

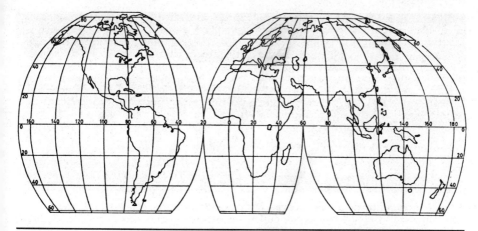

abbrev. to *ITCZ*. This boundary z was formerly known as the Intertropical Front (*ITF*), but the phenomenon is zonal rather than frontal in character, esp. over oceans, where the *ITCZ* is only weakly defined. The air-masses may be virtually stagnant, i.e. DOLDRUMS, and winds are gen. light or variable. The air is unstable, and convectional rainfall common. In this region shallow, slow-moving 'LOWS' develop; sometimes these move poleward out of the z, intensify and develop into intense tropical storms; see CYCLONE. Over a continental area, the front may be more sharply defined; temperature differences in the converging air-streams are small, but humidity may differ markedly; e.g. in W. Africa a dry continental air-mass from the Sahara may come into contact with a humid one from the Gulf of Guinea, causing a distinct front to develop.

intrazonal soil A s type not restricted to a latitudinal zone, but the result of partic. conditions or constituents. It includes; (i) *peat ss* (fen-peat, bog-peat, dry-peat, meadow-soils); (ii) *saline ss* (SOLONCHAK, SOLONETZ); (iii) *calcareous ss* (RENDZINA, TERRA ROSSA). These correspond resp. to (i) *hydromorphic* (continued water in the soil); (ii) *halomorphic* (the presence of soluble salts); (iii) *calcimorphic* (a limestone parent).

intrenched meander An INCISED M where valley sides are steep and symmetrical, where vertical erosion has been dominant; e.g. R. Wear near Durham. Ct. INGROWN M.
[*f* INCISED MEANDER]

intrusion, vb. to intrude, adj. **intrusive** (i) Penetration or injection of molten rock (MAGMA) into existing rocks, usually along a line of weakness. (ii) The particular type of solidified igneous rock mass thus formed; see BATHOLITH, DYKE, LACCOLITH, LOPOLITH, PHACOLITH, SILL. [*f in each case*]

intrusive rock A category of igneous r (as opposed to *extrusive r*), which solidified at depth among pre-existing rs.

invar An alloy of steel containing 36% nickel. It has an extremely low, almost undetectable coefficient of thermal expansion, and is used for surveyors' tapes, esp. for measuring base-lines of triangulation systems.

inversion of temperature An incr. of air t with height, so that warmer air overlies colder, contrary to the normal LAPSE-RATE. It can occur both at the surface of the earth (*ground-i* or *surface-i*) and at high altitudes. (i) A *ground i* may be the result of rapid heat-loss from the ground by radiation at night, esp. when air is calm and sky clear, or by warm air advection over a cold surface. I may be marked in a valley or basin, for radiation of heat into space on upper slopes is rapid, and cold dense air then drains down into the hollows. (ii) A *high altitude i* may occur when a cold air-mass undercuts a warm one (at a COLD FRONT), or a warm air mass over-rides a cold one (ahead of a WARM FRONT), or develops into an OCCLUSION. The subsidence and ADIABATIC heating of a high altitude mass of air may cause an i. This is esp. important in ANTICYCLONES; e.g. the Trade Wind i.

inverted (inversion of) relief The result of prolonged denudation, by which SYNCLINES may ultimately form high ground and ANTICLINES low ground. The crest of an anticline is gen. structurally weak, having been subject to tension, and a stream may develop along it,

Plate 37. INVOLUTION of gravels overlying chalk near Stockbridge, Hampshire, England, which were contorted by CONGELITURBATION during a PLEISTOCENE PERIGLACIAL phase. (*R. J. Small*)

forming a valley with infacing steep slopes; these gradually recede outwards, so reducing the area of the orig. adjacent synclinal valley. In due course the anticlinal valley will lie below the synclinal valley, and the remains of the syncline remain as a ridge or peak; e.g. mountains E. of the Bow R. in Alberta (Mt Eisenhower), the remains of a syncline; Snowdon in N. Wales is the remnant of a SYNCLINORIUM; Great Ridge on Salisbury Plain lies on a synclinal axis. [*f*]

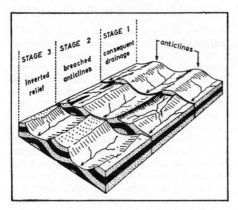

inverted (inversion of) structure When stratified rocks have been acutely overfolded or overthrust, the orig. older (lower) beds now rest upon younger (upper) beds; e.g. the Ardennes, where in places Cambrian rocks lie on top of Devonian; the Meuse valley in Belgium, where Devonian, Carboniferous Limestone and Lower Coal Measures have been thrust over younger coal-bearing Upper Coal Measures; Glacier National Park, Montana, USA, where Precambrian rocks rest on Cretaceous rocks, the result of the Lewis overthrust from the W.

involution (i) The results of frost-heaving and disturbance in upper layers of the soil. Broadly syn. with CRYOTURBATION. (ii) The modification of NAPPES after formation, either by re-folding together of 2 nappes, or penetration of the older by the younger nappe. See pl. 37.

ion An atom or group of atoms in an electrically unbalanced form, with either an excess (*positive i* or *cation*) or deficiency (*negative i* or *anion*) of electrons; e.g. hydrogen atom without its electron is a hydrogen ion. Almost all matter owes its existence to i s, held together by an ionic bond, as in the case of compounds, formed by combinations of elements. Is are partic. important in soil chemistry, since COLLOIDS are able to attract and hold is of calcium, magnesium and potassium (bases), and is of hydrogen. The degree to which the hydrogen is are held by the soil-colloids is the pH VALUE.

ionosphere The portion of the atmosphere above about 60 km characterized by distinctive layers (indicated by letters D to G), which reflect electromagnetic waves (including radio signals) back to earth, and the zone in which the AURORA occurs. In this zone the atmospheric gases are ionized by incoming solar radiation (ultra-violet and X-rays). The i includes the HEAVISIDE-KENNELLY LAYER at 100–120 km and the APPLETON LAYER (*c.* 300 km). The lower level of the i sinks during daytime to about 56 km, rising to about 100–105 km at night. The i is characterized by frequent disturbances, both regular (diurnal, latitudinal, seasonal), and random, assoc. with sunspot variations and geomagnetic storms. This may cause interruption to radio transmission. Constant research is in progress, using radio-echo techniques, rockets and satellites.

Iowan In N. America the earliest substage of the Wisconsin glacial advance, corresponding to Würm I. in the Alps and to the Brandenburg stage of the Weichsel in N. Europe. The drift consists of stony yellowish-brown clay, and has been identified in N. Iowa; a thin deposit, usually less than 6 m.

Ipswichian The last interglacial period in Britain, equivalent to the EEMIAN interglacial of N.W. Europe, and represented typically by former lake sediments at Bobbitshole, Ipswich.

irisation Colour sometimes seen on clouds, mingled or banded, through which the sun may be partially seen. Syn with *iridescence*.

Iron Age The culture period succeeding the BRONZE AGE, in Europe from *c.* 1500 B.C., in Britain from about 6th century B.C., characterized by the use of iron weapons and implements. First main impulse in Britain came in 5th century B.C., with invasion of Celts of the *Halstatt* culture; in 3rd century B.C. came Celts of the more advanced *La Tène* culture in E. and S. England, later in S.W. These introduced better weapons, slings, chariots, burial in round barrows, construction of hill-forts and lake-villages (Meare and Glastonbury, Somerset), and considerable technical improvements in working of iron, wood, pottery. In 1st century B.C. the Belgae introd. further improvements.

ironpan See HARDPAN.

isallobar A line on a map joining places where the same change in atmospheric pressure has taken place during a specif. time. They are plotted to reveal to the forecaster how a pressure system is developing and moving.

isanomal, isanomalous line A line on a map joining places with an equal difference (*anomaly*) from the mean or normal of any climatic element; specif. the difference between mean temperatures for individual stations and the avge of all available stations in the same latitude. Tables of standard distribution of temperature with latitude are available, or the following formula (by J. D. Forbes) can be used: $t = -17·8 + 44·9 \cos^2(\theta - 6·5)°C$, where θ is latitude and t is standard temperature. Anomalies are plotted and is are interpolated. Areas of high positive anomaly are *pleions* or *thermopleions*, of high negative anomaly *meions, antipleions* or *thermomeions*. On a map, areas of positive anomaly are conventionally tinted red, negative ones blue. The map emphasizes winter cold of continental interiors, summer heating and land-masses, and effects of the oceans.

island A piece of land surrounded by water, smaller than a continent; diminutive or poetic *isle, islet*. The largest is are Greenland (2·2 million km^2), New Guinea (820 000 km^2) and Borneo (750 000 km^2).

iso- From Gk. word meaning 'equal', used as a prefix to lines on maps linking points with similar values or quantities; gen. *isopleths, isarithms, isontic lines, isograms* or *isolines*. A formidable terminology has developed, which has by no means received universal acceptance. The following may be regarded as standard: ISALLOBAR, ISANOMAL, *isobar* (pressure), *isobase* (elevation or depression of land), *isobath* (depth), *isobathytherm* (temps. within sea water), *isobront* (places having thunderstorms at the same time), *isocheim* (winter temperature), *isocryme* (coldest period of time), *isoflot* (floristic features), *isogon* or *isogonic line* (magnetic variation) [*f* AGONIC LINE], *isograd* (rocks of same facies), *isohaline* (salinity), *isohel* (sunshine), *isohyet* (rainfall), *isohypse* (contour), *isokinetic* (equal wind speed), *isomer* (the mean monthly rainfall as a percentage of the average annual amount), *isoneph* (cloudiness), *isonif* (amount of snow), *isopach* (thickness of geological strata), *isophane* (departure from parallels of latitude at a constant rate of 1° of lat. to 5° of long.), *isophene* (flowering dates and other botanical and biological occurrences), *isophyte* (height of vegetation), *isopore* (annual change of magnetic variation), *isorad* (radiation from rock), *isoryme* (frost), *isoseismal* (earthquake intensity), *isostade* (significant dates), *isosteric* (density), *isotach* (equal distance travelled in a period of time), *isotherm* (temperature). As many stations as possible are plotted, their values noted, an isopleth interval chosen, and lines drawn through stations with selected values or interpolated proportionally.

isocline, isoclinal folding Tightly packed parallel overfolds, all limbs dipping at approx. the same angle, and in the same direction; e.g. in the S. Uplands of Scotland. If the folding is still more intense the folds are ruptured and become IMBRICATED. [*f*]

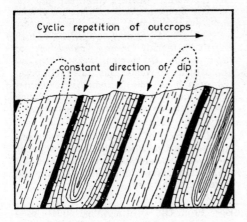

isometric Graph paper with 3 principal axes (vertical and 2 diagonals), dividing the surface into equilateral triangles. Can be used in producing simple pseudo-perspective block-diagrams. [*f*]

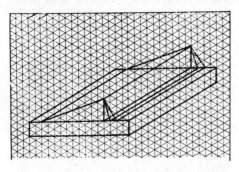

isostasy, adj. **isostatic** (From Gk., meaning 'equal standing'), Signifying a state of equilibrium or balance in the surface crust of the earth, the result of the necessity for equal mass to underlie equal surface area. Where mountains (of less dense SIAL) rise high above the mean surface of the GEOID (i.e. excess mass), their roots of similar sial must penetrate more deeply into the denser SIMA, exercising **i** compensation. Under the oceans deficiency of mass is compensated by higher density of sima. It is as though the continents are lighter 'rafts' of sial 'floating' on dense underlying sima. I involves: (i) vertical adjusting movements of sections of crust; e.g. as an ice-sheet melts, the continent rises; as a mountain mass is eroded, the continent rises; and as a great weight of sediment is deposited on a delta, there is sinking; and (ii) horizontal movement of sub-surface material to balance differences in mass. Cf. PLATE TECTONICS. [*f*]

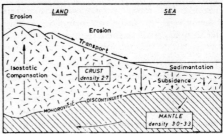

isosteric surface A s of constant density in the atmosphere.

isothermal layer A l of the atmosphere in which temperature remains the same for a considerable height. At one time this applied to the STRATOSPHERE, where it was believed these conditions obtained.

isotope Alt. form of an element, though with virtually identical chemical properties, but a different number of neutrons in its nucleus. The atomic number of the element and its is is the same, but their atomic weights differ. Oxygen has three naturally occurring is, $_8O^{16}$, $_8O^{17}$ and $_8O^{18}$. *Radiogenic is* are formed from the breakdown of a radioactive parent element, of great importance in finding the age of rocks; see LEAD-RATIO and RADIOACTIVE DECAY. The radioactive **i** of carbon, carbon 14 ($_6C^{14}$), can be used for dating organic material less than 50 000 years old (RADIOCARBON DATING).

isthmus Narrow neck of land between two seas or oceans; e.g. I. of Panama.

jade 2 distinct minerals: (i) hard translucent compact *nephrite* (an AMPHIBOLE); (ii) PYROXENE (*jadeite*, a bisilicate of calcium, magnesium, aluminium, etc.), greenish to whitish in colour, used for jewellery and ornaments.

jet A hard, dense, compact, black form of LIGNITE, capable of taking a brilliant polish. Found in UK in Upper Lias, notably in Yorkshire.

jet stream A high-altitude (about 12 000 m, in the TROPOSPHERE) W. air movement, with rel. strong winds concentrated in a narrow belt. In summer its mean speed is about 50–60 knots, in winter it may be stronger (95–120 knots); on occasion 200–250 knots have been recorded. Several streams are distinguished, including a *polar-front j s*, a *polar night j s*, and *arctic j s*, a *sub-tropical j s* [*f below*] between 20° and 30° latitude N. and

S. and an *equatorial j s* over the Pacific Ocean. The subtropical **j s** is rel. constant, the polar-front **j s** extremely variable. The position and nature of **j** ss have important climatological effects, esp. on FRONTOGENESIS (a **j s** may discourage CONVERGENCE and encourage DIVERGENCE), on bursting of the MONSOON, and on formation of a TORNADO. [*f*]

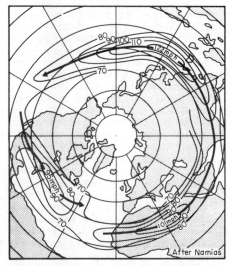

joint A surface of division or fracture in a rock, the result either of tearing under tension, or of shearing under compression, but involving little or no actual movement or displacement of the rocks (ct. FAULT). A **j** is usually transverse to the bedding; e.g. Checkerboard Mesa, Zion Canyon, Utah, where well-marked vertical **j**s cross well-defined bedding-planes, causing a 'checker board' effect. Other types of **j** occur in igneous rocks (e.g. granite) through tensile stresses caused during cooling and solidification, and in sedimentary rocks during consolidation (shrinkage cracks). When **j**s coincide with STRIKE, they are *strike-js*, when with DIP, *dip-js*. If 2 sets of **j**s develop, they are usually at right-angles. A dominant **j** is the *master-j*. J-ing is of value to quarrymen in extracting stone, forming planes of weakness along which forces of weathering and erosion operate; e.g. **j**s define rectangular blocks of a granite TOR. See pl. 54.

[*f* GRIKE]

joint-block removal The detachment of joint-bounded blocks by moving ice. **J-b r** is important in the formation of cirque headwalls, rock steps, and the 'stoss' faces of roches moutonnées (cf. PLUCKING). As the tensile strength of ice is low, it is likely that prior weakening of the rock, either by ice wedging or PRESSURE RELEASE, is essential.

joint-plane The plane of a joint; when 2 sets of vertical joints are at right-angles to the BEDDING-PLANE, the rock is divided into rectangular masses. See pl. 6, 69.

jökull (Icelandic) A small icecap, derived from Iceland; e.g. Vatnajökull, Heinabergs-jökull.

jungle Orig. waste or uncultivated ground (specif. in India); now used pop. for any tropical forest with thick undergrowth.

Jurassic The 2nd (middle) of the geological periods of the Mesozoic era, and the system of rocks laid down extending from about 180 to 135 million years ago. Derived from the type-region in the Jura Mtns N.W. of the Alps. Sediments were laid down in fluctuating, shallow seas, including clays and sands, and coral reefs were widespread, with periods of estuarine and fluviatile deposition. The result is a varied system of rocks, traceable in England in a diagonal band across the country from the Yorkshire coast (S. of the mouth of the Tees) to the Dorset coast.

Purbeck	limestones, clays
Portland	limestone and sandstone
Kimmeridge Clay	dark grey or black shaly clay
Corallian	limestone, with some clay
Oxford Clay	heavy clays
Great Oolite	oolitic limestone, fuller's earth
Inferior Oolite	oolitic limestone, with beds of clay and sandstone
Lias	mainly grey or blue clays

The **J** reveals a rich faunal life (incl. over 100 fossil zones), with ammonites, coral, echinoids, etc., while dinosaurs reached their max. development; first birds appeared in the Upper **J**, and a rich flora of conifers, ferns, etc. Sedimentary iron ores are widespread in the **J**: in UK in Humberside (Lower Lias), Cleveland Hills, S. Lincolnshire, Oxfordshire (marlstone of Middle Lias), in Northants. and S. Lincolnshire (Northampton Sands of Inf. Oolite); in Lorraine (France and Luxembourg) (*minette*).

juvenile water Hot, mineralized w liberated during igneous activity from the depths of the earth, not from the surface or atmosphere. Syn. with *magmatic w*.

kame From an old Scots word signifying an elongated steep ridge (from 'comb'). An undulating mound of bedded sands and gravel, deposited unevenly from melt-water along the face or front of a stagnant, slowly decaying ice-sheet. Its inner face represents a 'mould' of the former ice-contact slope. If deposition took place in a PROGLACIAL LAKE, materials are more clearly bedded and sorted, with a distinct flat top and FORESET BEDS; this is a *k-delta*. 'K-moraines' are found widely on the formerly glaciated lowlands of N. America and N.W. Europe. A belt has been traced in loops from Long I. as far W. as Wisconsin, and in Europe more or less parallel to and S. of the Baltic Heights. They are widespread among the bogs of central Ireland; one line can be traced from near Dublin into Galway, though cut through by the Shannon near Athlone. Ct. ESKER.

kame-and-kettle country A landscape of undulating KAMES, interspersed with shallow hollows or KETTLES; e.g. W. of L. Michigan, USA, in Wisconsin.

kame-terrace A t-like ridge of sand and gravel along the side of a valley, laid down by a melt-water stream occupying the trough between a glacier and its enclosing valley-wall; sometimes called an *ice-contact t*. K-ts resembles shorelines of former glacial lakes, but are less regular and have depressions where blocks of ice lingered before melting. Found commonly in valleys of the Appalachians in New England. [*f*]

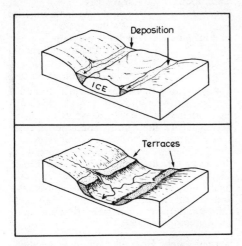

Kansan In USA a major glacial advance corresponding to MINDEL in the Alps. Its deposits consist of extensive, near level TILL, almost entirely clay, with little evidence of stones, sands and gravel. Covers much of Iowa, Nebraska and Missouri, though much dissected by post-glacial valleys. TERMINAL MORAINES at its S. margin and a few RECESSIONAL MORAINES have been distinguished indicating a steady recession of the ice, with no prolonged halt. The till was mainly brought from the Keewatin centre of ice-accumulation.

kantogram A BLOCK-DIAGRAM on which breaks of slope are emphasized.

kaolin A whitish clay produced by complex chemical breakdown of aluminium silicates, largely feldspars and esp. orthoclase, partic. in granite, the result of weathering (HYDROLYSIS), HYDROTHERMAL processes, and ascending gases and vapours from a magma reservoir (PNEUMATOLYSIS); the process is *kaolinization*. Strictly, *kaolinite* is the mineral hydrated silicate of alumina ($2H_2O.Al_2O_3.SiO_2$); k is the clay, incl. also particles of quartz and mica from decomposed granite. Its name originates from the Kauling ('high hill') range in Kiangsi province, China, where it has been worked for many centuries for pottery, hence alt. name 'china-clay'. Most k deposits are found *in situ*, and esp. where PNEUMATOLYSIS is the chief agent are of great thickness; e.g. near St Austell and gen. around Dartmoor and Bodmin Moor in Cornwall. Some k is transported by running water, and redeposited in thin beds as 'residual k' or 'sedimentary k'; e.g. 'ball clay' near Wareham, Dorset, though this is of lower grade. K is washed out from open pits by high-pressure jets of water, allowed to settle and dry, and removed in lumps, the waste forming conical spoil-heaps. K is used for pottery, as a filler and in paper-making.

kar (Germ.) See CIRQUE.

karaburan A strong dust-laden N.E. wind in the Tarim Basin of central Asia; *lit.* '*black buran*', in ct. to '*white buran*' of winter, which blows clouds of snow.

karren (Germ.) A limestone surface intersected with furrows; syn. with LAPIAZ.

Karroo, Karoo (i) The Gondwana series of rocks in S. Africa, ranging from Upper Carboniferous to Lias in age. (ii) A flat-topped terraced land-form, called after such a feature in Republic of S. Africa.

karst (It. *Carso*, Serbo-Croat *Kars*) (i) Proper name of an area of rugged limestone plateaus and ridges near Adriatic coast of Yugoslavia. (ii) Used (with lower-case k) for an area of limestone or dolomite with assoc. phenomena of SINKS, underground drainage and cave-systems, the result of carbonation-solution; e.g. parts of N. Pennines (Craven),

S.W. France (Causses), parts of Indiana and Kentucky. The original k phenomena were described in an area where features are emphasized by long aridity of summer and intensive but short-lived autumn and winter rain. In the Pennines these features are modified by a more constant humidity, and also in places by a cover of till. In tropical areas of high humidity, groundwater is plentiful, and, with its high content of organic acids from the rapid decay of vegetation, is partic. erosive, hence rapid and emphasized dissection occurs along well-marked joints, leading to 'COCKPIT K', '*cone k*' and 'tower k'. E.g. Jamaica, Cuba, S.W. China, N. Vietnam, Malaysia and Indonesia. See also POLYGONAL K.

karstification All processes, chemical and physical, responsible for a KARST landscape.

katabatic wind A cold w blowing downhill. Frequently occurs at night in valleys, caused by gravity flow of dense air chilled by radiation on upper slopes. The effect is emphasized when cold air blows down from an ice-cap; e.g. NEVADOS from Andes of Ecuador, outflowing winds from the Greenland and Antarctic ice-caps. Ct. ANABATIC.

katafront A COLD FRONT in which the warm air-mass is for the most part descending over the cold wedge. Ct. ANAFRONT.

kavir (Iran) Used as syn. with PLAYA.

K-cycle A concept of short-term cycles of development, designed as a framework for the understanding of sequences of soil formation and destruction and their geomorphological implications. Thus, where a number of buried soils is preserved, climatic changes are inferred as follows. During an 'unstable phase' (Ku) denudation occurs on the upper slope and accumulation on the lower; in a succeeding 'stable phase' (Ks) mature soil profiles are developed over the whole slope. In the next Ku phase, the soil on the upper slope is destroyed, while that on the lower slope is preserved by accumulation of overlying detritus. The whole process may be repeated any number of times.

Keewatin (i) The oldest Precambrian rocks near L. Superior. (ii) An ice-accumulation centre (or group of centres) in N. America, W. of Hudson Bay, from which Quaternary ice-sheets moved S. At the glacial max., the Keewatin ice-sheet probably merged with ice of other centres (*Patrician* and *Labradorian*) as a single *Laurentide* ice-mass.

Kegelkarst (Germ.) A form of tropical karst landscape, in which steep conical hills rise above 'cockpits' (deeply incised dolines, developed in relation to wide water-table fluctuations) whose floors are alluvium covered. K is well developed in Jamaica.

Kelvin scale, of temperature Syn. with the scale of ABSOLUTE T $0°K. = -273·1°C$ (usually approximated to $-273°C$), or $-459·4°F$. In 1967 an international conference recommended that K deg. or $K°$ should be abandoned and replaced simply by K.

Kelvin wave A tidal system which develops in a more or less rectangular area of sea (e.g. English Channel), in which the tidal range incr. on the right of the direction of the PROGRESSIVE W, and decr. on the left. This largely explains why the tidal range is less on the S. coast of England (about 4 m) than on the N. coast of France (11–13 m).

kettle, kettle-hole A circular hole in glacial drift, commonly water-filled, caused by the previous presence of a large detached mass of ice which subsequently melted; e.g. around Brampton to N.E. of Carlisle; near Lancaster; in Vale of Pickering. Numerous examples occur in the Kettle Moraine country of Wisconsin, USA. Hence KNOB-AND-KETTLE.
[*f*]

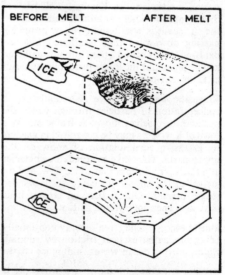

kettle-moraine An uneven area of morainic hills with numerous KETTLE-HOLES, so called after the Kettle Range in Wisconsin, USA.

Keuper A series of rocks, Upper Triassic in age, mainly red clays and marls, with bands of more compact sandstone. Beds of common salt and gypsum sometimes occur. Outcrops of the **K** are found in Midland England, Cheshire (where salt is extensively worked) and Lancashire.

Kew barometer A standard mercury b, which requires no adjustment for level of mercury in the cistern (as does the FORTIN type), so that pressure can be read directly from the top of the mercury column in the graduated glass tube.

key See CAY.

khamsin (Arabic) A hot S. SIROCCO-wind blowing in Egypt and S.E. Mediterranean. These are commonly preceded by a heat-wave and followed by a dust-storm. The wind is usually very dry, often dustladen, but sometimes humid when from the sea. It is assoc. with the front of a depression moving E. through the Mediterranean Sea.

killas A Cornish term for low-grade slates.

kilogram A metric unit of mass = 1000 grams (g) = 2·2046 lb.

kilometre Metric unit of distance: 1 km = 1000 m = 1093·61 yds = 0·621372 mi.

kimberlite An ultrabasic material (including varieties of MICA and OLIVINE), which fills volcanic PIPES in the Kimberley district of S. Africa, sometimes containing diamonds. At the surface it is weathered ('yellow ground'), passing downward into weathered but less oxidized 'blue ground', then into unweathered k. The pipes and the k. are probably the result of FLUIDIZATION.

Kinematic wave A wave represented by a concentration of particles through which the particles themselves move. A traffic jam is a form of **K w**. **K w**s may be generated in glaciers by excessive accumulation of snow on the upper parts. The wave proceeds downglacier at a rate up to 4 times that of ice flow, and on arrival at the snout will cause a spectacular surge forward. Gravel bars on stream beds have also been interpreted as **K w**s.

kinetic energy The 'free' energy continually being dissipated as heat friction by running water, moving ocean waves, sliding ice etc. In streams is defined by $Ek = \frac{MV^2}{2}$ where M is the mass of the water and V is velocity. Thus a large rapidly flowing stream generates more than twice the k e of an equally large stream flowing half as rapidly.

Klimamorphologie (Germ.) The study of the relationship between climate and morphological features. Hence CLIMATOMORPHOLOGY.

klint (Swedish, Danish) A steep cliff around the shores of the Baltic Sea, esp. in Denmark and Sweden; e.g. Möns K., a chalk cliff on island of Mön, rising 130 m from the sea.

klippe (Germ.) Part of an overthrust rock-sheet, isolated by denudation; a special case is a *nappe-outlier*. E.g. (i) zone of k in Switzerland, esp. Chablais Pre-Alps to S. of L. Geneva, eroded into a maze of limestone ridges and peaks over 2400 m, of great structural complexity. (ii) Hills in Assynt (Sutherland, N.W. Scotland), consisting of k of Lewisian Gneiss, Torridonian Sandstone and Cambrian sediments, lying W. of the main edge of the Ben More Thrust which carried the nappes W.-ward. (iii) Mountains in Glacier National Park, Montana, of Precambrian rocks, remnants of nappes carried E. by Lewis Overthrust of LARAMIDE age, now resting as isolated fragments on Cretaceous rocks (e.g. Chief Mountain). [*f*]

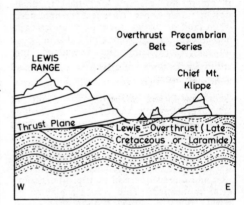

kloof (Afrikaans) A gorge or ravine, sometimes a pass, in S. Africa.

knap A hill-crest, or rising ground.

knick The sharp break of slope 'into which a boot may sometimes be placed' (L. C. King) at the junction of PEDIMENT and INSELBERG. The k may be produced by concentrated chemical weathering, or by lateral undercutting by water flowing over the upper pediment surface.

knickpoint (Germ. *Knickpunkt*) A break of slope in the long-profile of a valley, where a new curve of erosion, graded to a new sea-level (after a rel. lowering of former sea-level), intersects an earlier; the k recedes upstream as erosion proceeds. Sometimes the Germ. word is reserved for a rejuvenation feature formed in this way. The English word may refer to any break of slope in the long-profile of a stream,

Plate 38. A KOPJE, resulting from the disintegration of a small granite INSELBERG near Katsina, northern Nigeria. (*R. J. Small*)

which may also result from an outcrop of resistant rock ('hard bar'), or causes other than rejuvenation. [*f* REJUVENATION]

knob-and-kettle (USA) An area of irregular TERMINAL MORAINE, in which low knobby hills alt. with hollows, usually containing small lakes; see KETTLE.

knock-and-lochan (Scottish) A result of glacial erosion in an area of low relief, forming a hummocky landscape of water-filled hollows and marshy depressions separated by ice-worn knobs. The type-area is in N.W. Scotland and the Hebrides, eroded largely from Lewisian Gneiss. Equiv. broadly to American KNOB-AND-KETTLE.

knoll A small rounded hill. See REEF K.

knot (i) A speed of 1 nautical mi. (6080 ft 1·8532 km, in USA 6076 ft 1·852 km) per hour, so called because the speed of a ship used to be measured by counting knots, as a knotted rope (LOG-line) ran out, against a period of time indicated by a sandglass. (ii) A structural junction-area of 2 or more ridges of fold-mountains; e.g. k of Pasco in the Andes, Pamir k in central Asia.

kopje, koppie (Afrikaans) A small prominent, isolated hill in S. Africa, frequently an erosion remnant forming a castellated pile of rocks (*castle-k*). Used by M. F. Thomas to describe 'collapsed inselbergs', produced by subaerial disintegration, in contrast to TORS, formed by sub-surface weathering. See pl. 38.

Köppen climatic classification An empirical c of climatic types, devised by W. Köppen in 1918, and several times revised esp. in collaboration with R. Geiger. 5 major climatic groups were devised, based on climatic requirements of certain types of vegetation, lettered *A* to *E*. These were sub-divided according to features of rainfall régime and certain thermal characteristics, using lower case letters; e.g. *Csa*, warm temperate climate, with dry hot summer and warm moist winter (*humid mesothermal*), gen. known as Mediterranean type. A further category *H* (Highland) was added.

koum, kum (Turkestan) The sandy desert, with dunes and 'sand-seas', equivalent to ERG of the Sahara.

kratogen(ic) A rigid, rel. immobile part of the earth's crust, later modified to *kraton* and then to CRATON. The last is occas. used, but SHIELD is more pop.

kyle (Gaelic) A channel, sound or strait between the mainland and an island or two islands; e.g. K. of Loch Alsh, Ks of Bute.

laccolith, laccolite (from 2 Gk. words, 'rock-cistern'.) The result of an intrusive mass of MAGMA which has forced up or domed the overlying strata. A l is not necessarily fed from below by a PIPE, but sideways from a BYSMALITH, forming a tongue-shaped bulge. In the Henry Mtns ls cluster around the bysmalith '... like the petals of a lop-sided flower ...' In its simplest form, magma solidifies as a cake-like mass, but several ls may be formed one above another from a single intrusion (CEDAR-TREE L). E.g. Henry Mtns of S. Utah, where 5 groups of trachytic ls were injected among sandstones of age from Carboniferous to Cretaceous; some are simple (Mt Ellsworth), others are complex (Mt Ellen is thought to be a cluster of 30 individuals, Mt Hillers of one large and 8 small). These have been subject to denudation, and reveal stages of destruction and exhumation; the highest is Mt Ellen (3497 m, 11 473 ft). La Sal and Abajo Mtns are other ls in Utah, and there are several in Wyoming and S. Dakota (Belle Fourche valley). [*f*]

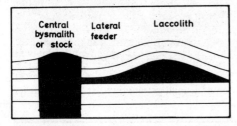

lacustrine App. to a lake, esp. to: (i) deposits of sediment therein; (ii) terraces around the margins of a lake whose surface was formerly at a higher level. L deposits often reveal clear layering or banding, each corresponding to one season's deposition. See L DELTA, L PLAIN, VARVE.

lacustrine delta A d built out by a stream into a lake; e.g. d of Rhône into L. Geneva, of R. Kander into L. Thun, Switzerland.

lacustrine plain The floor of a former lake, filled in by deposition of material by inflowing streams. The soils developed are usually favourable for agriculture, since they consist of fine, well-sorted materials, and the surface is level. Sometimes artificial drainage is necessary, because of the low-lying swampy character. L ps are widespread because of former numerous PROGLACIAL LAKES, now disappeared or represented only by fragments. E.g. Ls. Winnipeg, Winnipegosis and Manitoba, remnants of 'L. Agassiz'; Airedale in Yorkshire, which now contains a l p of at least 30 m thickness of deposits; l p between Derwentwater and Bassenthwaite in English Lake District.

lag (i) Coarse residual material (*l gravel*) left behind on the bed of a stream, or in a desert after finer material has blown away. (ii) The time that elapses between a change in conditions, and recording of the same by an instrument. Measuring instruments, esp. meteorological, do not respond immediately to changed conditions.

lag-fault A low-angled fault formed when rocks slide over one another, the uppermost of which lags behind the lower.

lagg (Swedish) A Fenland margin, in respect of vegetation, deposits and its gen. terrain.

lahar A mud-flow associated with volcanic activity in Java. Loose fine-grained volcanic ash is converted into a mobile paste by water from torrential rain, snow-melt or breaching of the walls of a CRATER-LAKE.

lagoon (i) A sheet of salt water separated from the open sea by sand- or shingle-banks; e.g. the Venetian Is (ii) The sheet of water between an offshore reef, esp. of coral, and the mainland. (iii) The sheet of water within a ring- or horseshoe-shaped ATOLL.

lake A water-filled hollow in the earth's surface. Some are saline inland seas; e.g. Caspian (437 000 km^2, 169 000 sq. mi.), Aral (63 300, 24 500), Dead (1048, 405). The largest named ls are: (*a*) *N. America:* Superior (82 410, 31 820), Huron (59 600, 23 010), Michigan (58 020, 22 400), Great Bear (30 200, 11 600); (*b*) *Africa:* Victoria (67 860, 26 200), Chad (fluctuating, about 51 800, 20 000), Nyasa (36 780, 14 200), Tanganyika (Malawi) (32 890, 12 700); (*c*) *Asia:* Baikal (29 990, 11 580), Balkhash (18 260, 7050); (*d*) *S. America:* Maracaibo (21 487, 8296), Titicaca (8290, 3200); (*e*) *Europe:* Ladoga (18 130, 7000), Onega (9840, 3800); (*f*) *Australia:* Torrens (6216, 2400), Eyre (fluctuating). Largest in England is Windermere, in UK Loch Lomond, in British Isles Lough Neagh (381, 147). The longest in the world, excl. Caspian, is Tanganyika (675 km, 420 mi.) At the other end of the scale are small sheets of water, *meres, ponds, tarns* (English Lake District), *llyns* (Wales), *loughs* (Ireland), *lochs* and *lochans* (Scotland), *étangs* (France), *stagni* (Italy). Ls may be classified according to origin of hollows in which they lie: (i) *erosion:* glaciation ('finger'- ls and tarns), solution (salt-meres), wind ('shotts' in N. Africa); (ii) *deposition* or *barrier ls*, enclosed by: rock-falls, sand bars, deltaic deposits, morainic deposits, ice-dams, vegetation dams (ls in W. European heaths and moorlands), calcareous dams (barrier of deposited calcite, as L. Plitvicka in Yugoslavia); (iii) *structural:* rift-valleys, hollows formed by crustal sagging, down-faulted basins; (iv) *volcanic:* crater-ls, lava-dams. The largest man-made ls are Bratsk in the Angara valley (USSR), Kariba (Zambia-Rhodesia) behind the Kariba Dam across R. Zambezi, Mead behind Hoover Dam across R. Colorado, USA. Deepest lake is Baykal in USSR (1940 m, 6365 ft), lowest surface Dead Sea (395 m, 1296 ft below sea-level), highest named surface L. Titicaca (3811 m, 12 506 ft above sea-level).

lake rampart A ridge of material on a l-shore, formed when ice on the frozen l exerted lateral pressure against the shore; clearly shown around the shores of Great Lakes in N. America. Not to be confused with a lacustrine terrace representing a former level of the l.

lalang (Malay) Thick coarse grass which rapidly covers abandoned plantations and clearings.

Lambert's Conformal Projection A type of CONIC P in which parallels are concentric circles, meridians are radiating, equally spaced, straight lines, with the scale true on 2 standard parallels, increasing N. and S. from them. This **p** gives true direction at every point, and is used for aeronautical charts, weather maps, and any type of distributional map where correct direction is important. A popular **p** for maps of USA, since there is little distortion for this shape and area, esp. when using parallels 29°N. and 45°N. Calculation is complex, and it is usually drawn from tables.

Lambert's Zenithal Equal Area Projection (L's 6th) In its polar form, meridians are represented by straight lines at true angles to each other, while parallels are circles of which meridians are radii. The radius of each parallel is calculated from $2a \times \dfrac{\sin z}{2}$, where $z=$co-latitude and $a=$earth's radius to scale. The p. can be applied to any part of the earth's surface, for deformation is symmetrical around the central point from which GREAT CIRCLES radiate. The equatorial and oblique cases are plotted from co-ordinates of grid intersections given in tables. Directions are correct at the central point, equal area, and there is little deformation for 30° from the central point.

laminar flow (i) The movement of a glacier, produced by a thrust along the line of slope caused by pressure resulting from the weight and solid character of ice higher up. The ice is fractured and sheared, and thrust along glide-planes. This may push the 'snout' of the glacier uphill for a short distance. (ii) Non-turbulent (*stream-line*) flow in a fluid; speed and direction of flow at any given point are steady.

lamination, lamina, *adj.* **laminar** The finest scale of stratification or sheeting in rocks, esp. metamorphic; sometimes defined as less than 1 cm in thickness, though there may be 40 or more layers in a cm. Each l represents a layer of original deposition in sedimentary rocks; each period of flood deposited a thin layer of mud, which compacted, and the next flood added another; or wind added successive layers of sand to dunes. In such rocks as shale, micas and clay minerals lie parallel with the bedding. E.g. Green River Shales of Wyoming, 800 m in thickness, in layers 0·17 mm thick; it is estim. that these took 6·5 million years to be deposited.

land breeze A cool night **b** blowing from the l, which has been cooled by radiation, towards the warmer sea, occurring partic. near the Equator, and elsewhere in calm, settled weather. During night pressure is slightly higher over l than over sea, so there is an airflow towards the latter, reinforced by the outward spread of cool denser air. In some islands these are so regular that fishing-boats go out at night with the l b, returning next afternoon with the SEA B. [*f*]

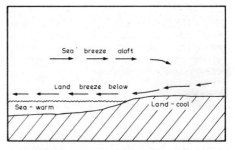

land-bridge (i) Geologically, a past or present land-link between continents; e.g. (past) across the Bering Strait; (present) isthmus of Panama. (ii) By biologists and anthropologists, former migratory route-ways by which terrestrial animals, incl. man, were dispersed.

land classification A c of l into categories, which can be used 'for a broad national policy of land-use planning and conservation of land resources' (L. D. Stamp). He proposed 3 major categories and 10 types, based on: (*a*) nature of *site* (elevation, slope and aspect); and (*b*) nature of *soil* (depth, texture and water conditions). The major categories were: I. good; II. medium; and III. poor. In 1974 a new l c grading was issued by the UK Ministry of Agriculture, Food and Fisheries, based somewhat subjectively on criteria of relief, slope, aspect, climate and wetness, depth, texture, structure and stoniness of soil, affecting range of crops, level and consistency of yield, and cost of production. The 5 grades are: (i) l of best quality, with high yields (2·8% of the total area of England and Wales); (ii) l with some minor limitations in range of possible crops (14%); (iii) l with moderate limitations, with restricted range of cropping (50%); (iv) l with severe limitations, mostly under grass, with occas. fodder crops (19%); (v) l with

174 LAND

severe limitations, under grass or rough grazing (14%). Maps on scale 1:63 360 were published in 1974. The term 'grade' is unfortunate, since it implies that grade 3 l is 'third-class l' which is adequate for cereals and incl. the country's best grassland. In developing countries (e.g. Nigeria, Uganda) l c studies carried out since the war by MEXE (Military Experimental Establishment) and other groups have been based on the delimitation and mapping of LAND SYSTEMS, the physical characteristics of which are tabulated in data storage systems. There is thus built up an 'inventory' of potential resources (soils, vegetation, surface and ground water, building and road construction materials etc.), of considerable value in the formulation of development schemes for agriculture, reclamation, irrigation, route construction etc.

landes (Fr.) (i) Proper name for sandy lowlands ('*Les L.*'), with dunes and lagoons, bordering S.W. shores of the Bay of Biscay. (ii) With lower case, used for other wastelands, as heathlands of Belgium.

landform The shape, form and nature of a specif. feature of the earth's land surface, the study of which was formerly called PHYSIOGRAPHY, in its modern scientific concept GEOMORPHOLOGY. Various classifications of ls have been devised. A. N. Strahler divided them into 2 main groups: *initial forms*, where orig. features produced by tectonic forces have been only slightly modified; and *sequential forms* where modifications are more pronounced and where initial forms may have been destroyed; the latter incl. (i) erosional types; (ii) residual types; (iii) depositional types. R. E. Murphy produced a classification using 3-fold categories: 7 types of structural region, 6 classes of topographical region, 5 classes of geomorphological process. Each is given a letter. Thus Labrador is *LHw*, (a Laurasian shield, of low tableland relief, recently glaciated); the Colorado plateau is *ATd* (Alpine system, high tableland, arid).

landscape Used orig. by artists to denote rural scenery (Dutch, *Landschap*), now gen. for the sum total of the aspect of any area, rural or urban. 'L Geography' developed as part of the regional view of the subject, stimulated by P. Vidal de la Blache in France, which examined both 'natural' and 'cultural' aspects of the face of the earth in terms of concrete reality. See NATURAL L.

landscape evaluation An effort to establish a scale of l values as a contribution to planning, development and CONSERVATION. The problem is to replace pure subjective judgement and personal bias by some scale of values, though personal observation from a chosen series of viewpoints ('view evaluation') is required. One suggested scale (by K. D. Fines) is from 0 to 32, with 6 categories: *unsightly*, 0–1; *undistinguished*, 1–2; *pleasant*, 2–4; *distinguished*, 4–8; *superb*, 8–16; *spectacular*, 16–32; e.g. 18 is the highest in UK (the Cuillins of Skye across Loch Coruisk), 12 the highest in Lowland Britain. A l e map has been made of E. Sussex, using a series of tracts (shaded on a CHOROPLETH basis) largely related to geological and physiographical features; 9 lowland l types, 3 highland l types and 6 townscapes were suggested.

Landschaft (Germ.) German word for landscape, which has come to have a specialized, though often diverse, meaning, sometimes in the sense of 'scene', sometimes with the character of a unit region. Translating the word simply as 'landscape' has led to confusion in certain theoretical discussions.

landslide, landslip A fall of earth and rock-material down a slope or mountain-side, the result of gravity and rain-lubrication; see MASS-MOVEMENT. The two words denote the same process; the former is more customary in the USA, the latter in UK.

land system The most widely employed and practicable method of dividing landscapes in land classification. L ss, comprising various land FACETS (valley floors, slopes, terraces, fans, plateaus etc), are virtually homogenous in terms of climate, relief, vegetation, soils and hydrological characteristics. They represent arguably the most basic method of physico-geographical regional differentiation yet devised.

lapiaz, lapiés (Fr.) A 'pavement' of limestone, fretted by carbonation-solution along well-marked joints, with ridges and furrows (equivalent to CLINTS and GRIKES resp.) *Lapié* (plur. *lapiés*) is sometimes used for individual furrows within the whole limestone pavement. Syn. with Germ *Karrenfeld*, *Karren*.

lapilli (It.) Small fragments of cinder, about the size of a pea, ejected from a volcano; sometimes defined as between 4–32 mm diameter.

lapse-rate The r of temperature decr. (*vertical temperature gradient*) in the atmosphere with incr. altitude. The ENVIRONMENTAL L-R is on average, about 0·6°C per 100 m, or 1°F for every 300 ft of ascent, but varies from place to place and with passing time (day to night, summer to winter, passage of different air masses etc.). This **l-r** seems to continue steadily as far up as the TROPOPAUSE, beyond which the temperature remains fairly constant in the lower STRATOSPHERE. This normal l-r may be interrupted by inversion, particularly at night when the earth is cooled by radiation, and heat is conducted to the ground from the overlying air. See DRY ADIABATIC L-R, SATURATED ADIABATIC L-R.

Laramide orogeny A mountain-building movement responsible for folding of the Rocky Mtns, which started in late Jurassic times and continued into the early Tertiary.

Late-Glacial The final stage of the DEVENSIAN glacial period. The L-G is usually divided into 3 'Zones'. Zone I was a very cold, dry period of considerable duration; Zone II (the ALLERØD interstadial) was relatively warm; Zone III, lasting only from 9800–9300 B.P., was again colder and more humid, with much summer melting of snow and rapid run-off and solifluction.

latent heat The amount of heat-energy expended in changing the state or phase of a body without raising its temperature, expressed in calories per g. In meteorology, the *l h of evaporation* indicates energy required or heat absorbed to convert water to water-vapour. When CONDENSATION occurs, the heat given out is *l h of condensation*. The *l h of fusion* represents the change from solid to liquid. The *l h of sublimation* represents the change from solid to gaseous state (Ice to water-vapour, 680 calories per g; ice to liquid, 80 cal per g; liquid at 15°C to water-vapour, 590 cal per g; liquid at 100°C, 540 cal per g).

latent instability In climatology, the state of the upper part of a conditionally unstable air mass, lying above the level of free CONVECTION. See CONDITIONAL I.

lateral dune A minor d flanking a main d within a pattern formed in a desert where some obstacle occurs. [*f* DUNE]

lateral erosion E performed by a stream on its banks, ct. vertical erosion on its bed. This is most marked on the outside of a bend or meander, so that the current impinges there, resulting in undercutting and BANK-CAVING. Gradually a river extends its l e, widening the swing of each meander and its overall valley; each meander gradually moves downstream. The valley-floor is bordered by a BLUFF, the outer limit of l e. The creation of a broad, nearly flat, surface in this way is *l planation*.

lateral moraine Rock debris lying on the surface of a glacier, forming a low ridge or band parallel to and near its edge. When the glacier melts, debris may remain as an embankment along the valley side; several, broadly parallel, indicate stages in the shrinkage of the ice. In a big glacier, a l m may be 30 m or more in height. Usually lines of **m** protect ice below from melting, and many so-called **m**s are actually only a thin veneer of debris. See pl. 39. [*f* MORAINE]

laterite Until recently used with varied, even vague, implications. *Sensu lato* a residual material formed through breakdown of rocks, partic. igneous, under humid tropical conditions though with a marked dry season, resulting in acute chemical weathering. Silica and alkalis are leached from the parent material, causing a concentration of sesquioxides of iron and aluminium in the lateritic horizon. The laterizing solutions may be carried down by ground-water to form a HARDPAN, esp. if they percolate into a layer of gravel (syn. with FERRICRETE). L may form an indurated, concretionary, slag-like layer, hard and massive; it may appear in a horizon as a series of nodules, or lenses, or as a hard cellular crust. When this is exposed by subsequent denudation, it forms a lateritic CUIRASS, 1–10 m thick, esp. on terraces and ancient peneplain surfaces. Some authorities apply the term to a non-consolidated though leached horizon, as *lateritic clay*, soft and plastic though hardening on exposure. Crushed lateritic gravels are used for road-making (e.g. murram in E. Africa), lateritic clays for brick-making (Lat., *later*, brick), and for building (Angkor Wat temples, Vietnam).

laterization The process of weathering under hot damp conditions (though with a marked dry season) to form LATERITE and, in the case of granitic rocks low in iron, BAUXITE.

latitude Angular distance of a point along a meridian N. or S. of the Equator (0°), the Poles 90°. A circle drawn around the earth parallel to the Equator is a parallel of l and every point on it has the same l. 1° of l at the Equator = 110·551 km (68·704 mi.), at 45° = 111·130 km (69·054 mi.), near the Poles 111·698 km (69·407 mi.) (Clarke spheroid) All parallels of l intersect all meridians of LONGITUDE at right-angles. [*f, page 176*]

latosol Coined in USA to denote *sensu lato* all tropical soils occurring in the savanna and rain-forest; these are coloured from brown through yellow to commonly red, are free-draining, with an acid reaction, rich in hydroxides of iron, aluminium and manganese, and heavily leached of soluble compounds, esp. bases. The product of constantly

176 LATOSOL

Plate 39. A high-level LATERAL MORAINE, marking several advances of the ice during the LITTLE ICE AGE in the Alps, above the Lower Arolla glacier, Switzerland. The effects of GULLY EROSION of the face of the moraine, following its exposure by the lowering of the glacier surface since 1850, are clearly seen. (*R. J. Small*)

high temperatures, humid conditions (though usually with markedly seasonal rainfall) and free drainage, which causes chemical disintegration of bed-rock to a consid. depth. They vary in texture from clay to loamy sand. Sometimes loosely called 'Tropical red and yellow earths'. In an effort towards precision, pedologists have defined 5 main divisions of ls. (i) weathered *ferralitic* soils; (ii) *ferruginous* soils; (iii) leached *ferralitic* soils; (iv) *basisols* (ls developed from basic rocks); (v) *humic* ls (found at high altitudes where oxidation of humus is less and therefore organic content is higher than in lowland ls).

Laurasia The N. primeval landmass; see PANGAEA.

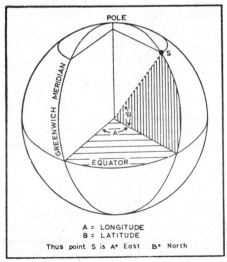

lava Molten rock (MAGMA) extruded on the surface of the earth before solidification. Classified as: (i) silica-rich ('acid'); (ii) basic (90% of total); (iii) intermediate. It may be glassy, PORPHYRITIC or VESICULAR. See AA, PAHOEHOE and PILLOW L. See pl. 24, 28, 40.

lavant See BOURNE.

lava tunnel, tube If the surface and sides of a narrow flow of basaltic l solidify to form a thick outer crust, hot l inside continues its forward movement, ultimately draining out of

Plate 40. A massive LAVA 'delta', which has escaped from a fissure in the side of the Boqueron volcano, San Benedicto Island, Mexico. (*US Navy*)

this crust to leave a tunnel, later occupied by an underground stream. If this water freezes in winter, it may not thaw out in summer because the basalt crust forms an insulating layer; e.g. in Modoc L. Beds National Monument, California.

law A rounded hill in Scotland, esp. near Firth of Forth; e.g. Traprain L.

lawn (i) An open area of grassland among woodland, notably in New Forest of Hampshire. (ii) Surface of terraces on limestone of the Isle of Portland, Dorset.

layer-tinting Distinctive shading or colouring of a map between pairs of contours in order to reveal the pattern of high and low land. The CONTOUR-INTERVAL at which each colour change is made must be carefully chosen. One alt. is to use a sequence of greens, yellows and browns, leading to red, purple and even white in high country. Another is to use one colour only, ranging from faintest to darkest possible density, as in purple-tinted RAF maps. A 3rd is to merge or grade successive tints to avoid the stepped appearance inherent in contour-filling. Line-shading in black can be used, but should be avoided if possible, owing to obscuring other detail and gen. lack of quality.

leaching Removal of soluble salts, esp. bases, from the upper layer (A and E HORIZONS) of the soil by percolating soil-water in humid climates. Much leached material is held in the B-HORIZON, a zone of accumulation. Soils gradually become mineral-deficient and 'sour'. Typical leached soils are PODZOLS and LATOSOLS. See ELUVIATION, ILLUVIATION.

lead A metal occurring as an ore in its sulphide form, *galena* (PbS), also in an oxidized form, *cerussite* ($PbCO_3$). It is mostly found in lodes, veins and 'flats' in sedimentary rocks, esp. in limestone.

lead (water) A stretch of open water between individual ice-floes in a pack-icefield.

lead-ratio In a piece of rock containing uranium and/or thorium, the l-r is the ratio of lead^{206} to uranium^{238}, of lead^{207} to uranium^{235}, or of lead^{208} to thorium^{232}, formed by the radioactive breakdown of some of the uranium/thorium. As the rate of breakdown is known (HALF-LIFE of $U^{238} = 4.51 \times 10^9$ years, of $U^{235} = 7.13 \times 10^8$), an estimate of the age of the rock may be obtained. See ISOTOPE, RADIO-ACTIVE DECAY.

league An obsolete British unit of length = 3 nautical mi.

leat (i) An open watercourse, commonly conveying water to a mine. (ii) A ditch which contours a hill-side, picking up minor streams, to feed a reservoir; e.g. a l contours the side of the Ogwen valley in N. Wales, delivering water into Llyn Cowlyd.

ledge (i) An underwater ridge of rocks. (ii) A projecting horizontal, shelf-like mass of rock on a mountain- or cliff-side.

lee, leeward The side away from, or protected from, wind (downwind); e.g. lee-side of a range of mountains.

lee depression A low pressure system formed in the l of a mountain range, the result of a dynamically created eddy as an air-stream crosses it; e.g. along E. margins of the Canadian Rockies; to E. of the S. Alps in New Zealand. In W. Mediterranean basin, a cold Polar Maritime air-stream, associated with a COLD FRONT, flows S. over the Maritime Alps towards the sea, where low pressure is intensified to form a distinct d.

lee-shore The s towards which wind is blowing; the danger s for shipping.

lee wave A w form in air motion caused by a relief barrier. When a gentle wind crosses a barrier, the effect on the leeward side is limited to simple LAMINAR FLOW (*streaming*). With a stronger wind, a STANDING W forms, and with still stronger winds a l w is formed which may develop into complicated TURBULENCE, sometimes called *rotor-streaming*. This may produce striking BANNER, LENTICULAR [and *f.*], and arched clouds.

length A basic unit of l is the *Ångstrom Unit* (Å U) = 10^{-10} m (10^{-1} im), or 3.937×10^{-9} in., used in atomic dimensions (an atom of most elements has a diameter of about 2 Å). See METRE, MICRON.

1 m = 1·09361 yds; 1000 m = 1 km = 0·62137 mi.; 1 yd = 0·9144 m; 1760 yds = 1 statute mi. = 1·6093 km; 1 cm = 0·3937 in; 1 in. = 2·5400 cm; 1 ft = 0·3048 m; 1 m = 3·281 ft.

lemniscate loop A type of curve (expressed by the formula $p = L \cos k\theta$, where L is the length of the long axis, and k is a 'dimensionless number' expressing the elongation of the l l) employed to express the form of large drumlins. In simple terms these are egg-shaped; the stoss end of the drumlin, experiencing greatest pressure from the moving ice, is comparable with the blunt end of the egg (which is first to emerge from the bird and is also subjected to maximum pressure).

lenticular cloud (*lenticularis*) (cloud species) A lens-shaped c, often associated with wind eddies over hills and mountains; e.g. Helm-c over the Crossfell escarpment in N. Pennines, lying along or just above the ridge; 'Long White Cloud' or 'North-West Arch' of S. Island, New Zealand. See HELM WIND. [*f*]

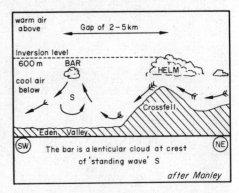

after Manley

leucocratic A category of igneous rocks comprising such light-coloured minerals as quartz, feldspar, white mica. Ct. MELANOCRATIC.

levante, levanter (Sp.) An E. or E.N.E. wind blowing between S.E. Spain and Balearic I., and in Strait of Gibraltar, when a depression is in the W. Mediterranean basin. Gen. a mild and humid wind, and can cause heavy precipitation and floods.

leveche (Sp.) A hot, dry S. wind of SIROCCO-type, blowing from Morocco across to coast of S. Spain. Occurs in the warm sector of a depression moving E.-ward through the W. Mediterranean basin.

levee, levée (orig. Fr.) (i) A natural bank built up by a stream along the edges of its channel, esp. during flood; when the water subsides bank remains. The bed of the river tends to rise above the surrounding FLOODPLAIN. If ls are breached during a subsequent period of high-water, widespread floods may occur; e.g. Mississippi, Hwangho. (ii) An artificial embankment constructed along a river to check flooding; e.g. lower Mississippi.

level Used in a variety of senses where a horizontal line or plane surface is implied. Specif. in a geographical sense: (i) A surveying instrument (ABNEY L, DUMPY L). (ii) A large tract of l land, usually drained; e.g. Bedford L. (E. Anglia).

levelling In surveying, the operation of finding difference in height between successive pairs of points, by sighting through an instrument, with peep-sight or telescope on which is mounted a spirit-level, on to a graduated measuring-rod. A line of levels can be run from the starting height or DATUM, and heights of points on the line are obtained. *Geodetic l* is of a high order of accuracy, used as the basis of a topographical survey. See ABNEY LEVEL, DUMPY LEVEL. [*f*]

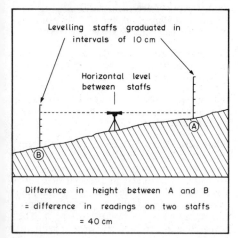

Lewisian A group of Precambrian rocks in N.W. Scotland, esp. in Sutherland and Ross, and in the Outer Hebrides. Specif. the lower of the 2 Precambrian systems (the other TORRIDONIAN), and consists mainly of gneiss, with some altered schists. The L occurs both as part of a BASAL COMPLEX and as a series of DYKES and SILLS; much altered by earth-movements and metamorphism.

liana, liane A woody climbing plant with roots in the ground and stem around a tree, found esp. in Tropical Rain-forests; e.g. rattan in Indonesia, so long and strong that it can be used as a rope.

Lias A Lower Jurassic formation, of clays, shales, limestones and sandstones, conspicuous by ammonite fossils. It can be traced across England from Dorset to the Yorkshire coast in a diagonal outcrop, and in S. Wales. Also used for limestones typically interbedded with shale or clay in other formations.

libeccio (It.) A strong W. or S.W. wind blowing towards the W. coast of Corsica. Max. frequency in summer, when it is sometimes boisterous and invigorating, although when it crosses Cap Corse it may reach the E. coast as a hot, desiccating squall. In winter the wind may bring rain or snow to W. mountain slopes.

lichenometry A method of time-estimation based on rate of growth of lichen on a stone, giving some indication as to how long the stone (e.g. in a MORAINE) has been lying in that position; has been used for periods up to 3 centuries. The method must be used with discretion, since growth rate is affected by so many variables, esp. of climate.

lido (It.) A BARRIER-BEACH, named after the L. at Venice; now used as a sophisticated bathing-beach.

lightning A visible electrical discharge in the form of a flash; see THUNDERSTORM. Many types are defined: *forked, sheet, ball, streak, chain, ribbon*. It may be within a single cloud, cloud to cloud, or cloud to ground.

light-year The distance l travels in 1 y at 186 326 mi. per second ($2 \cdot 998 \times 10^8$ ms^{-1}, i.e. 5 878 310 400 000 mi., or approx. $9 \cdot 7 \times 10^{12}$ km (6×10^{12} mi.). The nearest star (*Alpha Centauri*) is 4·29 l-ys away from the earth.

lignite Used gen. to denote types of low-grade coal, in structure and carbon-content between PEAT and sub-bituminous coal. Strictly, distinction should be made between l and BROWN-COAL; l is stonier, darker in colour, with a slightly higher carbon content, less moisture, and less evident vegetable structure.

limb, of fold The rock strata on either side of the axis, or central line, of a f. [*f* ANTICLINE]

limestone A rock consisting mainly (sometimes defined as 80%, by others 50%) of calcium carbonate ($CaCO_3$); i.e. *calcareous*; there are many varieties, defined by texture (e.g. *oolite*), mineral content (e.g. *dolomite*), origin (e.g. *coral*), age (e.g. *Carboniferous, Jurassic*). 3 major groups: (i) *organic* (shelly, reef, coral, algal, crinoidal, foraminiferal); (ii) *chemical* (oolitic, dolomite, tufa); (iii) *clastic (detrital)* (fragments of pre-existing carbonate rocks, shell-grits). (i) and (ii) are AUTOCHTHONOUS ls, (iii) is ALLOCHTHONOUS. Chalk is also a l, though as the latter is thought of as a resistant hard rock and chalk is manifestly different in appearance, this is often forgotten. See individual types of l.

limiting angle The range of angle within which certain forms occur or certain processes operate. In humid conditions the *upper* l a for the presence of a regolith cover (or the *lower* l a for an exposed face) is 40–45°. In periglacial environments in Canada the common occurrence of long slopes at 6–8° indicates that this is the *lower* l a for effective solifluction.

limnology The scientific study of fresh-water lakes and ponds, dealing with various physical, chemical and biological conditions and characteristics. One aspect of HYDROLOGY.

limnoplankton Microscopic organisms, vegetable and animal, that live in still water (lakes and ponds).

limon (Fr.) A superficial fine-grained deposit akin to LOESS, from which brown, loamy soils have developed. Probably of wind-blown origin, laid down during dry, steppe-like interglacial, or immediately post-glacial, times. In Europe wind blowing outwards from high-pressure areas over N. ice-sheets exercised its sorting effect and removed finer elements from the vast mass of materials deposited by fluvio-glacial action beyond the ice-margins, and transported them to the W. Occurs in Belgium and Paris Basin, and even in Brittany, on the highest interfluves and plateaus; thicknesses of 21 m have been found. A distinction is made between wholly wind-blown and wind-deposited l, and that reworked and redeposited by later river action.

limonite Yellowish brown ore of iron, various hydrated ferric oxides ($2Fe_2O_3 \cdot 3H_2O$), found notably in Lorraine. The end weathering product of iron-containing minerals.

limpo (Portuguese) A type of SAVANNA in Brazil, dominated by grasses, with rel. few bushes or trees.

lineament A major linear topographical feature, the result of some striking underlying factor; e.g. FAULT-LINE SCARP, DYKE, PEDIMENT, FISSURE ERUPTION, glacial TROUGH. A *megalineament* is on a world scale; e.g. RIFT-VALLEY, MID-OCEAN RIDGE, island ARC, FOLD MOUNTAIN range.

linear erosion E carried out along a distinct line, as by a river, or glacier in a valley.

linear eruption Syn. with FISSURE E.

line-squall A sharply-defined COLD FRONT, associated with very stormy conditions, along a line as much as 480 km in length, usually ahead of a COLD FRONT, with dark rolls of cloud, hail, thunderstorms and violent gusty winds, sometimes of gale force.

links Gently undulating sandy ground, with dunes, coarse grass and shrubs, near the seashore in Scotland and N.E. England; e.g. The Ls, Holy I., Northumberland.

lithification, lithifaction The conversion of an accumulation of loose sediments into a massive rock at temperatures and pressures normal to the earth's surface, through COMPACTION (consolidation) and CEMENTATION. Strictly l is the result, DIAGENESIS the process. METAMORPHISM is excluded.

lithogenesis The accumulation of sediment, sand and mud in the sea (notably in a GEOSYNCLINE), later compacted to form solid rocks.

lithology The study of rocks in connection with their physical, chemical and textural character (hence *lithological*). By some restricted to sedimentary rocks.

lithometeor An ensemble of solid, non-aqueous particles in the atmosphere; e.g. HAZE, DUST, smoke, DUST-STORM, SAND-STORM.

lithosol An AZONAL soil consisting of stony unweathered or part-weathered rock-fragments, scree, glacial till.

lithosphere Used in several senses; (i) (pop.) the CRUST of the Earth, hence *lithospheric plates* for individual masses; see PLATE TECTONICS; (ii) the SIAL and crustal SIMA layers above the MOHOROVIČIĆ DISCONTINUITY; (iii) (now gen. accepted) the SIAL, SIMA and upper MANTLE above the GUTENBERG CHANNEL. The l is the zone of earthquakes, since it is characterized by resistance to SHEAR-waves. Ct. underlying ASTHENOSPHERE, of weaker, hotter materials, with a marked reduction in shear-wave velocity.

litre A metric unit of volume; 1 kilogram of water at 4°C and 760 mm pressure = 1000·027 cm^3, usually accepted as 1000 cm^3. 1 l = 0·219976 UK gallons. 1 UK gallon = 4·546 l. In 1964 the international l was redefined as 1 dm^3, which gives a slightly different value for 1 UK gallon, though this is identical to 4 significant figures. I US gallon = 3·785 l. The correct SI value for all volume is the cubic metre (m^3).

litter The surface deposit or layer of vegetation, which may ultimately become HUMUS.

'Little Ice Age' (i) The period *c.* 3000–500 B.C. in E. Anglia, (though this varies with latitude, i.e. is *time-transgressive*), equivalent to the SUB-BOREAL, when climate became cooler than in the previous millennia. Glaciers which had disappeared from parts of Alaska and the Sierra Nevada again developed. This somewhat journalistic name has been dropped in favour of NEOGLACIAL. (ii) A much more recent advance of glaciers in the Alps during the period A.D. 1550–1850. Records show that villages were overwhelmed by **i**, summer pastures were no longer usable, and passes were blocked. Latterly there has been a gen. overall, though fluctuating, shrinkage of glaciers. In Alaska there seems to have been a maximum advance c. A.D. 1850. Since 1850 there has been a gen. rise of air temperature by 1°C in the N. hemisphere. See pl. 39.

littoral (i) App. gen. to the sea-shore. (ii) The zone between high- and low-water springtide marks (*enlittoral zone*). Other authorities, esp. in connection with living organisms, extend the zone more widely, some to 200 m depth, though the area between 60 and 200 m depth is usually specif. denoted as *sublittoral*.
[*f* ABYSSAL ZONE]

littoral deposit A d of sand, shells and shingle between high- and low-water marks, though sometimes incl. all shallow-sea ds, incl. offshore mud.

llano (Sp.) Tropical grassland or SAVANNA on the Guiana Plateau in S. America. Initially simply 'plain'; by extension, the vegetation found in such a situation, which reflects the summer rainfall régime (April to Oct.).

load Material transported by a natural agent of transportation, an integral part of denudation, used specif. of a river. Material is carried: (*a*) in *solution*; (*b*) in *suspension*; (*c*) by SALTATION (*bed-l* or *traction l*). The COMPETENCE or ability of a stream to move particles of a certain size is thought to be proportional to the 6th power of its velocity (SIXTH POWER LAW). A stream can carry a much larger l of fine material than of coarse. L is sometimes used of material carried by a glacier, the wind, or moved by waves, tides and currents.

loam soil A permeable, friable mixture of particles of different size, forming: (i) *sandy l*; (ii) *silty l*; (iii) *clay l*, according to proportion of constituents. Given a more precise definition in USA (7–27% clay, 28–58% silt, and 30–52% sand); particles of sand are 1·0–0·05 mm in diameter; silt, 0·05–0·005 mm; clay less than 0·005 mm (US definition).

lobe (i) A rounded tongue-like mass, used esp. of: (*a*) ice projecting from a larger sheet; (*b*) mass of wet clay moving down a steep slope on to a beach; (*c*) tongue of DRIFT projecting further than the main mass. (ii) In USA the land enclosed by an acute meander.

local climate The c of a small area which possesses marked contrasts with other areas nearby, resulting from minor differences of slope and aspect, colour and texture of soil, proximity of a water-surface, nature of vegetation cover, and effects of buildings. A large number of carefully sited recording stations is needed.

local relief The difference in altitude between highest and lowest points in a limited area. Syn. with *relative relief*.

local time Syn. with APPARENT T.

loch (Scottish) (i) A lake in Scotland; e.g. L. Lomond, Rannoch, Laggan. (ii) A long narrow arm of the sea on the coast of Scotland, e.g. L. Linnhe, Fyne, Hourn.

lochan (Scottish) A small lake, usually lying in a coire (CIRQUE); e.g. L. Meall an t-Suidhe on the slopes of Ben Nevis; L. Lagan in Black Cuillins of Skye.

lode (i) Artificial channel or watercourse, usually embanked (E. Anglia). (ii) Mineral vein, or closely parallel veins of ore, used esp. in Cornwall. Orig. from the fact that the miner could be led or guided in his search for ore by following a l.

lodgement till In a gen. sense, syn. with GROUND-MORAINE. Specif. material deposited subglacially (i.e. under actively moving ice). Particles of clay are plastered (or 'lodged') on the underside of the ice, so accumulating a thick deposit with no semblance of stratification or sorting, and with larger stones so arranged that their major axes are parallel to the direction of ice-flow. DRUMLINS are formed from l t. Ct. ABLATION TILL.

loess, löss (Germ.) Derived from the name of a village in Alsace, used to indicate local deposits by farmers and brick-makers and adopted as the name of a fine-grained, coherent, friable, porous, yellowish dust. Probably initially removed by wind, either from desert surfaces where loose material was unprotected by vegetation, or during dry inter- and post-glacial and fluvioglacial materials. L was studied by v. Richthofen in N.W. China, where a sheet, covering nearly 650 000 km², swathes the landscape to a depth of from 90–300 m. It occurs here at all elevations from near sea-level to 2500 m. It reveals innumerable vertical tubes, lined with calcium carbonate, thought to be the 'casts' of grass stems. It was deposited where vegetation had some binding effect, and possibly where incr. rainfall helped wash it down from the air. It is also found in Europe S. and W. of the terminal moraines of the glacial advance, from E. Germany, through W. Germany (Bördeland), Belgium and France (see LIMON). In Europe much may have been reworked and redeposited by running water. Also found in W. States of USA in the Mississippi-Missouri valleys and in Argentina. From it develop fine-textured, easily worked, deep, well drained soils, good for wheat and sugar-beet.

loessoïde Given by Dutch geologists to deposits in S. Limburg believed to be of loessic origin, but reworked and redeposited by stream-action, possibly with an admixture of residual materials formed by decomposition *in situ* of Upper Chalk. The resultant soils, *loessleem*, are extremely fertile.

log An apparatus for measuring the speed of a ship, which came into gen. use at the beginning of the 17th century. The *common l* was replaced by a mechanical or *patent l* at the beginning of the 20th century. The common l consisted of a *l-ship*, a flat piece of wood weighted with lead, secured to a *l-line* by a *bridle*, the l-line fastened to a *reel*. The l-line was divided into equal parts by knotted cords. The l-ship was thrown overboard, and a *l-glass* (sand-glass) was used to measure the number of knotted cords, hence KNOT.

logan A rocking-stone, after a granite block in the Land's End area of Cornwall, delicately balanced on its base. These rocks became individual masses by chemical weathering along joints in granite. See TOR.

logarithmic scale A s in which an increase of one unit represents a power increase in the quantity involved. When desired to plot rates of change (output, population), a vertical log-s is used, with horizontal linear s for time intervals; such semi-logarithmic graph-paper is available. Log-log graph-paper, with both horizontal and vertical log-scales, is used for some frequency graphs.

longitude Angular distance E. or W. of the prime (0° or Greenwich) MERIDIAN, measured to 180°. A circle drawn around the earth through each pole is a meridian of l, and cuts all parallels of latitude, incl. the Equator, at right-angles. All places on the same meridian have the same l. As meridians converge towards the Pole, the length of 1° of l becomes less. 1° of l at Equator = 111·320 km (69·172 mi.); at latitude 45° = 78·848 km (48·995 mi.); at 70° = 38·187 km (23·729 mi.); at Poles = 0. 15° of l = 1 hour in local time. [*f* LATITUDE]

longitudinal coast See CONCORDANT c [*f*].

longitudinal crevasse See CREVASSE.

longitudinal valley Used when a v is parallel to the gen. trend of mountain ranges (ct. transverse vs). E.g. *vaux* (sing. *val*) in Jura Mtns. Strictly should be reserved for a v developed along the STRIKE of the strata.

longitudinal wave A form of shock-w produced by an earthquake, which travels along the surface of the ground, its nature controlled by elasticity of the strata. It arrives after the *P*- and *S-ws*, which have taken more direct courses through the earth's mass. Subdivided into L_Q (*Love ws* or *Querwellen*) and L_R (*Rayleigh ws*), referred to as *Q* and *R* waves. [*f* EARTHQUAKE]

Longmyndian A series of mudstones and sandstones, once thought to be Cambrian, now known to be of Precambrian age. Outcrop typically in the Longmynd, in S. Shropshire.

long-profile, of a river The PROFILE of a r bed, from source to mouth

long-range forecast A f for a period greater than 5 days. These have been issued by the US Weather Bureau for some time; public forecasts of this kind for a month ahead were started in UK in Dec. 1963. Synoptic ANALOGUE methods, involving computer handling of vast quantities of data, are used.

longshore drift A d of material along a beach as a result of waves breaking at an angle. A breaker (SWASH) sweeps material obliquely up the beach, the BACKWASH drags some down again at right-angles, thus a net movement along the beach; e.g. to E. along S. coast of England, since dominant winds and waves are from the S.W. A l current will help this movement. [*f*]

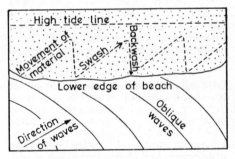

'long time-scale' A t-s of PLEISTOCENE glaciation, in which its onset is pushed back to 1·8 to 2 million years ago. Ct. 'SHORT T-S'. Much evidence for this proposed lengthened span of the Pleistocene is based on radiochemical analysis of CORE SAMPLES from deep-sea sediments of the Atlantic Ocean. One such study indicated that the avge rate of accumulation was 2·5 cm per millennium. The whole Pleistocene section, its start delimited by an abrupt change from warm to cold water planktonic foraminifera, totalled 3·0 m (= on this basis, 1·3 million years ago).

lopolith A large-scale saucer-shaped intrusion of igneous rock lying CONCORDANT with the strata, and forming a shallow basin; e.g.

Bushveld L. in the Transvaal, more than 480 km across; rhyolite plateau of Yellowstone National Park between Gallatin and Absaroka Mtns; Sudbury (Ont.); Duluth (Minn.) of gabbro. [*f*]

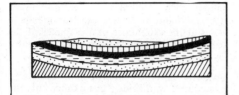

Lorenz Curve A c drawn using % values on each axis, designed to show, by means of its degree of concavity, to what extent a particular distribution (e.g. concentration or clustering of population) is uneven in cf. a uniform distribution (which would be shown by a straight line crossing the graph at an angle of 45°).

lough (Irish) A lake or arm of sea (equiv. to Scottish loch); e.g. L. Neagh, Derg.

'low', atmospheric A DEPRESSION, or low pressure system in the atmosphere, shown on a chart by closed isobars of diminishing values towards the centre. [*f* DEPRESSION]

low, beach A long shallow depression on a **b**, roughly parallel to the shoreline, separating 2 ridges (FULLS). Also known as a *swale*. E.g. on Lancashire coast near Formby; on Dungeness.

Lowestoft Till The thickest and most extensive glacial deposit of E. Anglia, characterized by many Jurassic and Cretaceous erratics from farther W., and formed during the ANGLIAN glacial period.

lowland A vague word with no precise meaning, gen. referring to land below 180 m, though distinguished rather by contrast with adjacent higher land.

loxodrome See RHUMB-LINE.

lunar day The period of time taken by the earth in rotating once in respect to the moon, i.e. between 2 successive crossings of the same meridian (24 hours, 50 mins). This is because while the earth is rotating once in 24 hours, the moon has its own orbital motion around the centre of gravity of moon and earth, and so crosses each meridian 50 minutes later. This is the cause of the interval of about 12 hours, 25 minutes between 2 successive high tides; i.e. any high tide is 50 minutes later than the corresponding tide of the previous day.

lune A portion of the surface of a sphere cut off by 2 semi-GREAT CIRCLES (i.e. half a great circle, with spherical distance of 180°). Area of l (where lunar angle is θ) =

$$\frac{\theta}{360} \times 4\pi R^2 = \frac{\theta \cdot \pi R^2}{90}$$

lutite A SEDIMENT or SEDIMENTARY ROCK consisting entirely of fine clay-particles (less than 0·002 mm in diameter); i.e. ARGILLACEOUS. E.g. MUDSTONE.

L-wave See LONGITUDINAL W (of an earthquake).

lynchet A man-made terrace on a hill-side, usually parallel to the contours. Ascribed to ancient cultivation practice (from Iron Age or earlier), constructed to provide a level, well-drained strip of land with a S. aspect, and to check soil-erosion. Ls are found esp. on chalk country in S. England, and on the N. sides of Yorkshire Dales. [*f*]

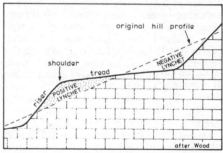

lysimeter An apparatus for measuring the quantity of water percolating through soil; it consists of a container filled with the material under examination, with measuring devices to assess the amount of water and dissolved materials which have passed through. Can also be used to measure the consumption of water by growing plants.

Maar (Germ.) A small, near circular sheet of water situated in an explosion-vent, the result of an eruption which has blown a hole in the surface rocks, surrounded by a low crater-ring of fragments of country rock, but accompanied by no extrusion of igneous rock. Derived from the Eifel Mtns, W. Germany; e.g. Merfeldmaar, Pulvermaar, Lachermaar.

macchia (It.) See MAQUIS.

machair A fine whitish shell-sand, found in broad, gently undulating tracts along the coast of W. Scotland and Hebrides, esp. in S. Uist and Tiree, where it affords light arable soils for crofters.

mackerel sky See CIRROCUMULUS CLOUD. Sometimes the same effect appears with small ALTOCUMULUS CLOUDS. Given because the cloud-pattern resembles the scales of a m.

macroclimate A broad large-scale climatic region, as distinguished orig. from MICROCLIMATE.

maelstrom (Dutch) (i) A whirlpool. (ii) A powerful eddy in a tidal current in a restricted irregular channel; e.g. the famous M of tidal origin in a channel between Lofoten Is, Norway.

maestrale (It.) It. name of the wind known as the MISTRAL.

mafic mineral Syn. with FERROMAGNESIAN M.

magma Molten rock material under the surface of the Earth at very high temperature, charged with gas and volatile materials, and under enormous pressure. The m is probably formed in local concentrations at a depth of 16 km or more, and cannot be regarded as a continuous layer; the fusion of its constituents may be due to local accumulation of radioactive heat. It consists chemically of a solution of a wide range of elements, mainly in oxide form, including silica and basic oxides. When it solidifies under the surface, INTRUSIVE (PLUTONIC) rocks are formed. If it reaches the surface, much gas and water is lost and it becomes LAVA, from which EXTRUSIVE, *eruptive* or *volcanic* rocks are formed on solidification. Hence *magmatic differentiation* or *segregation*, by which different igneous rocks are formed from a single m, largely through fractional crystallization. See also STOPING.

magmatic water See JUVENILE WATER.

magnesian limestone A l with a proportion of magnesium carbonate. With capitals, it is a stratigraphic formation of Permian age. See DOLOMITE.

magnesite Magnesium carbonate ($MgCO_3$), an ore of magnesium.

magnetic declination (or **variation**) The angle at any point on the earth's surface between the m needle of a compass (pointing to the M Pole and indicating the m meridian) and True North (or the geographical meridian), expressed in degrees E. and W. of True North. M d varies in different parts of the world, and also with time. In UK it was about $8\frac{1}{2}°$W. in 1972, decreasing by $\frac{1}{4}°$ every 4 years; i.e. if the rate is sustained m d will be zero in A.D. 2033. At present it is zero on a meridian passing approx. through Cincinnati, Ohio, the AGONIC LINE [*f*].

magnetic pole One of the 2 poles of the Earth's m field, situated in N. America and Antarctica, and indicated by a free-swinging m needle in a horizontal plane. The locations of these shift in a complex way, the N. one situated in Canada near Prince of Wales I. (about 73°N., 100°W.), the S. one in S. Victoria Land in Antarctica. Nor are these poles at the extremities of a diameter of the earth, for a line joining them misses its centre by about 1200 km.

magnetic storm A sudden and temporary, sometimes worldwide, disturbance of the Earth's m field, which can have serious effects on m surveys and short-wave radio. The cause is not fully known, but it is associated with active periods of sun-spots and solar flares, and with the AURORA.

magnetite An ore of IRON (Fe_3O_4); a valuable source of iron in N. Sweden and the Urals. Known as *lodestone*.

magnetosphere The zone of the Earth's magnetic field, extending far into space, incl. the EXOSPHERE and much of the IONOSPHERE. It incl. 2 belts of radiation, an inner at about 3000 km from the Earth, an outer, more intense, belt at 13–19 000 km, known as the *Van Allen* belts.

magnitude, of an earthquake A scale, devised by C. F. Richter in 1935, of earthquake shocks based on instrumental records, an index of earthquake energy at its source. Numbers range from 1 upwards, the largest so far recorded being 8·6. The m differs from *intensity* of an earthquake, related to surface effects of the waves; see MERCALLI SCALE.

mallee A dense scrubby thicket of dwarf eucalyptus, growing to about 2 m in arid parts of S.E. and S.W. Australia.

malpais (Sp.) An area of rough, barren lavasurface, so called because of the difficulty of crossing it.

mamélon (Fr.) Syn with CUMULODOME.

mammilated surface Rock ss smoothed and rounded by various agencies, esp. glacial action; e.g. in Canadian Shield, Finland, Sweden.

manganese A metal found mainly in concentrations of m oxide in bands and nodules in clay.

mangrove (i) A collective name for some genera and species (*Rhizophora* and *Bruguiera*), with ability to grow on tide-washed mud-flats in the tropics. They have short stumpy trunks, sometimes supported by a maze of aerial roots, or by roots which bend at right-angles ('knees'); others send out horizontal roots from which vertical ones grow up through the mud. These root-systems both anchor the m and act as aerating organs. (ii) A swamp or swamp-forest composed of a dense growth of ms; e.g. along coast of S. America near the Amazon delta; along edge of the Niger delta; along coasts of Sumatra and Borneo.

Manning equation A 'flow equation' designed to relate the velocity of a stream to factors of hydraulic radius, slope and channel roughness.

$$V = \frac{1 \cdot 45}{n} R^{\frac{2}{3}} S^{\frac{1}{2}}$$

in which V is mean velocity, R is hydraulic radius, S is slope and n is a roughness factor (the Manning 'n'). 'n' is determined empirically, and is influenced by channel sediment, curvature, bank and bed vegetation, rock outcrops etc. Some values for n are:

A straight channel with no bars 0.030
A curved channel with bars 0.040
A mountain stream with steep banks
   and a boulder strewn bed 0.050

mantle (i) A layer of ultrabasic rocks, density 3·0–3·3, 2900 km thick, lying between the CRUST and CORE of the earth (formerly called 'lower layer'). Its upper surface (up to 40 km) under the continents, (6–10 km under the oceans) is the MOHOROVIČIĆ DISCONTINUITY, its lower surface the GUTENBERG DISCONTINUITY (at 2900 km down). In the upper m between about 100–200 km down, is the ASTHENOSPHERE. (ii) The surface accumulation of soil and weathered rock; the equiv. of REGOLITH. [*f* ISOSTASY]

map projection The representation of the Earth's parallels and meridians as a net (GRATICULE) on a plane surface. Some ps are theoretically constructed on a developable surface, i.e. cone, cylinder or plane, capable of being laid out as a plane on to which the graticule is 'projected' geometrically or by calculation. Many CONVENTIONAL ps are not constructed in this way. See under names of individual ps.

maquis (Fr.) A low scrub of evergreen aromatic plants, found in Mediterranean lands: oleander, rosemary, heath, arbutus, lavender, myrtle, a profusion of interlacing creepers, vines, herbaceous and bulbous plants. It is partly a reflection of the Mediterranean climatic régime of summer aridity, partly the result of felling former evergreen oak forests and therefore a degeneration of the vegetation cover. It is usually found on siliceous soils, replaced by GARIGUE on limestone.

marble A crystalline limestone metamorphosed by pressure and heat, so forming a hard, patterned shiny rock; used for decorative purposes e.g. Carrara, Italy; Connemara (Ireland). Sometimes given loosely to any decorative stone that will take a polish; e.g. 'birds'-eye m', which is a crinoidal limestone.

'mares' tails' Long drawn-out wispy CIRRUS CLOUDS, indicating strong winds in the upper atmosphere.

marginal channel The c of a meltwater stream following the line of junction between ice and rock. Where the rock is resistant the channel will be cut wholly in the ice; and after deglaciation no evidence of its existence will remain. Where the rock is weak the channel will be cut partly in rock and partly in ice; and after deglaciation hill-side benches will remain. These benches will be discontinuous, and run at a low angle to the contours of the slope. A number of benches on a valley side may be used in the reconstruction of stages in ice withdrawal.

marginal deep A long narrow trough in the ocean floor, parallel and close to an island arc; e.g. the Sunda D. parallel to Sumatra and Java.

marginal depression A longitudinal channel developed at the base of a scarp or around an inselberg in some savanna regions. An **m d** is formed (i) by concentrated surface erosion, where run-off from the scarp encounters the upper pediment, (ii) by advanced chemical weathering, resulting from the approach of the water-table to the ground surface or the former presence of water holding detritus. R. W. Clayton has argued that **m d**s in Ghana result from the removal of rotted rock by springs working headwards up the pediments. Also referred to as '*linear depressions*' or '*scarp-foot depressions*'.

marginal sea A semi-enclosed s that borders a continent, and lies on a submerged portion of a continental mass, rather than within an ocean basin; e.g. S. of Okhotsk, S. of Japan, Yellow, S., Sulawesi (Celebes) S., Baltic S., North S., Mediterranean S., Red S., Persian Gulf.

marin A moist warm S.E. wind, blowing esp. in spring and autumn across the coast of S. France, the result of a depression in Gulf of Lions.

maritime air-mass See AIR-MASS.

maritime climate A climatic régime experienced on islands and near coasts (particularly W. coasts in mid-latitudes), usually with a small seasonal and diurnal temperature range, and with appreciable cloud and precipitation. It can occur in any latitude; e.g. Tropical M (Köppen's type *Af*), Humid Sub-tropical (*Cfa*), Cool Temperate M (*Cfb, Cfc*).

marl (i) A clay with an admixture of at least 15% calcium carbonate. (ii) A calcareous mudstone; (iii) Used loosely for any friable clay soil. (iv) In geology, proper name of several types of rock; e.g. Keuper M, Chalk-m.

marsh (i) *Coastal:* see SALT-M. (ii) *Inland:* area liable to temporary inundation, but usually wet and ill-drained; rivers frequently overflow and silt is deposited. There are extensive sheets of shallow water, with rushes, reeds and sedges, and occas. water-tolerant trees (e.g. alder). When drained, a m is usually fertile, with its mineral-rich soil; e.g. silt-lands of the N. Fen District, England; Pontine Ms Italy (both now largely drained). Ct. BOG, FEN, SWAMP.

marsh gas Methane (CH_4), a major constituent of natural g, also resulting from the decaying of vegetation in marshes. It occurs in coal-mines as *fire-damp*.

mascaret (Fr.) See BORE.

mass balance Calculation of the 'budget' or 'regime' of a glacier, by comparison of gross annual accumulation on the upper glacier and net annual ablation on the lower glacier. Glaciers with a '*positive m* b' (i.e. accumulation in excess of ablation) will thicken and accelerate in flow, producing a steep and advancing ice front. Conversely a '*negative mb*' results in glacier thinning and retreat. Where the 'total budget' (accumulation *plus* ablation) is large, the glacier will be very active, flowing rapidly and effecting much erosion.

massif (Fr.) A compact plateau-like mass of uplands, with clearly defined margins; e.g. M. Central of France. Applied to most Hercynian blocks in Europe. Also used of a distinct mountain group; e.g. Mont Blanc m.

massive App. to a rock markedly free from stratification, bedding-planes, jointing and cleavage; a thick uninterrupted stratum.

mass-movement, mass-wasting Downward movement of material on a slope under influence of gravity, usually lubricated by rain-water or snow-melt. (i) *Slow:* soil-creep, rock-creep, scree- (or talus-) creep, SOLIFLUCTION; (ii) *rapid:* EARTH-FLOW, EARTHSLIDE, SLUMP, ROCK-SLIDE, ROCK-FALL. See individual types.

material cycle The pathway of a partic. type of matter (e.g. water, carbon, nitrogen, sulphur) through the Earth's ECOSYSTEM. Sometimes referred to as *biogeochemical* or *nutrient c.*

mature An advanced stage in the development of a landscape, river, shoreline or soil. Much used for certain types of landscape, specif. of rounded hills and well developed valleys, following the writings of W. M. Davis, though there has been considerable criticism of such terminology.

maximum thermometer A t which registers the max. temperature recorded over an interval of time. One type consists of a metal rod within a sealed glass tube, resting on the mercury meniscus. As temperature rises this is pushed upwards, but it so fits into the tube that when the mercury falls again (with a drop in temperature), the marker is left at the highest point reached. This can be read later.

Meade's Ranch A geodetic station in central Kansas, approx. in the centre of USA, its latitude and longitude precisely fixed in relation to the Clarke Spheroid of 1866 (lat. 39° 13′ 26″·286, long. 98° 32′ 30″·506). The starting point for the geodetic triangulation of USA, Canada and Mexico. Positions of all other primary control points were related to M's R, a system known as the *North American Datum of* 1927. Topographical surveys on all scales are related to this.

meadow soil An INTRAZONAL *hydromorphic* s formed in flood-plain of a river where flooding occurs for a short part of the year, resulting in deposition of a thin layer of silt and mud, but with considerable growth of vegetation at other times. The A-HORIZON is dark, with much organic material, with an underlying GLEI horizon, where waterlogging prevents oxidation, and presence of ferrous salts gives the s a bluish-grey appearance, with occas. iron concretions.

meander A curved, loop-like bend or sinuosity in the course of a sluggish river, or of a valley (*valley-m*); derived from Maiandros R. in Asia Minor. Various types have been distinguished: INCISED, INTRENCHED and INGROWN. As the current flows round a bend, the curve is accentuated, since water impinges most strongly on the concave side (outside of

MEANDERING

channel occur. Gen. with an increase in the discharge of a stream, or a reduction in its load, the wider the **m-b** becomes. [*f*]

Plate 41. The MEANDERS of the north Tyne, near Simonburn, Northumberland, England. Note the POINT-BAR DEPOSITS on the inside of each meander bend. (*Aerofilms*)

the curve), causing max. erosion there, even undercutting (forming a *bluff* or *river-cliff*), while there is little erosion and usually deposition in the slack of the current on the inside of the bend (*slip-off slope*). Sometimes an initial slight bend in a river may be transformed into a **m**. Ms may develop until only a thin neck separates the stream on each side, and a 'cutoff' or OXBOW lake may be produced, with a remnant of the spur left as a M-CORE. Gradually ms move downstream The major problem is their initiation. The size and wavelength and the width of the belt appear to be related to such factors as the discharge and bed-load of the river, and to varying depths as the stream crosses an uneven surface, or from non-resistant to resistant rocks. The formation of a m lengthens the course of a river and so reduces its gradient and velocity, and enables a greater volume of water to be moved, thus possibly developing to a state of grade. See POOL AND RIFFLE. See pl. 41. [*f, opposite*]

meander-belt The flat floor of a valley between the outside limits of successive **m**s, within which migration of a stream and its

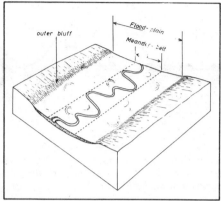

meander-core Land almost surrounded by the nearly complete circle of a river in an INCISED M. E.g. Palace Green, Durham, nearly surrounded by R. Wear; there are many examples in the Moselle valley. Some definitions limit **m-c** to isolated hill left when the river completely breaks through the 'neck' of the **m**; e.g. Wye valley near Redbrook; Dee in N. Wales to N. of Llantisilio.

meandering valley A sinuous valley, with steep crescentic slopes on the outsides of bends and gentler slopes on the insides, together with extensive valley-floor alluvium over which the *existing* stream has developed meanders much smaller than those of the valley as a whole. G. H. Dury has argued that **m** vs are ancient channels cut into solid rock at a time when climatic conditions produced a bankfull discharge some 20 times that of the present underfit stream. **M** vs are widely found

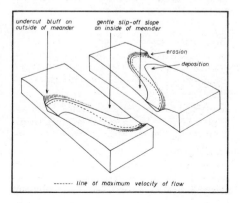

in S. England e.g. the R. Leach, Gloucestershire and the R. Evenlode, Oxfordshire. [*f*]

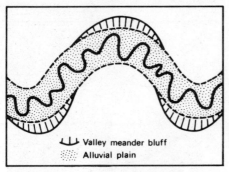

meander-scar A m abandoned by a river (see CUTOFF), filled in by deposition and vegetation, but still traceable.

meander-terrace An unpaired river-t, formed when a river is meandering freely, though eroding vertically. As it swings across the the valley, it removes part of a former higher level of its flood-plain, leaving a portion as a higher t. [*f*]

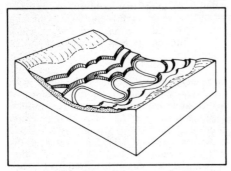

mean sea-level An avge level of the sea, calculated from a long series of continuous records of tidal oscillations (see ORDNANCE DATUM). **MSL** is calculated for UK at Newlyn, Cornwall. This is 40 mm below the Liverpool Datum used before 1921.

Mean Solar Time An avge or *mean solar day* of 24 hours, used because the interval between successive transits of the sun over the meridian (APPARENT T) is not uniform. See EQUATION OF T.

mechanical (or physical) weathering Disintegration of rock by agents of weather (frost, temperature change), without involving chemical change.

medial, median moraine A line of debris down the centre of a glacier, formed by merging of LATERAL MS of two confluent glaciers. **M ms** are wholly superficial formations comprising an ice ridge overlain by less than 1 m of protective rock debris. Following the melting of the glacier the **m m** is destroyed. E.g. Aletsch Glacier, Switzerland, where several **m ms** follow the curve of the valley. See pl. 42, 45. [*f* MORAINE]

median mass A high intermontane area within a zone of fold-mountains. As two forelands of a GEOSYNCLINE approached, the result of OROGENIC pressure from either side, their bordering portions may have been overthrust on to the margins of each foreland, forming fold-mountain ranges, but with a lofty, rel. unfolded part of the geosyncline remaining between as the **m m**. E.g. plateau of Tibet lying between Himalayas on S. and Kuen Lun on N.; Persian plateau between Elburz Mtns on the N. and Zagros Mtns on S. [*f*]

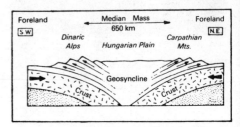

Mediterranean climate Sometimes called Western Margin Warm Temperate **c**, (*Csa* and *Csb* types on W. Köppen system). A warm temperate **c**, occurring on W. margins of continents in latitudes 30–40°, characterized by hot, dry, sunny summers, and moist warm winters, the result of alternation between gen. high pressure conditions in summer, and the passage of depressions assoc. with moist winds from the oceans in winter. Occurs in the New World as a rel. narrow coastal area, sharply defined inland by mountain ranges, in central California and central Chile; in S.W. of Cape Province (S. Africa); S.W. and S.E. Australia; and around the Mediterranean Sea, where total rainfall decreases progressively E. (Gibraltar 914 mm, Athens, 406 mm). A wide variety of **c**s is prevalent within the general type-area of the Mediterranean; ct. S.W. Spain with Corsica, Greece and Israel.

megashear A STRIKE-FAULT of continental dimensions.

megatherm One of a category of plants with the temperature requirement of each month with a mean of more than 18°C; typical of Tropical Rain-forest.

Plate 42. A MEDIAL MORAINE on the Upper Arolla glacier, Switzerland. The moraine is largely the product of FROST-SHATTERING of schists on the northern face of La Vierge, the small peak in the top right of the photograph. (*R. J. Small*)

190 MEGATHERMAL

Megathermal Stage See ATLANTIC STAGE, of climate.

melanocratic A category of IGNEOUS rocks consisting mainly of dark mafic ferromagnesian minerals; e.g. hornblende, olivine, augite, biotite (black mica). Ct. LEUCOCRATIC.

melt-water Water formed by melting of snow and ice; e.g. m-w stream issuing from the snout of a glacier.

Mercalli Scale (modified) App. to earthquake intensity, as given on a numbered scale ranging from I (only detectable by a seismograph) to XII (catastrophic, involving the total destruction of buildings). Ct. RICHTER SCALE.

Mercator Projection A CYLINDRICAL P with CONFORMAL properties, which Gerhard Mercator used for his world-map of 1569. Parallels are straight lines, drawn the same length to scale as the Equator, equally divided for meridians intersecting at right-angles. The distance between parallels incr. from the Equator to preserve correct ratio between latitude and longitude, giving great and increasing distortion in high latitudes. The parallel intervals are conveniently obtained from tables; their computation involves calculus. Its chief advantage is that lines of constant direction are straight, so that it is used widely for navigation to plot bearings (LOXODROMES) as straight lines (see GREAT CIRCLE [f]). One min. of latitude is given on the side of a chart, as a scale line equal to a nautical mile. See TRANSVERSE M P. [f]

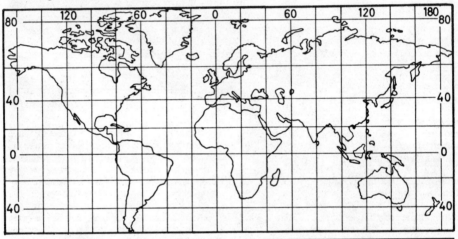

mercury The only metal liquid at ordinary temperatures; obtained mainly from red mercuric sulphide (*cinnabar*).

mere A small lake, esp. on a clay-covered plain; e.g. Breckland. Occurs specif. in Cheshire, the result of subsidence caused by removal of underground salt deposits.

meridian A line of LONGITUDE, a Great Circle passing from Pole to Pole, numbered to E. and W. from 0° (Greenwich, the Prime M) to 180°. A *m circle* is a GREAT CIRCLE consisting of a **m** and its complement; e.g. 0° and 180°.

meridian day Used for: (i) the **d** on which the INTERNATIONAL DATE-LINE is crossed; (ii) the actual **d** repeated on a ship crossing the International Date-Line when sailing in an E. direction.

meridional flow A type of atmospheric circulation in which N. to S. (meridional) movement of air is dominant. Ct. ZONAL F.

merokarst A region of KARST, where typical features of limestone solution are limited because of either: (i) its impurity (e.g. presence of bands of marl or sand); or (ii) thinness of the limestone strata, as in Avant-Causses (W. of the Grands Causses, S.W. France).

mesa (Sp.) (i) A flat-topped eminence (*lit.* tableland), commonly capped with a resistant rock-stratum, the remnant of denudation of a plateau in a semi-arid area; more extensive than, though similar to, a BUTTE. (ii) A tableland extending back from a scarp-edge. E.g. Arizona, Utah, and S. California in S.W. USA, where it forms a common place-name, as Checkerboard M. (Zion National Park, Utah), M. Verde (Colorado). [*f*]

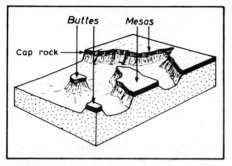

Mesolithic period (lit. 'middle stone', from Gk.) A culture p following the PALAEOLITHIC, from 10 000 to 4000 B.C., assocd. with the end of the Pleistocene and characterized by the use of microlithic implements, ranging from Mesopotamia to the Baltic and the New World. Other features involved development of fishing and domestication of the dog. In Britain lasted from the 8th to the 4th millennium B.C. During this time England was separated by sea from the continent of Europe (*c.* 5000 B.C.), and the moister ATLANTIC climatic stage began.

mesophyte One of a category of plants requiring a moderate or avge amount of moisture, incl. most trees; adj. *mesophytic.* Ct. XEROPHYTE, HYGROPHYTE, HYDROPHYTE.

mesosphere A zone in the atmosphere above the STRATOSPHERE between 50–80 km high. At its base (STRATOPAUSE) atmospheric temperature reaches a max. (about 77°C) (see OZONE) and then falls with incr. altitude to −100°C at the *menopause.* Above this lies the THERMOSPHERE. Near the top of the m can be seen NOCTILUCENT CLOUDS.

mesotherm One of a category of plants with mean temperature requirements of the warmest month over 22°C, of the coldest month between 6°C and 18°C; e.g. olive, the 'type-tree' of Mediterranean climate.

mesosystem Used by meteorologists to denote atmospheric systems of 'medium' size, with horizontal dimensions between 15–150 km, such as LEE WAVES, LEE DEPRESSIONS, LAND AND SEA BREEZES, THUNDERSTORMS, the 'eye' of a HURRICANE, convective systems. Akin is *mesometeorology,* in ct. to MICROMETEOROLOGY, SYNOPTIC meteorology, *macrometeorology.*

Mesozoic Derived from Gk. 'middle life', app. to the 2nd (hence sometimes Secondary) era of geological time subsequent to the Precambrian, and to rock-groups deposited during that time, lasting from *c.* 225 to 65 million years ago. Divided into 3 periods: (i) *Triassic*; (ii) *Jurassic*; (iii) *Cretaceous.*

mesquite A deep-rooted, drought-resistant leguminous shrub or stunted spiny tree, growing in arid parts of S.W. USA, Mexico and as far S. as Peru. It forms dense thickets.

metamorphism, adj. **metamorphic** The process by which an already consolidated rock undergoes changes in or modifications of texture, composition or structure, either physical or chemical. It involves: (i) THERMAL (CONTACT) M, the result of intrusion of a mass of molten rock (see AUREOLE); (ii) DYNAMIC M, small-scale break-up of rocks by localized stresses; (iii) REGIONAL M, large-scale, involving both heat and pressure, assocd. with OROGENIES: (iv) AUTOMETAMORPHISM; see PNEUMATOLYSIS, HYDROTHERMAL. METASOMATISM is a form of **m**, but involves the introduction of additional external material. M can produce **m** *rocks,* which tend to be compact and form masses involved within areas of mountain-building; e.g. much of the Highlands of Scotland, Hercynian uplands of Central Europe. A special case is *impact m*; see COESITE.

metasomatism Changes in a rock-substance as a result of processes in which a mineral is replaced by another introd. from external sources by percolating solutions, or by penetrating vapours of high chemical activity (PNEUMATOLYSIS). E.g. GRANITIZATION; the formation of dolomite from calcium carbonate by brine; PETRIFACTION (e.g. Petrified Forest of Arizona, formed by the exact replacement of wood by siliceous solutions).

meteor (i) A body of matter travelling through space, which becomes incandescent and visible when heated by friction with the atmosphere; hence pop. name of 'shooting' or 'falling star'. It usually becomes luminous at heights between about 150 km and 11 km disappearing at about 80 km. If not wholly consumed, it reaches the Earth as a METEORITE. (ii) In a meteorological sense (as by the World Meteorological Organization), any phenomenon (other than a cloud) in the atmosphere: HYDROMETEOR (water), LITHOMETEOR (solid particles), PHOTOMETEOR (luminous phenomena) and ELECTROMETEOR (manifestation of atmospheric electricity). M *symbols* are used on weather maps; see BEAUFORT NOTATION.

meteoric water W on the earth's surface derived from the atmosphere as rain or snow. Ct. JUVENILE W.

meteorite A mass of nickel-iron, silicate minerals and glassy material (TEKTITE) from outer space, which has survived its passage through the Earth's atmosphere and has landed on its surface; ct. METEOR. Several large and many small ms have been recorded as falling on the Earth; e.g. Great Meteor Crater in Arizona, a hole 1251 m across, 5 km in circumference and 174 m deep, was made by the impact of a m about 50 000 years ago. Recent gravimetric surveys have shown that below its floor lies a mass of metal weighing about 3 million tons, 250–275 m down. The Tunguska Crater in Siberia was made by a m in 1908. There are nine other authenticated m craters in the world.

meteorology The scientific study of physical phenomena and processes at work in the atmosphere. This is partic. applied to forecasting, the basis of which is construction of a SYNOPTIC CHART from simultaneous observations of atmospheric phenomena at many stations.

metre A unit of length in the metric system (now a standard SI unit) orig. taken to be 1/10 millionth of the quadrant of the meridian through Dunkirk. The standard m is the distance between two lines on a platinum-iridium alloy bar at 0°C (kept in Paris); it is now precisely defined in terms of the wavelength of the orange light emitted by krypton^{86}. 1 m = 10 decimetres = 100 centimetres (cm) = 1000 millimetres (mm); 1000 m = 1 kilometre; 1 m = 39·3701 ins. = 3·281 ft, = 1·094 yds; 1 cm = 0·394 ins.; 1 m^2 = 1550 sq. ins. = 10·7639 sq. ft; 1 m^3 = 35·317 cu. ft. See LENGTH.

metrication System of measurement based on a rationalized metric system, which UK is in process of adopting. See SI.

mica A group of silicate minerals, having a perfect cleavage, and splitting into thin tough lustrous plates. The main ms are *biotite* (dark), *muscovite* (light or transparent) and *phlogopite* (yellowish) (magnesium-rich).

microclimate The climate of the immediate surroundings of some phenomena on the earth, partic. around groups of plants. The dimensions of the space considered varies with the object; thus a larger scale is involved in studying trees than grass. *Urban m*s receive much attention, incl. effects of buildings on temperature, and air pollution. Hence *microclimatology*, scientific study of ms Ct. MICROMETEOROLOGY, LOCAL CLIMATE.

micro-erosion The study of small-scale e processes, esp. in creation of slopes; e.g. frost-heaving to depths of only 2–3 cm, wetting-drying cycles on the surface, the development of rain-rills.

microgranite A type of granite with medium- rather than coarse-grained texture, occurring in minor INTRUSIONS.

micrometeorology The detailed scientific study of lowest layers of the atmosphere, esp. from ground-level up to about 1·5 m.

micron A unit of length, 1-millionth of a metre (10^{-6} m = 3·937 × 10^{-5} in.) denoted by symbol μ.

micro-relief Slight irregularities of the ground surface (e.g. sheep-walks and terracettes, minor solifluction benches, rill channels and termite mounds) which are normally developed in the regolith rather than in bedrock.

microtherm A category of plants with mean temperature requirements of warmest month between 10° and 22°C, coldest month above 6°C; e.g. most temperate deciduous trees, such as oak.

mid-latitude A latitudinal zone, in its broadest sense between 23½° and 66½° (in both hemispheres). Incr. used as more specif. than TEMPERATE.

midnight sun In latitudes higher than 63½°, the s does not sink below the horizon from mid-May to late July in N. hemisphere, and mid-November to late January in S.

mid-ocean ridge An elongated arch or swell rising from the o floor; in the Atlantic O. (*Mid-Atlantic R.*, known as *Dolphin Rise* in the N. and *Challenger Rise* in the S.), and in Indian O., where a r can be traced from S. India to the Antarctic O. The rs are covered by 3000–3600 m (1700–2000 fthms) of water, and from them rise islands (Azores, Ascension, Tristan da Cunha). The rs are basaltic, probably volcanic in origin. Another theory believes the Mid-Atlantic R. is actually a double r separated by a gaping crack or rift, a possible indication of the moving apart of the continental mass on either side, i.e. CONTINENTAL DRIFT. See PLATE TECTONICS.

[*f, page 193*]

migmatite (Gk. *migma* (mixed) A rock formed by cooling of MAGMA injected into metamorphosed unmelted rock, part of *granitization* (see GRANITE). The m represents a stage in this process, whereby micaschists have been changed via ms into granite over a long period of time. Alt. layers or 'lenses' of schist and granite can be distinguished; e.g. islands off S.W. Finland. A banded *injection gneiss* is a m. See ICHOR.

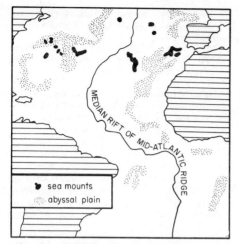

migration of divide The change in position of a **d**, the result of a more active river (with greater runoff or steeper slope) on one side cutting back more rapidly, and capturing an area formerly drained by the weaker stream. In a CUESTA, the steeper slope will usually erode more rapidly than, and thus at the expense of, the back-slope; the **d** at the cuesta crest is therefore continually shifting down-dip.

mile (i) *Statute mi.:* a linear measurement of 1760 yds = 5280 ft = 63 360 ins. = 880 fthms = 80 chains = 1609·3 m. (ii) *Nautical mi.:* the length of 1 minute of arc, or 1/21 600 of a mean Great Circle; i.e. 1 minute in latitude m = 6076·8 − 31·1.cos 2θ ft. This is standardized (in Great Britain) at 48°N. = 6080 ft = 1·1516 statute **mi.** = 1853·25 m; the USA (since 1954) and many other countries use the *International Nautical mi.* = 6076·1033 ft = 1852 m. 1 knot = 1 nautical *mi.* per hour. (iii) *Geographical mi.:* strictly the length of 1 minute of arc measured along the equator = 6087·2 ft; in practice, it is also taken as 6080 ft. Orig. the **mi.** was the Roman measurement of 1000 paces; hence the name (from *milia*).

military grid A **g** system used by British War Office until NATIONAL G was introduced for all official British maps. USA has a **m g** system; see UTM and UPS GS.

'millet-seed' sand Wind-borne **s**-grains in a state of constant movement, impacting against each other and rock-surfaces they meet; hence by ATTRITION each particle becomes more rounded. Not only are desert-**s**s of today like this, but grains in red Triassic sandstones of English Midlands and Precambrian sandstones of Charnwood Forest are similar.

millibar A pressure unit of 1000 dynes per cm², in recording atmospheric pressures as indicated by a barometer. 1000 mb. = 1 bar. No universal formula for conversion, except at constant temperature (0°C) and latitude (45°); 29 ins of mercury = 982 mb.; 30 ins = 1016 mb.; 31 ins = 1049 mb.; 1000 mb. = 29·531 ins (750·1 mm). On SI scale, 1 mb. = 100 newtons per m² (10² N m⁻²).

Millstone Grit A hard, coarse-grained sandstone, sometimes in massive beds, which occurs under the Coal Measures at the base of the Upper Carboniferous. Occas. layers of shale, thin seams of coal, and bands of ironstone are incl. Laid down under shallow-water marine conditions, probably as a delta-deposit. Found in central Pennines (where it forms gritstone moorlands), and in Northumberland; in places it is up to 1500 m thick, and forms prominent edges, as around Kinder Scout (between Manchester and Sheffield).

mima mound A type of large earth hummock, up to 2 m high and 3–20 m in diameter, found in the W. USA. **M m**s have been attributed to burrowing gophers, but most are due to periglacial activity. Possibly **m m**s represent former frozen ground-ice cores, around which surface run-off has eroded channels in thawed soil.

Mindel The 2nd of the 4 periods of fluvio-glacial deposition on the Alpine Foreland which A. Penck and E. Brückner correlated with periods of glacial advance during the Quaternary Glaciation. The Younger Deckenschotter (upper outwash terrace) is associated with this glacial advance. This 4-fold concept of glacial advance has been accepted elsewhere, although research has now shown the picture to be more complicated in many areas, sometimes more, sometimes fewer, glacial periods. The **M** corresponds with Elster glaciation in N. Europe and with Kansan in N. America. [*f*]

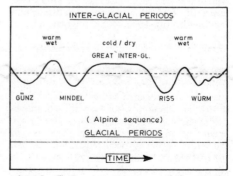

mineral (i) In gen. terms a substance obtained by mining: coal, oil, a metallic ore. (ii)

194 MINERAL

Scientifically an inorganic substance with specif. chemical composition; mixtures of **m** particles comprise rocks. Nearly all **m**s are crystalline. Some are simple in composition, consisting of a single element (e.g. diamond, carbon), most of 2 (e.g. pyrites, of iron and sulphur, FeS_2). **M**s have various properties: crystal form, hardness, specific gravity, colour, lustre and transparency, streak, cleavage, fracture, striations. Over 2000 **m**s are known, though only a few are important as rock constituents.

mineral spring A **s** of water containing an appreciable amount of **m** salts in solution, incl. iron compounds (Harrogate), hydrogen sulphide, magnesium chloride and sodium chloride (Droitwich). May sometimes be of medicinal value, and may lead to development of a spa, as Bath, Carlsbad (Karlovary), Vichy.

mineral water Orig. **w** impregnated with **m** substance, drunk locally for medicinal purposes or bottled; e.g. alkaline Vichy **w** from France. Now covers artificial imitations of such natural **w**, and in gen. carbonated drinks.

minette (Fr.) An iron ore, carbonate or silicate, with a metal content of 24–40% and 1·7 to 1·9% of phosphorus, occurring in strata of Middle Jurassic age in French Loraine and S. Luxembourg.

minimum thermometer A **t** that automatically records lowest temperature attained over a period of time. One type consists of an alcohol **t** with a dumb-bell-shaped marker kept just beneath the meniscus of the alcohol by surface tension. When temperature falls, the marker is drawn to the level of the lowest temperature attained, where it remains.

minute A unit equivalent to: (i) 1/60 of an hour; (ii) 1/60 of a degree of latitude and longitude; (iii) 1/60 of an angular degree, horizontal or vertical. This is not an SI unit (ct. SECOND), but is internationally recognized.

Miocene The 3rd geological period of the Cainozoic era from approx. 26 to 7 million years ago. No rocks of **M** age have been found in Britain, where it was a time of earth-movements and erosion, though MOLASSE of **M** age is widespread in Central Europe. The max. of the Alpine orogeny occurred during the **M**. (ii) In USA, internationally and incr. in UK, regarded as an epoch within the TERTIARY (see table), preceded by OLIGOCENE and succeeded by PLIOCENE.

miogeosyncline A GEOSYNCLINE which experienced little VULCANICITY during its infilling by sedimentation. Ct. EUGEOSYNCLINE.

mirage An optical illusion caused by refraction of light by the atmosphere, partic. in hot deserts, when the layer of air near the ground is greatly heated by conduction, hence becomes less dense, so that rays of light from the sky may be bent upwards; thus the sky may be seen by refraction, giving the impression of a shimmering sheet of water. A **m** may be seen over a road-surface on a hot day; known as an *inferior m*. In a *superior m*, where light rays are bent down from a warm layer of air resting on a cold one (e.g. in high latitudes), a sharply defined double or inverted image of a distant object, such as a ship, may be seen.

mire Soft, spongy waterlogged ground. Used in place-names; e.g. Great Close M., W. Yorkshire; Foxton M., Dartmoor.

misfit river A **r** much too small (*underfit*) for its present valley, because: (i) its headwaters have suffered capture and its size thereby reduced; e.g. R. Meuse in Lorraine; (ii) a change of climate has occurred and its volume reduced; e.g. most English rivers; (iii) a valley has been enlarged by glaciation to a broad U-shape; e.g. small R. Ogwen in the wide Nant Ffrancon, N. Wales.

[*f* CAPTURE, RIVER]

Mississippian Formerly in USA the lower of divisions of the Carboniferous period (and system). Now gen. regarded in USA as a period and system in its own right, 5th in order of age (*c*. 345–310 million years ago), and characterized by widespread deposits of Carboniferous Limestone. Succeeded by the PENNSYLVANIAN.

mist Obscurity of ground layers of the atmosphere, the result of condensation of water droplets, with visibility between 1 and 2 km. Really a form of FOG, with not such restricted visibility. Ct. HAZE.

mistral (Fr.) A strong, cold, dry N.W. or N. wind, blowing from the Massif Central of France towards the Mediterranean Sea, felt esp. over the Rhône delta and Gulf of Lions. Cold air is funnelled down the lower Rhône valley from the winter anticyclone over Central Europe towards low pressure over the W. Mediterranean basin. Avges about 60 km per hour, but 137 km has been recorded. See also BORA.

mixing ratio The **r** of weight of water-vapour in a 'parcel' of the atmosphere to total weight of the air (excluding water-vapour), stated in g of water-vapour per kg of dry air. Ct. SPECIFIC HUMIDITY.

mizzle Very fine rainfall, in the form of a misty drizzle. Cf. Dutch dialect, *miezelen*.

model (i) In its simple conventional form, a m can provide a 3-dimensional reproduction of the landscape, with length and breadth to scale, though of necessity altitude is usually exaggerated. The material may be of superimposed layers of paper, card, pulp, hardboard or wood, moulded plaster or potter's clay, or various plastics (which may be produced by vacuum-forming); these are *hardware ms*. The *m* may be prepared from topographical maps and/or air photographs. Apart from showing land-forms, ms are widely used in civil engineering projects, geological studies etc. A *working scale m* enables processes to be reproduced, to enable their results to be studied; e.g. tidal-m (Southampton Water at the University of Southampton, Mersey at the Hydraulics Experiment Station at Wallingford), a river m (US Corps of Engineers Research Establishment at Vicksburg, Tennessee). (ii) "The identification and association of some supposedly significant aspects of reality into a working system which seems to possess some special properties of intellectual stimulation" (R. J. Chorley). This brings together certain aspects of reality into a clear-cut theoretical m forming a bridge between observation and theory, and providing a working hypothesis against which reality can be tested. They may be graphical, mathematical, theoretical (conceptual), experimental or natural; they may be *iconic* (presenting the same properties, though reduced in scale), *analogue* (in which actual properties are represented by different, though analogous, properties), according to the degree of abstraction, generalization and presentation of information, or *symbolic*, in which the orig. properties are represented by symbols.

model building The abstraction of those parts of complex reality relevant to the problem under discussion, their sorting, sifting, and presentation to form a m of reality.

moder Organic matter present in some soils, intermediate between MULL and MOR, containing a richer fauna than MOR.

mofette (Fr.) A small hole in the Earth's surface, from which issues carbon dioxide, with some oxygen and nitrogen, and occas. water-vapour; e.g. Phlegraean Fields near Naples; Auvergne in central France; and Java. It indicates a late stage in minor volcanic activity.

mogote A steep-walled limestone residual hill, rising above an alluviated plain. (cf. TURMKARST).

Mohorovičić Discontinuity The d between the Earth's crust and MANTLE, so-called after A. Mohorovičić, who discovered it in 1909 while studying a Balkan earthquake. The d affects the speed at which earthquake waves travel. American scientists made preliminary trials towards drilling a hole through the crust into the mantle ('Operation Mohole', off W. Mexico, abandoned 1966). Soviet scientists made similar drillings off the Kuril I. Both US and Russian drills reached BASALT. The **M D** lies at a depth of up to 40 km under the continents, but at only 6–10 km under the oceans. [*f* ISOSTASY]

Mohs' Scale See HARDNESS S. Named after F. Mohs (1773–1839), a German-Austrian mineralogist.

moisture index An i devised by C. W. Thornthwaite (1948) as a basis for climatic classification according to water-balance evaluations of energy (POTENTIAL EVAPOTRANSPIRATION, P/E), and m. It indicates a deficiency or surplus of precipitation. $M\ I = 100\ \frac{(P-PE)}{PE}$. The MI has a value of -100 when precipitation is 0, and may exceed $+100$ when precipitation greatly exceeds evaporation. Using this i, Thornthwaite devised a quantitative climatic classification, with types ranging from A (perhumid) to E (arid), with subdivisions.

Molasse A soft greenish sandstone, with conglomerate and marl, mainly of Miocene age. Its constituents were worn from the Alpine ranges of Europe during and after the orogenic max. (ct. FLYSCH, laid down before the max.), and deposited under continental or shallow fresh-water conditions. Used with l.c., incr. applied to all deposits of similar origin and character.

mollisol See ACTIVE LAYER.

Mollweide Projection An EQUAL-AREA P, in which the central meridian is a straight line half the length to scale of the Equator. If this central meridian is 0° longitude, the area bounded by parallels 90°E. and W. will represent a hemisphere πr^2, with radius $(r) = \frac{1}{2}$ central meridian. Area of hemisphere $= 2\pi R^2$, where R is radius of the Earth, therefore $\pi r^2 = 2\pi R^2$, therefore $r = \sqrt{2}R$. If R is unity (i.e. scale 1/250 million), $r = 1.414$; length of the central meridian $= 2r = 2.828$; length of Equator $= 4r = 5.657$. The parallels are drawn as straight lines at right-angles through the central meridian; their distances apart are obtained from the following table, and are spaced more closely towards the Poles:

(*Pole to Equator* = 1)
Distance from Pole
10°	=	0·137
20°	=	0·272
30°	=	0·404
40°	=	0·531
50°	=	0·651
60°	=	0·762
70°	=	0·862
80°	=	0·945
90°	=	1·000 (i.e. Equator)

Each parallel within the hemispherical circle is divided equally according to the meridian interval (e.g. by 9 to give 20° intervals), and the same divisions, the parallels outside the circle, will give points of intersection of the outer meridians. Smooth ellipses are drawn from Pole to Pole, through each parallel–meridian intersection. The **p** is equal area, since the distance apart of each parallel was calculated to make this so. The linear scale is true only on parallels 40° 40′ N. and S., increasing poleward and decreasing equatorward. The **p** is quite good for a world map of distributions centred on longitude 0°, but if drawn with a central meridian through N. America, Asia is cut in two. Within the inner hemisphere shape is good, but distortion incr. towards the margins in high latitudes, though less so than on the SINUSOIDAL P. The **M P** is used in INTERRUPTED [*f*] form, and its high latitude portion in GOODE'S HOMOLOSINE P.

molybdenum A metal occurring in ores *molybdenite* (MoS$_2$) and *wulfenite*.

monadnock A residual hill or erosional survival standing above the gen. denuded level, named after Mount M. (965 m) in New Hampshire, USA.

monoclinal fold An asymmetrical **f**, the result of compression in the crust, with one limb markedly steeper than the other. Ct. HOMOCLINE, UNICLINE.

monocline The bending or flexing of strata along a line, through tension in the crust; strata are near horizontal (though at different levels), except along the line of flexure. A **m** may turn into a fault along its length or at depth.   [*f, opposite*]

monoglaciation A glacial period in which only one major advance of an ice-sheet occurred, in ct. to the multi-advances of the Pleistocene ice-sheets.

monolith A single block of stone, used of rock buttresses on mountains; e.g. the M., Cwm Idwal (N. Wales).

monsoon (from Arabic '*mausim*' for 'season') The seasonal reversal of pressure and winds over land-masses and neighbouring oceans. Orig. referred merely to winds, specif. in the Arabian Sea, but now pop. used both for 'the rains' accompanying inflowing moist winds, and for various climatic types. In part, the result of much more marked differential heating and cooling of land by comparison with neighbouring oceanic areas, in part the result of poleward shift of hemispheric wind-belts in summer. The Asiatic **M** is considerably affected by the interrupting effect of the Plateau of Tibet on the Upper Westerlies (incl. sub-tropical JET STREAM), which in summer shift N. of the Plateau, permitting the N.-ward surge of the S.W. **M**.

monsoon forest A tropical **f** which experiences a marked seasonal drought. It may occur in true monsoonal lands (Vietnam, Burma, Indonesia, India and N. Australia), and on margins of the equatorial climatic belt, where it forms a transition zone between rain-**f** and tropical grasslands. Luxuriant growth occurs during the wet season, and a markedly deciduous habit of leaf-shedding during the dry. Dominant trees are teak, bamboo (Asia), acacia (Africa) and eucalyptus (Australia).

montaña (Sp.) The forested slopes of the E. Andes in equatorial latitudes.

monte An area of low XEROPHYTIC scrub in foothills of the Andes in Argentina.

Moon The earth's sole satellite, revolving around it in a plane inclined at approx. 5°09′ from the plane of the ECLIPTIC, in a period of 27 days, 7 hours, 43·25 minutes (*sidereal month*). It also has an apparent motion in the Celestial Sphere, rel. to the stars, from W. to E., and completes one revolution rel. to the sun in a mean time of 29 days, 12 hours, 44 minutes (*lunar* or *synodic month*, from one new **m** to the next). A LUNAR DAY is of 24 hours, 50 mins. This time varies slightly

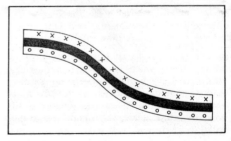

because of the eccentricities in the orbits of both the **m** and the earth. The **ms** min. distance from the Earth is 348 292 km (216 420 mi.), 398 579 km, max. (247 667 mi.) The **ms** mass is approx. 1/81 of that of the earth, its diameter approx. ¼. The **ms** DECLINATION ranges from between about 23° 27′ N. and S., plus about 5° 09′ N. and S. (the inclination of its own orbital plane), i.e. a total possible of 57° 12′, though this occurs only once each 18·5 years. It has no water or atmosphere. The phases of the **m** are: (1) 1st quarter (a semi-circle with the bow facing W.) (QUADRATURE); (ii) full moon (OPPOSITION); (iii) 3rd quarter or last quarter (a semi-circle with the bow facing E.) (QUADRATURE); and (iv) new **m**, invisible except for a faint light reflected from Earth, or earthglow (CONJUNCTION). The **m** is *gibbous* when more than a half-circle is visible. From new **m** to full **m** is *waxing*; from full **m** to new **m** is *waning*. It rotates on its axis once during each revolution in its orbit, and therefore the same face is always turned to the Earth. 41% of the Earth's surface cannot be seen from the **m**. A major result of the **ms** presence is its contribution to the gravitational forces mainly responsible for the TIDES. *Note.* Man first landed on the **m** on July 20–21, 1969.

moor, moorland Strictly an upland area of siliceous rocks, such as Millstone Grit, where acid peat (derived from sphagnum, cottongrass, purple moor-grass) has accumulated under damp conditions; the lit. sense of Germ. *Moor* or *Hochmoor*. More gen. it includes: (1) unenclosed upland waste ('the moors'); (ii) undulating 'upland heath', dry and covered with ling and heather (N. York Ms, Scottish grouse-ms); (iii) lowland marsh (Sedgemoor in Somerset).

moorpan See HARDPAN.

mor A 'raw' humus, markedly acid (pH less than 3·8), formed on heaths, moors and in pine-woods, the result of slow and only partial decay of vegetation because of cool, moist conditions and poverty of soil organisms.

moraine (Fr.) (i) Masses of clay and stones carried and deposited by a glacier; (ii) arrangement of this material to form a landform. M was used by Fr. peasants in the Alps in the 18th century for any bank of earth and stones, and gradually became accepted in Alpine literature. Frost action is potent on rock-buttresses and mountain-slopes above a glacier, and blocks fall on to the ice, some carried on the surface, others sinking in. Material is also picked up from the underlying valley-floor through PLUCKING and ABRASION by the ice. Ultimately much material is deposited in humps and mounds: LATERAL, MEDIAL, PUSH, RECESSIONAL, STADIAL and TERMINAL MS. [*f*]

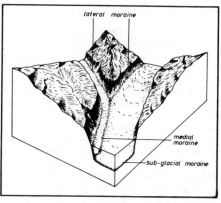

morass An area of waterlogged ground, swamp, marsh or bog.

morphochronology The dating of MORPHOLOGICAL features.

morphogenesis, adj. **morphogenetic** The origin of forms, applied partic. to landforms. A m description of a landform is one in which origin and development are considered. In a *m region* partic. climatic and erosional conditions predominate, giving a regional character which cts. with those of other areas subjected to contrasting processes and conditions. Various classifications have been devised; e.g. by L. Peltier, who distinguished 9 regions, based on annual temperatures and precipitation, plus morphological character: glacial, periglacial, boreal, maritime, selva, moderate, savanna, semiarid, arid.

morphographic map A small-scale **m** showing physiographic features by standardized pictorial symbols, based on the appearance if viewed obliquely from the air. Such ms have become well-known through the work of A. K. Lobeck and E. Raisz in USA.

morphological map A m designed to depict objectively the detailed surface forms, usually of a comparatively small area. Breaks and changes of slope, slope gradients etc. are determined by field survey, and shown in symbolic fashion. M ms are essentially 'nongenetic' (ct. GEOMORPHOLOGICAL MAP), but are intended to assist in the interpretation of relationships between landforms and soil and vegetation distributions.

morphological region In geomorphology, a distinctive unit, of various scales or orders, demarcated according to form, structure and evolution. D. L. Linton suggested a hierarchy of incr. size and complexity: *site, stow, tract, section, province*, and *continental subdivision*. On a m map m types are distinguished by a range of symbols and tints.

morphology The scientific study of form in various connections; e.g. landforms.

morphometry The exact measurement of external features of landforms; a mathematical treatment of them. This basis (e.g. for stream characteristics, drainage basins or slopes) is becoming prevalent in geomorphology. Relationships between precisely measured aspects of landscape are revealed, partic. by the use of statistical methods. While some authorities doubt the real value of some techniques, a morphometric approach ensures an objectivity which was lacking, and the incr. precision of research is a corrective to the subjective knowledge of landscape on which many older theories were based.

morphostasis Used in GENERAL SYSTEMS THEORY to describe a counteracting process to DEVIATION. Any tendency to change a system is counteracted, and steady state characteristics are perpetuated.

mortlake See OXBOW.

mosaic A composite vertical photograph, made by joining individual overlapping photographic prints.

moss Waterlogged boggy land, orig. derived from extensive development of sphagnum and other ms. Used in several ways: (i) an area of siliceous rocks, with layers of peat, characterized by sphagnum and cotton-grass; e.g. Featherbed M. in S. Pennines, N.E. of Glossop; (ii) a lowland m of basic fen-peat; e.g. S. Lancashire Ms (Chat M., Risley M.); (iii) coastal marshes; e.g. Solway M.

moss forest A type of f found esp. on mountains in the tropics (Sri Lanka, E. Africa), and in warm temperate regions (Olympic F., N.W. of Washington State, USA), where trees are covered with thick layers of dripping m. The result of heavy precipitation throughout the year.

mother-of-pearl cloud A somewhat rare, high altitude (about 24 km), c-form, delicate, lenticular in structure, occas. visible in winter in high latitudes during low pressure atmospheric conditions. It reveals an iridescence which can persist after sunset. Scientifically a *nacreous c*.

mottled zone A widely recognized 'horizon' lying between an upper 'zone of induration' (e.g. LATERITE) and an underlying PALLID ZONE in well-developed tropical weathering profiles. The m z may reach 10 m in thickness, and comprises mainly clay with many quartz grains. On exposure it may harden and form BREAKAWAYS along valley edges.

moulin (Fr.) Circular sink-hole ('glacier mill') in the surface of a glacier, sometimes penetrating to the bed-rock beneath, worn by swirling melt-water falling down a CREVASSE. There is commonly a loud roaring noise, prob. the origin of the name. Large pot-holes beneath glaciers have been attributed to the impact of moulin streams, but it is doubtful whether a hole in a glacier would remain stationary for a sufficiently long period for a cavity to be worn in the rock floor.

mound A low hill, either natural or artificial. The latter is common in areas which were liable to flooding, hence refuge-ms; e.g. *terpen* in Friesland, *wierden* in Groningen, *vliedbergen* in Zeeland (Netherlands).

mountain A markedly elevated landform, bounded by steep slopes and rising to prominent ridges or individual summit-peaks. There is no specif. altitude, but usually taken to be over 600 m (2000 ft) in Britain, except where eminences rise abruptly from surrounding lowlands, e.g. Conway M. In such a case, the term Mount is sometimes used; e.g. Mt Caburn in Sussex, an outlier of S. Downs. The world's highest: Everest (8848 m±3 m, 29 028±10 ft), K2 or Godwin-Austen (8610 m, 28 250 ft), Kanchenjunga (8586 m, 28 168 ft). The highest in N. America is Mt McKinley (6191 m, 20 330 ft); in S. America, Aconcagua (6906 m, 22 834 ft); in Africa, Kilimanjaro (5895 m, 19 340 ft); in Europe, Mont Blanc (4811 m, 15 782 ft); in Australia, Mt Kosciusko (2227 m, 7308 ft); in UK Ben Nevis (1343 m, 4406 ft); in England, Scafell Pike (977 m, 3206 ft); in Wales, Snowdon (1085 m, 3560 ft).

mountain-glacier See GLACIER.

mountain wind See ANABATIC W.

mouth, of river The junction of a tributary with its main stream; outfall of a main stream into the sea or a large lake.

mud, oceanic Bathyal deposits of Blue, Coral, Green and Red Ms deposited on the CONTINENTAL SLOPE, derived from clay particles worn from the land, larger than the very fine ooze on the ocean-floor.

mud-flat Area of fine silt, usually exposed at low tide but covered at high, occurring in

Plate 43. Boiling MUD-POT at Krisuvik, south-west Iceland. (*Eric Kay*)

sheltered estuaries or behind shingle-bars and sand-spits. In British waters these **m-fs** are colonized by eel-grass (*Zostera*), marsh-samphire (*Salicornia*), perennial rice-grass (*Spartina townsendii*). Some plants help trap particles of mud and bind it with their roots. In tropical waters, **m-fs** are commonly colonized by MANGROVES.

mud-pot Pool of boiling **m**, usually of sulphurous quality, sometimes brightly coloured, which bubbles away in an area of minor volcanic activity. They may be up to 10 m in diameter. E.g. Yellowstone National Park, Wyoming; Iceland; N. Island, New Zealand. See pl. 43.

mudstone Used to describe argillaceous non-fissile strata, initially of the Silurian system, but used of similar rocks elsewhere.

mud-volcano Ejection of hot water and **m** from a volcanic vent, building a small short-lived cone; e.g. near Paterno in E. Sicily; Krafla in Iceland; the S. of the N. Island of New Zealand. Sometimes natural gases, assoc. with oil-deposits, escape through soft water-logged deposits, building small cones; e.g. Baku, Caucasus, USSR.

mulga Dense thicket of spiny acacia-scrub (*Acacia aneura*) on the margins of the desert of central Australia.

mull (Swed.) (i) Mild HUMUS (pH 4·5–6·5) derived from leaf-mould, found in deciduous forests as a surface layer. (ii) The surface horizon where soil is mixed with humus, giving a granular crumby texture. (iii) Headland in Scotland; e.g. M. of Galloway.

multivariate analysis Quantitative methods of examining and evaluating variables in any problem, espec. where more than 2 variables are involved. E.g. relationship between DISCHARGE and LOAD of a river involves dimension and nature of the load, water temperature, volume, velocity, etc. If sufficient statistical information is available, such methods as multiple correlation and multiple regression may be used with a computer.

murram Reddish lateritic material, widely used for road surfacing in Africa. When rolled out hard it makes a tolerable surface, though can be dangerous when wet.

Muschelkalk (Germ.) The middle of the 3 series of the TRIASSIC system, consisting of fossiliferous limestone (containing ammonoids and crinoids). Occurs in central Germany; e.g. in Teutoburger Wald and Thuringia where it forms a low out-facing CUESTA; Luxembourg and French Lorraine.

muskeg Waterlogged depressions in the sub-arctic zone of Canada, largely filled with sphagnum moss, with scattered lakes and groups of tamarack and fir-trees on slight eminences. There are festoons of meandering, though virtually stagnant, streams. Sometimes pools are covered with moss, with an appearance of solidity, but liable to collapse under the unwary. Very difficult country to cross, and in summer mosquito ridden.

nab (i) A spur, in N. England, esp. in N. York Moors; e.g. N. End, Glaisdale; N. Ridge, near Helmsley. (ii) A headland; e.g. Long N., White N., Cunstone N. near Scarborough.

nacreous cloud See MOTHER-OF-PEARL C.

nadir The point on the CELESTIAL SPHERE opp. the ZENITH; applied gen. to 'the lowest point'.

nailbourne See BOURNE.

naled See AUFEIS.

nano-relief Small-scale surface relief features of an order of magnitude less than that of MICRO-RELIEF e.g. minute rill channels a few cm across.

nappe (Fr.) (i) An overthrust mass of rock in a near-horizontal fold, in which the reversed middle limb has been sheared out as a result of the enormous pressure; actually the hanging wall of a very low-angled THRUST-FAULT. As a result, rocks have been forced for many km from their 'roots', so covering underlying formations. Some writers denote an unbroken recumbent anticline, but this is not correct. A portion of a **n**, surviving after denudation, is a KLIPPE. E.g. in the Alps, a series of **n**-remnants may be distinguished, forming distinctive relief regions: (*a*) *Pre-Alps*, S. of L. Geneva; (*b*) *Helvetic* (Helvetian) ns, Helvetides or High Calcareous Alps, six individual ns; (*c*) six ns of *Pennine Alps:* three Simplon–Ticino ns, Grand St Bernard, Monte Rosa and Dent Blanche ns. Remnants of the last 2 form high peaks of the Pennine Alps. (ii) Esp. in France, used widely for any overlying, covering sheet; e.g. lava flow; equivalent to Germ. *Decke*. [*f*, *opposite*]

narrows A constricted section of a river, strait, valley or pass. 'The N.' is a common place-name; e.g. The N. between Staten I. and Long I., separating the Lower and Upper Bays in New York Harbor.

National Grid The **g** based on the TRANSVERSE MERCATOR PROJECTION, used on current UK Ordnance Survey maps. The axes of the N G are 2°W and 49°N., intersecting at the True Origin, from which the FALSE ORIGIN is transferred 400 km W. and 100 km N. The N G is drawn on the metric system, with: (i) 500-km squares designated by a letter; (ii) within these are 100 km squares designated by a second letter (A to Z, excl. I); (iii) within these are 1-km squares, with every 10 km accentuated, which appear on 1-in., 1:50 000 and 1:25 000 maps. Thus a single reference system is provided for UK, correct on the 1-in. scale to 100 m with still greater precision on large-scale plans. Thus the N G reference of Southampton University is SU 427153. G-lines correspond with sheet-lines, so that 10 km **g** on the 1-in. and 1:50 000 map is an index of the 1:25 000 series. Each of these, together with the 1/2500 and 1/1250, is numbered and identified by the **N G** reference of its S.W. corner. [*f* GRID]

national park An area set aside for preservation of scenery, vegetation, wild-life and historic objects, 'in such manner and by such means as will leave them unimpaired for future generations', both for scientific purposes and public enjoyment. The status of **n p**s varies from country to country. In USA the concept began with the designation of Yellowstone in 1872; the N. P. Service under the Department of the Interior was established in 1916. By 1974 there were 35 **n p**s, 86 **n** monuments and 61 'preserved places', totalling 23 million acres, and incl. Yosemite, Katmani in Alaska with active volcanoes, Death Valley, Grand Canyon, and Statue of Liberty. Canada has spectacular **n p**s (e.g. Banff and Jasper), and Japan has Fuji-Hakone. Manu N. P. in Peru is a rain-forest. In Africa

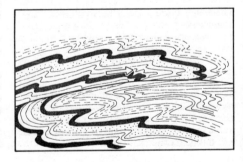

the main problem has been the preservation of animal life, as in Kruger (Transvaal) and Serengeti (Tanzania). In England and Wales n ps were first established in 1949, and now comprise: Lake District, Snowdonia, Exmoor, Brecon Beacons, Yorkshire Dales, N. York Moors, Peak District, Dartmoor, Northumberland and Pembrokeshire coast (total 13 618 km², 5258 sq. mi.) While these include inhabited countryside, they are under planning safeguards. In Scotland 5 'n p direction areas' are subject to special planning control: Loch Lomond, Trossachs; Glen Affric, Glen Cannich, Strath Farrar; Ben Nevis, Glencoe, Black Mtn; Cairngorms, Loch Torridon, Loch Maree, Little Loch Broom.

natural history An old term, popular in the 19th century, wide in its scope and application, implying the study of n objects: animal, vegetable and mineral. Some would restrict it to living creatures, but a n historian is interested in all his physical environment (incl. rocks and minerals; e.g. A. Harker, *The Natural History of Igneous Rocks*). The term seems to be growing in pop. again, implying a science student with wide interests, rather than a narrow specialist.

natural landscape The l as unaffected by man, in ct. to cultural l. Regarded as syn. with *physical l*, i.e. concerned with relief and n vegetation. But so little of the Earth's *l* has been unaffected by man that it is better not to make any distinction, and to refer to n and cultural elements in the l as a whole.

natural region (i) Unit-area of the Earth's surface, with certain uniform and distinctive physical characteristics (structure, relief, climate, vegetation), the basis of regional division. (ii) A geographical r, a sum total of all significant characteristics, in contrast to political or administrative rs. The concept of a r has been the subject of much debate, for while rs delimited in some way are the basis for much geographical teaching, many writers wonder whether they exist other than as the result of applying subjective criteria formulated in the minds of geographers themselves. There is a current trend against using 'n' in connection with rs, unless in very gen. terms. The concept of 'major' n rs was formulated in 1905 by A. J. Herbertson, revised in 1913. These have been the basis of many textbooks and much teaching.

natural vegetation The primeval plant-cover, unaffected either directly or indirectly by Man. *Climatic* CLIMAX VEGETATION is now preferred. Some would distinguish between *actual n v* (the v that would exist were man removed from the scene). Very little of the Earth's present v cover remains unaffected in this way; some writers have doubted whether the concept is of any value, in view of its present small extent and doubtful definition. Now usually taken more widely to mean all v not deliberately organized or included in farming activity, and includes 'wild' or SEMI-NATURAL V. See CULTURAL V.

nature reserve An area preserved so that botanical and zoological communities may survive. This does not necessarily mean an untouched WILDERNESS, but may involve careful control to maintain a partic. environment; e.g. Wicken Fen in Cambridgeshire, where a high water-table is maintained. Regulations are usually imposed to limit public access or disturbance. The N. Conservancy in the UK has demarcated numerous n rs; e.g. Old Winchester Hill, Hants., Cwm Idwal, N. Wales. In UK these incl. 131 national n rs, 26 rs run by local authorities, 13 forest n rs, 600 rs privately owned by voluntary bodies (e.g. National Trust, Council for Nature, County Naturalists' trusts, etc.), and 2600 sites of SPECIAL SCIENTIFIC INTEREST. 13 n rs have been established in N. Ireland.

nautical mile See MILE.

naze A promontory or headland, cognate with 'nose', NESS, NAB and Nase; e.g. the N., Essex; Carr Nase, near Filey, Yorkshire.

neap tide When earth, sun and moon are in quadrature (i.e. at right-angles, with the earth at the apex), t-producing forces do not reinforce each other and the tidal range is reduced, producing ts of low amplitude (i.e. high low ts and low high ts), about the time of the first and last quarters of the moon. N ts occur about every 14·75 days and are about 20% lower than mean ts. Ct. SPRING T.

neat line The boundary, usually a GRID L or GRATICULE, which encloses the detail of a map.

Nebraskan In USA, an early main advance of the continental ice-sheet during the Pleistocene glaciation, corresponding to GÜNZ in the Alps. Ice moved from the Keewatin accumulation centre, and advanced to just S. of the present Mississippi–Missouri junction. The N. drift has been identified in Nebraska, S. Iowa (up to 45 m deep), N. Missouri and W. Illinois, but lies on the surface only in part of Nebraska, and is elsewhere identified only when revealed by post-glacial denudation. The N. period was followed by the Aftonian interglacial.

neck (i) A mass of solidified lava which fills the pipe or vent of a volcano; this may be later exposed by removal of surrounding material by denudation; e.g. Castle Rock, Edinburgh. (ii) An isthmus or promontory, a 'n of land'.

needle-ice A small spike of **i**, (PIPKRAKE), formed just below the surface of soil or loose weathered material. Its presence contributes to downhill movement of particles.

neese A prominent ridge or spur, cognate with NOSE, NESS, NAB and NAZE; e.g. Gavel N., Great Gable, Cumbria.

negative anomaly, of gravity A g measurement below that computed for an ideal globe, implying downward penetration of less dense SIALIC rocks to a greater depth than elsewhere, as beneath rift-valleys of E. Africa and deeps parallel to the Indonesian arcs. Sometimes a *Meinesz zone*, after V. Meinesz, who studied **n** as from a submarine in 1926.

negative landform A rel. depressed or low-lying **l**, incl. valleys, basins, plains and ocean-basins, in ct. with upstanding or POSITIVE LS.

negative movement, of sea-level A change in rel. **l** of land and **s**, causing net lowering of actual **s l**. This may be the result of either: (i) a world-wide or EUSTATIC fall in the **l** of the water (e.g. during the Quaternary glaciation, when abstraction of water in the ice-sheets was equivalent to 90 m fall); or (ii) a more local uplift, warping, tilting, or ISOSTATIC recovery of the land.

'negro head' Name given on the Great Barrier Reef of Australia to a mass of coral, darkened by the growth of lichens, which has been broken off the outer part of a coral reef and hurled on to the reef-flat behind by storm-waves.

Nehrung (pl. **Nehrungen**) (Germ.) A long sand-spit on S. coast of Baltic Sea, formed across the seaward side of a shallow embayment (HAFF), the result of deposition of sand by longshore currents moving E. and N. in the Baltic; e.g. Kurische N., Frische N. (names now altered to Russian and Polish forms), 80 and 56 km long resp. though only 0·4 km to 3 km wide. Lined by almost continuous tracts of sand-dunes up to 60 m high, mostly 'fixed' with marram-grass or planted with pines.
[*f, opposite*]

nekton Plant and animal organisms that actively swim in surface waters of the ocean, as distinct from floating and drifting. Ct. PLANKTON, BENTHOS.

Neogene The younger of the 2 divisions of the Tertiary system/period, as defined by International Geological Congress, and used by European and American geologists. It has not found favour in UK and there is doubt as to the time-period and rocks it covers. Some use the term to indicate all time and rocks from Miocene onwards, others just the Miocene and Pliocene. A convenient term for a period of erosional geomorphology. See table, TERTIARY.

Neoglacial, Neoglaciation Used incr. in USA in place of 'LITTLE ICE AGE' to refer to renewal of glacier growth in the Sierra Nevada and Alaska, after shrinkage or disappearance during the milder HYPSITHERMAL phase.

Neolithic (lit. 'new stone', from Gk.) A culture period, following the MESOLITHIC, from the latter part of the 4th millennium B.C. until the onset of the BRONZE AGE. Characterized by the addition of polishing and grinding of stone tools (notably FLINT) to the earlier percussion and pressure flaking methods. Its characteristic feature was the beginning of the domestication of animals, cultivation of crops and the making of pottery, and in Britain by the construction of long barrows, megalithic tombs and great religious sanctuaries; e.g. Woodhenge and the first part of Stonehenge.

Neozoic (Gk. *neos, zoe*, 'new life') Not gen. in use, but sometimes denoting a combination of Mesozoic and Cainozoic, as ct. with Palaeozoic ('old life').

nephanalysis A generalized cloud chart made from large quantities of data transmitted from a meterological satellite, in the form of either a facsimile image or in a 5-figure code.

nephoscope An optical instrument for measuring direction and speed of movement of clouds.

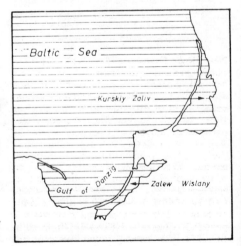

neritic Related to shallow water, hence *n deposits*, organic material derived from remains of shellfish, sea-urchins and coral, found in LITTORAL and shallow water zones along the coast.

ness Syn. with NAZE; e.g. Orfordness, Dungeness.

neutral coast A coastline where there has been no change in rel. level of land and sea; includes stationary cs of sedimentation (MUDFLATS, MARSHES, DELTAS), CORAL CS and LAVA CS.

neutral stability An unsaturated column of air has n s when its ENVIRONMENTAL LAPSE-RATE is equal to the DRY ADIABATIC LAPSE-RATE, and a saturated column has n s when its environmental lapse-rate is equal to the SATURATED ADIABATIC LAPSE-RATE; i.e. the 'parcel' of air is in equilibrium with its surroundings. Syn. with *indifferent equilibrium*.

Nevadian (Nevadan) orogeny A period of earth-movement in N. America, in late Jurassic-early Cretaceous times, incl. folding movements in W. part of the N. American cordilleras, with emplacement of massive BATHOLITHS of granitic rock in the Sierra Nevada. It lasted until, or even overlapped, the LARAMIDE O in the E. part of the Cordilleras of N. America.

nevados, nevadas (S. Am.) Cold down-valley wind blowing from Andean snow-fields to high valleys of Ecuador. A type of KATABATIC WIND, strengthened by chilling of air in contact with a snow surface, resulting in a gravity flow downhill.

névé (Fr.) Usually regarded as syn. with FIRN (Germ.); the latter is preferred by most glaciologists. Some differentiate between the two, firn regarded as actual snow, n the *accumulation area* of firn, or *firn-field*.

New Red Sandstone S rocks deposited during Permian and Triassic times; the line between Palaeozoic and Mesozoic rocks is drawn in the middle of the N R S. Occurs widely in N.W. England (St Bees Head and Eden valley); Devonshire; Durham; and parts of Scotland (incl. S. part of Isle of Arran).

newton The SI unit of force. That force which applied to a mass of 1 kg gives it an acceleration of 1 m per second per second ($=1$ kg m s^{-2}). 1 BAR$=10^5$ N m^{-2}. 1 mb$=10^2$N m^{-2}. 1 lb force per sq. in.$=6894\cdot8$ N per m^2. 1 N per m$^2=1\cdot4504 \times 10^4$ lb per sq in. In 1971 the PASCAL was internationally recognized as the basic SI unit of pressure$=1$ N per m^2 (N m^{-2}).

niche Used ecologically for the partic. position or specif. environment occupied by an organism, for which it is best suited.

niche-glacier A small CIRQUE G, lying high in the mountains in a steeply sloping hollow, bench or gully, probably developed from a compacted snow-patch through NIVATION.

nickel A metal occurring as a sulphide-ore in conjunction with copper and iron, resistant to corrosion and non-tarnishable.

nick-point See KNICKPOINT.

nife The mass of nickel-iron (Ni-Fe) of density about 12·0, believed to comprise the CORE of the earth, lying beneath the MANTLE below a depth of about 2900 km from the surface.

nightglow The very faint light emitted continuously by the upper atmosphere, even on the darkest moonless night, the result of radiation emitted by the ionized products of daytime ultraviolet waves. This is additional to direct light from the stars and from zodiacal sources.

nimbostratus (*Ns*) A low thick cloud, from which continuous rain falls; dark grey in colour and nearly uniform in texture, and occurs partic. above a warm frontal surface in a low-pressure system, having progressively thickened from ALTOSTRATUS.

nimbus A gen. term for clouds from which rain is falling, attached to other cloud-types; e.g. CUMULONIMBUS, NIMBOSTRATUS. Not used by itself in the international system of cloud classification.

nip A distinct break of slope at the higher edge of a beach, clearly rel. to high-water mark, and usually just above the level of the highest tides.

nitrogen cycle The complex of n involving the atmosphere, soil, plants and animals. N is released to the atmosphere from the soil through n-fixing bacteria.

nivation (i) Rotting or disintegration of rocks beneath and around the margins of a patch of snow by chemical weathering and alt. FREEZE-THAW, sometimes called 'snow-patch erosion'. Melt-water by day trickles into cracks and re-freezes at night, causing shattering. The slight hollow in which the initial snow-patch accumulated is progressively enlarged, representing an early stage in the creation of a CIRQUE. (ii) In a more gen. sense, includes the work of snow and ice beyond or outside true glacial limits.

nivation cirque A large nivation hollow, resembling a true glacial cirque in scale and general form. N cs possess steep frost-shattered headwalls, sometimes more than 100 m high, and in plan are semi-circular with a preferred orientation facing N.E. in Britain. However, **n cs** do not possess basin-shaped floors and rock lips; and moraine-like mounds are the result of sliding of debris from the headwall across snow-drifts (cf. PROTALUS RAMPART). Glacial striae and polished rock surfaces, features diagnostic of glacial erosion, are absent from **n cs**. [*f*]

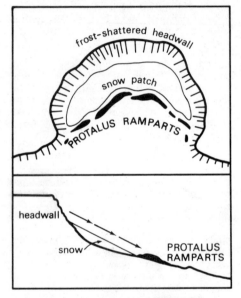

niveo-fluvial A term used by M. P. Kerney *et al.* to describe an assemblage of periglacial processes which operated in Britain during Zone III of the LATE-GLACIAL period. A relatively mild climate, with heavy winter snowfall on high ground but effective summer thaw, produced many fluctuations of temperature about 0°C, rapid solifluction and powerful episodic run-off. This resulted in heavy and selective 'erosion' in some places e.g. the scarp face of the E. N. Downs near Wye. It is argued that this **n f** erosion formed several spectacular escarpment coombes during a period of only 500 years.

noctilucent cloud A luminous bluish or silvery-white c seen (though rarely) after sunset on summer evenings in the STRATOSPHERE at 50–80 km, believed to consist of either dust, cosmic dust, or ice-crystals. Most frequently recorded in mid to high latitudes.

nocturnal A star-clock developed at the beginning of the 15th century for determining time during night at sea. By measuring the angle between the plane of the local meridian and a line through a circumpolar star (usually 'pointers' of the Plough) and the Pole Star, SIDEREAL TIME can be measured. By means of another dial, sidereal time was converted to APPARENT TIME. The medieval **n** was frequently a decorative work of art.

nonconformity, adj. **nonconformable** A type of UNCONFORMITY where igneous rocks are denuded and then overlain by material which compacts to form sedimentary rocks; e.g. near Cape Town, a dark sandstone overlies nonconformably pale-coloured granite.

non-instrumental diaries Source material of past climatic phenomena recorded by contemporary observers; e.g. analysis of Hudson's Bay post journals for dates of freeze-up and ice break-up on rivers and lakes.

non-sequence In a CONFORMABLE series of rocks, a brief gap in the normal sequence, during which no deposition took place.

Nordic Projection A map **p** designed by J. Bartholomew, 'an oblique area-true **p** designed to give optimum representation to Europe, and to routes across the Atlantic, Arctic and Indian Oceans'. The major axis is a GREAT CIRCLE touching 45°N., the lesser axis the Greenwich meridian.

normal erosion Orig. used for river **e** under conditions of a mid-latitude climate, in contrast to desert and glacial **e**, which were termed 'special **e**'. It is now realized that **e** as found in mid-latitude areas may in fact be abnormal, certainly when considered in respect of the earth as a whole, and in relation to past geological conditions.

normal fault A **f** resulting from tension, when the inclination of the **f**-plane and direction of downthrow are in the same direction; i.e. the HANGING WALL [*f*] has been depressed rel. to the FOOT WALL. A vertical **f** is regarded as a **n f**. The term can be misleading if it is thought to imply that this kind of **f** is most commonly found; this is far from the case. [*f, page 205*]

Northeast Trades The TRADE WINDS of the N. hemisphere.

norther A cold dry N. winter wind bringing low temperatures to Texas and Gulf coast of USA. A form of polar outbreak, whereby a

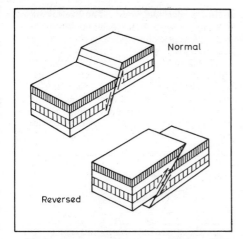

cold air-mass moves S. across the N. American continent, unimpeded by a transverse relief barrier. A similar wind, *norte*, affects the coasts of Mexico and Central America. N is used for cold polar winds in many parts of N. hemisphere (California, Portugal).

northing The distance on a map from the origin of a GRID in a northerly direction, providing the second half of a grid-reference; ct. EASTING.

Nor'wester A FÖHN-like wind in S. Island, New Zealand.

nose A prominent spur or ridge, cognate with NESS, NAZE, NAB, NESSE; e.g. Wrynose Fell, on N. side of Wrynose Pass, Cumbria.

nubbin Used by A. L. Washburn to describe a type of small rounded or elongated 'earth hump' found in areas such as N. E. Greenland, and believed to result from the activity of NEEDLE-ICE.

notch (i) Undercutting of a cliff at high-watermark, which produces a **n** at its foot. (ii) A pass, esp. in the Appalachians; e.g. Pinkham N., White Mtns, New Hampshire. See pl. 44.

nuée ardente (Fr.) A mass of hot gas, superheated steam and incandescent volcanic dust, which moves down the slopes of a volcano during an eruption as a glowing avalanche; particles slide over each other, with a high internal velocity. E.g. during eruption of Mont Pelée in the W. Indies in 1902, St Pierre was obliterated, destroying its 25 000 inhabitants. Another **n a** rolled down the flanks of Lassen Peak in 1915, where a 'Devastated Area' was created as forests were swept away by 'the Great Hot Blast'.

nullah (Indian) The bed of a stream which flows only occas. following sporadic though intensive downpours of rain. Welded tuffs are formed from the deposited dust.

numerical forecasting The f of weather by solving, usually with the aid of a computer, sets of equations relating to associations of observed atmospheric phenomena. Various models are used, and no personal or subjective judgement by the forecaster is required.

nunatak (Eskimo) A rock-peak projecting prominently above an ice-sheet; orig. used in Greenland, where they are numerous along the coastal rim; also common in Antarctica.

Plate 44. NOTCH formed by marine undercutting of upraised coral at Mombasa, Kenya. (*R. J. Small*)

oasis A place in a desert where sufficient water is available for permanent plant growth and human settlement, varying from a group of palms around a spring to an area of some hundred km². An o may occur wherever the WATER-TABLE approaches or reaches the surface, as in a DEFLATION hollow (e.g. district of the Shotts, in Algeria and Tunisia), and esp. in an ARTESIAN BASIN (Libya). Sometimes water must be pumped from depth; e.g. S. Algeria, in the Mzab, where are 3300 wells, from which water must be laboriously raised to supply irrigated gardens.

oblate A flattening or depression of opp. parts of a sphere, hence an *o spheroid*, shape of the Earth.

oblique aerial photograph A p taken from an aircraft with the camera pointing down at an angle. It combines the familiar ground-view with the pattern obtained from a height. In low undulating areas it may be used for map-making (e.g. Canadian Shield), but since scale decreases away from the foreground, this presents difficulty.

oblique fault A FAULT with strike oblique to the strike of the beds it traverses.

obsequent fault-line scarp A f-l s facing in a direction opp. to that caused by the initial earth-movement. After a f-l s has developed, denudation may progress so far that resistant strata on the higher side are removed, exposing less resistant underlying strata. The strata on the DOWNTHROW side will now be more resistant, and incr. stand out as the gen. surface is lowered, so that a f-l s develops along the line of the same fault, but facing in the opp. direction; e.g. Mere Fault on N.W. of Vale of Wardour, Wilts.

[*f* FAULT-LINE SCARP]

obsequent stream Introd. by W. M. Davis to describe ss which flow in a direction opposite to the original CONSEQUENT drainage to join a SUBSEQUENT S. Often used for a *scarp s* or *anti-dip s*, one flowing in the opp. direction to the dip. This transference of meaning can be confusing, esp. if consequent drainage, initiated perhaps on a marine platform, does not conform to the dip. There is still a difference of opinion; many French and American geomorphologists use the orig. Davisian concept as 'anti-consequent', while some English authorities prefer the 'anti-dip' meaning. [*f* SUBSEQUENT STREAM]

obsidian Extrusive igneous rock, with 65% or more silica, in appearance like a mass of dark glass, rhyolitic in composition.

occlusion, occluded front The overtaking of a WARM F by a COLD F in an atmospheric depression, which ultimately lifts the WARM SECTOR off the surface, so forming an occluded front. The cold air in the rear of the depression has now come up against the cold air against which the warm front was orig. formed. If the overtaking air is colder than the air in front, it is a *cold o*, if not as cold it is a *warm o*, as in [*f*] below. If there is no marked difference in temperature, it is a *neutral o*. [*f*]

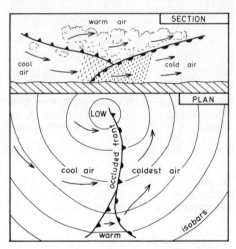

ocean A main water area of the globe, lying in a basin; os occupy in all 71% of its total surface.

	Area		Greatest depth	
	1000 km²	1000 Sq. mi.	m	ft
Pacific	165 384	63 855	11 033	36 198
Atlantic	82 217	31 744	8378	27 498
Indian	73 481	28 371	8047	26 400
Arctic	14 056	5427	5449	17 880

This excludes shallow marginal seas, such as Caribbean and Mediterranean. The first 3 include various sectors of Southern O, sometimes defined as o areas S. of 40°S. Some authorities speak of 6 oceans: N. and S. Atlantic, N. and S. Pacific, Indian and Arctic.

oceanic climate See MARITIME C.

oceanography The scientific study of phenomena associated with oceans. It incl.: (i) *physical o*, extent and shape of ocean basins, structure and relief of their floors, movements of sea-water, its temperature and salinity; (ii) *biological o*, study of life in the oceans.

ocean weather station Fixed meteorological ss in the N. Atlantic, at present 4 in the W. manned by US ships, 5 in the E. by European

Plate 45. The pattern of OGIVES on Morsárjokull, southern Iceland. Note that the curved bands of dark and debris-stained ice are developed on the two glacier-tongues (separated by the prominent MEDIAL MORAINE), indicating that each is flowing to some extent independently of the other. (*C. A. M. King*)

countries in rotation. Each has a code-name and fixed latitude and longitude; e.g. *B* (*Bravo*) at 56° 30′ N., 51° 00′ W.; *K* (*Kilo*) at 45° 00′ N., 16° 00′ W.

offshore App. to movement away from the shore, as o wind, o current, or to accumulation of material away from the shore, as o BAR [*f*].
[*f* COASTLINE]

ogive One of a series of bands of light and dark ice, arcuate downstream, on the surface of a glacier fed by an ice-fall. The o shape results from more rapid flow of the centre of the glacier; at the edges it is retarded by friction. The dark ice bands represent re-freezing of meltwater containing dirt. The light bands contain air-bubbles; this white ice was probably formed from winter accumulation of clean white snow. The bands of dark ice are bubble-free, and represent partial melting and refreezing during passage during summer down the ice-fall. The os form surface ridges, up to 1 m high, owing to the fact that white bubbly ice reflects more insolation, while the dark bands suffer great melting because of their increased conductivity. In USA called *Alaskan bands* or *Forbes bands*. See pl. 45.

oil-pool An accumulation of PETROLEUM held in the pores of sedimentary rock. Though o is widespread, it occurs in concentrations in favourable structural locations; e.g. above a salt-dome (Texas and N. Germany), in a symmetrical anticline (o-dome) above water, in any stratigraphic or lithogical trap.

oil-sand S impregnated with hydrocarbons (bitumen); e.g. tar-sands in N. Canada, which cover a large area of the Athabasca valley 160 km N. of Edmonton, believed to comprise the largest o reserves in the world.

oil-shale Ss containing hydrocarbons (*kerogen*), which can be distilled to give o. Occur in the Carboniferous rocks of the Midland Valley of Scotland in West Lothian.

okta The proportion of the sky cloud-covered, expressed in 8ths of the total, and indicated on a weather map by a proportionally shaded disc. 0 okta=clear sky, 8 oktas= total cloud cover. First adopted in UK on 1 Jan., 1949.

Older Drift Older of the main groups of Pleistocene glacial deposits, more weathered than, and overlain in part by, YOUNGER D. More complex subdivisions have been made, with several periods of glacial and interglacial deposition distinguished. In Europe O D denotes that laid down during earlier glacial advances, lying S. and W. of the Baltic Heights, the main terminal moraine of WEICHSEL glaciation. The O D has been much weathered, eroded, resorted and redeposited by post-glacial rivers.

'oldland' An area of ancient rocks reduced by denudation to subdued relief.

Old Red Sandstone Series of red, brown and white ss, with some conglomerate, marl, shale and limestone, of Devonian age. These rocks are non-marine, probably deposited in large inland lakes. Found in Herefordshire, S. Wales, Midland Valley of Scotland, N.W. Scotland (Caithness), and Orkneys. See pl. 74.

Oligocene (i) In UK until recently the 2nd of the geological periods of the Cainozoic era, and system of rocks laid down from about 38–26 million years ago. (ii) In USA and incr. internationally and in UK, the 3rd series/epoch of the TERTIARY system/period within the CAINOZOIC era; see table, TERTIARY. In England, O rocks occur only in the Hampshire Basin, the result of deposition under freshwater or brackish conditions, incl. limestones, clays and marls, and in Devonshire between Newton Abbot and Bovey Tracey (clays, gravel, sands and lignite). The O is more widespread in Europe, occurring in Paris Basin, N.E. Belgium and N. Germany.

oligotrophic Applied to an environment such as a lake, poor in nutrients and hence having poverty of living organisms, plant and animal. This is the result of the surrounding area, where there is little arable agriculture, scanty population and rugged relief, hence negligible contributions of lime, fertilizers, humus, human waste etc. E.g. Wastwater and Ennerdale Water, English Lake District. Ct. EUTROPHIC.

olivine Chief constituent of a group of rock-forming minerals, basic and ultrabasic in character, consisting of silicates of magnesium and iron, usually dark green in colour. It forms much of the SIMA layer in the crust. An older name is *chrysolite*. Rocks in which o is the most important mineral incl. *peridotite*, *serpentine*.

Omega system Used by US Navy, an electronic s to establish position (within about 1 km) anywhere on the Earth's surface. This depends on the transmission of long-wave radio signals by 6 stations; a ship can use any 2 to establish position and cross-check on a 3rd.

onion-weathering See EXFOLIATION.

onset and lee The result of abrasive effect of a glacier on the upstream side of a rock, and plucking effect on the downstream side. See ROCHE MOUTONNÉE.

oolite A sedimentary rock, usually calcareous, consisting of rounded grains (*ooliths*) from 0·25 to 2·00 mm in diameter. Often likened to fish-roe, hence the name. A grain commonly has a nucleus of a quartz or shell particle, with surrounding concentric calcareous structure. O as a formation name refers to Inferior and Great O outcrops of Jurassic age, which can be traced across England from Dorset to N. Yorkshire.

ooze Fine-textured PELAGIC DEPOSIT, which accumulates on the ABYSSAL zone of the ocean-floor. (i) *Calcareous o* (PTEROPOD, GLOBIGERINA); (ii) *siliceous o* (RADIOLARIAN, DIATOM). O accumulates slowly, at about 2·5 cm in 20 000 years in the Pacific Ocean. Total thickness is great; echo-sounding indicates a layer of about 3700 m, which must have taken 300–400 million years to accumulate.

open system An approach to a GENERAL SYSTEMS THEORY that is increasingly applied to geography. The o s is characterized by supply and removal of energy and material across its boundaries (ct. CLOSED SYSTEM). Under such conditions the s regulates itself by homeostatic adjustments and attains a *steady state*. Throughout time the s attains a constant magnitude and is said to behave equifinally, for different initial conditions can produce similar end-results. Application of the o s concept enables geomorphology to free itself from many constraints, and helps to solve the problem of GRADE. In a closed s grade may be regarded as a hypothetical end-stage in the denudational process, but an o s approach demonstrates how a provisional equilibrium can be established at a much earlier stage. In meteorology an example of an o thermodynamic s is a precipitating cloud i.e. movement out of the s (the cloud). In geomorphology an example is a drainage basin.

opisometer Small serrated wheel linked to a calibrated dial, used for measuring distances on a map (along a road, railway or river) by running the wheel along the partic. line.

opposition When three heavenly bodies are on a common line, the two outer bodies are in opposition to each other and the central body. Thus when sun and moon are in **o** to the Earth, the result is the phase of full moon and SPRING TIDES. Ct. CONJUNCTION, QUADRATURE.

orbit, of the Earth The path through space of the E in its annual journey around the sun. This is an ellipse, with max. distance 152 million km on 4 July (APHELION), and min. distance 147 million km on 3 Jan. (PERIHELION). The mean velocity at which the E travels this orbit is about 106 000 km per hour, varying according to the partic. portion of the orbit.

order, of streams See S ORDER.

Ordnance Datum (OD) Mean sea-level calculated from hourly tidal observations at Newlyn, Cornwall between 1915 and 1921, from which heights on official British maps are derived.

Ordovician 2nd of the geological periods of the Palaeozoic era, and the system of rocks laid down approx. 500–440 million years ago. Formerly regarded as Lower Silurian (hence index letter *b* is used for both **O** and Silurian on Geological Survey maps), but in 1879 a separate system was proposed by C. Lapworth. In Britain the rocks occur in S., central and N. Wales and in neighbouring parts of Shropshire (an area inhabited by a tribe, *Ordovices*, hence name); in English Lake District; S. Uplands of Scotland (a narrow belt continued into Ireland from near Belfast Lough); and S.W. Ireland from Dublin to Waterford. Rocks include shales, flagstones, grits and slates, interbedded layers of volcanic tuff and lava (the period was characterized by widespread vulcanicity, which compose mountainous country in Snowdonia and English Lake District (Borrowdale Volcanic Series).

Series	N. Wales	English Lake District
4. Ashgillian	Blue-grey slates	Coniston Limestone
3. Caradocian	Rhyolitic lavas, with slates	
2. Llandeilian	Volcanic ashes, agglomerates, limestones, flags	Borrowdale Volcanic Series
1. Arenigian or Skiddavian	Arenig Series (flags, grits, shales)	Skiddaw Slates

Note: A series, the Llanvirnian, is sometimes intercalated between 1. and 2., assimilating the upper Arenigian and the lower Llandeilian.

---

ore A mineral containing metal, in sufficient concentration to be economically workable.

organic Living organisms and their remains. Hence **o** deposits: shelly limestone, pelagic oozes, coral, coal and lignite, chalk.

organic weathering Breakdown of rocks by plants such as moss and lichen, causing humic acids to be retained in contact with rock; a tuft of moss may lie in a small hollow, slowly enlarged by rotting beneath it. Vegetation can also have a mechanical effect, the result of penetrating and expanding force of roots. Worms, rabbits and moles have a considerable effect on surface soil.

orientation The setting of a map or a surveying instrument in the field so that a N. to S. line on map or drawing sheet is parallel to the N. to S. line on the ground. Medieval maps were drawn with E. at the top, hence the term.

origin, of a grid A point from which a **g** system is laid out, at the intersection of the central meridian and a line drawn at right-angles (*true o*). To avoid negative values, this is transferred to a point beyond the gridded area (FALSE O). [*f* GRID]

origin, seismic The point in the Earth's crust at which an earthquake shock originates as a result of tectonic movements. Also known as S FOCUS. [*f* EARTHQUAKE]

orocline A bent or flexed (in plan) orogenic belt; e.g. Alaskan O., Baluchistan O.

orocratic period App. to a **p** of crustal earth movement and vulcanicity, the geological duration of an OROGENY. Ct. PEDIOCRATIC P.

orogeny Major phase of fold-mountain building. Hence *orogenesis*, process of fold-mountain building; *orogenic*, forces which cause this process. The main **o**s which affected Europe: (i) Charnian (late Pre-cambrian); (ii) Caledonian (Silurian-Devonian); (iii) Hercynian (Carboniferous-Permian); (iv) Alpine (mid-Tertiary). The chief American **os**: (i) Laurentian; (ii) Algomanian; (iii) Killarnean (all Precambrian); (iv) Taconian (Ordovician-Silurian); (v) Acadian (Devonian-Lower Carboniferous); (vi) Appalachian (Permian, early Triassic); (vii) Nevadian (late Jurassic-early Cretaceous); (viii) Laramide (late Cretaceous-early Tertiary); (ix) Cascadian (Pliocene-Pleistocene). Nos. (vii) and (ix) are not **os** in the true sense of folding, since they involved mainly emplacement of batholiths, *en masse* uplift and vulcanicity (Sierra Nevada, Cascades).

orographic precipitation Caused by ascent (hence cooling) of moisture-laden air over a mountain range. Heavy **p** falls partic. on windward slopes of mountains facing a steady wind from a warm ocean; e.g. Vancouver airport on the flat delta of Fraser R. receives 635 mm (25 ins.) of **p**, Vancouver city 1500 mm. (59 ins.), the mountains behind Vancouver at 1200 m altitude about 2286 mm (90 ins.). But Kamloops, E. of the Coast Ranges in a valley, receives only 250 mm (10 ins.); see RAIN SHADOW. Gen. the **o** factor emphasizes the rainfall caused by other means; a relief barrier causes an intensification of depressional **p** (as W. Britain), since ascent 'triggers off' CONDITIONAL INSTABILITY, causes convergence and uplift, and retards the rate at which a depression moves (prolonging the period of actual **p**).

orography In gen., description or depiction of relief; more specif. relief of mountain systems.

orthodrome Section of a GREAT CIRCLE on the Earth's surface; i.e. the shortest distance between any 2 points. Sometimes called a *geodesic line*. All **os** are straight lines on a GNOMONIC PROJECTION. Ct. LOXODROME.
[*f* GREAT CIRCLE]

orthogonal Lit. right-angled. Applied to PROJECTIONS in which all parallels and meridians intersect at right-angles. E.g. O *Map of the World*, based on the PETERS' P.

orthography The spelling of names on a map, according to some accepted usage.

Orthographic Projection A category of **p** in which a global hemisphere is projected on to a perpendicular plane as if from an infinite distance. Only at the centre of the **p** is scale accurate, and error increases rapidly outwards. Little used except for astronomical charts or pictorial world maps. It can be drawn in a polar, equatorial or oblique plane, but the largest area that can be shown is a hemisphere. [*f*]

orthomorphic projection Category of **p**, syn. with CONFORMAL, meaning lit. 'true shape'.

os, (pl. **osar**) (Swedish *ås*, plur. *åsar*) An ESKER *sensu stricto*.

oscillatory wave theory of tides The concept that the ocean surface can be divided into 'tidal units', each with its node. In each unit the water is set oscillating, varying with rel. position of earth, moon and sun, together with shape, size and depth of the water body within the unit; a gyratory movement is imparted by rotation of the earth. From the nodal points, where the height of water remains virtually level, the tidal rise incr. outward. See AMPHIDROMIC POINT, CO-TIDAL LINE. [*f* AMPHIDROMIC POINT]

outcrop That part of a rock occurring at the surface of the Earth, though covered with superficial soil, vegetation and buildings. Also used as a vb., *to o*. Ct. EXPOSURE.

outfall Narrow outlet of a river or drain; e.g. of the Fleet into the Thames.

outlet glacier A **g** emerging from the edge of an ice-cap on a high plateau; e.g. N. Norway, Iceland. See pl. 5.

outlier (i) *Erosional*: outcrop of newer rocks surrounded by older, the result of its separation by erosion from the main mass of which it forms a detached portion, as in *f* below; e.g. *butte témoin*. (ii) *Structural*: outcrop of newer rocks let down between faults or in a syncline, which has survived when adjacent portions of the same rock have been removed by denudation. Ct. INLIER. [*f*]

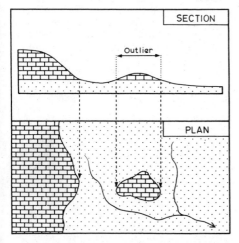

outwash Clay, sand and gravel washed out by melt-water streams from glaciers and icesheets. These deposits form extensive plains, or more limited fans. Hence *o plain, o apron*. In N. European Plain, sheets up to 75 m thick were laid down beyond main terminal moraines, cut into individual blocks by valleys of post-glacial rivers, and now form heathlands. Most glacial forelands and terraces of rivers crossing them are covered with **o** resorted and redeposited. [*f* ESKER]

overbank deposit Sands and silt deposited adjacent to, though outside, a CHANNEL during periods of river flood. This may accumulate to form a LEVEE.

overcast Used as a noun for a complete cloud cover, esp. in aviation.

overflow channel A c by which a lake has overflowed during a former high water-level. Applied partic. to a c draining a lake dammed between a pre-glacial watershed and an icesheet; e.g. Newton Dale, cutting across N. York Moors, which drained water from PROGLACIAL LAKES on their N. flanks S.-ward into the Vale of Pickering; Forge Valley, through which the Derwent flows S., eroded when an ice-sheet blocked its orig. E. course to the N. Sea. When L. Bonneville occupied much of the Great Basin in USA, and its height was 300 m above the present Great Salt L., water spilled over through the Red Rock Pass into Snake R. valley. The Great Basin is now an area of inland drainage. [*f, opposite*]

overfold An asymmetrical ANTICLINE completely pushed over, or overturned, by compressional forces. [*f*]

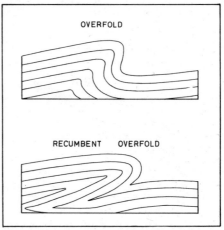

overland flow The proportion of total rainfall which is not intercepted by vegetation and does not infiltrate the soil, but which runs over the ground surface as sheet or rill wash. The amount of **o f** increases with rainfall intensity, and is at a maximum during convectional storms in semi-arid and tropical areas. However, **o f** is rare in humid temperate regions, where vegetation cover is continuous and rainfall intensity low. O f is sometimes referred to as '*Horton o f*', after the model of run-off and infiltration proposed by R. E. Horton.

overthrust The result of very intensive folding, causing the upper limb of a fold to be forced along a horizontal or gently inclined plane over the lower limb (cf. NAPPE). This may involve fracturing of the rocks. E.g.

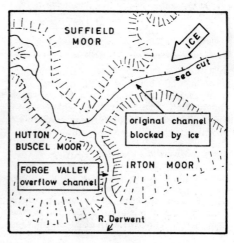

Lewis O. in Glacier National Park, Montana, where Precambrian rocks were thrust E. over Cretaceous rocks; 4 thrust-planes of Caledonian folding in N.W. Scotland: Glencoul, Ben More, Moine and Sole. [*f*]

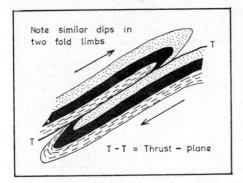

oxbow Surviving portion of an acute MEANDER, formed by river current ultimately cutting through the meander-neck, abandoning a small crescent-shaped lake. Also called 'cutoff' or 'mortlake'; found on most rivers in their flood-plain tracts; e.g. Mississippi, Trent. [*f*]

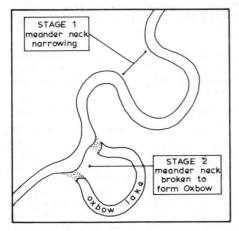

oxidation A process of chemical weathering, involving combination with oxygen. The results are commonly shown when the rocks contain iron. The ferrous state changes into the oxidized ferric state, forming a yellow or brown crust which readily crumbles.

ozone Allotropic triatomic form of oxygen (O_3), very faintly blue in colour, found in minute quantities in the Earth's atmosphere. The **o** layer (*ozonosphere*) has 2 definitions: (i) the layer between 10–15 km in which concentration of **o** is appreciable; and (ii) the zone of max. concentration at 20–25 km. O is believed to result from photochemical changes through the absorption of ultraviolet light radiation by oxygen; this causes a temperature rise at the top of the layer at 50 km, indicating the STRATOPAUSE. An important result is that this layer absorbs most of the sun's shorter ultraviolet waves, forming a screen (or 'hot layer') over the lower stratosphere, though allowing the longer u.-v radiation to pass to the Earth's surface.

Pacific type, of coast See CONCORDANT C.

pack-ice Masses of **i** floating on the sea, specif. when floes have been driven together to form a mass, more (*close p*) or less (*open p*) coherent, but covering the entire sea-surface, with little or no open water except for occas. channels (LEADS). They may be up to 5 m thick, beyond which heat is transferred from underlying depths beneath the insulating ice-layer, precluding further freezing. Frequently individual floes are driven together, their margins buckled and irregular PRESSURE RIDGES formed.

padang Land in S.E. Asia that is gen. treeless, with scrub and heath-type vegetation, commonly developed on sandstones with leached soils.

pahoehoe (Hawaiian) A newly solidified lava-flow with a wrinkled, 'ropy' or 'corded' surface. The liquid lava is at a high temperature, gases escape quietly, and the flow congeals smoothly; ct. jagged 'block' lava (AA), PILLOW LAVA. See pl. 46.

paintpot Descriptive term applied, esp. in Yellowstone National Park, Wyoming, to hole from which hot coloured mud is ejected, in an area of minor vulcanicity.

paired terrace See ALLUVIAL TERRACE.

palaeobotany The study of plant life of past ages, mainly from the evidence of fossils.

Palaeocene Used in 3 contexts. (i) The earliest part of the Eocene, lasting from *c.* 70 to 60 million years ago. (ii) Strata intermediate in age between Cretaceous and Tertiary systems. (iii) By international usage, the 1st. of the epochs of the Tertiary period.

palaeoclimate, palaeoclimatology The climate, or the study of a climate, of some period in the geological past. Climates have varied considerably, with major glacial, desert and pluvial episodes.

Palaeogene, Paleogene The older of 2 divisions of the TERTIARY (see table) system/period, as defined by the International Geological

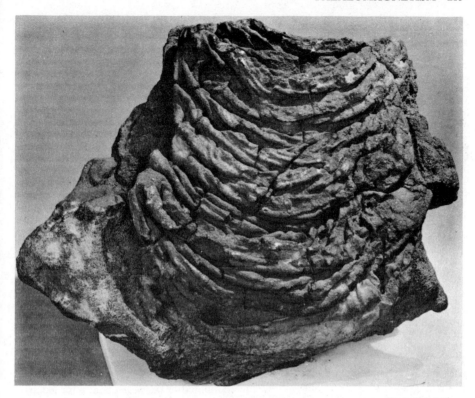

Plate 46. PAHOEHOE. (*Institute of Geological Sciences*)

Congress, and used by American and European geologists (though less in UK) for the 3 oldest series/epochs of the Tertiary (PALAEOCENE, EOCENE, OLIGOCENE).

palaeogeography The reconstruction of the pattern of the Earth's surface during its geological history.

Palaeolithic age (Gk. *palaios* ancient, *lithos*, a stone) The Old Stone A, the earliest period of human pre-history, coinciding with the greater part of the Pleistocene Ice-age. Divided into: *Lower P* (c. 1–2 million years ago–120 000); *Middle P* (120 000–40 000); Upper **P** (40 000–10 000). A series of culture-periods has been distinguished, related to the various glacial and interglacial phases: *Pre-Chellean, Chellean* (or *Abbevillian*), *Acheulean, Mousterian, Aurignacian, Solutrean, Magdalenian* and *Azilian*, named after places in France. The earliest implements were roughly fashioned palaeoliths of flint, but gradually these became more highly fabricated ('blade industries'), and at some stages bone was used. Cave-paintings are a significant feature, esp. in S.W. France, as at Les Eyzies in the Vézère valley.

palaeomagnetism Igneous rocks when cooled preserve a certain magnetization, the result of the presence of iron, accordant with the direction of the earth's magnetic field at the time. Similarly do sediments containing detrital magnetite, in which grains have oriented themselves. A wide range of samples is taken, structural complications eliminated, and statistical analysis applied. This fossil magnetism thus provides a record by which the position of the magnetic poles at various times may be located, and has strengthened in partic. arguments that CONTINENTAL DRIFT has occurred. **P** information from rocks of the same age indicates that continents must have moved rel. to one another in order to harmonize the direction of magnetization in differing areas. The most palpable indication of change is given when a complete reversal of direction of the Earth's magnetic field is deduced; this has occurred several times since the Cretaceous, with at least 2 reversed and 2 normal epochs.

palaeontology The scientific study of FOSSILS.

palaeorill The eroded channel of a former minor stream, its course indicated by a basal layer of stream-bed gravel; with the disappearance of surface flow, the channel became filled with sandy downwash from adjacent slopes. A considerable throughflow in this sand may develop, thus keeping it moist. Cf. PERCOLINE.

palaeosol An ancient soil profile, preserved by overlying younger deposits e.g. a 'warm' interglacial soil covered by 'cold' periglacial head. Ps often contain charcoal and pollen grains, which enable accurate dating and thus the reconstruction of Pleistocene climatic phases from p sequences.

Palaeozoic (Gk. 'old life') The first (hence 'Primary') era of geological time subsequent to the Precambrian, and rock-groups deposited during that time. Divided into six periods (seven in USA), with corresponding systems of rocks, and lasted from c. 500–600 million to c. 225 million years ago. See individual periods. Boundary between Upper and Lower P is put at c. 400 million years ago.

Paleocene See PALAEOCENE.

paleoecology Used esp. in USA; the study of the ECOLOGY of past geological times, partic. in relation to changing climates, using RADIOCARBON DATING, POLLEN ANALYSIS, fossils and the landforms with which the climate is associated.

pali ridge A sharp r at the intersection of adjacent valleys on the flanks of a heavily eroded volcanic dome, specif. in Hawaii.

Palisadian In USA block-faulting movements which affected E. parts of that country in late Triassic-early Jurassic times.

pallid zone A widely recognized 'horizon' lying between the overlying MOTTLED ZONE and underlying partially decomposed bedrock in well-developed tropical weathering profiles. The p z may be up to 60 m thick, but is usually less than 25 m. It comprises highly decomposed kaolinitic clay and quartz sand, from which iron minerals have been removed (possibly upwards to form laterite sheets). The p z has been widely noted in arid W. Australia, where it is a 'fossil' feature related to early Tertiary humid weathering.

palsa (Swedish) A conical or elongated mound containing thin ice lenses and occurring in bogs in periglacial areas. Ps may be several m high, and are 100 m or more in length. Most ps are cored by peat rather than mineral soil; the mode of formation of the constituent ice lenses is problematical.

palynology The science of fossil pollen analysis. Pollen grains, blown by wind and deposited in peat-bogs, are partic. resistant to decay. By counting and identifying these grains, a picture of vegetation during past ages may be obtained. A boring is made in a peat-bog, a core brought up, and samples studied under the microscope. Various pollen grains have characteristic and recognizable shapes, and plants responsible can be identified. A 'pollen count' is made, and the contributory proportion of various plants, esp. trees, can be calculated. This affords valuable indications of climatic change during the QUATERNARY.

pamir Poor grassland on high plateaus of central Asia; used as a proper name for a mountain complex in Tadzhikistan, where meet the Hindu Kush, Karakoram and Tien Shan.

pampa (Sp.) Orig. given by Sp. settlers to the extensive, monotonous plain in Argentina and adjacent parts of Uruguay. This comprises a basin of sedimentary deposits, with superficial loess and alluvium. At the arrival of Europeans, the p was covered with tall bunchgrass (*pasto duro*). Gradually the term p was transferred to this vegetation; p (or pampas) is now sometimes used as syn. with temperate grassland. Much has been ploughed and replaced by European meadow-grasses (*pasto tierno*), alfalfa and wheat, and a large part is occupied with rearing cattle on large *estancias*.

pampero (Sp.) A dry, bitterly cold wind, bringing low temperatures in winter to parts of Argentine and Uruguay, the result of a movement N. of a polar air-mass. Associated with the cold front of an E.-moving low-pressure system, and blows from a S. or S.W. direction.

pan See HARDPAN.

pancake ice Small near-circular pieces or 'cakes' of newly formed ice on the surface of the sea.

pancake karst A type of K developed in limestone with closely spaced joints or bedding planes, giving the impression of a pile of ps; e.g. in S. Island, New Zealand.

panfan The endstage of desert geomorphological development, when ridges have been worn down and destroyed, PEDIMENTS are extensive, and basins filled in.

Pangaea A concept put forward by A. Wegener of a single large landmass of SIAL in

Precambrian times, surrounded by a primeval SIMA-floored ocean (*Panthalassa*). He suggested that the land-mass, **P**, broke up to form a S. (GONDWANALAND) and a N. (LAURASIA) land-mass, between which lay TETHYS.

panorama An outline sketch of a piece of country as viewed from some prominent point, covering a considerable horizon distance, and emphasizing foreground, middle-ground and background detail. An essential part of field-sketching. Various geometrical methods can be used. A **p** can be drawn in the field (preferably), or from a contour map.

panplain A plain formed by coalescence of several flood-plains, each created through lateral erosion by its respective river. Hence *panplanation*. Introd. in ct. to PENEPLAIN, formed by degradation of divides between rivers.

pantanal A category of SAVANNA along the flood-plain of the Paraguay R. and its tributaries in Brazil, flooded for several months during summer, but with continuous drought for the rest of the year. A varied vegetation complex, with grasses on lower areas and clumps of trees on higher mounds.

pantograph An instrument for enlarging or reducing a map. It consists of four metal arms of equal length, loosely jointed, with one end fixed to a weighted stand. A cross-bar is moved along parallel to two of the sides, in order to fix the scale-factor. Tracing points are attached at the opposite end to the fixed one and in the cross-bar. If line-work is carefully traced round by the point on the cross-bar, the outer point will draw the same pattern on an enlarged scale, and vice versa.

papagayo (Sp.) A dry, strong, cold N.E. wind, bringing low temperatures and clear weather in winter to coastlands of Mexico. Caused by air moving from the high plateau towards low pressure over the Caribbean Sea and Gulf of Mexico.

parabolic dune A sand-**d**, curved in pattern, but with the curve in the opp. direction to a BARKHAN; i.e. convex down-wind and the 'horns' trail upwind. These occur esp. on sandy shores, after a 'blow-out' has formed by deflation caused by eddying; the face of the **d** is steep, and it may migrate inland. E.g. in Landes of France and W. Denmark. **P ds** are found under similar 'blow-out' conditions in arid interior plateaus; e.g. Mongolia, Tarim Basin. If the **d** moves rapidly forward, its 'horns' will be drawn out parallel to each other, forming a 'hairpin **d**'.

paradigm A type of large-scale MODEL, pattern or exemplar, stable in character, enabling the effective and selective presentation of theoretical and methodological belief in some aspect of scientific enquiry. This involves the handling of a vast field of data matrices (storage, retrieval, analysis and display), possible only with an electronic computer.

parallel drainage A pattern of **d** in which streams and their tributaries are virtually **p** to one another. [*f*]

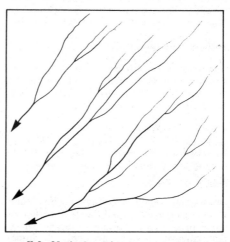

parallel of latitude Line on a map joining all points of the same angular distance N. or S. of the Equator, and parallel to it. See LATITUDE. All **p**s are 'small circles', except for the Equator which is a GREAT CIRCLE.

parallel retreat of slope R of a s under the attack of weathering and erosion without any appreciable change of gradient. The concept of **p r** is partic. associated with W. Penck, and the idea that **s** forms may remain fairly uniform throughout long periods of erosion has become incr. accepted. **P r** is a fundamental process in L. C. King's theory of formation of a PEDIPLAIN and in development of INSELBERGS.
[*f*]

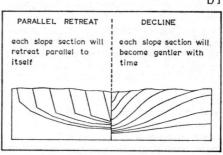

'parallel roads' Horizontal terrace-like features occurring **p** to one another on each side of a valley. These **p rs** may mark former shorelines of lakes formed by ice-damming at several levels; e.g. Glen Roy, near Fort William, Scotland.

paramo (Sp.) High, bleak plateau in the Andes, between the tree-line and permanent snow-line, covered with pasture and tundra vegetation.

parent material Weathered rock-**m** upon which soil-forming processes operate to create soil. The C-horizon of an *in situ* soil.

parhelion A kind of 'sun-image' forming a spot on or near a solar HALO where the light is intensified; sometimes white, sometimes prismatically tinted, with the red nearest the sun, esp. where 2 haloes intersect. Known also as *mock-sun*. Similarly *paraselene* or *mock-moon*.

park savanna(h) S grassland, with trees (acacia, baobab, palm) scattered sporadically among tall grasses, and along water-courses or around water-holes.

paroxysmal eruption Volcanic e with violent explosive activity, usually after a period of quiescence of a volcano, during which subterranean pressures accumulated. Sometimes referred to as a 'Vesuvian' or 'PLINIAN' E after that of Vesuvius in A.D. 79. A **p e** occurred in 1883, when Krakatoa, in the Sunda Straits between Java and Sumatra, blew up.

partial drought In British climatology, a period of 29 consecutive days, some of which may have slight rain, but during which the daily average does not exceed 0·25 mm. Ct. ABSOLUTE D, DRY SPELL.

particulate matter Solid and liquid **m** which occurs in the atmosphere (smoke, dust, pollen, liquid drops as in clouds and fog), some of which are pollutants (esp. products of industry).

pascal (PA) Internationally recognized in 1971 as the SI unit of pressure. 1 Pa=1 NEWTON per m^2=N m^{-2}. 1 Bar=10^5 Pa, 1 mb.=10^2 Pa, 1 lb. force per sq. in.=6894·8 Pa.

pass (i) Col, notch or gap in a mountain range, affording a routeway across; e.g. Great St Bernard P., Khyber P. (ii) Channel of a Mississippi distributary. (iii) Narrow channel through a coral BARRIER REEF; e.g. on Great Barrier Reef of Australia.

passive glacier A g with low rates of ALIMENTATION and ABLATION, the result of light snowfall and low summer melting, as in central Antarctica and Greenland, where only 5–15 cm of snow falls annually. Little ice and debris are transported and the rate of flow is extremely slow. Cf. *active g*, with heavy winter snowfall and warm summers causing rapid ablation, therefore considerable transportation power.

Pastonian An early pre-BEESTONIAN temperate phase in the Pleistocene of Britain, represented by tidal deposits at Paston, Norfolk.

patana (Sinhalese) A coarse grassland on the uplands of Sri Lanka above about 1800 m; e.g. Horton Plains.

paternoster lake One of a string or series of ls in a glaciated valley, separated by morainic dams or rock bars, partic. where the valley is stepped. Derived from their resemblance to a string of beads. E.g. the ls in many parallel valleys of Sweden, draining E. to the Baltic Sea. [*J*]

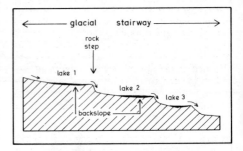

patina Coloured film on the surface of a rock or a pebble, formed by exposure to weathering.

patterned ground Well defined circles, polygons, nets, steps and stripes, characteristic of areas at some time subject to intensive frost action. Material is sorted either into polygonal forms of varying dimensions, with stones round the perimeter and finer material in the centre, or into 'stone stripes'. **P g** also incl. non-sorted forms: polygons outlined by vegetation zones and step-like forms on slopes. There is no certainty as to their mode of origin, though involving various freeze-thaw processes, contraction following temperature change, moisture controlled movements and solifluction. **P g** features probably have several

Plate 47. A limestone PAVEMENT (foreground) above Malham Cove, Yorkshire, England. In the background the CARBONIFEROUS LIMESTONE forms a prominent SCAR. On the pavement itself CARBONATION SOLUTION has etched joint-lines into deep furrows or GRIKES, separated by upstanding CLINTS. (*Eric Kay*)

origins, the object of much research, both in field and laboratory. [*f*]

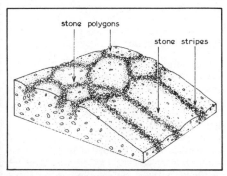

pavement (i) Bare rock surface, produced by weathering, wind erosion, or glacial scouring. See DESERT P, GRIKE. (ii) A rel. smooth and level mosaic of coarse angular or rounded particles, set on an underlay of sand, silt or clay, sometimes cemented. They have such local names as *gibber*, *hammada*, *reg*, *serir*, *gobi*, and occur mainly in hot deserts. See pl. 47. [*f* GRIKE]

pays (Fr.) A small region in France with a distinctive unity based on geology, relief and land-use which distinguish it from its neighbours, though gen. it has no administrative significance. Some **p**-names are derived from feudal administrative units or families (e.g. *Valois*); from striking physical characteristics (e.g. *Champagne humide*); from types of land-use (e.g. *Ségalas*, *P. noir*); from a near-by urban centre (e.g. *Bordelais*, *Lyonnais*). The origin of many is lost in antiquity, others have been coined recently, sometimes by geographers. Hence *paysage*, the whole landscape.

peak Prominent pointed mountain summit, often incorporated in a proper name; e.g. Pike's P., Long's P. in the American Rockies. But the P. in S. Derbyshire, hence the P. District, is a flat-topped plateau.

peat Partially decomposed vegetable matter, dark brown or black, accumulated under waterlogged (ANAEROBIC) conditions. It may be either acid in reaction (*bog p*, *moss p*), or neutral or alkaline (*fen p*); see BOG, FEN. Formed chiefly under temperate or cold conditions, and has accumulated more rapidly in the past (e.g. in ATLANTIC STAGE). Acid **p** is mainly used for fuel (e.g. Ireland), both in domestic hearths and in thermal generators, or for animal litter; fen **p** is used for horticulture and for dressing lawns and racecourses.

PCSM Indices In synoptic climatology a classification devised by H. H. Lamb of daily weather-types or -sequences experienced over the British Isles, summarized as Progressive, Southerly, Cyclonic and Meridional.

pebble Small water- or wind-worn stone, larger than GRAVEL, smaller than a COBBLE; sometimes defined as 10–50 mm in diameter; another definition is 4–64 mm.

pedalfer A leached soil from which base compounds (esp. calcium) have been removed, leaving aluminium and iron, found in humid climates (with over 600 mm of precipitation), esp. when assoc. with high temperatures. Used broadly of one of the 2 major divisions of ZONAL SOILS, esp. in USA (cf. PEDOCAL). A line drawn N. to S. through USA along the Mississippi broadly separates **p**s to the E. from pedocals to the W.

pediment (i) *Erosional*. A gently sloping rock-surface, bare or with but a thin veneer of debris, stretching away from the foot of a ridge or mountain in a semi-desert or desert region. Usually regarded as the product of denudation under arid, sub-arid or savanna conditions. Its upper edge runs abruptly, with a marked change of slope (though sometimes masked by a fan), into the mountain-front; its lower edge slopes gently away under the sands and gravels found in a desert basin. One school of thought regards it as the product of lateral planation by streams, SHEET-FLOOD, rills and downwash, resulting from episodic rainstorms. A second school regards a **p** as resulting from SLOPE RETREAT; a steep mountain-front with an angle of slope of 30° or more, retreats under the attack of weathering and erosion without having its gradient altered to any marked extent. The **p** therefore is a low-angle (6°–7°) slope developing independently at the foot of a steep, parallel-retreating slope, commonly of granite; it is a basal slope of transport over which weathered material derived from the steep slope above is carried by occasional rainstorms. Some refashioning of the **p** by lateral planation may occur, but only as a subsidiary process moulding a feature whose origin is the result of other factors. Some authorities claim that the production of features which are genetically **p**s is a much more widespread process.

Ps are widely found in S.W. USA (esp. Arizona), in the Kalahari Desert in S.W. Africa, and in the savanna lands gen. of Africa. The term *pedimentation* is used for the collective processes which produce a **p**. See INSELBERG, PEDIPLAIN. (ii) *Depositional*: formed by the coalescence of low-angle WANING (or wash) SLOPES near the base of a weathering slope. See FREE FACE (and *f*). [*f*]

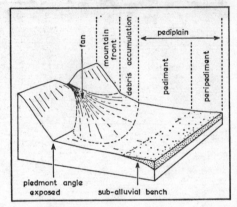

pediocratic period P of crustal calm during geological time between 2 OROGENIC **p**s; e.g. most of the MESOZOIC.

pediplain Multi-concave surface resulting from the coalescence of several large-scale adjacent PEDIMENTS [*f*]. Widespread erosion surfaces are thought to have been formed in this way; e.g. in Africa, the 'Gondwana **p**' at over 1200 m above sea-level probably dates from Jurassic times, and the 'Africa **p**', at about 600 m, is possibly of early Tertiary age. Alt. spelt 'pediplane', but it is best to reserve this for the vb. 'to pediplane' and for the process of 'pediplanation'. See pl. 48.

pediplanation cycle The formation of PEDIPLAINS, over very long periods of geological time (up to 100 million years) by the twin processes of scarp retreat and pedimentation (L. C. King). In youth, the land-surface is dissected by streams incising lines of geological weakness (e.g. fracture zones); steep valley-side slopes (scarps) and low-angled basal slopes (PEDIMENTS) are thereby initiated. In maturity scarp retreat reduces the interfluves to numerous towering BORNHARDTS. In old age pediments are extended to dominate the landscape, and bornhardts are reduced by weathering to piles of boulders (KOPJES).

pedocal Soil rich in calcium carbonate which has not been much leached, characteristic of areas with less than 600 mm of precipitation. Ct. PEDALFER.

Plate 48. The so-called African PEDIPLAIN (of early Tertiary age), at 1500 m above sea-level south-east of Nairobi, Kenya. In the left of the photograph a bold scarp leads up to the Gondwana pediplain (of Jurassic age), represented by the summits of the Machakos Hills at over 2000 m. (*R. J. Small*)

pedogenesis The development of soil.

pedology The scientific study of soils, their origin and characteristics, and their utilization.

pedon Used mainly by American pedologists as a generic name for a small volume of soil, which can be rel. through groups and orders into major Soil Orders, the basis of the American Soil Survey classification known as the SEVENTH APPROXIMATION.

pegmatite An igneous rock characterized by large coarse crystals, tightly interlocking, commonly found as dykes or veins, esp. on the margins of BATHOLITHS. Individual crystals may be several cm or even m in size; a crystal of a lithium mineral 12 m in length was found in the Black Hills of S. Dakota. The commonest type is a granite-**p**, chiefly of quartz and feldspar. Some **p**s contain rare minerals, concentrated during final stages of slow magmatic cooling; e.g. of such elements as boron, fluorine, uranium, niobium, tantalum, lithium, beryllium, thorium. **P**s may be simple or complex in composition. See pl. 49.

pelagic App. to the water of the sea, and to organisms living there, independent of shore or sea-bottom.

pelagic deposit Material deposited as an ooze on the abyssal floors of the oceans, derived from floating organisms which die and sink. See individual oozes: GLOBIGERINA, PTEROPOD, RADIOLARIAN, DIATOM, RED CLAY.

Peléan eruption Volcanic e accompanied by clouds of incandescent ash (see NUÉE ARDENTE), from the type-example of Mont Pelée which erupted in 1902, wiping out St Pierre, Martinique.

'Pelé's hair' Glass-like 'hairs' or 'threads', extrusions from basaltic lava, produced by explosions or the bursting of bubbles of contained gas in a pool of liquid lava; after Pelé, the Hawaiian goddess of fire.

'Pelé's tears' Small drops of basaltic volcanic glass, thrown out and solidifying during eruptions of lava.

pelite Rock of fine particles of clay; e.g. mudstone; i.e. an ARGILLACEOUS rock. Cf. PSAMMITE, PSEPHITE.

peneplain An almost level plain, the product of long-sustained denudation. The vb. is 'to peneplane', and the process 'peneplanation', never 'peneplaination'. Introduced by W. M. Davis in 1889, and quite adequate for planation surfaces of whatever origin. However, because Davis assoc. partic. processes with the production of such features, others with different processes in mind have introd. other terms, notably PEDIPLAIN and PANPLAIN. Some would exclude a surface of marine, as distinct from sub-aerial, planation.

Plate 49. PEGMATITE. (*Institute of Geological Sciences*)

penetrometer A device for measuring the force required to cause soil penetration by a pointed rod or cone (*cone p*) to a given depth at a specified rate. Soil strength, as thus determined, may be an important control of erodibility and thus of relief development.

peninsula An elongated projection of land into the sea or a lake; e.g. S.W. P. of England, Dingle. P. of S.W. Ireland, Iberian P. May be used adjectivally, as Peninsular Italy, P. Europe.

penitent rocks An unusual landform developed by deep weathering of schistose rocks (which weather rapidly along major planes), where planes are but slightly inclined from the vertical. Weathering is mainly subaerial, but leaning **p rs** are exposed by subsequent exhumation. Found notably in Rhodesia and Zambia.

Pennsylvanian Formerly in USA the upper of the 2 epochs of the Carboniferous, equiv. to the UK Upper Carboniferous (Coal Measures) system. Now has the status of a period in USA, 6th in order in the Palaeozoic, *c.* 310–280 million years ago.

perception The stimulus-response aspect of processes of decision-making with locational implication in the study of spatial/environmental patterns. Perceptual studies are esp. concerned with identifying social and cultural determinants of image-making in response to both the physical and social environment. 'Ps form keystones in our world views, symbolic schema and cosmologies.' Geographers are concerned with **p** of the quality of landscape, effects of environmental stress, areal images, HAZARDS, etc. in understanding human behaviour. Results of such research can be expressed qualitatively, or measured and expressed quantitatively and/or cartographically.

perched block A boulder perched in a delicate state of balance, usually left by an ice-sheet or glacier, or having fallen by gravity (e.g. the Bowder Stone, Borrowdale, Cumb.), or having been thus left by weathering *in situ*. Sometimes called a *pedestal rock*.

perched water-table An independent and isolated area of ground-water, above the w-t proper and separated from it by unsaturated rocks; i.e. it occurs in the VADOSE zone. [*f*]

River 'perched' on impermeable bed
Main water-table
Perched water-table
PERMEABLE SAND
IMPERMEABLE BED ROCK
Lens of impermeable clay

percolation The downward movement of water through pores, joints and crevices within the mass of soil and rock. It can carry down soluble minerals, hence LEACHING.

percoline A line of concentrated water seepage laterally within the soil, constituting a major component of THROUGHFLOW. Ps, which may coincide with underground 'pipes' (possibly initiated by rodent burrowing), can form dendritic patterns representing an extension of the surface drainage network.

pereltok A temporarily frozen layer, developed at the base of the active layer and remaining unthawed for one or two summers. Ps may easily be confused with the true surface of the PERMAFROST.

perennial stream A s that maintains its flow throughout the year, in ct. to an intermittent one.

perforation hypothesis An hypothesis of kame development in which it is inferred that debris collects within pools on the surface of stagnant ice. As the pools warm, they melt through the ice, depositing their debris contents on the underlying ground surface. The surrounding ice is in turn removed by ablation, leaving the pool deposits as upstanding hillocks and mounds.

pergelation Formation of PERMAFROST, or, in a more limited sense, local freezing of the soil.

pergelisol See PERMAFROST.

perhumid The highest category (*A*) in C. W. Thornthwaite's classification of climatic regions according to his MOISTURE INDEX.

pericline A small ANTICLINE which pitches along its axis in each direction from a central point, forming an elongated dome; e.g. Kingsclere and Shalbourne ps to W. of Basingstoke; Pays de Bray in N. France. From their nature, ps are partic. vulnerable to breaching along their axes, as has happened in these cases. Used commonly in UK, rarely in USA (sometimes called *centrocline*).

peridotite Coarse-grained, plutonic igneous rocks, consisting almost entirely of olivine, composed of ferromagnesian silicate. Commonly dark green in colour.

perigean tide When the moon is at its nearest position to the Earth (PERIGEE), high ts are higher and low ts are lower than usual. If a p t coincides with a SPRING T, the tidal range is greater still.

perigee The point in the orbit of a planet when at its min. distance from Earth. Now used strictly with ref. to the moon; when in **p** it is 348 292 km from Earth. Ct. APOGEE.

periglacial App. to an area bordering the edge of an ice-sheet, to the climate of that area, to physical processes involving FREEZE-THAW activity and to their results. Some confusion may be caused by the fact that climates along the edge of ice-sheets vary considerably, nor are all present or past ice-fronts marked by what are now gen. understood to be **p** conditions or processes. See pl. 11, 37.

perihelion The nearest point of a heavenly body in its orbit around the sun; the Earth at **p** (3 Jan.) is at a distance of 147·3 million km. Ct. APHELION.

period A time-interval in the geological record, the corresponding division of rocks being a SYSTEM. A **p** forms a division of an ERA, and is divided into EPOCHS. There are 2 usages: (i) The gen. UK practice recognizes 15 ps since the beginning of the Palaeozoic: Cambrian, Ordovician, Silurian, Devonian, Carboniferous, Permian, Triassic, Jurassic, Cretaceous, Eocene, Oligocene, Miocene, Pliocene, Pleistocene, Holocene (or Recent). (ii) The usual American practice is to demote the last 6 of these to epochs; the order is Cambrian to Devonian (as above), Mississippian, Pennsylvanian, Permian, Triassic, Jurassic, Cretaceous, Tertiary, Quaternary (the last 2 within the Cainozoic era), hence 12 ps in all. (iii) Incr. by UK and international usage as (ii), but with Carboniferous for Mississippian–Permian, hence 11 ps in all.

periodicity A time variant of a function consisting of a single period or combination of periods. E.g. in averaging a large number of meteorological elements, certain periodically recurring phenomena may be distinguished, such as daily or seasonal temperature and pressure changes. Efforts are made to develop long-range weather forecasting by this means, though as yet with limited success. Cf. SINGULARITY, ANALOGUE.

periodogram A deviational graph in connection with the determination of climatic cycles; amplitudes of temperature in the form of deviations are plotted as ordinates and time-units as abscissae.

peripediment Area of alluviation in a desert basin, adjacent to a PEDIMENT [*f*].

permafrost Permanent freezing of soil, subsoil and bedrock, sometimes to great depths; borings indicate that p exists in the Siberian TUNDRA to 600 m below the surface. Thawing may occur at and near the surface during summer, while the frozen ground below still forms an impermeable layer. The limit of p accords gen. with the $-5°C$ annual isotherm. See PERIGLACIAL.

permafrost table Defined by S. W. Muller as the 'more or less irregular surface which represents the upper limit of the PERMAFROST' i.e. the dividing line between the permafrost and the active layer.

permanent hardness, of water The result of the presence of dissolved magnesium and calcium sulphate, which cannot be removed by boiling.

permeable rock A r which will allow the passage of water, either because of its POROSITY (e.g. sand, sandstone, gravel, oolitic limestone, but not clay) or of its being PERVIOUS (chalk, Carboniferous Limestone, quartzite). Hence *permeability*.

Permian 6th of the geological periods of the Palaeozoic era (7th in USA) and the system of rocks laid down during that time, from about 280 to 225 million years ago. Sandstone strata were laid unconformably upon worn-down Carboniferous rocks, except possibly in the coalfields of the English Midlands. The lower part of the P. contains fossils related to Palaeozoic fauna, while the upper in places passes indistinguishably into TRIASSIC rocks. Hence *Permo-Trias* is often used of the transition system of rocks (New Red Sandstone). Other rocks of **P** age incl. breccias, Magnesian Limestone in N.E. England, and marls, sometimes containing salts. The **P** period was a time of widespread glaciation in the S. continents. Derived from province of Perm in the USSR, where marine rocks of this age are well developed.

persistence When meteorological conditions exist for more than a normal length of time, they are persistent; e.g. a persistent anticyclone. Some conditions are notably persistent; once the conditions exist, they are likely to remain for some time. See BLOCKING HIGH.

perspective projection Category of MAP P based on geometrical p from a given point through the surface of a globe. Includes ZENITHAL (azimuthal), CONICAL and CYLINDRICAL PS.

perturbation In meteorology, any disturbance in the steady state of a system. In partic., applies to departures from ZONAL FLOW and similar wave disturbances in atmospheric circulation.

pervious rocks One of the 2 groups of PERMEABLE RS. They are traversed by joints, cracks and fissures through which water can flow; e.g. chalk, quartzite, Carboniferous Limestone. Some writers regard permeable and p as syn., others refer to *primary* and *secondary* degrees of permeability respectively.

Peters' Projection A p announced in 1973, a modified version of CYLINDRICAL EQUAL AREA P, in which latitudes 46°N. and S. are used as standard parallels. Sometimes called *Orthogonal Map of the World*. The variable displacement of the parallels involved destroys the property of equal area.

petrifaction The conversion of matter of vegetable or animal origin into stone by percolating solutions of silica or calcium carbonate, which slowly replace the original tissue and structure; e.g. Petrified Forest, Arizona, where tree-trunks were buried under sediments including a layer of siliceous bentonite. The silica was carried down in solution and the trunks were converted into crystalline silica in the form of agate and other minerals.

petrofabric analysis The measurement of the degree and nature of the orientation of mineral grains, notably in metamorphic rocks in an area of JOINTING or CLEAVAGE, i.e. under stress conditions. This **a** may help to interpret the structure of an area.

petrogenesis A branch of PETROLOGY which deals with the origin of rocks, specif. igneous rocks.

petrology The scientific study of the chemical and mineral structure and composition of rocks.

p-forms A series of 'plastically moulded' features of glacially abraded rock-surfaces. **P-fs** include channels on steep rock faces ('cavettos'), grooves on open flat surfaces, curved and winding channels, and bowls and potholes. Some **p-fs** may result from the impact of basal ice containing rock debris; others have been attributed to the flow of water-soaked mobile ground moraine, squeezed out by overlying ice, and to subglacial meltwater streams under great hydrostatic pressure and moving at high velocities; in the latter the rare process of CAVITATION occurs.

phacolith, phacolite Concordant intrusion of igneous rock, lying near the crest of an anticline or base of a syncline in folded strata, in the shape of a lens; e.g. Corndon Hill, Shropshire, where a lens of igneous rock, forced into an upfold of Ordovician rocks, has been exposed by denudation.   [*f*]

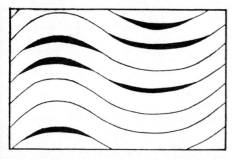

phanerozoic Lit. 'obvious' or 'plainly visible' life. Applied to the time-span (*eon*) during which remains of plants and animals have accumulated abundantly within sediments. The term has been transferred to geological time and to the stratigraphic systems from the Cambrian onwards. Hence *P time-scale*. Ct. CRYPTOZOIC.

phenocryst Large, conspicuous crystal in a porphyritic igneous rock, sometimes in a ground-mass of glassy or fine-crystalled material; e.g. feldspar in Shap Granite, and in granite of Haytor, Dartmoor.

phenology The scientific study of effects of seasonal climatic change upon recurring natural phenomena, such as bird migration and flowering of vegetation.

phonolite A fine-grained extrusive igneous rock, consisting essentially of alkaline feldspar, with the mineral nepheline. So named because of its ringing note when struck by a hammer. The Devil's Tower, Wyoming, is a mass of **p**, 213 m high, with the rock in large pentagonal columns.

phosphate rock A variety of **p**, including phosphorite, phosphatic limestones and leached guano accumulated from sea-bird droppings (e.g. Nauru). Phosphorus is also obtained from the mineral apatite and from basic-slag out of blast-furnaces. One of the 3 main elements needed by plants, and used as a fertilizer in the form of super-phosphates; i.e. **ps** treated with sulphuric acid.

photic zone The shallow-water **z** of seas and oceans, between 0–60 m in depth (0–200 ft), in which penetrating light has sufficient intensity to allow photosynthesis by algae. Ct. APHOTIC Z.

photometeor A luminous phenomenon in the atmosphere produced by reflection, refraction, diffraction and interference of light; e.g. BROCKENSPECTRE, HALO, IRISATION, RAINBOW, FOG-BOW.

photo-relief Shading on a map to give the impression of a photograph of a relief model, usually with the appearance of the source of light from the N.W. (USA, *plastic relief*).

phreatic eruption A volcanic explosion, usually violent, resulting from the rapid conversion of groundwater into steam, gen. from magmatic contact.

phreatic water See GROUND W.

pH value The quantitative degree or scale of soil acidity and alkalinity, measured in terms of the negative index of the logarithm of the concentration of the hydrogen-ion activity in soil-colloids. In pure water, 1 part in 10 millions (10^{-7}) is dissociated into hydrogen ions, and pH is 7; a neutral state in the scale of acidity. If a strong alkali, such as caustic soda, is dissolved in water, the solution is markedly alkaline, and an infinitesimal part (10^{-14}) is dissociated into hydrogen ions, hence pH is 14. With hydrochloric acid, 1 part in 1000 is dissociated, i.e. 10^{-3}, pH of 3. A neutral soil has pH value of about 7·2, an acid soil of less than 7·2, and an alkaline soil of more than 7·2. An alkaline mineral soil has a pH of 10, an acid sulphate soil of 2; under humid condition the usual range is from 5 to 7.

physical geography Aspects of **g** concerned with the shape and form of the land-surface, the configuration, extent and nature of seas and oceans, the enveloping atmosphere and processes therein, the thin layer of soil, and the 'natural' vegetation cover; i.e. Man's physical environment, though normally independent of his effects.

physical geology 'The nature and properties of the materials composing the earth, the distribution of materials throughout the globe, the processes by which they are formed, altered, transported and distorted, and the nature and development of the landscape' (L. D. Leet and S. Judson). This covers one major division of g, the other HISTORICAL G. P g is sometimes divided into *structural g* and *dynamic g*. It includes much of the content of GEOMORPHOLOGY, but is wider in scope.

physical landscape See NATURAL L.

physical weathering Disintegration of rock by agents of weather (frost, temperature change), without involving chemical change.

physiognomic Applied specif. to the life-form characteristics of vegetation and its structure (hence '*formation type*'), as distinct from its floristic composition and (in attempted classifications) from associated climatic criteria. A. W. Küchler devised a successful p classification of world vegetation.

physiographic pictorial map Depiction of relief on a m by systematic application of a standardized set of conventional pictorial symbols, based on the simplified appearance of physical features, as viewed obliquely from the air at an angle of about 45°. The American cartographer, Erwin Raisz, was a supreme exponent of this art.

physiography Orig. 'a description of nature', or of natural features in their causal relationships; then almost syn. with physical geography; gradually, esp in USA, became limited to the study of landforms. By some authorities has been superseded by GEOMORPHOLOGY; by others, regarded in wider terms as an integration of geomorphology, plant geography and PEDOLOGY.

phytogeography Plant geography; the geographical description of plant distribution.

phytoplankton Microscopic floating plant-life in the sea.

piedmont Area near the foot of a mountain range; used widely as an adjective, as with *p-plain, -fringe,* -GLACIER. As a proper name, it denotes: (i) the plateau lying E. of the Appalachians; (ii) province in N. Italy at the foot of the Alps.

piedmont angle The sharp angular junction between the mountain front and PEDIMENT [*f*] at its foot.

piedmont glacier A mass of ice on the flanks of a mountain range, formed by coalescence of parallel valley-gs. P gs were common in the Quaternary Ice-Age, when they spread on to the Bavarian and Swiss Forelands, N. Italian Plain, and Lannemezan in N. Pyrenees. Present examples are the Malaspina G. in S. Alaska, covering 4000 km^2; Frederikshaab G. in W. Greenland; Butterpoint G. in S. Victoria Land (Antarctica).   [*f*]

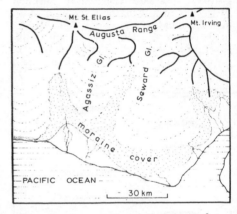

pie-graph A pop. descriptive name for a DIVIDED CIRCLE g.

piezometric surface The level to which water rises in a well sunk into an AQUIFER.

pike A mountain peak, esp. in English Lake District; e.g. Scafell P., Langdale Ps.; and in Pennines; e.g. Buckden P.

pilang An area of close acacia-forest in Indonesia.

pillow lava L which has solidified in the appearance of a pile of pillows, probably under water, partic. in the case of basic l as basalt. The form results from sudden cooling of the outer skin, forming a rounded cushion-like mass which partially flattens under its weight.

pinch-out Where a rock-stratum thins out in a horizontal plane and disappears.

pingo (Eskimo) An isolated dome-shaped or conical mound of earth or gravel found in

Alaska, Arctic Canada, Greenland and Siberia. Autumn freezing traps a layer of water between the newly frozen surface and underlying PERMAFROST, and hydrostatic pressure resulting from the expansion of water on freezing may raise a 'blister' on the surface. Many have been recently formed, though some large ones, from evidence of RADIOCARBON DATING, were formed some 5–6000 years ago. American workers distinguish between '*closed system ps*', where the layer of water trapped is an isolated body [*f*], and '*open system ps*', where the hydraulic head is rel. to an adjacent slope. The latter are common in eastern Greenland, where plentiful water from summer snow-melt on the mountains moves downwards towards the coast over the zone of permafrost. **Ps** range in size from mounds of 6 m to hills of over 90 m in height with a basal diameter of 0·8 km. The top of the **p** may collapse, leaving a 'crater' partly filled with a shallow pond. Other workers call these features *hydrolaccoliths* or *cryolaccoliths*, restricting **p** to a small isolated mound with an existing ice-core, but there appears to be no unanimity. See pl. 50.
[*f*]

Thick lens of ice forming pingo

Upper soil layers slide back

Ice melts leaving central depression

pinnate drainage A **d** pattern in which the main tributaries receive a number of closely spaced affluents at acute angles, like a feather.
[*f, opposite*]

pipe (i) The vent of a volcano which leads to its crater. (ii) In chalk, vertical joint enlarged by carbonation-solution, and filled with sand and gravel. (iii) A cylindrical mass of mineral ore or diamantiferous rock; e.g. diamond-bearing **ps** of Kimberley (S. Africa).

containing ultrabasic rock (KIMBERLITE) in which diamonds occur. These were formed by FLUIDIZATION.

pipkrake (Swedish) A needle, wedge or lens of ground-ice in the soil, formed under PERIGLACIAL conditions.

piracy, of streams See CAPTURE, RIVER.

pisolith, pisolite A coarse-grained oolite, of rounded grains about the size of a pea (1–10 mm), hence the name; e.g. Jurassic Pea Grit.

pitch (i) The direction in which the AXIS of a fold dips. (ii) A solid hydrocarbon, formed in distillation of coal-tar. (iii) Used loosely for asphalt or bitumen.

pitchblende A complex oxide of uranium (*uraninite*): black, with pitchy lustre, containing small amounts of other elements—lead, thorium, radium, helium. Found in Zaire, Canada (Gt Bear Lake, Gt Slave Lake districts), USSR. The other main uranium mineral is *carnotite*, found esp. in Jurassic sandstone of Utah and Colorado.

pitchstone A waxy-looking mass of solidified lava. When acid lava is extruded on the surface, usually forming a glassy mass, it may sometimes contain a higher proportion of water than usual, so losing its glassy lustre. E.g. in islands of W. Scotland, esp. Arran, Mull, Eigg. The highest point on the last, the Sgurr of Eigg, is made of **p**.

piton (Fr.) (From a Fr. word for a metal spike

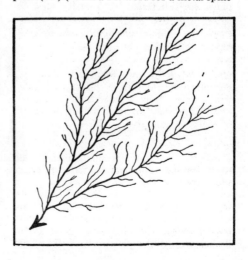

Plate 50. A large ice-cored mound or PINGO in sediments at Wollaston Peninsula, Victoria Island, northern Canada. Note the partial collapse of the roof of the dome, and clear evidence of cracking and other disturbances on the flanks of the pingo. (*A. L. Washburn*)

or bolt with a ring-head, used in mountaineering) (i) Applied to a sharp-pointed peak. (ii) Used in KARST terminology (*Karst à pitons*) for sharp projections, the result of the acute dissection of limestone under sub-tropical conditions.

pitted outwash plain P of fluvioglacial deposition, dotted with KETTLES; found extensively in Wisconsin, Minnesota and adjacent parts of Canada.

placer Mass of sand or gravel containing particles of gold, tin or platinum, eroded from exposed veins, washed down by a river, and laid down as an alluvial deposit; e.g. tin in Malaysia, gold in S.W. USA.

plagiosere A series of communities which forms through disturbance of natural CLIMAX VEGETATION by human interference leading to a *plagioclimax*. Many English uplands, formerly wooded, have been turned into grassland as a result of clearing, and have been maintained by subsequent grazing. The decline of grazing intensity on hill-slopes has led to bracken colonization and re-establishment of SUBSERES.

plain Continuous tract of comparatively flat country, sometimes gently rolling or undulating, with no prominent elevations or depressions. Some would restrict it to an area of horizontal structure, others to any flat area at low elevation. Used in various contexts; see COASTAL P, UPLAND P, FLOOD-P. A common place-name element; e.g. Salisbury P., Great Ps. of N. America, N. European P.

planation The denudation of rocks to produce a fundamentally flat surface. Various types have been distinguished: peneplanation (see PENEPLAIN), pediplanation (see PEDIPLAIN) and panplanation (see PANPLAIN). *P surface* is to be preferred to erosion surface when used to describe this feature, since an erosion surface can be of any form, flat or otherwise. See pl. 51.

plane, of the ecliptic The p of the apparent path of the sun. See ECLIPTIC.

[*f* AXIS, OF EARTH]

Plate 51. The PLANATION of steeply dipping CARBONIFEROUS LIMESTONE near Paviland, south Wales. The PLANATION SURFACE, at about 60 m above sea-level, may have been formed by marine action during Pliocene times. (*Eric Kay*)

planet One of the 9 solid heavenly bodies revolving in the same direction around the sun: these are:

		mean distance from sun		time of 1 revolution in orbit	
		(*million km*)	(*million mi.*)	earth-years	earth-days
(i)	Mercury	58	36	—	88
(ii)	Venus	108	67	—	225
(iii)	Earth	150	93	1	0
(iv)	Mars	228	142	1	322
(v)	Jupiter	777	483	11	315
(vi)	Saturn	1426	886	29	167
(vii)	Uranus	2869	1783	84	6
(viii)	Neptune	4495	2793	164	288
(ix)	Pluto	5900	3666	247	255

plane-table Small drawing-board fixed on a tripod, used for topographical surveying in conjunction with an ALIDADE. See RESECTION.

planetary winds Pop. name for the gen. ATMOSPHERIC CIRCULATION [*f*], comprising latitudinal wind-belts (equatorial easterlies or Trades, mid-latitude westerlies and polar easterlies). The zonal longitudinal component of the circulation and the vertical movement are brought about mainly by thermally controlled pressure distribution (see HADLEY CELL); the latitudinal meridional component is the result of deflection due to rotation of the Earth (see CORIOLIS FORCE), producing WESTERLIES, which intensify aloft into JET STREAMS. The idealized pattern is further complicated by imposition of perturbations, from major 'waves' in the Westerlies (with wavelengths measured in thousands of km) to smaller moving depressions and local convection 'cells', valley ws., etc. These disturbances result from: (i) *thermal effects:* differential heating and cooling of land and sea, seasonal variations in heating, responsible for some MONSOON ws and local ws such as sea breezes, and often considered as the main cause of major waves in the upper ws; (ii) *orographic effects:* influence of relief on local ws, and influences of major mountain belts on gen. pattern of latitudinal ws, held by many to be of greater importance than thermal effects.

Planetesimal Hypothesis A theory put forward in 1904 by F. R. Moulton and T. C. Chamberlin to account for the origin of the Earth and other planets. The main principle is that they were formed from coalescence of small particles (planetesimals), rather than from a gaseous, then molten, mass, which slowly cooled. This theory has been revived in a modified form by some modern cosmologists.

planetoids See ASTEROIDS.

planèze (Fr.) A triangular wedge of lava on the slopes of a dissected volcano, narrowing upwards to a broken projection on a crater rim. These occur where the lava flows protect an otherwise unresistant cone; e.g. around Plomb du Cantal (1858 m.) in Central Massif of France, where many ps have proper names (P. de St Flour, Limon, Carlades).

planimeter An instrument for measuring areas on a map. Several types vary in complexity from a simple form of tracer-bar (*hatchet-p*) to delicate instruments fitted with recording dials (*wheel-p* or *polar-p*). A pointer is traced round the perimeter of the area to be measured. Some models record the area in cm², which are converted according to the scale of the map; others have variable tracer-arms set to any scale-value. Hence *planimetry:* measurement of areas.

planina, pl. **planine** (Serbo-Croat.) Long limestone ridges, with comparatively level flattened crests, in Yugoslavia, trending broadly N.W. to S.E. in conformity with the geological 'grain' of the Dalmatian area. The high **p** rise to 900–1500 m.

planisphere A star-chart of one of the celestial hemispheres.

plankton Minute plant (not incl. seaweed) and animal organisms that float or drift in the ocean. Ct. NEKTON, BENTHOS.

planosol A soil in which clay HARDPAN has developed midway between the surface and bedrock, found on an upland surface in a humid climate. The shallow A-horizon is leached, but as the claypan develops the soil becomes more or less permanently saturated and reveals GLEI features.

plant productivity, index of An effort to determine the potential growth of vegetation in terms of climatic conditions in which it is developing, depending on length of growing season, annual and seasonal precipitation, temperature of warmest month, temperature range, amount of solar radiation, etc. Various empirical indices have been calculated on a basis of *potential p p* (i.e. what vegetation could be supported purely in climatic terms). E.g. by S. Paterson; on his **i** a forested area near Portland (Maine) is 3·1, one near Miami (Fla.) is 22·3, one near Belem (Amazon Basin) is 118·0. He drew maps of the continents indic. 6 zones plotted from his **i**, ranging from *A* (over 50, diverse tropical rainforest) to *F* (under 0·25, ice-caps, tundra fringe, desert).

plastic deformation, of ice Under certain conditions, **i** near the base of a thick glacier becomes **p**, as a result of pressure causing intermolecular movement. This results in *creep*, an important contributory process to glacier flow.

plastic relief (i) The depiction of **r** by means of HILL-SHADING, producing the effect of a 3-dimensional **r** model. (ii) A 3-dimensional model of **r** moulded in plastic.

plateau (Fr.) An upland with a more or less uniform summit level, sometimes bounded by slopes falling steeply away, sometimes rising by steep slopes to mountain ridges. (i) *Tectonic p* (Arabia; Deccan; Meseta); (ii) *residual p*, formed from ancient fold-mountains by denudation and sometimes later uplifted (Middle Rhine P.; Allegheny-Cumberland P. of Appalachians); (iii) *intermont p* enclosed within fold-ranges (British Columbia; Great Basin of Utah and Colorado; ps of Ecuador and Bolivia; Anatolia and Armenia in S.W. Asia; Tibet); (iv) *volcanic p* (Antrim; Abyssinia; Columbia-Snake; N.W. Deccan). Some ps may be extensively dissected, leaving ACCORDANT SUMMITS as indications of the orig. p-surface. Used very widely, both in gen. terms and as part of a proper name.

plateau basalt See BASALT, BASIC LAVA.

plateau-gravel Sheet of small stones, sand and grit, often compacted, capping a plateau or its dissected remnants. Poor acid soils develop on **p-g**. Gravels of different origins are marked on maps of the Geological Survey of UK. They are found over parts of S. England up to about 120 m. E.g. in Southampton area between Test and Itchen valleys.

Plate Carrée Projection See EQUIRECTANGULAR P.

plate tectonics The seismologically and tectonically active major rifts in the Earth's crust, comprising a kind of continuous network dividing crust and upper MANTLE into rigid ps, separated by fault-lines, ocean rifts, rises and deep TRENCHES (the SUBDUCTION ZONE). This new geophysical concept utilizes both sophisticated observational evidence and models based on computerized calculations. Various p systems have been postulated: 6 major (Eurasian, American, African, Pacific, Indian and Antarctic) and a dozen smaller ones (e.g. Caribbean). These ps comprise both sialic continental 'rafts' and simatic ocean floors, so that the time-honoured concept of CONTINENTAL DRIFT may in this light be no longer strictly correct. The movement of ps rel. to each other, the result of thermal convection, may go far to explain the creation of fold-ranges, island arcs, batholithic intrusions and volcanic eruptions along their contact margins. 'This great unifying concept draws sea-floor spreading, continental drift, crustal structures and world patterns of seismic and volcanic activity together as aspects of one coherent picture' (E. R. Oxburgh).

platform (i) Any level terrace or benchlike surface; specif. the product of wave erosion; as WAVE-CUT P. (ii) A more extensive block forming a continental basement; e.g. Russian P., which extends eastward from the Baltic Sea, buried progressively more deeply.

platinum A rare metal with a white lustrous appearance, found in nature alloyed with other rare metals (osmium, iridium, palladium).

playa (Sp.) (i) A basin of inland drainage containing a shallow, fluctuating lake, saline or alkaline, surrounded by sheets of crust or mud. (ii) The lake itself in such a basin; e.g. in Nevada and Arizona.   [*f, opposite*]

Playfair's Law In areas of uniform bedrock and structure which have been subject to river erosion for a long period of time: (i) valleys are proportional in size to the streams which they contain; and (ii) stream junctions in these valleys are accordant in level.

pleion An area with a high positive ANOMALY in respect of some climatic element, notably temperature (THERMOPLEION).

Pleistocene (Gk. lit. 'most recent'). Used with several meanings: (i) The last 1–2 million years (by some authorities, the last 600 000), including Recent or Holocene, and coinciding with Man's appearance. (ii) The last 1–2 million years but excluding Recent or Holocene (post-glacial); i.e. coinciding with the last Ice Age. (iii) In UK, the first of 2 periods into which the Quaternary era is divided, the other being Recent or Holocene. (iv) In USA, and incr. in international and British usage, the first of 2 epochs into which the Quaternary period (within the Cainozoic era) is divided. See table, CAINOZOIC. In E. Anglia the base of the **P** is taken as the Red Crag, reflecting climatic deterioration before the Ice Age. Recent work has shown great difference of opinion as to the length of the **P**. Some workers have pushed back the onset of the 1st glacial period to 1·8 to 2 million years ago ('LONG TIME-SCALE'), others limit its onset to 600 000 years ago (one of the 'SHORT TIME-SCALES'). While many workers still regard the **P** as syn. with the Quaternary glaciation, others believe that the classic glacial sequence occurred at the end of a long period of time (all of which they would incl. in the **P**), during which there was little evidence of glaciation, though with rapid alternations of climate; at least 12 named stages have been postulated, incl. 6 warm, 6 cool, from the *Ludhamian* to the *Devensian* (UK usage). See pl. 20, 28, 37.

Plinian eruption A PAROXYSMAL volcanic E, so-called because observed and described by the younger Pliny, during the Vesuvius e of A.D. 79.

plinthite See DURICRUST.

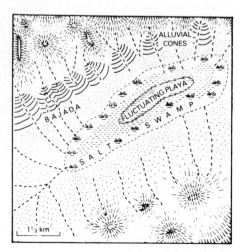

230 PLIOCENE

Pliocene (i) In UK the 4th of the geological periods of the Cainozoic era, lasting about 11 million years, and the system of rocks laid down during that time; i.e. shelly sands and gravels known as Coralline Crag in E. Anglia. In USA and incr. in Europe and UK the 5th epoch of the Tertiary system within the Cainozoic era (see table, TERTIARY). Until recently the boundary of the **P** in Britain was taken to be the top of the Cromer Forest Bed Series, but this has been shown to be an interglacial deposit. The underlying Weybourne Crag may be representative of the 1st glacial period. The base of the **P** is that of Coralline Crag, its top is the base of Red Crag. There is still argument about the position of this **P**-Pleistocene boundary.

plottable error The smallest distance that can be shown on any map, depending on its scale. Based on the fact that the finest possible line on a map is 0·25 mm, **p e** on a map of scale 1:10 000 is 2·5 m.

ploughing block An isolated boulder, with an elongated depression on the upslope side, and bounded by a 'rucked up' ridge downslope. **P bs** move slowly under the influence of frost disturbance, though sometimes the furrow and ridge appear to have been formed by a rapid slipping movement.

plucking The pulling away of masses of rock on the floor below a glacier, by means of ice freezing on to protuberances and detaching them as it moves; one of 2 main processes of glacial erosion. Syn. with EXARATION.

plug A more or less cylindrical mass of solidified lava, occupying the vent of a dormant or extinct volcano, exposed by denudation; e.g. Arthur's Seat and Castle Rock, near Edinburgh; Rocher St Michel and Rocher Corneille in Auvergne (Central Massif of France). See pl. 52.

plug-volcano A mass of viscous acid lava squeezed out of a vent in molten form, and solidifying as dacite, rhyolite, trachyte; e.g. Mont Pelée 'spine', formed in 1902; Santiagito plug in Guatemala, formed in 1922–4; Mt Lassen, Cascades, USA, a dacite plug with

Plate 52. A prominent volcanic PLUG, the Rocher St Michel at Le Puy, southern France. The ash cone, which once surrounded the volcanic vent, has been completely eroded away by recent erosion. (*Eric Kay*)

an estim. volume of 4.4 km^3. The last is not typical of **p-vs**, since it had an eruption in 1914 and new craters were formed near its summit.

plume Used by geophysicists in PLATE TECTONICS to indic. vertical convection movement of magma from a 'hot spot' at the boundary between the Earth's CORE and MANTLE, causing first a swelling, then a rupture in the rocks. E.g. volcanic domes in such Saharan plateaus as the Tibesti are thus formed; there is probably a **p** in the Hawaiian area, where chains of volcanoes indicate that a plate has moved over it.

plum-rains See BAI-U.

plunge-pool A hollow eroded by the force of falling water at the base of a waterfall (*p flow*), partic. by eddying and CAVITATION; e.g. Horseshoe Falls, Niagara, Canada.

plutonic rock Intrusive igneous **r**, which cooled slowly at considerable depth in the Earth's crust, of large-crystalled coarse texture; e.g. granite, syenite, diorite, gabbro, peridotite. Hence *pluton*, esp. in USA for any large intrusive igneous mass.

pluvial, pluviose Rainy, or app. to rain. Used in connection with a past rainy period, in ct. to dry periods; e.g. ATLANTIC was a **p** period between 5500 and 2500 B.C., with drier preceding BOREAL and succeeding SUB-BOREAL.

pluviometric coefficient A value arrived at by expressing the mean monthly rainfall total for a given station as a ratio of the hypothetical amount equivalent to each month's rainfall were total rainfall to be equally distributed throughout the year. E.g. suppose mean annual rainfall = 31 ins., and mean Jan. rainfall = 4·0 ins. If uniformly distributed, Jan. avge = $\frac{31}{365}$ of 31 ins. = 2·63 ins. Then **p** c = $\frac{4·00}{2·63}$ = 1·52. See EQUIPLUVE.

pneumatolysis Chemical changes produced in rocks by heated gases and vapours other than water (HYDROTHERMAL) (notably when MAGMA is cooling and solidifying), both within the igneous material and in surrounding country-rock. New minerals are formed; e.g. tourmaline, kaolin, fluorite, topaz.

pocket beach See BAY-HEAD B.

podzol, podsol (Russian) A soil formed under cool, moist climatic conditions with a vegetation of coniferous forest or heath, esp. from sandy parent materials, resulting in intense leaching of base-salts and iron compounds. The thin A_0-horizon consists of raw humus, beneath which is the partly leached A_1-horizon, and the heavily leached, ash-grey A_2 or E-horizon; the B_1-horizon contains an accumulation of partly cemented leached materials; the B_2-horizon contains sesqui-oxides, clay and humus. Widely spread in the heathlands of W. Europe and coniferous forests of N. Europe. The process is *podzolization*.

point A narrow projection of land into the sea, usually low-lying; e.g. P. of Ayre (the N. tip of I. of Man); St Catherine's P. (I. of Wight); Start P. (Devon).

point-bar deposits Alluvium, sand and gravel deposited on the shelving SLIP-OFF SLOPE on the inside of a river meander. See pl. 41.

polar air-mass An **a-m** which originated in middle latitudes (40–60°), either over the ocean (*p maritime*) (*Pm*) or over a continental interior (*p continental*) (*Pc*). Air masses originating near the Poles are unfortunately not called *p*, but Arctic (*A*) and Antarctic (*AA*), a source of possible confusion. The latter were coined after **p** became well established.

polar distance The complement of the DECLINATION of a heavenly body, i.e. 90°-declination angle. Named *N.P.D.* (North Polar Distance) or *S.P.D.* (South P.D.), according to whether measured from N. or S. Pole.

polar front A **f** or frontal zone in the N. Pacific and N. Atlantic Oceans, along which meet P Maritime and Tropical Maritime air.

polar outbreak The penetration of a cold air-mass (*Pc*) from middle and high latitudes into lower latitudes, bringing exceptionally cool, even cold, weather, accompanied by strong winds. This is partic. common when there are no protective transverse barriers of mountains, as in N. America. The winds are so distinctive as to be given names: *norther, norte, papagayo, friagem, pampero, southerly burster*.

polar projection A map **p** centred on the N. or S. Pole; e.g. polar cases of the various

ZENITHAL PS. The diagram shows ZENITHAL EQUIDISTANT P. [*f*]

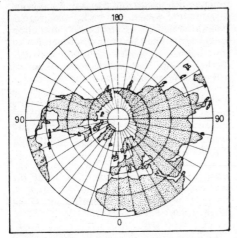

polar vortex A W. circular low pressure circulation in middle and high latitudes in both hemispheres, appreciably stronger in winter, when the airstream circulation may intensify in the STRATOSPHERE to form the polar night JET, of 150 knots plus.

polar wind A wind blowing away from the N. or S. polar regions.

polder (Dutch) A unit of land near, at or below sea-level, reclaimed from the sea, lake or river flood-plain by endyking and draining, often kept clear of water by pumping. (Vb. to *empolder*.) The largest scheme in the Netherlands is the partial reclamation of the Zuider Zee; about 2227 km², or 60% of the total, will be reclaimed in 5 ps, leaving the fresh-water IJssel Meer. [*f*]

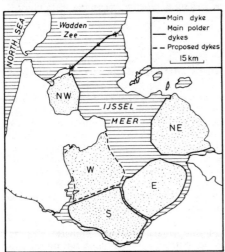

pole (i) Geographical **p**, N. and S. extremities of the Earth's axis. (ii) MAGNETIC P. (iii) CELESTIAL P.

polje, pl. **polja** (Serbo-Croat.) A large depression in limestone KARST, elliptical in plan and flat floored, often covered with TERRA ROSSA, sometimes with a marsh, or small intermittent or permanent lake. E.g. Yugoslavia, where most are elongated N.W. to S.E., parallel to the gen. 'grain' of the country. Some are large; e.g. Livanjsko **p**, 64 km in length and 400 km² in area.

pollen analysis See PALYNOLOGY.

Polyconic Projection A CONIC P, on which parallels are non-concentric circles, each drawn on its own radius = R. *cot. lat.* The central meridian is a straight line divided truly, all other meridians are curves. Scale along each parallel is correct, but incr. on meridians with incr. distance from the centre. Even so, scale error is less than 1% within 900 km of the central meridian. It is easy to draw, using tables giving coordinates of parallel-meridian intersections. Neither equal area nor conformal, but suitable for maps of countries of great longitudinal extent; e.g. Chile. Also used for large-scale topographical maps in USA, since the scale error within the standard 15 minute quadrangle is small. The INTERNATIONAL MILLION MAP is drawn on a modified polyconic **p**, on which meridians are straight, not curved, with 2 standard meridians on each sheet, and only the top and bottom parallels are standard. Sheets can be fitted along each margin.

polycyclic App. to geomorphological features which have been subjected to several partial cycles of erosion. In partic. used of a stream whose course reflects base-levelling to more than one former sea-level. Some authorities use *multicyclic*, but **p** is gen. preferred.

polygenetic With many origins, used of: (i) geomorphological features which have undergone several differing processes of erosion; (ii) mountains formed at different times by various forces (folding, faulting, vulcanicity). (iii) soils of complex origin; (iv) pebbles in a conglomerate.

polygon, soil A surface formation developed under PERIGLACIAL conditions, through the sorting of fine and coarse weathered materials by freeze-thaw. See PATTERNED GROUND.

polygonal karst A form of KARST developed specif. in humid tropical areas, with a cellular network-plan; each cell or basin (formed by superficial erosion rather than by collapse) drains centripetally towards an internal drainage system. Various types have been idenitified: *pinnacle, conic, tower, linear*, etc., according to the character of residual remaining eminences. Cf. COCKPIT KARST.

Polyhedric Projection P used for a large-scale topographical map, whereby a small quadrangle on the spheroid is projected on to a plane trapezoid. Scale is true either on the central meridian or along the sides.

Ponente A W. wind on the coast of Corsica and Mediterranean France, usually cool and dry.

ponor, pl. **ponore** (Serbo-Croat) A deep vertical shaft in limestone country, leading from the surface into an underground cave system; a stream may fall into it, permanently or intermittently. Syn. with *aven* (Fr.), and in speleological parlance with 'pot', 'pot-hole' or 'swallow-hole'. E.g. Gaping Gill, W. Yorkshire, 111 m (365 ft) deep; Gouffre de Pierre St Martin in the Pyrenees, 346 m (1135 ft) deep.

pool and riffle A pattern of alt. deeper **p**s and shallow gravel bars (RIFFLES), which apparently develops in a stream even with a straight channel and uniform cross-section. The distance between the **p**s is related to the width of the stream (approx. 5–7 times). If the bed is readily erodible, the channel gradually becomes sinuous, as **p**s develop near the outer concave bank, **r**s on the inner side where they become POINT BARS. This may help to account for the initiation of MEANDERS. (See KINEMATIC WAVE.)

pore water pressure The pressure exerted on soil and rock particles by water contained within interstices. Under conditions of saturation particles are forced apart by **p w p**, and rapid mass failure may ensue. Disastrous slips, such as that of the coal spoil heap at Aberfan, S. Wales in 1966, are often induced by exceptionally high **p w p**, related to the sub-surface accumulation of water when springs and seepages are impeded.

pororoca (S. Am.) The BORE on the estuary of R. Amazon.

porosity The nature of a rock with open texture, and coarse grained, widely spaced constituents; e.g. sand, gravel, sandstone, conglomerate, oolitic limestone. The **p** of a rock can be stated as a ratio or percentage, in terms of total volume of pore spaces related to total volume of rock. An evenly graded rock, of uniform spherical particles, has a **p** of 27%; loose gravel 45–47%; sandstone 5–15%. It can be measured by a *porosimeter*, which injects liquid or gas at a specific pressure into a sample of rock. Most porous rocks are PERMEABLE, except clay (where pore-spaces are minute and are often sealed with water held by surface tension) and un-jointed chalk. Sometimes referred to as *primary permeability*.

porphyry A HYPABYSSAL rock containing large crystals (PHENOCRYSTS) in a fine-grained ground-mass. Hence *porphyritic*.

positive anomaly, of temperature Where the mean **t** at a meteorological station is higher than that for all stations in that latitude. See ANOMALY.

positive landform An upstanding **l** such as mountain, hill and plateau, in ct. to NEGATIVE forms.

positive movement, of sea-level A change in rel. level of land and sea, resulting in a net rise of **s-l**, in ct. to a NEGATIVE MOVEMENT. The result of: (i) subsidence of land as a result of earth-movements or ISOSTATIC loading; (ii) actual rising of water-level in the sea on a world-wide and uniform scale (EUSTATIC movement), as the return of water to the oceans as Quaternary ice-sheets melted.

post-glacial App. to time since the Quaternary ice-age. As this ended at different times in various areas, according to their positions rel. to centres of ice-dispersal (i.e. *time-transgressive*), no single date can be ascribed to the beginning of **p-g** time, and glaciation is still present at high latitudes and altitudes. An arbitrary date for E. Anglia would be 15 000 B.C., for S. Scandinavia 8000 B.C., for S. Finland 6500 B.C. In geological terms, the beginning of **p-g** time is the division between Pleistocene and Holocene or Recent. See pl. 56, 71.

pot Used in a placename as syn. with CIRQUE, notably in Mourne Mtns of Ireland; e.g. P. of Legawherry, cut into N.-facing slopes of Slieve Commedagh.

potamology The scientific study of rivers.

potamoplankton Minute organisms, animal and vegetable, that live in slowly flowing rivers and streams.

potash Gen. applied to salts of potassium (notably *sylvite* or **p** chloride, *carnallite* or **p** magnesium chlorate, and *kainite*, a mixture of **p** chloride and magnesium sulphate). Usually found in assoc. with common salt on the site

234 POTASH

of a past salt-lake as in Stassfurt (E. Germany) and Alsace. P is also recovered from present salt-lakes, as Dead Sea and S.W. USA.

potential energy The energy 'stored' in bodies of water (e.g. rivers, lakes, ocean waves) standing higher than the natural rest level. In streams p e is defined by $E_p = W_z$ where W is the weight of the water and z the height above base-level (head). P e is released as free energy when the water actually moves downhill (or the ocean wave breaks on impact with the shore) (ct. KINETIC ENERGY).

potential evapotranspiration index Empirical i devised by C. W. Thornthwaite (1948) as the basis for his 2nd scheme of climatic classification. Based on the moisture i, rel. the amount of precipitation required by e, assuming vegetation to be present, to actual water available. Abbrev. to P.E.I; the same initials are sometimes used for PRECIPITATION EFFICIENCY I with which it should not be confused.

potential instability, of an air-mass Orig. condition of an a-m which when lifted over a relief barrier, or over a mass of cooler air at a FRONT, becomes conditionally unstable. See CONDITIONAL INSTABILITY.

potential model A mathematical construction that measures the force exerted by any defined phenomenon on a point in space by ref. to the same phenomenon located at all other points on the spatial domain under study.

pot-hole (i) H in the bed of a stream, formed by the grinding effect on bed-rock of pebbles whirled round by eddies. (ii) Used pop. though incorrectly, to denote vertical cave-systems, esp. by speleologists, hence 'pot-holing'. See pl. 53.

pound A British unit of mass: (*a*) *avoirdupois:* 7000 grains; 27·692 cu. in. of water at 4°C; 453·592 grams (g); 2240 lb. = 1 British long ton. (*b*) *Troy:* 5760 grains; 373·2418 g.

powder snow Dry, loose s crystals accumulated under conditions of low temperature. As much as 750 mm of this s may be equivalent to 25 mm of rain, as ctd. with wet slushy s where 100–150 mm may melt to form 25 mm

Plate 53. A POT-HOLE being actively drilled in the bed of the Afon Rheidol at Devil's Bridge, Wales. (*Eric Kay*)

of water. P s is found gen. in Antarctica, N. Canada and Siberia, and in the Alps, etc. forms an ideal ski-ing medium.

pradoliny (Polish) A large-scale glacial overflow channel; syn. with URSTROMTAL.

prairie An extensive area of grassland, occurring in middle latitudes in the interior of N. America. The use is complicated by Fr. word *prairie*, lit. a meadow. In its modern world-scale usage, equated with STEPPE. In its natural form, developed under a rainfall total of 250–500 mm, it presented a continuous cover of tufted grass, coarse and hardy, bluish-green in spring, yellow and straw-like in summer, with bulbous and herbaceous plants. Little survives today, since it has been ploughed-up for wheat, or replaced by cultivated grasses. But survives as a broad land-use category in Canada and USA.

prairie soil A s type akin to CHERNOZEM, with a similar profile except that it lacks deposition of calcium carbonate in the B-HORIZON. Occurs in areas formerly under p grasses, now growing grain and fodder crops. Also known as *brunizem* or *brown steppe s*.

Preboreal The climatic phase immediately following the Quaternary glaciation; e.g. in E. Anglia until about 7500 B.C. Characterized by dry, cold conditions, with a birch-pine flora. Sea-level was at least 60 m below the present, and the British Isles were joined to Europe.

Precambrian All geological time, and the assoc. rocks (occupying about a fifth of the Earth's surface area) before the CAMBRIAN period of the Palaeozoic era (i.e. before *c.* 570–600 million years ago) and extending back for at least 4000 million years. Sometimes referred to as CRYPTOZOIC EON; see also ARCHEAN. Some 30 stratigraphic names are used for different parts of the P in Canada, the Baltic region and Siberia, from *Algomanian* to *Uriconian*. This results from the difficulty of correlation of rocks on a local basis, though as radioactive dating of P rocks extends, a scheme of world-wide application is developing. During the P numerous orogenies occurred; e.g. Laurentian, Algomanian and and Killarnean in N. America.

precession of the equinoxes The change which takes place in the rel. positions of ECLIPTIC and EQUATOR, the result of the axis of rotation of the Earth describing a slow, slightly conical rotation, caused by gravitational forces acting between the Earth and sun and the Earth and moon; this causes the Celestial N. Pole to appear to sweep in a complete circle in the heavens in 26 000 years. Similarly, positions of the EQUINOXES move round the ecliptic once in 26 000 years.

precipice A picturesque term for a high, steep and abrupt face of rock.

precipitate A rock which results from the compaction of ps from solution, mainly of calcium and magnesium salts; e.g. chemically formed limestones.

precipitation The deposition of moisture on the Earth's surface from the atmosphere, including dew, hail, rain, sleet, snow. The main problem is the growth of water drops from the minute droplets (less than 15 mm × 10^{-3}) in non-precipitating clouds, either via ice-crystals forming and growing from supercooled droplets, or by coalescence.

precipitation-day A day (24 hours) with at least 0·25 mm of p. In a country with an apprec. part of p in the form of snow (as Canada), this is more accurate than RAIN-DAY.

precipitation-efficiency (or p-effectiveness, or p-evaporation) index An i devised by C. W. Thornthwaite for the basis of his first climatic classification system from the formula $i = 11 \cdot 5 \left(\frac{p}{T-10}\right) 10/9$ where p = monthly mean precipitation in ins. and T = monthly mean temperature. 5 major regions are distinguished based on humidity rather than on thermal character. A THERMAL EFFICIENCY I was used to subdivide these. The abbr. term P.E.I. should not be confused with POTENTIAL EVAPOTRANSPIRATION I, which is abbr. in the same way. Cf. MOISTURE INDEX.

pressure In climatology, an abbr. for ATMOSPHERIC P.

pressure gradient See BAROMETRIC G.

pressure melting point The temperature at which ice is on the verge of melting, such that exertion of pressure on the ice will induce melting. P m p is normally 0°C at the surface of glaciers, but at depths will be fractionally reduced by pressure (at -30 m p m p will be $-0.0192°C$). Most Alpine glaciers are at p m p throughout the ice thickness, but Polar glaciers, influenced by very low atmospheric temperatures, will be well below 0°C.

pressure-plate anemometer A simple instrument for indicating wind-force, consisting of a wooden base and upright support, with a metal plate suspended from a knife-edge. The angle from the vertical to which the plate blows is observed, and the velocity read off from tables. Not very accurate, esp. when wind is squally.

236 PRESSURE

Plate 54. The effects of PRESSURE RELEASE on exposed granite at Haytor, Devon, England. The removal by denudation of strata formerly covering the granite has led to the formation of numerous JOINTS, which because of their broadly horizontal character give to the rock an appearance of bedding ('pseudo-bedding'). (*R. J. Small*)

pressure release The outward expanding force of p released within rock masses as a result of 'unloading' (e.g. removal by denudation of overlayers), causing pulling-away of outer layers of rock. This may occur partic. following deglacierization. It is esp. evident in massive unjointed rocks; e.g. granite. The 'domes' of Yosemite Valley, California are probably the result of this which has caused a succession of curved 'rock-shells', 25 mm to 6 m or more in thickness, to pull away along DILATATION joints. See pl. 54.

pressure ridge A r of floating ice, formed where one floe has been squeezed against another.

pressure system An individual atmospheric circulation s of high or low p. See ANTICYCLONE, DEPRESSION, RIDGE OF HIGH PRESSURE, COL, SECONDARY DEPRESSION, WEDGE.

pressure tube anemometer A self-recording instrument measuring direction and velocity of wind. It consists of a prominently placed vane, with a shaft going down to a recording unit; the front end of the vane has an opening, kept facing into the wind, communicating changes of pressure (and thus velocity) to the recorder.

prevailing wind A w which blows most frequently from some specif. direction; e.g. S.W. ws are p ws over much of W. Britain. Not to be confused with DOMINANT W.

Primärrumpf (Germ.) W. Penck's name for an initial, slowly and progressively uprising, flat-topped domed surface, or 'primary peneplain', in ct. to the ENDRUMPF.

Primary First used in the early 19th century for the period of time and associated rocks now referred to as Precambrian. Then was transferred to Lower Palaeozoic, finally to the whole Palaeozoic era. Now replaced by the last by most authorities, though some retain it to conform with Secondary, Tertiary and Quaternary.

primary (P-) wave A compressional vibration produced by an earthquake; this is similar to a sound-w, with each particle displaced by the w in its direction of movement; a 'push-w.' Mean velocity through $SIAL = 5.57$ km s^{-1}, through $SIMA = 6.50$ km s^{-1}
[*f* EARTHQUAKE]

principal meridian Central m on which a rectangular GRID is based. Used specif. of 32 systems employed for the US LAND SURVEY.

prisere Collection of SERAL COMMUNITIES which lead to development of a CLIMAX COMMUNITY. These may develop from conditions liable to drought (*xeromorphic*), as bare earth (*lithosere*) and sand-dunes (*psammosere*), or under fresh (*hydrosere*) or salty (*halosere*) water conditions. The emphasis is on development of a community under conditions which in the first instance are unsuited for vegetation.

prismatic compass A magnetic c, with a small magnet which swings on a central pivot, with a circular card graduated clockwise from N. A peep-sight with slot and a prism is on one side, on the other a vertical hair-line in a hinged glass lid. If a sight is taken on an object, the angle of bearing can be read through the prism.

probability In the statistical analysis of data (climatic, population), the **p** of an event occurring is assessed by dividing the number of occurrences by total number of cases or trials. P is normally expressed as a percentage. The occurrence of an absolutely certain event is 100% probable (sometimes written 1·0). If a penny is tossed, the likelihood of its coming down heads is 50%, not exactly so, since the coin may not be exactly true and there is the outside chance that the coin will land on its edge.

process elements In a MODEL in physical geography, **p** es form measurable contributions of various forces involved. E.g. in a model of beach development, **p** es incl. height, period and direction of DOMINANT WAVES, resulting in creation of a set of RESPONSE ES. These may be analysed statistically.

process lapse-rate The r of decrease of temperature of a small parcel of air as it rises, in ct. to ENVIRONMENTAL L-R. See DRY ADIABATIC L-R, SATURATED A L-R.

profile The outline produced where the plane of a section cuts the surface of the ground, as river **p**, coast **p**, dune **p**. See also COMPOSITE, PROJECTED, SUPERIMPOSED P. [*f*]

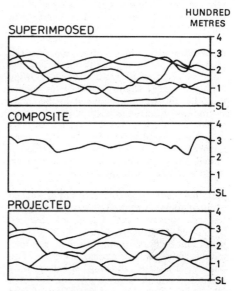

profile of equilbrium (i) Of a *river*: a long-**p** adjusted to prevailing conditions so that at all points the stream is in a state of GRADE. This **p** is gen. a flattened parabolic curve, concave upwards. Recent work has questioned this assumption, which is largely based on the concept that long-**p** is the only significant variable in the achievement of grade. (ii) Of a *shore*: a sloping shore where the mud, sand and shingle which accumulates is more or less balanced by amount removed. This balance is temporary, easily disturbed by exceptionally high tides, strong onshore winds and destructive waves. After a storm is spent, processes of accumulation labour to restore the **p of e**, the net result of the avge set of conditions along that stretch of coast.

proglacial lake An area of water, dammed during a glacial period between the edge of an ice-sheet and some divide; e.g. L. Agassiz in N. America (Ls. Winnipeg, Winnipegosis and Manitoba are remnants); L. Lapworth in W. Midlands of England; L. Humber in Vale of York; Ls. Eskdale, Glaisdale and Wheeldale on N. side of N. York Moors.

progradation Extension of land into the sea by deposition and accumulation, of river-borne sediment and material moved along the coast by longshore drift. Ct. RETROGRADATION.

progressive wave A w propagated in a channel of theoretically infinite length; its w-length is defined as the distance between adjoining crests, and its period as the time it takes to move one w-length. This concept has been applied: (i) to tides (below); and (ii) to wind-generated ws in the open ocean.

progressive wave theory, of tides The old theory of ts, involving the formation of two tidal ws in the S. Ocean, one following (though lagging slightly behind) the moon, the other on the opp. diameter of the earth. From these, branches passed N. into Atlantic, Indian and Pacific Oceans, and successively into their marginal seas. This has been replaced by the STATIONARY W theory, though a **p w** does move up a narrow estuary, in extreme cases forming a BORE.

projected profile One of a series of **p**s, spaced at equal intervals, plotted on a single diagram, but incl. only portions of each not obscured by higher intervening forms. This gives a panoramic effect, with distant sky-line, middleground and foreground; the effect of an outline landscape drawing showing only summit detail. A *multiple p p* consists of a series of individual **p**s cut out of thin board and mounted to form a 3-dimensional model.
[*f* PROFILE]

projection See MAP PROJECTION.

promontory Projection, headland or cliff protruding boldly into the sea, usually with rocky cliffs, STACKS and offshore rocks. See pl. 19.

proportional dividers Instrument to copy detail from a map where enlargement or reduction is required. Two bars fit diagonally across each other, held by a screw in a sliding groove; the screw is set to the proportion required, according to a scale-line, and screwed down tight. Any distance stepped off with the **d**s at one end will give equivalent distance at the other, incr. or decr. proportionally according to scale-setting.

protalus rampart A moraine-like accumulation of coarse angular debris, resulting from the sliding of frost debris across perennial snow-banks. In plan **p r**s form parallel to the slope nourishing them (e.g. the steep headwall of a NIVATION CIRQUE will give rise to a concave-in-plan debris rampart, whereas in a GLACIAL CIRQUE the moving ice will form a convex-in-plan moraine ridge).
[*f* NIVATION CIRQUE]

Proterozoic (i) The younger of 2 Precambrian eras, and by some authorities syn. with Algonkian. (ii) By US Geological Survey, all Precambrian time and rocks. (iii) By some authorities, the 3rd era of Precambrian time, succeeding Eozoic and Archaeozoic.

Protozoic Obsolete term for Older or Lower Palaeozoic era; i.e. Cambrian, Ordovician and Silurian periods and systems, as ctd. with also obsolete Deuterozoic.

provenance Partic. used in connection with sedimentary rocks to describe the origin of the constituent materials, or the source area from which the sediments came.

psammite Rock made of particles of sand 'cemented' together, hence *psammitic*. An ARENACEOUS rock. Ct. PELITE, PSEPHITE.

psephite A coarse, compacted, fragmental rock, with usually rounded particles larger than sand grains; e.g. CONGLOMERATE.

psychrometer Type of HYGROMETER, using wet- and dry-bulb thermometers. The *Assman P* has an electrically driven fan which forces a current of air past the wet bulb to ensure max. evaporation. In another form, the *sling p*, thermometers are whirled round to ensure max. ventilation. The aim is to ascertain RELATIVE HUMIDITY.

Psychozoic Used by some to indicate the era of geological time since man appeared on the Earth, Gk. *psyche* (soul or mind); really syn. with Quaternary.

pteropod ooze Calcareous deep-sea o, formed of shells of minute conical molluscs (*pteropods*). Of limited occurrence, found in rel. small patches only in N. and S. Atlantic Oceans.

puddingstone Pop. name for CONGLOMERATE.

pumice The scum, containing bubbles of steam and gas, on the surface of a LAVA flow, which solidifies to form a spongy vesicular, or cellular rock, so light that it may float. Its chemical composition is similar to rhyolite, of fine-grained texture.

puna (S. Am.) High intermont plateau at about 3600–4800 m between W. and E. Cordillera of the Peruvian and Bolivian Andes with a sparse cover of coarse grass and XEROPHYTIC plants. Used also: (i) for the vegetation itself; (ii) as a proper name for such a region in Bolivia.

push-moraine Mounds of sand and gravel near the margins of ice-sheet movement, pushed into ridges by the ice when it advanced. The materials are thus folded and contorted by pressure. Known in Dutch as *stuwwallen*, in Fr. as *moraines de poussée*. They occur in the Netherlands just S. of the IJssel

Meer in parts of Veluwe, Gelderland and Overijssel, and in neighbouring parts of W. Germany. The product of the SAALE glacial period, known in the Netherlands as *Drenthian*. [*f* ESKER]

push-wave (P-wave) See PRIMARY W. [*f* EARTHQUAKE]

puszta (Hungarian) Type of temperate grassland in the plains of Hungary.

puy (Fr.) Small volcanic cone. In Auvergne some are of ash and cinder, some are dome-shaped of silica-rich lava, others have double cones. In all, about 70 rise from the plateau-surface at 830–990 m; the largest is P. de Dôme (1465 m, 4806 ft), of trachyte.

pyramid[al] peak Sharp p formed when 3 or more CIRQUES develop, cutting into the orig. upland, with prominent faces and ridges; e.g. Matterhorn. [*f*]

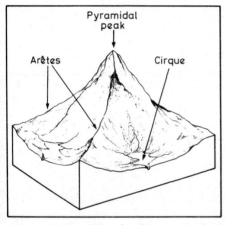

pyrites, pyrite Abbr. for iron **p**, or iron sulphide, FeS_2, yellow mineral with metallic lustre, mined chiefly for sulphur content. Sometimes called 'fool's gold'. Also refers to other sulphides; e.g. copper **p**, but if used alone it denotes iron **p**.

pyroclast Fragmental volcanic material: lava, cinders, ash and dust, consolidated and compacted into TUFFS; hence *pyroclastic*.

pyrometer An instrument used to measure the temperature of molten LAVA.

pyroxene minerals A group of rock-forming silicates, incl. augite, jadeite (m of which jade is composed), and other complex ms.

quadrangle A map of a piece of territory, bounded by parallels and meridians, which fits into a uniform topographical map series. The term is used in USA for the National Topographic Map Series, ranging from R.F. 1:24 000 to 1:1 million. The 15-minute qs of 1:62 500 scale are $17\frac{1}{2}$ ins. in N. to S. dimension, though width varies from about 15 ins. in the S. to 12 ins. in the N. because of progressive convergence of the meridians.

quadrant (i) Instrument formerly used for measuring angles and altitudes, consisting of a graduated arc with an eye-piece. Now obsolete and replaced by the sextant. (ii) The quarter of a circle; an arc subtending 90°.

quadrat A unit area selected for sampling purposes in order to obtain an accurate statistical description of vegetation. For sampling ground species, 1 m^2 might be sufficient, but for woodland a larger area would be necessary.

quadrature A situation when sun, Earth and moon (or other planet) are at right-angles, with the Earth at the apex, which occurs in the case of the moon twice each synodic month. The tide-producing gravitational effects of sun and moon are in opposition, and range of the tides is reduced; hence NEAP tides, with low high tides and high low tides. Cf. CONJUNCTION, OPPOSITION.

quagmire A BOG which shakes under the weight of a man or animal.

quantification A process dealing with quantities; use of statistical methods in presenting, explaining and solving a problem in quantifiable concepts and relationships, in handling data, and in producing objective systems of classification. In recent years **q** has been applied to virtually every aspect of research in physical geography. It should be regarded as a supplement to, not a replacement for, any qualitative or descriptive work, and 'high-power' methods should not be applied to 'low-power' data.

quarry A place where stone is excavated from an open surface-working.

quartz Crystalline form of silicon dioxide (SiO_2), with bright lustre, specific gravity 2·65, and hard enough to scratch glass (No. 7 in the hardness scale). It occurs commonly as clear and transparent, but there are varieties with shades of colour; e.g. rose **q**, smoky **q** (*cairngorm*), amethyst, citrine.

quartzite Rock composed almost entirely of quartz recemented by silica, hard, resistant and impermeable. Some qs are metamorphic in origin, where sandstones have been

240 QUARTZITE

recrystallized by heat into a mosaic of quartz grains, others are sedimentary. Qs of Cambrian age are found in the Welsh borders, Midlands (Hartshill quarry), and N.W. Scotland. Much of the main range of the Canadian Rockies between Banff and Jasper (Alberta, BC) are of q. In the Ardennes, qs are of Lower Palaeozoic age.

quasi-equilibrium A state of apparent EQUILIBRIUM attained by streams, a better term to use than GRADED, since it allows for temporary changes, adjustments and minor oscillations on either side of a mean state.

Quaternary (i) For long regarded as the 4th era in the logical sequence Primary, Secondary, Tertiary, Q; i.e. post-Pliocene, starting near the onset of the last Ice Age. Divided into 2 periods, Pleistocene and Holocene (or Recent). (ii) In USA and now incr. accepted elsewhere, the 2nd (younger) of the 2 periods in the Cainozoic era (the first being Tertiary), and divided into 2 epochs, Pleistocene and Holocene. See Table, TERTIARY.

quebracho forest An evergreen f of gnarled bushy trees with very hard wood, found in the Gran Chaco of Paraguay–Argentina.

quicksand A thick mass of loose sand and mud impregnated with water, which may swallow a heavy object such as an animal.

race (i) Rapid flow of sea-water through a restricted channel, usually caused by marked tidal differences at either end; e.g. Pentland Firth; R. of Alderney (Channel I.). (ii) Strongly flowing offshore current swirling round a headland or promontory; e.g. off Portland Bill, coast of Dorset. (iii) Narrow channel leading water from a river to the wheel of a water-mill (*head-r*) and from the mill (*tail-r*).

radar meteorology The use of **r** in the measurement of upper atmospheric winds (see R WIND) and screen detection of cloud and precipitation in both horizontal (*PPI*=plan position indicator) and vertical (*RHI*=range height indicator) planes.

radar winds An upper atmosphere airstream detected (range, elevation, azimuth) by means of reflections from a **r** target on a free balloon.

radial drainage A pattern of streams flowing outward down the slopes of a dome- or cone-shaped upland. Volcanic cones afford examples of **r d**, their sides scored with channels of streams. English Lake District has near **r d** from the W. to E. axis of its elongated dome; d goes N.W. via Derwent, N. via Calder, N.E. via Eamont, S. via Kent, Leven, Crake, Duddon, S.W. via Esk, Irt and Ehen. An example of SUPERIMPOSED D, which developed on a cover of rocks later removed by denudation. [*f*]

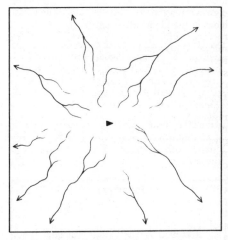

radian The angle subtended by an arc of a circle equal in length to its radius $=57\cdot2958°$; π radians $=180°$; $90°=1\cdot571$ r.

radiation (electromagnetic) The process by which energy is propagated or transferred through a medium by means of wave motion. In meteorology this involves: (i) radiant energy poured out by the sun into space (see INSOLATION, SOLAR CONSTANT); (ii) loss of heat into space in the form of long-wave **r** (with wavelength exceeding 4 microns) from the surface of the Earth (*terrestrial* or *thermal r*), cooling is greatest esp. on a clear night. The *r balance* (*net r*) is positive by day, negative by night, highest in low latitudes, greater over sea than land. An annual positive balance obtains over the globe as a whole, but a negative balance in the atmosphere of the same amount results in an overall equilibrium in the earth/atmosphere system.

radiation fog Shallow layer of white **f**, formed during settled weather in low-lying areas, esp. in spring and autumn. The surface of the ground, rapidly cooled by **r** at night, cools the layer of air resting on it, which flows down into hollows by gravity, and is cooled to dew-point, so causing condensation. The formation of such **f** is favoured partic. by a surface layer of moist air, clear skies permitting max. radiation, and calm air, common in fine weather in early summer; as the sun rises, **f** is dissipated. Under cold anticyclonic conditions in late autumn and winter, **r f** may be thicker and more persistent under an INVERSION layer, and around large towns SMOG may develop.

radioactive decay The progressive breakdown of unstable 'parent-elements' by emission of particles from nuclei of individual atoms, to form stable 'daughter' ISOTOPES; e.g. uranium^{238} to lead^{206}, uranium^{235} to lead^{207}, thorium^{232} to lead^{208}. Hence age of rocks containing **r** elements can be determined. Other radiometric methods of dating include potassium^{40}/argon^{40} and rubidium^{87}/strontium^{87} ratios. See also HALF-LIFE, RADIO-CARBON DATING, RADIOACTIVE DECAY, ISOTOPE, LEAD RATIO.

radiocarbon dating Carbon^{14}, a radioactive ISOTOPE of carbon with half-life of 5570 years, created in the upper atmosphere by bombardment of nitrogen by neutrons liberated by cosmic radiation. Carbon^{14} oxidizes to carbon dioxide, and enters the Earth's CARBON CYCLE. It is absorbed by living matter; e.g. in plants by photosynthesis. After the death of an organism, or its burial under sediments, carbon^{14} not only ceases to be assimilated, but the content diminishes at a known rate. The age of a buried piece of wood, a bone in a tomb, peat in a bog, and shells on the ocean-floor can be obtained by finding the proportion of radiocarbon in the total carbon it contains. It can be used to measure ages up to 50 000, sometimes as much as 70 000 years ago, though beyond 30 000 years accuracy rapidly diminishes. Developed by Dr W. F. Libby, of Chicago in 1947. Recent research has shown that there is an unexplained discrepancy between **r d** and **d** by DENDROCHRONOLOGY; for a sample of wood dated by C^{14} as 4000 years old, the tree-ring age was 4600. See SUESS EFFECT.

radiolarian ooze Siliceous deep-sea **o**, formed of skeletons of radiolaria, with lattice-like structure. Found in an elongated band in mid-Pacific Ocean, and in S.E. of the Indian Ocean.

radiosonde (Fr.) A self-recording and radio-transmitting instrument, carried by a hydrogen balloon to high altitudes, from which meteorological data are sent back by radio-signals.

rag, ragstone Hard rubbly rock outcrop in certain geological formations; e.g. Coral R. (Jurassic), Kentish R. (part of the Hythe Beds within Lower Greensand).

rainbow Arc of multi-coloured light, caused by refraction and internal reflection of the sun's rays by drops of rain, when the sun is behind the observer and rain is in front. The light entering each drop is reflected at its far side, broken up into the colours of the spectrum. In the primary bow, red is on the outer side, violet on the inner. The angle which the radius of the bow subtends at the observer's eye varies from about 41° for the red end of the spectrum to 43° for the violet. This means for an observer at sea-level a **r** can only be seen when the sun's altitude is less than about 42°; the lower its altitude, or higher the observer, the more bow he sees (from an aircraft he may see a complete circle). The larger the rain-drops, the more vivid the colours; in a fine-dropped fog, the bow is white (FOG-BOW). A secondary fainter bow of about 50° angular radius is sometimes observed, the result of a double reflection within each raindrop; colours are reversed compared with the primary bow, i.e. red on the inner side, violet on the outer.

rain-day A period of 24 hours, commencing at 09·00 hours, with at least 0·25 mm of rainfall in UK. The US Weather Bureau uses phrases such as a 'day with measurable precipitation'. Ct. WET-DAY.

rain-drop erosion The displacement of soil particles by the impact of large rain-drops, partic. under intense convectional precipitation and bare earth conditions. The impact of the rain has several effects incl. (i) detachment of soil particles; (ii) sealing of the soil surface by the forcing into voids of the finest particles, thus increasing surface run-off. **R-d e** is very effective in tropical and semi-arid environments.

rain factor An index to express a relationship between precipitation and temperature, to give some indication of climatic aridity to be used in the delimitation of climatic regions; devised by R. Lang.

$$rf = \frac{\text{annual precipitation in mm}}{\text{mean annual temperature in } °C}$$

rainfall When minute droplets of water are condensed from water-vapour in the atmosphere on to nuclei, they float in the atmosphere as clouds. If droplets coalesce, they form large drops, which, when heavy enough to overcome by gravity an ascending air-current, fall as rain. For condensation and precipitation to occur, ascent of an air-mass is essential; this is brought about in 3 ways, hence 3 main types of **r**: (i) CONVECTIONAL; (ii) OROGRAPHIC; (iii) CYCLONIC or frontal. **R** is measured in a RAIN-GAUGE. Records of **r** are given as monthly and annual means. See also INTENSITY OF R, DISPERSION DIAGRAM. *Note:*

242 RAINFALL

1 in. of $r = 100.9$ tons of water per acre = 14 460 000 gallons per sq. mi = 160 000 hl km^{-2}. The current standard period in UK for which statistics are given is 1916–50.

rain-forest Dense forest of HYGROPHILOUS trees growing in conditions of heavy, well distributed rainfall; found in: (i) tropical latitudes (e.g. Amazon Basin, Zaïre Basin, Borneo, New Guinea); (ii) warm temperate latitudes (e.g. S. Brazil, Florida, coastlands of Natal, central and S. China, S. Japan, E. Australia, N. Island of New Zealand).

rain-gauge A meteorological instrument, comprising a funnel resting in a collecting vessel, used to measure rainfall. The standard **r-g** has its rim 30 cm from the ground, with interchangeable apertures from 150 to 750 cm^2. The vessel is emptied periodically into a measuring cylinder. The g must be carefully sited, if possible twice as far from the nearest building as this is high. Self-recording **gs** are used at many stations, such as the HYETO-GRAPH, tipping-bucket and tilting-siphon types.

rain-shadow The markedly drier (lee) side of a mountain; see OROGRAPHIC PRECIPITATION. The term introduces confusion if it is thought that dryness is simply due to blocking of or sheltering from air-masses; when heavy rain occurs on the windward side, the moisture content of the air-stream is lower on the leeward. E.g. Alberta in **r-s** of Rockies; Deccan in **r-s** of W. Ghats; E. Scotland in **r-s** of mountains of W. Scotland.

rain-spell A period of 15 consecutive days, each with at least 0·25 mm of rainfall.

rain-wash The movement of loose surface material down a slope, esp. in semi-arid areas with scarce vegetation protection, caused by heavy rainfall and resultant runoff not confined to precise channels.

raised beach A coastline, sometimes backed by a cliff and fronted by a wave-cut platform covered with ancient **b** material, standing above present sea-level as a result of a NEGATIVE MOVEMENT. Such features below about 40–45 m (130–150 ft) are **r bs**, above this marine terraces, platforms or erosion surfaces. Around Britain have been identified: (i) pre-glacial 3 m (10 ft) **b** (e.g. Holderness, English Channel, I of Man, N. Wales); (ii) 30 m (100 ft) **b** (e.g. W. Isles and W. Scotland opposite Skye); (iii) 15 m (50 ft) **b**, W. Scotland between 14 and 20 m (45 and 65 ft) above O. D.; (iv) 8 m (25 ft) (e.g. W. Scotland, esp. Arran); (v) **r bs** a few m above present high-tide level (e.g. Gower Peninsula, S. Wales). The order listed is broadly in decr. age; no. (iv) probably of early post-glacial age. Recent work has thrown doubt on grouping of **r bs** by height alone because of isostatic recovery since glacial times and complex sea-level changes. [*f*]

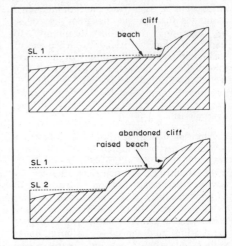

raised bog A thick accumulation of peat, mostly sphagnum moss, acid in character, forming a lens-shaped sheet in a shallow basin, as much as 5 m thick; e.g. central Ireland, **R b** is one of 3 main varieties of BOG, the others *valley bog* and BLANKET-BOG.

rake A sloping terrace on a mountain or rock-face; e.g. Lord's R. on Scafell, Jack's R. on Pavey Ark, Langdale, English Lake District.

Randkluft (Germ.) Gap between a FIRN-field and surrounding rock-face in a CIRQUE. Ct. BERGSCHRUND, where both sides of the gap are of ice. The **r** develops because of melting caused by radiation of heat from the rock-face.

random sample Method of overcoming the problem of coverage of a large mass of data, or a large area, in research. In ct. to *purposive sampling*, where one chooses typical **ss** subjectively, **r** sampling is such that every item has an equal chance of selection, and every item selected is independent of all others. In a *simple r s* the objects are listed, each assigned an index number, and a **r s** of index numbers obtained by ref. to a table of **r** numbers, as in *Cambridge Elementary Statistical Tables*. When a geographer wishes to obtain a **r s** of a spatial variable, either continuous (height of land, RELATIVE RELIEF, slope angle, soil features) or discontinuous (land use), he may superimpose a GRID over a map of the specif. phenomena, and select a **r** series of grid intersections. If the number of items involved is so large as to be virtually infinite, *area sampling* must be used; e.g. a botanist studying a large COMMUNITY of plants may select QUADRATS. If there is marked spatial clustering of phenomena, **r s** may lead to a biased **s**, and it is necessary to make a *stratified r s* in which total phenomena are broken down into classes ('strata') before the **s** is taken.

random walk network A method of SIMULATION used in geomorphology to investigate stream pattern development. On a sheet of squared grid paper stream sources may be arbitrarily located at equal intervals. The 'flow' of these streams 'down' the paper may be determined by the throw of dice, the use of random number tables etc. E.g. with dice, throws of 1 and 4 could indicate direct flow down the paper for one unit square; 2 and 5 a movement 45° to the right for one unit square; 3 and 6 a movement 45° to the left for one unit square, It is significant that the patterns thus generated closely resemble in certain respects actual stream patterns, showing the role of randomness or chance, as opposed to geological and other controls, in the operation of fundamental geomorphological processes. [*f*]

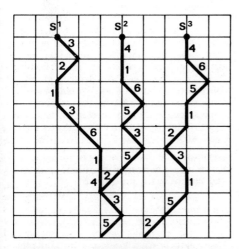

range (i) Line of mountains. (ii) An open area, usually unfenced, used for grazing, as High Plains of USA. (iii) The difference between the max. and min. of a series of numerical values, esp. climatic elements, such as seasonal temperature; e.g. Verkhoyansk (Siberia), Jan. mean −50°C (−59°F), July mean 15·5°C (60°F), hence **r** = 65·5°C (119°F). (iv) Limit of habitat of a plant or animal. (v)

The *tidal-r* between the highest high and lowest low spring tides.

rank Category of coal, from lignite to anthracite, according to its chemical and physical composition, the result of progressive metamorphism.

rapid(s) Area of broken, fast flowing water in a stream, where slope of the bed increases (but without a prominent break of slope which might result in a waterfall), or where a gently dipping bar of harder rock outcrops; e.g. Nile Cataracts; Istein **Rs.** on Rhine below Basle; rs in Iron Gate of the Danube.

rare earths Widely distributed, though extremely scarce, oxides of metallic elements with atomic numbers from 57 to 71, of atomic weight 138·92 to 175·0, from lanthanum to lutetium. These minerals and their compounds are difficult to separate.

rating curve A graph plot used in the determination of stream discharge at an artificially constructed flume or in a stable natural channel. Field observations are made of (i) stream velocity measured by current meter, and (ii) variations in cross-sectional area, especially those revealed by the height or 'stage' of the stream. When discharge (calculated from velocity and cross-sectional area) and stage are plotted on logarithmic graph-paper, a 'straight-line' relationship is usually revealed. The graph can then be used to 'read' discharge from the changing 'stage' of the river, as revealed by a 'Stage-recorder'. [*f*]

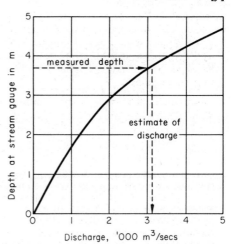

rattan Species of climbing palm in Indonesia with thorny stems, used for ropes, nets, and weaving baskets.

ravine Narrow steep-sided valley, larger than a gully or a cleft.

rawinsonde (radar wind-sounding) (Fr.) Hydrogen balloon equipped with self-recording and transmitting meteorological instruments, carrying a radar target to enable its course to be followed and plotted, the apparatus to be recovered after descent by parachute. Ct. RADIOSONDE.

ray-diagram A type of **d** to illustrate dynamic concepts radiating from a point; e.g. winds (see WIND-ROSE). Often known as *star-graphs, star-diagrams, clock-graphs*.

reach (i) Specif. section of a river. (ii) In navigation, a straight section between bends.

Réaumur scale An obsolete thermometric **s** in which melting point of ice is 0°R and boiling point of water

$$80°R \frac{R°}{80} = \frac{F° - 32}{180} = \frac{C°}{100}$$

Recent (i) The period/system since the end of the Pleistocene (i.e. post-glacial). This varies with position rel. to the main centre of ice-dispersal; for E. Anglia *c.* 9–11 000 B.C.; S. Scandinavia *c.* 8000 B.C.; S. Finland *c.* 6500 B.C. A gen. figure is 10 000 B.C. Syn. with Holocene **R** rocks incl. alluvium, peat, sand-dunes, shell-beds, coral. (ii) In USA **R** is a series epoch, the younger part of the Quaternary. See table, TERTIARY.

recession col A form of WIND-GAP resulting from the beheading of a dip-slope valley by escarpment recession.

recessional moraine A terminal **m** which is the result of a brief pause in the retreat, or even a slight readvance, of an ice-sheet. Syn. with *stadial m*. E.g. as the ice-sheet of the last major glaciation in N. Europe (WEICHSEL) retreated, it paused at 3 stages, leaving *Brandenburg, Frankfurt, Pomeranian* **r** ms.
[*f* ESKER]

recharge area The surface **a** over which water is received (rainfall, snowmelt), which then penetrates into an AQUIFER; e.g. hill-slopes on the flanks of an ARTESIAN BASIN.

reciprocal bearing (*negative, reverse* or *back* **b**) A line drawn 180° from any **b**.

reconnaissance mapping A type of exploratory map, produced rapidly yet efficiently, using such techniques as cameras, phototheodolites, car and sun compasses, terrestrial photogrammetry, TELLUROMETERS, air photographs. The essence is speed and rel. cheapness in an area where detailed accurate surveys (which may follow the **r m**) are not available, as in deserts, polar areas. The map should nevertheless be accurate within certain limits, even if thin in some detail.

rectangular drainage A **d** pattern in which tributary junctions are gen. at right-angles, and all streams, major and minor, exhibit sections of approx. the same length. May be controlled by a rectilinear joint pattern. Ct. TRELLIS D. [*f*]

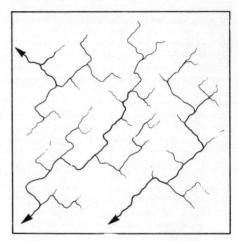

rectilinearity A form of slope which is 'straight in profile'. The **r** may represent a small part or large proportion of an actual slope profile. A number of distinct **r**s may occur, though commonly the main rectilinear segment occurs between an upper convex element and lower concave element, giving a 'convexo-rectilinear-concave' profile overall. In this instance the **r** will be the steepest part of the slope, tending to develop at the angle of rest of debris weathered from bedrock (cf. REPOSE SLOPE).

recumbent fold An OVERFOLD forced over into a near horizontal position, with its axial surface nearly horizontal.
[*f* OVERFOLD]

Red Clay Fine-grained PELAGIC deposit of hydrated silicate of alumina coloured by iron oxide, found on the ocean floor (ABYSSAL ZONE); derived from volcanic and meteoritic dust, material carried by icebergs, and insoluble relics of marine life. The most widespread of pelagic deposits.

red earth A tropical zonal soil, produced by chemical weathering under high temperatures and humidity, with well defined seasonal rainfall. It consists of a loamy mixture of clay and quartz, coloured by iron compounds.

This may be as much as 15 m thick, and covers extensive areas of savanna in Brazil, Guyana, E. Africa, S. Deccan, Sri Lanka, Burma, Vietnam. But the term is loose and unscientific, and has been replaced by 3 groups: (i) *ferralitic* soils; (ii) *ferrisols*; (iii) *ferruginous* soils. See LATOSOL.

Red Mud A m of terrigenous origin, stained by ferric oxide, found on the continental slope.

reef (i) Mass of rock with its surface at or just above low-tide mark. (ii) Specif. a mass of CORAL; see CORAL R. (iii) Vein of metal or ore, esp. gold. (iv) Bed of CONGLOMERATE containing gold (BANKET); e.g. the Rand, Transvaal.

reef-flat Platform of coral fragments and coral-sand accumulated on the inner side of a coral r, on which dunes collect and vegetation (esp. palms) is established.

reef knoll A k made of compacted material that orig. as a r, exhumed or exposed by later denudation, so that it stands out as a rounded or conical k, e.g. along S. or downthrow side of Mid-Craven Fault, between Settle and Appletreewick, Yorkshire (High Hill, Scaleber, Burns Hill, Cowden Pike and at least 5 others). Farther W., in Bowland Fells, the K. Series of r-limestones form isolated ks rising approx. 600 m above the gen. level (K. Hill, Crow Hill, Worsaw K., Sykes K., Twisted K.).

re-entrant (i) Prominent indentation into a landform, esp. in an escarpment where a transverse valley occurs. (ii) Angular inlet into a coastline.

reference net A system of squares, lettered along one side, numbered along the other, to facilitate the location of a place on a map. Thus a place may be in square P4. E.g. Ordnance Survey 1-inch series, 4th (Popular) edition, had 2-inch squares, but this gives only approx. location of a point within 4 sq. mi., and can refer only to a single sheet. Commonly used in the index or gazetteer of an atlas. Ct. GRATICULE, GRID.

refraction, wave As a wave approaches a shore and the water shallows, its speed is reduced. This shallowing is more rapid in front of a promontory than in the deeper water of a bay. The wave is therefore bent or refracted from the bay on to the side of the headland, thus accentuating the erosive processes there. R also occurs on a straight coast should a wave advance obliquely as the result of wind direction, so that ultimately it breaks parallel to it.

reforestation, reafforestation Planting or restocking with trees of a former forested area (which has been cleared or burnt); an integral part of forestry practice.

reg (Arabic) A stony desert, esp. in Algeria, where sheets of smoothly angular, wind-scoured gravel cover the surface. Strong mineralized solutions drawn to the surface by capillarity evaporate to form a 'cement', which binds the gravel into a hard sheet. See DESERT PAVEMENT.

regelation Pressure within an ice-mass converts ice-grains into molecules of water; these move to points where pressure is less (i.e. downhill) and then re-crystallize and re-freeze, causing a gradual movement of and within the ice-mass; a contributory cause of glacier-flow.

regelation layer A thin band of distinctive ice, usually less than 3 cm in thickness, found at the base of glaciers. The **r l** is evidently formed by thawing of basal ice, at PRESSURE MELTING POINT, as it is forced over the upglacier slope of rock obstructions. Refreezing of this thawed water, to give the **r l**, occurs as pressure is reduced on the downglacier side of obstacles. The process of regelation leads to (i) small-scale freeze-thaw 'weathering' below the ice, and (ii) entrainment of small particles within the basal ice; it is therefore a factor, of disputable importance, in glacial erosion.

régime (Fr.) (i) Seasonal fluctuation in volume of a river (more recently used of a glacier). (ii) *Climatic r*: seasonal pattern of climatic changes.

regional metamorphism Metamorphic action on a major scale of both pressure and heat on the rocks, associated with earth-movements and vulcanicity, resulting in both mineralogical and structural changes. Syn. *dynamothermal m*.

register mark A small cross at each corner of a map to be printed in more than one colour. The accuracy of printing of each colour is checked by the coincidence of crosses on the printed map.

regolith Mantle of more or less disintegrated, loose, incoherent rockwaste overlying the bed-rock, together with superficial deposits of alluvium, drift, volcanic ash, loess, wind-blown sand and peat. Also incl. the soil layer. Derived from Gk, *regos*, a blanket, *lithos*, a stone.

regosol One of the AZONAL group of soils, derived from freshly deposited alluvium, dune-sands and mud-flats. Used as one of the

2 divisions of Azonal, the other being stony mountain soils (LITHOSOL).

regur (Indian) See BASISOL.

rejuvenation Revival of erosive activity, esp. by a river, because of: (i) fall in sea-level; (ii) local movements of land-uplift; both result in a change in BASE-LEVEL (*dynamic r*). These initiate a new cycle, causing KNICKPOINTS, TERRACES and INCISED MEANDERS. R also occurs without change in base-level, as when following river CAPTURE or an incr. in precipitation; both may give incr. discharge and therefore greater eroding power to a drainage system (static **r**). [*f*]

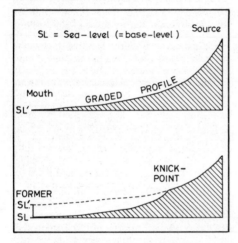

rejuvenation head See KNICKPOINT.

relative age, of rocks Determination of whether one **r** is younger than another, thereby producing a tabulated list of **r**s in order of age. (i) In all beds of sedimentary **r**s, the oldest is at the bottom and each overlying bed is progressively younger (*law of superposition*), unless inverted by earth movements. (ii) Any sedimentary **r** is younger than the fragments of which it is composed. (iii) **R**s containing the same fossil assemblages are similar in age. (iv) An intruded igneous **r** is younger than the **r**s which it cuts across (*law of cross-cutting relationships*).

relative humidity (i) The water-vapour present in a mass of air expressed as a percentage of the total amount that would be present were the air saturated at that temperature; (ii) the ratio of the air's VAPOUR PRESSURE to the saturation vapour pressure. **R h** varies with: (*a*) ABSOLUTE HUMIDITY; e.g. saturated air (100%) at 20°C (68°F) contains 17·117 g cm^{-3}, but if air at that temperature contained only 8·262 g, **r h** $\frac{8·262}{17·117} \times 100 =$ approx. 48%; (*b*) *temperature*; if a mass of air saturated (100%) at 4·4°C (40°F) is warmed, at 10°C (50°F) **r h** falls to 71%, at 15·5°C (60°F) to 51%, at 32°C (90°F) to 19%.

relative relief The relation of altitudes of the highest and lowest points of land in any area, sometimes called *local r* or *available r*. The difference between these points is the *amplitude of r r*. Various maps have been devised, usually dependent on gridding the area, finding a value for amplitude in each grid-square, and producing an ISOPLETH or dot map to depict the distribution of these values.

relaxation time The time required for the re-establishment of a state of equilibrium in a system, after a previous state of equilibrium has been disturbed by changes in the factors controlling the system. E.g. a climatic change may entail a change in the equilibrium angle of slopes in a particular area from 32° to 20°. This will require a lapse of time, during which as the slope angles are progressively changed a temporary 'inequilibrium' will obtain. The duration of **r t** can vary immensely, according to the 'scale' of the change in controlling factors and the resistance of rock materials to change. A beach profile will adjust very rapidly to a change in wave type; an African pediplan formed in granite and gneiss, and uplifted by 500 m, will remain in an 'unadjusted' condition for tens of millions of years.

reliability diagram A **d** given in a map margin to indicate date/quality of material on which it is based.

relict landform A landform produced by, or bearing the clear imprint of, processes no longer operative upon it. E.g. in the English chalk, dry valleys formed by periglacial meltwater streams are **r l**s, as are chalk scarp slopes fashioned mainly by frost action and solifluction during the Pleistocene. Owing to the magnitude of recent climatic changes perhaps the majority of landforms are to some extent 'relict'.

relict (relic) mountain A surviving upland mass in an area of denudation (MONADNOCK, INSELBERG); obsolescent.

relief The physical landscape, the actual configuration of the Earth's surface, used in a rather loose sense of differences in altitude and of slope, of inequality of surface, of shapes and forms. It should not be confused with TOPOGRAPHY, though American usage of *topographical relief* is permissible.

relief map A m which shows surface configuration by any of the following methods, some used in combination: (i) SPOT-HEIGHTS; (ii) CONTOUR-LINES and FORM-LINES; (iii) LAYER-TINTING; (iv) HACHURES; (v) HILL-SHADING (or PLASTIC R or photo-**r**); (vi) cliff- and rock-drawing; (vii) PHYSIOGRAPHIC PICTORIAL symbols.

relief rainfall See OROGRAPHIC PRECIPITATION.

relief model Reconstruction of the surface features by means of a 3-dimensional **m**, usually with some VERTICAL EXAGGERATION, known in USA as *terrain m*. A m may be made from cardboard (each contour drawn on card, cut around, and stuck over each other in exact position); plaster; a mixture of sawdust, plaster, paste and glue; sheet metal hammered to shape; vinylite and other plastics, pressed over a mould, or by a vacuum-forming process. The **m** can be painted and lettered.

rémanie (Fr.) Materials in the Earth's crust that have been (lit.) 'rehandled'; e.g. boulders in glacial drift; pebbles in conglomerate; masses of country-rock involved in the engulfment or emplacement of a batholith, caught up and solidified within a lava-stream; a fossil removed by natural agencies of denudation from a bed and redeposited in another.

remote sensing The gathering, retrieval and storage of mass data by aerial survey, airborne electronic scanning devices, and increasingly by orbital satellites and spacecraft, these sensors operating at considerable distances from the source. While a limited amount of data can be processed manually (e.g. air-photo interpretation), the vast increase in data requires computer techniques. The *Earth Resources Technology Satellite* (*ERTS-1*), launched in 1972, views the Earth from nearly 1000 km up, using 4 multispectral scanners and 3 TV cameras, each on a different wavelength band. Each portion of the Earth's surface is scanned once in 18 days. Different surfaces reflect different proportions of wavelengths, providing a *spectral signature* which enables bare ground, types of vegetation, etc., to be identified. Relief features can be determined by laser profiling and radar scanning. See pl. 55.

rendzina (Polish) An INTRAZONAL soil, dark coloured, with A-horizon of friable, almost granular, loam, lying on a B-horizon containing chalk or limestone fragments, which in turn rests on solid rock. It has developed where grassland formerly dominated, notably on chalk Downs and Wolds. **R**s also found in USA in E. Texas and 'Black Belt' of Alabama. Some pedologists refer to TERRA ROSSA as 'red **r**.'

renewal of exposure The continual re-exposure of a rock face to weathering processes, as the weathering products are removed by denudational processes such as gravity fall, creep, rainwash etc. **R of e** is most effective on steep slopes (e.g. FREE FACES), and helps to determine the rapidity of slope recession.

repose slope Used by A. N. Strahler to describe a slope whose steepness is controlled by the angle of repose of the superficial debris layer. **R** ss are usually steep (in the order of 30° on resistant bed-rock), and maintain their angle as they retreat. Cf. BOULDER CONTROLLED SLOPE.

---

representative fraction The ratio which the distance between 2 points on a map bears to the corresponding distance on the ground, expressed as a **f**.

R.F. 1/to	Miles to 1 inch	Inches to 1 mile	Km to 1 cm	Cm to 1 km
million	15·78	0·0634	10·0	0·1
633 600	10·0	0·1	6·336	0·1578
500 000	7·891	0·127	5·0	0·2
253 440	4·0	0·25	2·534	0·395
250 000	3·945	0·245	2·5	0·4
126 720	2·0	0·5	1·267	0·789
100 000	1·578	0·6336	1·0	1·0
63 360	1·0	1·0	0·6336	1·578
50 000	0·789	1·267	0·5	2·0
25 000	0·395	2·534	0·25	4·0
10 560	0·167	6·0	0·1056	9·468
10 000	0·158	6·336	0·1	10·0
2 500	0·0395	25·34	0·025	40·0
1 250	0·0198	50·69	0·0125	80·0

Plate 55. REMOTE SENSING, a satellite photograph of the Bay of Bengal, showing a major atmospheric disturbance. (*National Oceanic and Atmospheric Administration*)

resection In making a PLANE-TABLE survey, the position of the observing station on the map can be fixed by drawing rays from observed points. The plane-table is orientated by compass before observations are taken. The three rays intersect to form a TRIANGLE OF ERROR, which can be eliminated to fix the position accurately. [*f*]

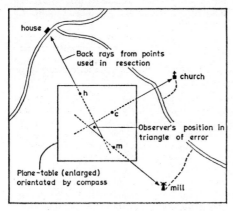

resequent drainage A pattern of d in which a stream lies more or less along the line of a former longitudinal (synclinal) consequent stream in an area of ancient folding, following a long period of denudation. It is not a SECONDARY CONSEQUENT STREAM, as often defined. E.g. in Hampshire Basin, R. Ebble occupies a syncline in the Chalk; it may have migrated from its orig. synclinal position to follow a course along either the neighbouring Bower Chalke anticline or the Vale of Wardour anticline, and has now returned to its orig. line. R streams are common in the Ridge and Valley section of the Appalachians.

resequent fault-line scarp A feature formed when erosion along an OBSEQUENT F-L S produces an escarpment facing in the orig. direction of downthrow.

residual debris Highly decomposed granite, in which protracted weathering of the ferromagnesion and feldspar minerals has produced an admixture of clay (mainly kaolinite) and stable quartz particles. R d is characteristic of deep weathering profiles in tropical environments.

residual deposit The residue of rock disintegration by weathering *in situ*. Hence *r soil* (sometimes *sedentary soil*), also formed *in situ*; e.g. Clay-with-Flints on Chalk, TERRA ROSSA in hollows on limestone.

resistant rock A r which can withstand weathering and erosion, because of its hardness, compactness, cementation, absence of jointing, and the nature of its chemical composition. This is usually considered in rel. terms, as when denudation picks out resistant and unresistant rs by differential weathering and erosion.

resource management Responsible and efficient m of both environmental and cultural rs implying a close and integrated relationship between the ecological basis and the operative socio-economic system involved. It can be applied specif. to *primary r m*, involving on the one hand factual surveys of geology, geomorphology, hydrology, climatology, soil and vegetation, on the other to the most efficient usage of this physical environment in the light of its land-use capability. This may range through rapid exploitation, controlled and regulated use, protection, preservation and even sterilization; e.g. in respect of a forest: clear felling, selective logging and replanting, maintenance as a recreation resource, or as a wilderness.

response elements In a MODEL the features produced by the combined operation of a series of PROCESSES. E.g. in a model of beach development, r es are the gen. form of the beach, its gradient, size and distribution of material of which it is composed, etc. These may be analysed statistically.

response time The time-delay in the reaction of glaciers and ice-sheets to short-term climatic changes, causing significant alterations in accumulation/ablation ratios. In small glaciers the **r t** may be up to 30 years; in the Antarctic ice-sheet it may be 5000 years. During the past few years glaciers in the Alps have begun to advance; this may be related not to present temperature, but to changes in winter snowfall during the 1950s.

resurgence The emergence of an underground stream from a cave, usually near the point where an impermeable stratum intersects the surface. In limestone water works its way through the mass of the rock, now vertically, now horizontally, issuing near its base. A **r** is larger than a spring, involving a considerable stream flowing strongly; e.g. R. Axe out of Wookey Hole in the Mendips; R. Aire from under Malham Cove; Peakshole Water from Peak Cavern at Castleton in Derbyshire; Echo R. in Kentucky emerging from the vast Mammoth Cave system to join Green R., hence Ohio. Syn. with VAU-CLAUSIAN SPRING.

retarding basin Construction of a dam across a river in order to reduce its flood flow and therefore its CREST; e.g. on the Mississippi. Excess water may be discharged into a floodway in a waste or 'back-swamp' zone of a river flood-plain.

retrogradation Esp. of shoreline studies, erosion or cutting back of a beach by wave action; hence 'retrograding shore-line'. Ct. PROGRADATION.

reversed drainage See CAPTURE, RIVER.

reverse(d) fault A f caused by compression, where older beds on one side of a f-plane are thrust over younger beds on the other; i.e. the HANGING WALL has been raised rel. to the FOOT WALL, resulting in crustal shortening. One effect is that landslips are common along a **r f**, because of overhanging strata.
[*f* NORMAL FAULT]

reversing thermometer A mercury **t** with a constriction in its glass tube. When reversed or inverted, the mercury column breaks, leaving the previous reading. Used for taking temperatures at depth in the sea.

Rhaetic Series of shales, marls and limestones, sometimes ascribed to the top of the TRIASSIC system, more usually regarded as transitional between the latter and the Lias at the base of the JURASSIC.

rheidity The relationship between the resistance of a solid to viscous flow (*viscosity*) and its resistance to elastic deformation (*elasticity*). This has important aspects in folding of rocks and movement of ice in a glacier. **R** involves a time factor. A substance undergoing deformatory flow is a *rheid*.

rheology (Gk. *rheo*, flow) Study of flowage of materials, partic. that of plastic solids. See RHEIDITY.

rhexistasy A condition of 'biological inequilibrium' (ct. BIOSTASY). In tropical rainforests **r** may be induced by Man's activities or natural catastrophies such as lava flows. As vegetation is destroyed it is not regenerated owing to the operation of rapid erosion, stripping soil and removing plant nutrients which were previously recycled by the forest.

rhinn Rugged ridge in W. of S. Uplands of Scotland; e.g. R. of Kells, Rs. of Galloway.

rhombochasm A parallel-sided gap (RIFT-VALLEY) in the sialic crust, floored by dense simatic material (not basaltic lava), which has probably moved there through rock-flowage (RHEIDITY).

rhumb-line A **l** of constant bearing, i.e. it cuts all meridians at a constant angle; syn. with *loxodrome*. Shown as a straight line on MERCATOR projection.   [*f* GREAT CIRCLE]

rhymite Syn. with VARVE.

rhyne, rhine Drainage channel in the Somerset Levels, intermediate in size between small ditches and rivers such as Parrett and Yeo or larger artificial channels or drains (e.g. King's Sedgemoor Drain), into which they discharge. These channels also serve as field boundaries.

rhyolite An igneous rock, consisting of alkali-feldspars and quartz, commonly with ferromagnesian minerals; it was extruded, and is fine-crystalled or even glassy in structure. It may contain porphyritic crystals of quartz, and reveal distinct banding ('banded **r**'.) as a result of its flowing as a molten MAGMA (Gk. *rheo*=flow). Rs of Ordovician age are found in English Lake District (e.g. W. face of the Pillar Rock, Ennerdale), and Snowdonia; e.g. upper parts of Glyders (huge blocks on Glyder Fach summit are **r**), and parts of Snowdon. Most of Yellowstone National Park, Wyoming, consists of a **r** plateau; the rock is distinctly yellowish, hence the name.

rhythmite Alt. name for VARVE.

ria (Sp.) (i) Funnel-shaped coastal indentation formed by submergence due to rise in sea-level affecting an area where hills and valleys meet the coast at right-angles. It

250 RIA

Plate 56. A RIA formed by the POST-GLACIAL rise of sea-level, which has flooded a small coastal valley at Boscastle, Cornwall, England. (*Eric Kay*)

decr. in width and depth inland. The stream which flows into its head, responsible for eroding the orig. valley, is too small for the present size of the inlet. E.g. N.W. Spain (R. de Vigo, de la Coruña, del Ferrol); S.W. Ireland (Dingle Bay, Kenmare R., Bantry Bay); W. Brittany (Rade de Brest, Baie de Douarnenez). (ii) Used more widely for the submergence of any land margin dissected more or less transversely to the coastline; e.g. S. coast of Devon and Cornwall (Tamar estuary with Plymouth Sound, and Fal estuary with Carrick Roads). See pl. 56. [*f*]

Richter Scale S of earthquake magnitude which has replaced the Rossi-Forel and Modified Mercalli Ss of earthquake intensity, devised in 1935 by C. F. Richter, seismologist at the California Institute of Technology. The s is based simply on instrumental records, making allowance for distance from EPICENTRE. The s assigns largest numbers to largest earthquakes from 0 to over 8·0; 8·9 has been recorded on 3 occasions. It cts. with Mercalli S, based on observed effects on buildings.

ridge A long narrow upland, with steep sides, but no very specif. application.

ridge, of high pressure An elongated region of high atmospheric **p** between two areas of low pressure, wider than a WEDGE. It brings fine, though short-lived, weather, often within a gen. period of rainy conditions. 'The borrowed day, too good to last'.

ridge-and-ravine topography A characteristic landform type in tropical forest environments, in which a high and uniform drainage density results in reduced slope lengths. To quote A. Young: 'the entire surface is made up of valley slopes with fairly narrow but smoothly curved convex crests above long 30–40° segments'. Also used by J. T. Hack to describe Appalachian landscapes, where fluvial dissection of this nature has produced accordant interfluves and hilltops i.e. 'apparent' peneplains.

ridge and valley Type of relief characterized by a close pattern of nearly parallel rs and vs; the type-region is R. and V. region in the Appalachians (E. USA), lying between the Allegheny-Cumberland plateau on W. and the Blue Ridge on E. The rs consist of resistant sandstones, quartzites and conglomerates, vs of weaker shales and limestones.

Ried (Germ.) A marshy flood-plain, specif. of the Rhine, consisting of marshland and backwaters, with damp pasture, clumps of willows, alders and poplars, often flooded in spring and early summer. Though much has been drained, there are still considerable remaining areas in the section of the Rhine valley between Basle and Strasbourg.

riegel A pronounced irregularity or 'rock-step' in the long-profile of a glaciated upland valley. Rs are usually separated by near-level 'treads' or rock basins, either alluviated or containing lakes. The r may be associated with a V rather than U-shaped cross-profile, indicating modification by fluvial activity or a form of modified glacier flow (see EXTENDING FLOW). Many rs coincide with hard rock barriers, which are abraded on their upper surfaces and plucked on their downvalley faces; thus they resemble large-scale ROCHES MOUTONNEES.

riffle A depositional bar in a stream, over which the water flows rapidly with a 'rippled' surface. Rs form, with deeper intervening pools, POOL AND RIFFLE sequences, partic. in 'gravel-bed' streams.

rift-valley A narrow trough between parallel FAULTS, with THROWS in opp. directions, so forming a long steep-sided, flat-floored v. There may be a series of step-faults on either side of the trough, or the sides may be clean-cut as a result of the downthrow along a single major fault on either side. Its exact origin is a matter of argument. (i) *Tension* in the crust may pull the 2 sides apart, leaving the centre to subside. (ii) *Compression* from either side may thrust masses on either side higher than the central block, which was also forced down. (iii) There may be gentle upbending of strata, so that a gaping crack developed along the crest of the swell. E.g. line of the Jordan V. (floor of the Dead Sea 750 m below the surface of the Mediterranean Sea), Gulf of Akaba, Red Sea, Abyssinia, E. Africa to the Zambezi, length 4800 km, middle Rhine r-v between Vosges and Black Forest; Midland V. of Scotland between 2 boundary faults. Ct. GRABEN. Now believed

Plate 57. Looking south along the major escaprment, approximately 600 m high, forming the margin of the East African RIFT VALLEY near Limuru, Kenya. The flat floor of the rift stretches for some 30 km to the base of the far (western) scarp. (*R. J. Small*)

that the MID-OCEAN RIDGES are r-vs. See pl. 28, 57. [*f*]

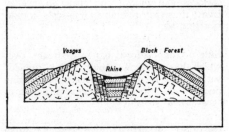

right ascension One of 2 references (the other DECLINATION) which enables a heavenly body to be exactly located on the CELESTIAL SPHERE, equivalent to longitude for terrestrial reference. For all heavenly bodies the **r a** and declination are tabulated in *Nautical Almanac*. The lines of **r a** (*hour-circles*), pass through the Celestial Poles, and cut the Celestial Equator and all parallels of declination at right-angles. **R a** is reckoned E.ward, starting from the point of the spring EQUINOX on the Celestial Equator, known as the 'First Point of Aries'(♈); analogous to 0° Greenwich meridian. So called because when this point was chosen 2000 years ago, it was situated in the constellation of Aries. Since then, the point has moved along the Celestial Equator (see PRECESSION OF THE EQUINOXES) into the next constellation (Pisces), though still termed First Point of Aries. **R a** is therefore the arc of the Celestial Equator intercepted between the First Point of Aries and the hour-circle through the body, measured in time E.ward from 0 to 24 hours. [*f*]

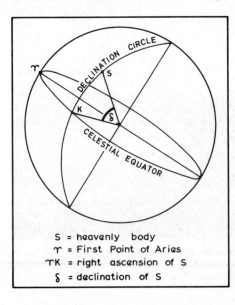

rill erosion Uneven removal of surface soil by the formation of small-scale **r**-channels; known in USA as *rilling*. If this continues, the rs coalesce into large gullies (*concentrated wash*) leading to GULLY E.

rimaye (Fr.) Fr. form of Germ. BERGSCHRUND.

rime Accumulation of white opaque granular ice-particles, formed when fog composed of super-cooled droplets is driven by a slight wind against objects with temperatures below freezing (telegraph poles and wires, trees and rock buttresses). When particles coalesce, pop. known as 'frost-feathers'.

ring-dyke A **d** in a zone surrounding, in more or less arcuate form, a circular or dome-shaped igneous intrusion. MAGMA forming the intrusion seems to have exerted pressure upward and outward, causing fractures; these are filled with magma, which solidifies as **ds**. If the fractures are vertical in section, **r-ds** are formed. E.g. Mull, Skye, Arran, Ardnamurchan and N. Ireland. Ct. CONE SHEET.

rip Turbulence and agitation in the sea, caused by: (i) meeting of 2 tidal streams; (ii) a tidal stream suddenly entering shallow water; (iii) return of water piled up on the shore by strong waves, esp. when they break obliquely across a longshore current. Hence *r-tide*, *r-current*. [*f*]

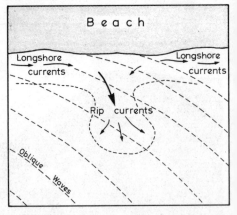

ripple marks Familiar patterns on sand or mud made by winds, waves or tidal currents. Their crest-lines may be straight, sinuous or lobate, the last *linguoid* (pointing down-current) or *lunate* (up), their cross-sections symmetrical or asymmetrical. They may be preserved on a surface of deposition, now compacted rock (sandstone, shale) covered and preserved by sediments, and later 'exhumed' by removal of sediments to reveal the orig. patterns.

rise An elongated, gently sloping elevation rising from the sea-floor, though its crest lies far from the surface; e.g. Dolphin Rise in N. Atlantic Ocean.

Riss The 3rd of 4 main glacial periods of the Quaternary glaciation identified and named by A. Penck and E. Brückner in 1909 in the Alpine Foreland, where they recognized distinctive fluvioglacial gravels on the High Terrace, assoc. with moraines and drumlins nearer the main Alpine ridges. The **R** is equated with *Saale* in N. Europe, and *Illinoian* in N. America, prob. the period of max. glaciation. [*f* MINDEL]

river Water flowing in a definite channel towards the sea, a lake, a desert basin, a main **r**, a marsh. The longest are:

	km	mi.
Nile	6649	4132
Amazon	6276	3900
Mississippi–Missouri	6111	3860
Ob	5570	3461
Yangtse	5520	3430
Hwangho	4672	2903
Zaïre (Congo)	4667	2900
Amur	4509	2802
Irtysh	4421	2747
Lena	4269	2653
Mackenzie	4240	2635
Mekong	4184	2600
Yenisey	4132	2566
Parana	3942	2450
Volga	3690	2293
St Lawrence	3057	1900
Rio Grande	3034	1885

For features assoc. with **r**s, whose names are prefaced with **r** (e.g. *r-cliff, r-terrace*), see under specif. term.

river profile Outline of the shape of a **r** valley: (i) *longitudinal*, from source to mouth, the *long-p*; (ii) *transverse*, across the valley at right-angles to the **r**, the *cross-p*. See also THALWEG and PROFILE. [*f* KNICKPOINT]

roadstead, roads Anchorage outside a harbour, off a coast, or in an estuary or bay, with some degree of protection against wind and heavy seas; e.g. Cowes Roads, Carrick Roads. Fr. form *rade*; e.g. Rade de Brest.

Roaring Forties The uninterrupted ocean S. of latitude 40°S., where N.W.–W. (or 'Brave West') winds blow with strength and constancy; an area of gales, stormy seas, overcast skies, and damp, raw weather associated with a constant W. to E. procession of low pressure systems.

roche moutonnée (Fr.) A glacially-moulded mass of rock, with a smooth, gently sloping, rounded upstream side (the result of abrasion), and a steep, rough, irregular downstream side (the result of plucking). Orig. so called because of its resemblance either to recumbent sheep or to wigs smoothed down with mutton grease at the end of the 18th century. Common features in most glaciated areas; e.g. Pass of Llanberis, N. Wales. [*f*]

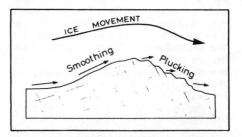

rock (i) An aggregate of mineral particles, forming part of the LITHOSPHERE. In pop. use, hard, consolidated, compact and massive, but in correct sense incl. sand, gravel, mud, shells, coral, clay. **R**s may be classified: (i) *by mode of formation*, including (*a*) IGNEOUS; (*b*) SEDIMENTARY; (*c*) METAMORPHIC; (ii) *by age*, i.e. tabulation according to their rel. ages, forming a convenient time-chart or chronology. See STRATIGRAPHY. (ii) Used as a place-name where a **r** is a striking feature; e.g. R. of Gibraltar, Spider R. (Arizona), Ship R. (New Mexico).

rock cycle (i) The cycling of **r**-forming minerals from IGNEOUS to SEDIMENTARY **r**, then METAMORPHIC **r**. The last return to the molten interior of the earth as LITHOSPHERIC PLATES descend, are incorporated in magma, and return to the surface to form a new igneous **r**. (ii) The sequence of: extrusion and intrusion–denudation – sedimentation – accretion – lithification–metamorphism–subduction–melting–extrusion and intrusion.

rock-drumlin A smooth elongated hummock of glacial clay, but with a mass of rock as its 'core', round which material is plastered.

rocketsonde A payload of self-recording and transmitting meteorological instruments carried by a rocket. It is ejected at a given altitude, various observations telemetered by radio, and then descends by parachute, tracked by radar and recovered to yield photographic records.

rock-fall Free fall of rocks or boulders down a steep mountain side. Ct. ROCK-SLIDE.

rock-flour Fine r-material produced by the grinding effect of glaciers. A stream flowing from a glacier-snout is milky white because of its suspended load of r-f. So too are lakes into which glacial streams flow; e.g. S.E. end of L. Geneva; L. Louise in Canadian Rockies.

rock glacier Defined by R. P. Sharpe as a glacier-like tongue of angular rock waste usually heading in cirques or other steep-walled amphitheatres and in many cases grading into true glaciers. The surface of a r g comprises concentric ridges of large blocks, with finer detritus at depth in association with an ice-core. The latter (in some cases the remnants of a former glacier buried beneath ablation moraine or landslide debris) experiences very slow creep, in the order of 1 m/year.

rock pavement See RUWARE.

rock-salt A clear, white, grey, yellow or brown mass of sodium chloride, crystallized in cubic form, sometimes with calcium and magnesium sulphates and chlorides. Formed esp. during desert conditions of Permian and Triassic, on dried-up beds of inland seas and lakes; e.g. near Strassfurt (E. Germany); near Cracow (Poland); in Cheshire; and widely in USA.

rock-slide A mass of r which slides *en masse* down a gentle hill-side over a bedding-plane or a fault-plane; e.g. in the Madison R. valley, S.E. Montana, on 18 Aug., 1959, when a vast r-s ('triggered-off' by an earthquake) blocked the valley, ponding up a lake. A section of the Dorivarz mountain slid into the valley of the Zeravshan R. on 26 April, 1964, in Soviet Uzbekistan, the result of an earthquake, and ponded up a lake whose waters threatened Samarkand; engineers lowered the natural dam and partly diverted the river. Ct. R. FALL, r-avalanche.

rock step See RIEGEL.

roddon A raised bank in the Fen District of E. Anglia, the LEVEE of a former river built up by silt. It was emphasized by compaction, oxidation and gen. lowering of the peat-lands between the rivers. A r stands above the gen. level of the surface.

roller Pop. word for an ocean swell, the result of a long FETCH, which forms immense breakers on exposed coasts, even in calm weather; e.g. on Atlantic islands (St Helena, Ascension, Tristan da Cunha).

rond See BROAD.

roof pendant A mass of country rock which projects downwards into the top of an igneous intrusion; e.g. older rocks penetrating a BATHOLITH, and completely surrounded by igneous rocks.

root Used in a tectonic sense in connection with NAPPE structures in highly folded mountain ranges where the 'core' of a RECUMBENT FOLD turns steeply downwards. Hence in the Alps the 'r zone' of the nappes of Bernese Oberland and Pennine Alps lies in N. Italy, each vertical nappe r separated by metamorphosed sediments.

ropy lava See PAHOEHOE.

Rossby wave Large-scale undulation in the upper-air W. wind flow in high latitudes; typically 3 to 6 waves form a complete pattern around the globe.

Rossi-Forel Scale A s of intensity of earthquake shocks, devised in 1878 by M. S. de Rossi and F. A. Forel, and used until 1931, when modified MERCALLI SCALE was introduced. The latter has been superseded for many purposes by RICHTER SCALE.

rotational slip A downhill movement *en masse* on a s-plane of solid material, such as rock or ice, which pivots about a point. The mass is left with a marked back-slope facing uphill. E.g. the Warren, Folkestone. [*f*]

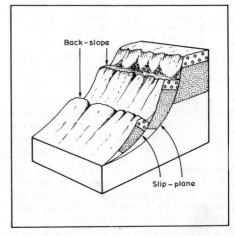

rotation of Earth The movement of the E on its polar axis from W. to E.; this results in an apparent daily r of sun, moon and stars from E. to W. The period of r, measured as 2 successive transits of a meridian by a star, is 23 hours, 56 minutes, 4·09 seconds (SIDEREAL DAY). The avge interval of time between 2 successive transits of the sun across the meridian (i.e. period of **r** of the E rel. to the sun) is 24 hours; see MEAN SOLAR TIME, EQUATION OF TIME. At the Equator, velocity of the Es rotation is about 1690 km per hour, at 60°N. and S. about 845 km per hour, at the Poles zero. To find the velocity per hour in any latitude, divide the length of that parallel ($2\pi R \cos \theta$) by 24.

rotor streaming In a LEE WAVE a form of turbulent air motion or closed eddy in a vertical plane. See LENTICULAR CLOUD, HELM WIND.

roughness An expression of the degree to which a stream channel is marked by irregularities, which impede stream flow (hence '*channel r*'). Factors affecting **r** include the size and angularity of bed material, the presence of ripples and sand-bars, channel sinuosity, bank vegetation and man-made obstructions. The effect of **r** on flow is proportionately less in large streams, or streams at high discharge. Within a stream the influence of **r** is reduced downstream, resulting in an increase of mean velocity from source to mouth.

roundness An index of roundness of particles and pebbles, expressed by the formula
$$R = \frac{2r}{L} \times 1000$$
where r is the smallest radius of curvature measured in the main plane of the particle, and L is the major axis of the pebble.

rubber, natural A substance derived from latex, the milky fluid which issues from cuts in the bark of certain species of trees, the most valuable being *Hevea brasiliensis*, originating in the Amazon basin.

rudaceous rocks A category of detrital sedimentary rs of coarse variable texture, both compacted (BRECCIA, CONGLOMERATE, TILLITE) and unconsolidated (SCREE, GRAVEL, GLACIAL TILL).

ruderal vegetation See CULTURAL VEGETATION.

run Used in E. USA for a small stream and its valley; e.g. Bull R., Virginia.

runnel See SWALE.

running mean A statistical procedure which endeavours to smooth out irregularities in a time series of values (e.g. annual precipitation), producing a moving avge, which may be plotted for each time-interval on a graph. E.g. in a series $x_1 \ldots x_n$, on a 5-year moving avge, the value for year 3 would be
$$\frac{x_1 + x_2 + x_3 + x_4 + x_5}{5},$$
while the value for year 4 would be
$$\frac{x_2 + x_3 + x_4 + x_5 + x_6}{5}$$

runoff Surface discharge of water from rainfall or snow-melt down a slope, also incl. surface discharge of underground streams. The proportion of precipitation that becomes **r** depends on: (i) slope; (ii) nature of the rock and soil; (iii) presence or absence of a vegetation cover; (iv) rate of evaporation. On a steeply inclined bare rock surface, **r** comprises 100% of precipitation; on a nearly level surface of sand, **r** is virtually nil.

Plate 58. Low domes or 'whalebacks' of granite, called RUWARE, representing exposed parts of the BASAL SURFACE OF WEATHERING near Kano, northern Nigeria. (*R. J. Small*)

ruware A low, rounded and in plan often elongated exposure of rock (sometimes 'whaleback'), rising a few m above the surrounding plain (underlain by alluvium and/or weathered rock *in situ*) in tropical areas. Referred to by L. C. King as a ROCK PAVEMENT. Widely regarded as an exposure of a domical rise in the BASAL SURFACE OF WEATHERING; thus represents an early stage in the formation of SHIELD INSELBERGS. An alt. view is that rs result from the destruction of inselbergs and the coalescence of surrounding pediments. See pl. 58.

Saale Early part of a major glacial period in N. Europe, corresponding to the Riss in the Alps and the Illinoian in N. America, occurring about 150 000 years ago. The S. icesheet extended farther S. than those of other glacial periods; its S. limit is represented by a discontinuous series of low sand-hills extending from near Utrecht to near Krefeld in Germany. Known as *Drenthian* glaciation in the Netherlands.

sabkha A salt-encrusted flat, on the shores of a lagoon or shallow sea; EVAPORITES are commonly found (halite, gypsum, carbonates). They occur on the coast (Persian Gulf), or in a DEFLATION hollow where the water-table is near the surface and strong solutions rise by capillarity to form evaporites. Some ss are 'fossil', with the minerals replaced by METASOMATISM.

saddle A broad col in a ridge between two summits; e.g. the Saddle between Dodd and Red Pike, Buttermere, English Lake District.

saddle-reef A lens-shaped mass of ore-bearing rock between beds near the axis of an ANTICLINE; a type of PHACOLITH.

saeter, sete (Norwegian) (i) An upland pasture in Norway. (ii) Farm high in the mountains of Norway used in summer following snow-melt.

sagebrush Scrub vegetation in semi-desert areas, of greyish heathlike shrubs (*Artemisia* spp.), esp. in Great Basin of Utah, Arizona, Nevada and Mexico.

St Elmo's Fire Small electrical brush discharge, seen playing around the masts and spars of ships, with the appearance of luminous flames. Esp. common in the DOLDRUMS. The same type is experienced in high mountains, making a hissing sound and tingling a climber's skin; this is rarely visible. A sign of stormy weather and usually assoc. with the passage of a FRONT.

salar A basin of inland drainage in S.W. USA, usually containing a salt-lake or salt-flat; syn. with PLAYA.

salina (Sp.) A PLAYA with a high salt content; used in the deserts of S.W. N. America.

saline soils A group of INTRAZONAL ss characterized by a considerable proportion of salts, esp. sodium. They occur widely wherever there is strong evaporation, both in hot deserts and in cool temperate continental interiors with high summer temperatures. Strong s solutions rise by capillarity, and salts form a greyish surface-crust, below which lies a granular salt-impregnated horizon. Syn. with *solonchak*. E.g. in Great Basin of Utah, esp. around Great Salt Lake; around Caspian Sea; in Tarim Basin. Many salt-crusts are the result of shrinkage and drying-out of former extensive salt-lakes in basins of inland drainage; s ss occur around their margins. Such a s is of little use for irrigation, unless desalinized; e.g. some areas around the Caspian Sea now grow cotton. Constant irrigation can incr. salinity of s, necessitating expensive systems of drains through which water can be periodically flushed.

salinity In sea-water, the proportion of salts in pure water, in parts per thousand by mass. The mean figure for the sea is 34·5 parts per thousand, written 34·5‰. In actual salts, made up of sodium chloride (23), magnesium chloride (5), sodium sulphate (4), calcium chloride (1), potassium chloride (0·7); minor ingredients incl. salts of bromine, strontium, carbon, boron, silicon, phosphorus, fluorine and others, incl. 'trace elements' of importance to marine organisms. S can be expressed in terms of actual elements; on an avge, sea-water contains 18·98 grams per kilogram of chlorine (55% of the total salt content), 10·56 of sodium (31%), 1·27 of magnesium, 0·88 of sulphur, 0·40 of calcium, and 0·38 of potassium. The proportions of constituents remain constant from place to place, despite changes in total s. The standard method of determining s is by precipitating halides by adding silver nitrate. Surface s varies according to temperature (causing evaporation and concentration), supplies of additional fresh water from rivers, rainfall, melting ice and snow (causing dilution), and degree of mixing by surface and sub-surface currents. In the open ocean, differences are small, from about 37‰ near the tropics, 35‰ near the Equator and in mid-temperate latitudes, to 34‰ towards the Poles. In partially or wholly enclosed seas the variation is greater; Red Sea is about 40‰ (high summer temperatures and evaporation, few inflowing rivers), Baltic off Bornholm only about 8‰ (low evaporation, many inflowing rivers).

salt, common (sodium chloride) (NaCl) Occurs in solution as brine, as sheets and crusts around margins of s-lakes (e.g. Dead Sea, Great Salt Lake of Utah, Aral Sea), in s-domes, and as deposits esp. in Permo-Triassic rocks (e.g. Cheshire, Stassfurt in E. Germany, Texas and neighbouring US states). See also ROCK-S.

saltation The process by which solid material moves along the bed of a stream in hops. Also used for similar movement of sand grains in deserts.

salt-dome (sometimes **salt-plug**) A roughly circular mass of solid s, from 90 m to 1·6 km in diameter, extending vertically to great depths, sometimes as much as 13 km. It was forced up by slow flowage from deeply buried deposits of rock-salt. Commonly associated with deposits of gypsum, anhydrite and petroleum, and appears to be crowned with a limestone cap-rock. See DIAPIR.

salt-flat A horizontal stretch of s-crust, representing the bed of a former s-lake, temporarily or permanently dried; e.g. Bonneville S.-fs. W. of Salt Lake City, USA, where motor speed-trials are held as the surface is so level and firm; around L. Eyre in Australia.

salting Slightly higher areas of a SALT-MARSH (still sometimes inundated by tides), where grass is present and there is little bare mud.

salt-lake Highly saline lake, located in a basin of inland drainage, in an area with high temperatures and evaporation rates. Water entering brings in some saline material, left when the water evaporates; e.g. Great Salt Lake (Utah) (220‰); Dead Sea (238‰); L. Van in Asia Minor (330‰).

salt-marsh Coastal marsh along a low-lying shore, enclosed by a shingle-bar or a sand-spit, or in the sheltered part of an estuary. Fine silt and mud are deposited by tides in backwaters, added to by alluvium brought down by rivers. Vegetation gradually spreads, and helps the process of accretion: eelgrass (*Zostera*), marsh samphire (*Salicornia*), rice-grass (*Spartina townsendii*). These form increasingly dense communities which help to trap silt; hummocks of vegetation, then more continuous areas develop, and whole surface is raised naturally, while tidal waters flow in increasingly restricted channels. Gradually other plants are established, and the marsh may turn into a SALTING, esp. if Man helps by dyking or building wickerwork fences. E.g. along Norfolk coast; marshes of the rivers of S. Suffolk and Essex; Romney Marsh behind Dungeness (now reclaimed); Solway marshes; and extensive areas between the Frisian Is and mainland coasts of Netherlands, W. Germany and Denmark (*Watten* or *Wadden*).

salt-pan A small basin containing a s-lake, surrounded by and lined with a solid deposit of s; e.g. the SHOTTS on the plateaus in the Atlas Mtns.

samun, samoon (Persian) A warm, dry descending wind in Persia, of same nature as the FÖHN.

sand Small particles, mainly quartz, with diameter of between 0·02 to 2·0 mm. The gen. category is subdivided into: (i) *fine* (0·02 to 0·2 mm); (ii) *medium* (0·2 to 0·5 mm); (iii) *coarse* (0·5 to 1·0 mm); (iv) *very coarse* (1·0 to 2·0 mm). Also applied to soils which consist of more than 90% s.

sandbank Accumulation of sand in the sea or river, usually exposed at low water.

sandr, sandur (pl. *sandar*) (Icelandic) An OUTWASH PLAIN of sand.

sandstone A sedimentary rock, consisting mainly of grains of quartz, often with feldspar, mica and other minerals; consolidated, cemented and compacted. S can be classified according to the 'cementing' material which binds the individual grains: (i) *calcareous*; (ii) *siliceous*; (iii) *ferruginous*; and (iv) *dolomitic*. Colour varies from dark brown or red through yellow to grey and white, mainly due to iron content and its degree of oxidation or hydration; some ss have a greenish shade due to the presence of glauconite or reduced iron compounds. Sands were laid down before compaction in: (*a*) shallow seas; (*b*) estuaries and deltas; (*c*) along low-lying coasts; (*d*) hot deserts. Ss are commonly laminated and sometimes show FALSE-BEDDING, others are FREESTONES. S is widely spread in space and time; e.g. Torridon S. (Precambrian) in Wester Ross; Bala S. (Ordovician) of Caer Caradoc in Shropshire; Old Red S. (Devonian) of Herefordshire, S. Wales, central and N.W. Scotland; various sandstones in Millstone Grit (Upper Carboniferous); Coal Measure S. (Upper Carboniferous) in most coalfields; New Red S. (Permo-Triassic) of N.W. England (St Bees S.), S part of I. of Arran; Bunter and Keuper Ss (Triassic) of Midlands, Cheshire, Merseyside; Hastings S., Lower and Upper Greensand (Cretaceous) of Weald; various rocks of Eocene age (Thanet S.,

Bagshot S.) in London and Hampshire Basins; younger Tertiary sands.

sandstorm Storm in a desert or semi-arid area, in which the wind carries clouds of sand, usually near the surface and rarely above 15 m. The erosional effect of sand in a storm is potent, akin to 'sand-blast'.

Sangamon interglacial I period in N. America, corresponding to Riss-Würm i in the European Alpine Foreland.

Sanson-Flamstead Projection (also known as **Sinusoidal P**) An equal area p, on which the Equator is taken as standard parallel and drawn as a straight line to scale $= 2\pi R$, and central meridian drawn as a straight line to scale $= \pi R$; each is divided truly (i.e. by 36 for 10° intervals). Parallels are drawn as straight lines through the points of division on central meridian; each is made its true length $= 2\pi R \cdot \cos \theta$, and divided truly (for 10° intervals $= \dfrac{2\pi R \cdot \cos \theta}{36}$). Meridians are sine-curves drawn through corresponding points on each parallel. Can be used for a world-map, but extensive 'shearing' occurs in high latitudes because meridians are very oblique to the parallels. Useful for maps of S. continents, esp. when they fall about central meridian, since shape is good. The **p** may be 'INTERRUPTED', and is used in GOODE'S HOMOLOSINE P for tropical latitudes, the rest drawn on MOLLWEIDE. The S-F P is a partic. case of BONNE, with the Equator as the standard parallel. [*f*]

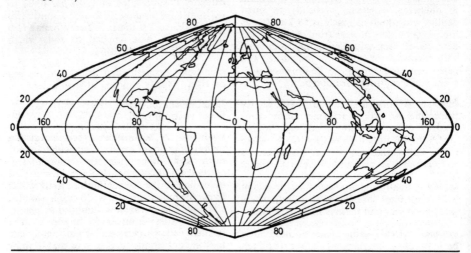

Santa Ana (Sp.) A hot, dry wind blowing from N. and N.E., descending from the Sierra Nevada across deserts of S. California; a wind of FÖHN type, though often laden with dust. It can do damage to fruit orchards through desiccation and withering of buds or blossom.

sapping (i) Syn. with glacial PLUCKING. (ii) See SPRING-SAPPING.

saprolite Residual weathered cover, the result of rotting of bedrock *in situ*. Partic. evident in warm humid climates, where hydrolysis is effective, and here may be as much as 90 m thick.

sapropel, adj. **sapropelic** A sludge or ooze which collects in freshwater swamps and in shallow sea-basins, rich in organic matter. The decomposition and putrefaction of the organic content is believed to be the basis of origin of petroleum and natural gas, esp. when organic matter is derived from fatty algae. Derived from Gk. *sapros* (rotten) and *pelagos* (sea).

saprophyte A plant (such as fungus) that lives on decaying vegetable matter. Ct. EPIPHYTE.

sarn (Welsh) A natural 'causeway' (though there are claims that it is of artificial origin), notably in Cardigan Bay where there are at least 5. Formed of loose stones and boulders, sections are exposed at very low tide; e.g. about 14·5 km of the total 34 km of S. Badrig. Obviously the result of rise of sea-level, possibly inundating a low watershed of glacial clay, washing away finer materials, developing like a storm-beach as material accumulates.

Plate 59. SARSEN blocks of resistant Tertiary sandstone, occupying the floor of the 'Valley of Stones' near Dorchester, southern England. The blocks, remnants of a former DURICRUST layer, have been transported from adjacent INTERFLUVES by SOLIFLUCTION to the valley bottom. (*R. J. Small*)

sarsen A large block of siliceous sandstone, found mainly on the chalk Downs, remnants of a now vanished cover of early Tertiary rocks. The precise constituents of a s vary. They may be relatively homogeneous or conglomeratic (comprising mainly broken flint gravel). They are often concentrated in isolated valleys in England, esp. in Wiltshire, having been moved downhill by periglacial SOLIFLUCTION. They have been used as building stone, as gate-posts, and in prehistoric times in megalithic monuments; e.g. Stonehenge (partly) and Avebury. In N. France they have been important in an area otherwise devoid of building stone; e.g. in construction of *pavé* roads. Often called 'greywethers', because of their likeness to sheep seen from a distance. Conglomeratic forms in the Chilterns are called Hertfordshire Puddingstone. See pl. 59.

sastruga, pl. sastrugi (Russian) A sharp ridge on a snow-field or ice-field, caused by scouring and furrowing erosion of wind laden with ice-particles. Partic. evident to polar travellers, since they interfere with the smooth running of a sledge, partic. if the journey lies transverse to the gen. pattern (a result of the direction of prevailing winds).

saturated (or **wet** or **moist**) **adiabatic lapse-rate** The r of decrease in air temperature by dynamic cooling, the result of expansion within a vertically ascending s 'parcel' of air. This is less than DRY A L-R because of release of latent heat. The actual r varies with the amount of water-vapour condensed, which varies with the amount of water-vapour present, which depends on temperature. It can lie anywhere between about 0·4°C and 0·9°C per 100 m. An air-mass of about 27°C (80°F) may contain so much water-vapour, and therefore releases so much latent heat, that s a l-r may be as low as 0·4°C per 100 m (2°F per 1000 ft). In a very cold air-mass there may be so little water-vapour that s a l-r differs little from dry a l-r.

saturation, saturated (i) The state of the atmosphere when it can hold no further water-vapour; i.e. as many molecules of water enter the air as leave. The amount of water-vapour that can be held by the atmosphere varies with temperature and pressure. If s air is cooled, a certain amount of water-vapour may condense. For condensation to occur nuclei are required, unless the water-vapour molecules coagulate under very high VAPOUR PRESSURE. If nuclei are not present, the atmosphere may become SUPERSATURATED at temperatures below the theoretical DEW-POINT. S must always be considered with respect to the physical condition to given 'parcels' of air. (ii) A rock holding the max. amount of water in its interstices is said to be in a state of s; hence the *zone of s*.

See WATER-TABLE. (iii) Used in another sense by petrologists to indicate minerals which can crystallize out from magma in the presence of an excess of silica; e.g. feldspar, mica, amphiboles. If a rock is over-saturated, excess silica occurs in the form of free quartz, if under-s it has a high proportion of olivine and/or feldspathoids (leucite, nepheline, lazurite).

saturation deficit Amount of water-vapour required to bring non-saturated air at a given temperature and pressure to the point of saturation. Syn. with *vapour-pressure d*.

saturation overland flow Run-off occurring when rainfall does not exceed the INFILTRATION CAPACITY. The soil may already be saturated, as on lower slopes where the permanent water-table is at ground level.

savanna, savannah, savana Open tropical grassland, with tall grass, scattered trees and bushes usually of a XEROPHILOUS character. In some areas grass forms a discontinuous cover, separated by bare ground; it may be only 0·3 m high, or as much as 4 m (e.g elephant grass, a stiff yellow straw crowned with silvery spikes). Patches of low thorny scrub are common. Clumps of trees—various palms, baobab, ceiba, euphorbia and acacia—grow in hollows where ground-water approaches the surface. Many trees are wedge- or umbrella-shaped as a result of strong winds. As the desert border is approached, grass becomes shorter, more tufted, with bare sand between clumps; equatorward the trees become taller and more numerous, merging into light forest and then rain-forest. The vegetation pattern is the result partly of a marked precipitation régime, partly of soil conditions, partly of extensive fires (some claim that s is a result of man's interference with forest). During the dry winter, grass is parched and trees bare, but with rains in summer there is a short-lived luxuriance. The s covers a large area in Africa, S. America on the Guiana Plateau (*llanos*) and on the Brazilian Plateau (*campos*) to N. and S. of the rain-forest, and N. Australia.

scabland American term for a landscape in parts of the Columbia-Snake Plateau in N.W. of USA, where the basalt surface was extensively eroded by glacial floodwater, and bare rock has been exposed or covered with angular debris derived from its own disintegration; there is a thin soil cover, and scanty vegetation.

scalded flat Applied in Australia to: (i) a lowland covered with salt-impregnated soil; (ii) specif. a sub-type of red-brown soil.

scale The proportion between a length on a map and the corresponding length on the ground. It may be: (*a*) expressed in words; (*b*) shown as a divided line; (*c*) given as a REPRESENTATIVE FRACTION.

scar A rock-face, partic, in N. England, where it refers to a limestone cliff, in some cases outcropping across country for some distance; e.g. Attermire S., Langcliffe S., Gordale S. One of the massive resistant bands in the Carboniferous is known as Great S. Limestone. See pl. 47.

scarp, scarp-slope, scarp-face Steep slope of a CUESTA [*f*]; ESCARPMENT is preferable.

scarp-foot spring A s that breaks out at or near the foot of an ESCARPMENT, esp. where chalk lies on clay (S. Downs), or limestone and sandstone lie on clay (Cotswolds).

scarp retreat See PARALLEL RETREAT OF SLOPES.

scarth Bare rock-face, esp. in English Lake District; e.g. S. Gap above Buttermere.

scatter-diagram A d giving a graphic indication of the amount of correlation between 2 sets of statistical data, one set plotted as ordinates, the others as abscissae. If when the values are plotted they tend to be grouped along a diagonal line, some degree of correlation is manifest.

Schattenseite (Germ.) Shady side (N.-facing in N. hemisphere) of a deep valley; syn. with UBAC. Characterized by thick growth of conifers in the Alps, in ct. to settlements, terrace cultivation and pasture on S.-facing sunny side. Ct. SONNENSEITE.

schist Medium-grained rock affected by regional metamorphism, causing re-crystallization; usually with a foliated, sometimes a wavy, texture. Flakes of such 'platy' minerals as mica are usually visible. This texture is independent of the bedding-planes of the orig. rock. Varieties of s incl. quartz-s, metamorphosed from sandstone; hornblende- and biotite-s from basalt and gabbro; mica-s (Fr. *schistes lustrés*) from phyllite.

Schuppenstruktur (Germ.) Germ. name for IMBRICATE STRUCTURE [*f*], though used in English geological literature.

scirocco See SIROCCO.

sclerophyll An evergreen tree or shrub with small hard leaves, hence *sclerophyllous*. Found mainly in lands with a Mediterranean climate, with long hot, dry summers, during which leathery leaves resist TRANSPIRATION. E.g. OLIVE, cork oak (*Quercus suber*), Holm oak (*Q. ilex*), Aleppo pine (*Pinus halepensis*), lavender (*Lavandula latifolia*).

scoria (i) Coarse clinkery mass of volcanic rock, of slaggy nature, usually basic in composition, fine-grained but cellular in texture, hence *scoriaceous* as a result of former abundant gas-bubbles and steam-blisters. It results from rapid cooling of the surface of a lava stream which contained gas and steam. (ii) The accumulation of similar clinkery material blown out of a volcano as PYROCLASTS.

'Scotch mist' Fine drizzling rain among British hills, when clouds lie at or near to the ground.

scour (i) The powerful erosive effect of a tidal current, removing deposits on the sea-bed close inshore. The importance of tidal s in forming gentle submarine erosion surfaces is becoming incr. apparent. A tidal current of 4½ knots can scour shingle at a depth of 59 m off Hurst Castle Spit in Hampshire, while 'sand waves' have been observed 164 m down on the CONTINENTAL SHELF off S.W. England. A tidal s may be felt partic. in the mouth of a bottle-necked estuary; e.g. Mersey. (ii) Powerful and concentrated erosive effect of a river current; esp. on an outside curve. Concrete training-walls are built to direct the s so as to assist in the regularization of a shipping channel; e.g. along Rhine below Basle; Danube.

scree Slopes of angular rock-debris on a mountainside, lying at an angle of rest $c.$ 35°, which remains remarkably uniform. The material is mainly formed as the result of frost, hence it occurs strikingly at the foot of steep rock buttresses, on which frost weathering is potent; sometimes a distinct s-cone tapers outwards from the base of a buttress or the foot of a rocky gully. In USA TALUS is used, sometimes syn., though by other authorities s is defined as all loose material lying on a hill slope, while talus accumulates specif. at the base of cliffs. Some refer to s as the ingredient, talus as the whole slope or feature. See pl. 60.

screen, meteorological White-painted wooden box on legs, 1·25 m above the ground, with louvred sides for ventilation, in which meteorological instruments are placed, so as to give shade readings unaffected by direct sunshine and strong winds. Other types are now used, known gen. as *thermometer ss.*

scrub Vegetation association in a semi-arid climate, or on poor sandy or stony soils, characterized by stunted trees, bushes and brushwood. (i) Tropical and semi-desert type (MULGA, SPINIFEX, CHAÑARAL, ACACIA); (ii) warm temperate type (MAQUIS, CHAPARRAL, GARIGUE, MALLEE BRIGALOW, SAGEBRUSH). Also used loosely of any rough vegetation on heathland. Plants are mainly XEROPHILOUS

Plate 60. Massive SCREE accumulations, formed by long-continued CONGELIFRACTION of FREE FACES at Mela Svelt, western Iceland. (*Eric Kay*)

in character, incl. cacti, thorny aromatic shrubs, small gnarled evergreens, saltbushes, mesquite, creosote, sharp spiny grasses.

scud *Stratus fractus*; tattered, ragged masses of cloud, driven by the wind beneath main cloud-layers of nimbostratus; gen. a bad-weather cloud and sign of wind and storm.

sea (i) The salt waters of the Earth's surface, as in 'land and s'. (ii) Proper name for any specif. area of water, usually on the margins of continents, in ct. to an ocean; e.g. North S., Mediterranean S., China S. A few large inland bodies of water are also so-called; e.g. Dead S., Caspian S., Aral S.

sea breeze A local b blowing from the s during the afternoon towards a low pressure area (produced by heating and convectional uplift) over the land, esp. in equatorial latitudes. Felt for only short distances inland, but considerably ameliorates stagnant, humid heat and freshens the air. It blows only during periods of calm, settled weather when not masked by the Trades. Ct. LAND B. [*f*]

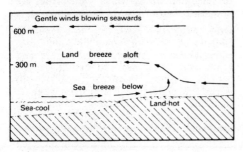

sea-level See MEAN S-L.

sea-marsh See SALT-M.

sea-mile See MILE.

seamount An isolated peak rising from the floor of an ocean, the summit of which is 900–1800 m (sometimes as much as 3000 m) below the water surface. Esp. numerous in the Pacific Ocean, where about 10 000 (incl. GUYOTS) are estimated. (A s has a prominent crest, a guyot is flat-topped.) Both are usually volcanoes, formed by sub-oceanic eruptions. See SUBMARINE RIDGE [*f*].

seascarp An ESCARPMENT (300–500 m high) on the floor of the ocean, the result of large-scale faulting movement, notably off W. coast of S. America; e.g. Mendocino, Murray, Clipperton Ss. They are thought to be the result of tectonic activity (possibly trans-current faults) on the margins of the continental and oceanic plates.

season One of the distinct periods into which the year may be divided, partic. in terms of duration of daylight and climatic conditions, as a result of changes in duration and intensity of solar radiation. If ss are defined astronomically, spring is from Mar. 21 (when the apparent sun is at the First Point of Aries; see ZODIAC) to June 22 (at the First Point of Cancer); summer from June 22 to September 23 (at the First Point of Libra); autumn from September 23 to December 22 (at the First Point of Capricorn); and winter from December 22 to March 21. Because the Earth's orbit is elliptical and the orbital rate of movement changes, ss are of unequal length. For N. hemisphere, spring=93 days; summer=94 days; autumn=90 days; winter=89 days. In Europe a 4-fold division, essentially reflecting the life cycle of cultivated plants, is usual: dormancy (winter), sowing (spring), growth (summer), harvest (autumn). These periods can seldom be precisely defined in terms of climatic factors, and the main contrast is between extremes of summer and winter in daylight hours and temperatures, July and January as indicative. Low latitude ss may be differently defined; e.g. in India by rainfall and drought, or by the monsoons.

seat-earth By miners and others, the layer of material on which coal-seams rest; this is a fire-clay (used for fire-resistant bricks) or sand (see GANISTER). Roots of swamp-forest from which coal was derived grew in this layer, and their fossil remains are commonly found.

Secant Conic Projection Inaccurate name sometimes given to C P WITH 2 STANDARD PARALLELS. Strictly a secant is a straight line cutting the circumference of a circle at 2 points, and a true S C P would have its 2 standard parallels separated by this secant distance, whereas the conic with 2 standard parallels actually has these separated by the arcuate distance. [*f*, *page 263*]

second (i) 1/60 of a minute of arc in the measurement of latitude, longitude, and angles gen. (ii) 1/60 of a minute, 1/3600 of an hour (the SI unit in the measurement of time).

Secondary The second of the eras of time and corresponding group of rocks subsequent to the Precambrian. Syn. with Mesozoic, which has superseded it in gen. use.

secondary consequent stream A tributary to a SUBSEQUENT S, flowing parallel to the main c s. A s c s is initiated after the formation of subsequent ss, but in a direction consistent with that of the orig. c proper. In areas of

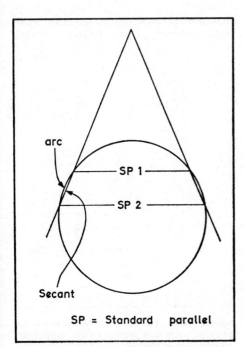

folded rocks, it is also a s draining the flanks of an ANTICLINE, leading into a synclinal depression, occupied by a *primary* or *longitudinal c*. [*f* SUBSEQUENT STREAM]

secondary depression A small area of low atmospheric pressure on the margins of a main **d**. It may be just a bulge in the isobars, or an individual system with closed circular isobars. It may deepen to become more intense, with lower pressure and associated with more stormy, rainy weather than the primary. The s **d** travels around the primary, in the N. hemisphere in an anti-clockwise direction. Most s **d**s form at FRONTS, though non-frontal ones may form in any large unstable air-mass.

secondary growth, of vegetation Specif. in tropics, a quick-growing type of vegetation (notably bamboo) which takes over formerly forested land, cleared by slash-and-burn, cultivated for a few years and allowed to revert. Used for any type of natural or semi-natural vegetation which takes over from a cleared area, esp. scrub and heath.

section Vertical cut through soil and/or rock, natural or artificial, or a representation of such a cut. Used only when details of the underlying strata are shown; the surface outline alone is properly a PROFILE (though SOIL PROFILE seems to contradict this). Geological ss may be: (*a*) *diagrammatic* (small-scale and generalized); (*b*) *semi-diagrammatic* (with an accurately plotted surface profile but diagrammatic representation of the strata); (*c*) *accurate*, with no vertical exaggeration which would falsify the dip of the strata.

secular An adj. implying a long period of time; i.e. a s change is very slow, virtually imperceptible.

sedentary soil A s formed *in situ* from underlying parent rock, in ct. to one derived from transported parent material. Not now used by pedologists.

sediment Deposited particles or grains of rocks. Sometimes extended to incl. all residual and detrital material laid down by rivers, wind, ice and sea. Hence *sedimentation*, the deposition of s. In its broadest sense, sedimentation includes a consideration of the rocks from which s is derived, processes involved in transportation of the fragments, and actual settling of material, whether: (i) in deep calm water of an ocean basin or GEOSYNCLINE; (ii) in a shallow marginal sea; (iii) under deltaic conditions; (iv) on a river FLOOD-PLAIN; (v) in a lake; (vi) in a desert.

sedimentary rock A **r** consisting of sediments, laid down in layers and/or 'cemented' (DIAGENESIS, LITHIFICATION). Main types: (i) *mechanically formed* (CLASTIC): (*a*) arenaceous (sand, sandstone, conglomerate, grit); (*b*) argillaceous (mud, clay, mudstone, shale); (*c*) rudaceous (breccia, conglomerate, tillite, scree, gravel, glacial till); (ii) *organically formed*: (*a*) calcareous (coral limestone, crinoidal limestone, shelly limestone); (*b*) ferruginous (ironstone); (*c*) siliceous (diatomaceous earth); (*d*) carbonaceous (peat, brown-coal, lignite, cannel coal, bituminous coal, anthracite); (iii) *chemically formed*: (*a*) carbonates (travertine, dolomite); (*b*) silicates (sinter, flint, chert); (*c*) ironstone (limonite, haematite, siderite); (iv) *formed by desiccation: evaporites*; (*a*) sulphates (anhydrite, gypsum); (*b*) chlorides (rock-salt).

sediment discharge rating In hydrology, the ratio between the total **d** of a stream and the **d** of s carried by it.

sediment yield The sediment load transported by a stream, expressed in terms of weight of sediment per unit area of drainage basin (e.g. s **y** may be as high as 800 tons per square mile). S **y** is an expression not only of stream capacity, but of intensity or erosion within the stream catchment. S **y** is particularly high in areas transitional between desert soils and grassland (annual rainfall 400 mm) and low in extreme deserts (annual rainfall less than 50 mm) and thick woodlands (annual rainfall over 1000 mm).

'seeding' of clouds The stimulation of CONDENSATION, then PRECIPITATION by dropping from aircraft particles of 'dry ice' (solid carbon dioxide), silver iodine, volcanic dust, etc. into supercooled cs. These may be seeded from the ground by coke-burning braziers from which silver iodine suffuses. Conditions must be suitable for precipitation before s is effective as a 'triggering' agent, and the proportion of successful attempts is not high, though research is proceeding. S of cs seems most practicable on high plains adjacent to mountain ranges (e.g. in Midwest USA), where orographic ascent causes SUPERCOOLING.

seepage (i) Slow sinking of surface water into the soil and subsoil. (ii) Oozing out of water along a fault or joint-plane, though not strongly enough for a spring. (iii) 'Show' of mineral oil at the surface, often a guide to prospectors. *Seep* is used as a vb., and in USA as a noun ('a seep').

seepage step A small scarp, usually 2 m or less in height, developed within the regolith by the seepage of water, along either a geological junction (e.g. that of the Barton Sand and Clay in the New Forest, Hampshire) or the soil-bedrock junction. S ss run broadly parallel to the hill-slope contours, but in detail show crescent-shaped sections produced by local collapses.

segetal vegetation See CULTURAL VEGETATION.

segregation In an ecological sense, spatial separation of people or institutions into distinct areas. Whereas ecological concentration or centralization implies separation of functions that operate in relation to the whole area, s stresses the separation of one function from all other functions. The opp. situation, breakdown of s, is *ecological desegregation*.

seiche A short-term standing-wave oscillation in the surface of a lake, the result of changes in atmospheric pressure and wind direction. The oscillation may be periodic, determined by the physical characteristics of the enclosing basin. Occas. a s may be caused by seismic forces; e.g. earth tremor.

seif-dune (Arabic) A longitudinal **d** forming a steep-sided ridge, often many km in length, aligned across the desert in the direction of the prevailing wind, which sweeps through depressions between the parallel lines. Sand supply is less than in BARKHANS and extensive 'sand-seas'. The orig. of a **s-d** is not clear, but it may be formed by coalescence of lines of small crescentic **d**s, so that the wind is funnelled between, sweeping away their 'tails', and leaving a pair of ridges. [*f*]

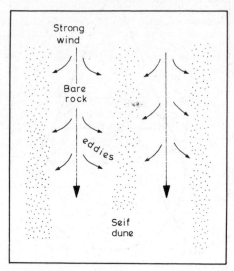

seismic focus The point in the Earth's crust at which an earthquake shock originates; also known as *s origin*. [*f* EARTHQUAKE]

seismic wave See TSUNAMI.

seismology Scientific study and interpretation of earthquakes; hence *seismologist*; *seismograph* (pendulum-based instrument for recording seismic waves); *seismogram* (the record obtained); *seismometer* (delicate instrument for receiving seismic impulses). About 600 observatories in the world make seismographic observations.

selenomorphology Study of land-forms on the surface of the moon.

selva (Portuguese) TROPICAL RAIN-FOREST in the Amazon Basin, and used gen. for any area of such vegetation.

semi-artesian well A w driven through overlying impermeable beds in a structural basin, in which water rises under hydrostatic pressure but does not reach the surface; e.g. at Bovington Camp, Dorset, a bore was driven through Tertiary beds into Chalk to a depth of 221 m, and water rose to within 28 m of the surface.

semi-desert, semi-arid climate Transition zone of climate and vegetation between savanna grassland and true desert, and farther from the Equator between true desert and Mediterranean vegetation. Also occurs in continental interiors; e.g. parts of Great

Basin and Colorado Plateau of USA. S-d is characterized by scrubby XEROPHILOUS bushes, coarse grass and bare sand.

semi-diurnal tide 2 tidal rises and 2 falls occurring during each lunar day at intervals of approx. 12 hours, 25 minutes; each high t attains more or less the same level, as does each low t. [*f* TIDE]

semi-natural vegetation Plant-association that is 'wild' in that it has not been planted by Man, yet owes its characteristics to his direct or indirect influence. Much of what was thought to be n v has been shown to be s-n; e.g. if rain-forest is cleared and then neglected, it will be rapidly covered by secondary forest. SAVANNA may be the result of destruction of forest by SHIFTING CULTIVATION, HEATH the result of clearance of light deciduous forest.

send, of waves See SWASH.

senile river A r in the 'old-age' stage of its cycle of development, a concept derived from W. M. Davis's CYCLE OF EROSION.

sensible temperature Sensations of heat and cold felt by the human body, involving not only temperature, but humidity and wind. A wet bulb thermometer reading gives an accurate indication; if it reads much above 24°C sustained manual labour is difficult. Damp raw cold is much more trying than dry cold; very low ts can be endured when air is dry and no wind. Dry heat mitigates high ts by accelerating evaporation from the skin surface. See also WIND CHILL.

sequence Used by some authorities in place of the provocative term CYCLE, in considering the succession of denudational changes which may affect the landscape. See SEQUENTIAL LANDFORM.

sequential landform In some classifications of ls, divided into INITIAL and s; the latter incl. those where modifications by subsequent denudation are so pronounced that only vestiges remain of initial structural forms.

sérac (Fr.) Pinnacle in an ice-fall on a GLACIER, the result of intersecting CREVASSES. See pl. 35.

seral community A plant c which forms a stage in development of a CLIMAX c, and itself only temporary, affording conditions more favourable for colonization by successively more demanding cs.

sere A developmental sequence of a plant COMMUNITY: *hydrosere* (in water); *xerosere* (orig. under dry conditions); *lithosere* (on a bare rock surface); *psammosere* (on sand).

series Stratigraphical unit, a division of a system; corresponding in the time-scale to an EPOCH; e.g. Carboniferous Limestone is a s within the Carboniferous; the Chalk is a s within the Cretaceous.

serir (Arabic) Stony desert of Libya and Egypt, where sheets of angular gravel cover the surface; syn. with REG in Algeria.

serozem (Russian) See GREY EARTH.

serpentine A common rock-forming mineral of complex chemical composition, containing hydrous magnesium silicate (*chrysotile*). Of variegated shades of green, sometimes with streaks of red and attractive markings resembling the skin of a serpent, with a greasy or soapy feel. Sometime occurs in large masses, as near the Lizard Peninsula, Cornwall, where it is worked for ornamental purposes.

set (i) The direction of a wind, tide or current. (ii) Clearly definable collection of things or items, known as the *elements* or *members* of that particular s, listed individually (finite) or defined collectively (infinite).

settling velocity A steady speed attained by an object descending through a liquid or gas; e.g. dust from a volcanic eruption falling through the atmosphere, particles in suspension in a river, lake or sea (partic. important in the consideration of sedimentation).

sextant An instrument used for measuring angles, esp. in navigation for obtaining altitude of the sun or a star. The observer looks through the eye-piece, and can see both the horizon and the mirrored image of the sun at the same time, then reads off the angle.

shade temperature Air t as indicated by a thermometer sheltered from the sun's rays, from radiation from neighbouring objects, and from precipitation. This may be achieved by using a SCREEN. This is the t quoted in climatic records and textbooks, unless 'sun t' is specif. used.

shakehole Steep-sided hole or depression, found commonly in limestone, sometimes in Millstone Grit, usually the result of collapse of an overlying cave roof, leaving debris in the bottom of the hole. It is usually dry, as ct. a SWALLOW-HOLE with a descending stream. More accepted scientific term is *collapse doline*, in USA *collapse sink*.

shake-wave An *S*-wave in an earthquake shock, similar to a light-wave, where each particle is displaced by the wave at right-angles to its direction of movement. Also known as *transverse wave*. Mean velocity through $\text{SIAL} = 3.36$ km s^{-1}, through $\text{SIMA} = 3.74$ km s^{-1}. [*f* EARTHQUAKE]

shale An ARGILLACEOUS sedimentary rock formed from deposits of fine mud, compacted by compression of overlying rocks, and laminated in thin layers which easily split apart. The lamination is gen. that of the bedding. Used widely in geological proper names; e.g. Ludlow S. and Wenlock S. of Silurian age; Yoredale S. of Carboniferous age; Alum S. of Upper Lias of Jurassic age.

shamal (Arabic) Strong, dry, dust-laden N.W. wind, blowing across the plains of Iraq in summer, producing a hazy sky.

shatter-belt A zone in which rock has broken through earth-movements into angular fragments (FAULT-BRECCIA) along the line of a FAULT which is not clean-cut. It forms a distinct line of weakness on which weathering and erosion can concentrate.

shearing The cracking and breaking of a rock at a fault or thrust-plane through compression or tension, with resultant slipping; hence *shear-fault*, *shear-plane*, *shear-cleavage*. The volume of the rock remains the same but its form is altered as the 2 adjacent portions slide past each other. Sometimes crushing and shattering takes place along the line of s.

shear moraine A transverse debris ridge, developed on the glacier surface in close proximity to the snout. The debris is released by melting-out of sediment-rich ice-bands, intersecting the ice-surface vertically or at a steep angle, and believed to be shear-planes up which sediment has been 'dragged' from the glacier bed. On melting out the debris protects the ice from ablation, forming an ice-cored s m. S ms are regarded by some as syn. with THULE-BAFFIN MORAINES.

[*f, opposite*]

sheet erosion, sheet wash SOIL E in which soil is gradually removed by surface runoff in a thin s over a large area, as it moves down gentle slopes. It is a slow process, less spectacular than other forms of erosion, but cumulatively very serious.

sheetflood, sheetflow Water flowing down a slope in thin continuous sheets, rather than concentrated into individual channels. Such flow occurs before RUNOFF is sufficient to promote concentrated flow and also after torrential rainfall when existing rills cannot carry inc. runoff.

sheeting The process of splitting-off of shells of rock in a massive, little jointed rock, probably the result of outward expanding force of PRESSURE RELEASE, causing curved rock-shells to pull away from the mass along DILATATION JOINTS. This causes a doming or rounding effect; e.g. granite domes of Yosemite Valley, California. Sometimes confused with EXFOLIATION.

sheet lightning The effect of a flash of l in a cloud discharge or cloud-to-cloud discharge, where the cloud causes illumination to be diffused in a sheet-like appearance.

shelf A ledge or platform, esp. CONTINENTAL s. Hence *shelf-sea*, syn. with EPICONTINENTAL SEA.

shelf-ice Floating ice-mass formed by coalescence of glaciers along the margins of Antarctica, ending in an ice-cliff; sometimes called *barrier ice*.

shell-sand A s consisting of comminuted fragments of shell; e.g. along W. coast of Skye at heads of the sea-lochs; MACHAIR of the Outer Hebrides, esp. on S. Uist and Tiree; on beaches of Brittany.

shield Rigid mass of Precambrian rocks, the 'nucleus' of a continent, which has remained rel. stable since an early period in the Earth's history. It may have undergone gentle warping, but otherwise little disturbed by later orogenies. (i) Baltic S. (Fennoscandia); (ii) Siberian Platform (Angaraland); (iii) Chinese Platform; (iv) Deccan; (v) Arabia; (vi) continent of Africa; (vii) W. Australia; (viii) Canadian (Laurentian) S.; (ix) Guiana Plateau; (x) Brazilian Plateau; (xi) Antarctica.

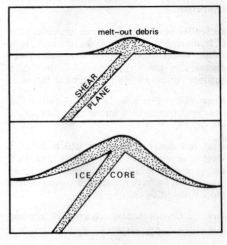

shield inselberg Used J. Budel (Germ. Schildinselberge) to describe domed INSELBERGS formed by lowering of the upper wash surface and exposure of high points of the BASAL SURFACE OF WEATHERING. Ct. ZONAL INSELBERG. [*f*]

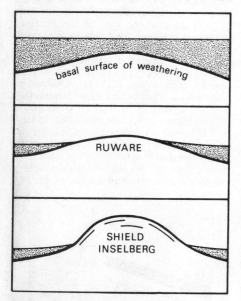

shield volcano Volcanic cone of basic lava, with a low angle of slope and large basal diameter; e.g. Mauna Loa, largest in the world, with a base 480 km in diameter on the ocean floor, total height from floor to summit about 9750 m (32 000 ft), height of summit above sea-level 4171 m (13 680 ft), 113 km in diameter at sea-level, with a broad shallow crater 16 km in circumference. The angle near its cone is about 10°, but near its base only about 2°. Mauna Loa is active, and large-scale lava flows occur.

shift The max. or total displacement of rocks on opp. sides of a FAULT, far enough from it to be outside the dislocated zone. In a fault which also involves bending or flexing of strata, the actual movement of rocks (SLIP) along the fault-line will be only part of the total displacement.

shingle Mass of water-worn rounded stones; usually limited to beach-deposit.

shoal (i) Shallow area in a sea, river, lake. (ii) Bank of sand, mud, pebbles responsible for that shallow area. (iii) As a vb., esp. in navigation, to shallow gradually.

shore (i) The area between the lowest low-water spring-tide line and the highest point reached by storm-waves. Sometimes divided into: (i) BACKSHORE, covered with beach material; FORESHORE, to edge of low-tide terrace. Beyond is *offshore*, a shallow area with bars and troughs permanently covered with sea-water. (ii) In nautical terms simply the land as seen from the sea. [*f* COAST]

shore-face terrace Area of deposition at the outer edge of a marine ABRASION PLATFORM, where wave-worn material is deposited in deep, calm water.

'short time-scale' A t-s of the PLEISTOCENE glaciation, in which its onset is put at *c.* 600 000 years ago. Ct. 'LONG TIME-SCALE'. This dating is largely based on the astronomical theory of periodical perturbations in the orbit of the Earth to explain climatic fluctuations (F. E. Zeuner), not on radioactive dating of Pleistocene deposits.

shott (Arabic) (i) Shallow fluctuating salt-lake in a hot desert. (ii) Hollow or depression in which such a salt-lake lies; e.g. S. el Jerid, S. Melrir, on either side of the Algerian-Tunisian border. Alt. *chott*.

shoulder (i) A bench on the side of a valley. Partic. developed on sides of a valley which has been deepened by erosive action of a glacier; the s occurs at the marked change of slope between the steep-sided inner valley and more gentle upper slopes above the level of glaciation. See ALP. (ii) Short rounded spur on a mountain-side.

SI The *Système Internationale d'Unités*, a rationalized and simplified metric system coming into international use; 23 countries, including the UK, have agreed the system (1970), which may become universal. Nearly all quantities required are based on 6 primary units: (*a*) *length*: METRE; (*b*) *mass*: KILOGRAM; (*c*) time: SECOND; (*d*) *electric current*: ampère (A); (*e*) *temperature*: the degree KELVIN (K); (*f*) *luminous intensity*: candela (cd). In addition, there are coherent derivatives e.g. NEWTON (*N*), PASCAL (*Pa*), and other units in common use (e.g. nautical mile and knot for the UK, °Celsius) are incl. The metre and kilogram are basic units, as opposed to centimetre and gram; multiples and sub-multiples of these are used: 10^{12} (prefix *tera*, T); 10^9 (*giga*, G); 10^6 (1 million) (*mega*, M); 10^3 (*kilo*, k); *10^2 (*hecto*, h); *10^1 (*deca*, da); *10^{-1} (*deci*, d); *10^{-2} (*centi*, c); 10^{-3} (*milli*, m); 10^{-6} (*micro*, u); 10^{-9} (*nano*, n); 10^{-12} (*pico*, p); 10^{-15} (*femto*, f); 10^{-18} (*atto*, a). Preference is expressed for multiples or sub-multiples separated by the factor 1000; values indicated by * are non-preferred; i.e. 1 cm is

not used, but expressed as either 10 mm or 0·01 m.

sial, adj. **sialic** Surface granitic rocks of the continental crust, composed largely of (*si*)lica and (*al*)umina. Density of 2·65 to 2·70. Avge velocity of earthquake P-waves = 5·57 km s^{-1}, of S-waves 3·36 km s^{-1}. [*f* ISOSTASY]

sidereal day Interval of time between 2 successive transits of a star over the same meridian, i.e. 1 rotation of the earth rel. to the stars = 23 hours, 56 minutes, 4·0996 seconds of solar time. A **s d** is nearly 4 mins. shorter than a MEAN SOLAR D, and stars appear to rise nearly 4 mins. earlier every night.

sidereal year The period of time in which the earth makes a complete revolution of the sun with ref. to the stars = 365·2564 mean solar days = 365 days, 6 hours, 9 minutes, 9·54 seconds.

siderite (i) Ferrous carbonate (FeCO$_3$), ore of iron, found in Coal Measures and in Jurassic limestones; alt. *chalybite*. (ii) An 'iron METEORITE' consisting of iron-nickel alloy.

side slip The sliding of a glacier margin past adjacent bedrock and/or lateral moraine. S s is measured by the insertion of a stake into the ice edge, and careful observation of its displacement from fixed points off the glacier. Measurements of the Tsidjiore Nouve glacier, Switzerland, in the summer of 1975 showed s s to average 6 cm/day. However s s is variable through time, being more rapid during warmer periods and/or rainy spells.

sierra (Sp.) High range of mountains with jagged peaks projecting like teeth of a saw; e.g. S. da Guadarrama, S. Nevada. Used widely in Spain, S.W. USA, Mexico and Latin America.

sieve-map A series of maps drawn on transparent paper, each depicting the distribution of some factor; if superimposed the **s ms** will 'sieve out' suitable or unsuitable areas.

silcrete Hard silicified sandstone; e.g. a SARSEN. Gen. a silicified DURICRUST.

silicon An element (atomic no. 14), second most abundant (to oxygen) in the crust, comprising about 28% by volume. Combines with oxygen to form s dioxide (SiO$_2$) (*silica, quartz*), and with other oxides as a large group of rock-forming silicates, including the feldspars, hornblende, the micas, olivine, pyroxenes, amphiboles. Hence *siliceous*, containing *Si*.; vb. *to silicify*, make siliceous;

silicification, replacement by silica. Cf. PETRIFACTION.

sill (i) A near-horizontal or tabular sheet of igneous rock, solidified from MAGMA intruded between bedding-planes; i.e. concordant with structure. E.g. Great Whin S. in N. England extends over about 3900 km^2 from Northumberland coast to W. edge of the Pennines, varying in thickness from 1 m–75 m. Consists of dolerite. Other examples are Salisbury Crags near Edinburgh; Drumadoon in S. Arran; Palisades [*f*] along W. bank of Hudson R., New Jersey, USA. (ii) A submarine ridge separating one ocean basin from another, or from a sea; e.g. in Strait of Gibraltar; in Strait of Bab-el-Mandab, over which water is 365 m deep, ct. over 2100 m deep on either side in the Red Sea and Indian Ocean. (iii) A submarine s or threshold near the mouth of a FJORD, usually of rock, sometimes with a capping of morainic material. See pl. 61. [*f*]

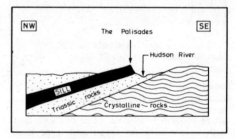

silt Material laid down in water, strictly with particles of diameter 0·002 to 0·02 mm; coarser than those of clay, finer than sand. Used of *s-soil*, containing over 80% s.

Silurian The 3rd geological period system of the Palaeozoic, succeeding the Ordovician and preceding the DEVONIAN. Dated c. 440–400 million years ago. The UK Geological Survey included Ordovician rocks in the S until the end of last century (still the practice in continental Europe). Rocks of this age occur in Shropshire, central and S. Wales, S. part of English Lake District, S. Uplands of Scotland, N.E. Ireland. The accepted series are: (i) *Llandovery*; (ii) *Wenlock*; (iii) *Ludlow* of shales, sandstones, flags, limestones.

silver A precious metal, occurring in native form in association with others such as gold, and as a mineral, silver sulphide or *argentite* (Ag$_2$S), found among most lead, zinc and copper ores.

silver thaw Descriptive name in USA for GLAZED FROST.

sima, adj. **simatic** Dense (2·9–3·3) rocks of (*si*)lica and (*ma*)gnesia underlying less dense

Plate 61. The exposure of the hard dolerite of the Whin SILL at Bamburgh Castle, Northumberland, England. (*Eric Kay*)

continental SIAL masses, and forming the floors of much of the oceans. Avge velocity of earthquake P-waves = 6·5 km s^{-1}, of S-waves = 3·74 km s^{-1}. [*f* ISOSTASY]

similar triangles A method of reduction or enlargement of some narrow strip on a map, such as section of road, railway or river. [*f*]

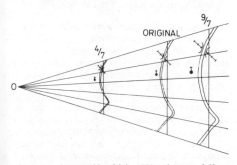

simoom, simoon (Arabic) Very hot, swirling, intensely dry, sand-laden wind, experienced in summer in N. Sahara. Usually assoc. with a low pressure system passing to the N.

Simple Conic Projection See CONIC P.

Simple Cylindrical Projection See EQUIRECTANGULAR P.

simulation The operation of a MODEL by manipulation of its constituent elements, subjectively by a scientific worker, objectively by a computer; the s may be projected indefinitely into the future. See RANDOM WALK NETWORK.

singularity In climatology, the tendency of a type of weather to recur with reasonable regularity each year around the same date.

sinkhole, sink (i) Hole, shaft or funnel-shaped hollow in limestone or chalk country, corresponding to a PONOR in karst terminology. The result of: (*a*) solution under the surface: *solution-s*; (*b*) collapse of rock above a large cavern system: *collapse-s* (in UK SHAKEHOLE). (ii) Hollow down which surface water disappears; i.e. syn. SWALLOWHOLE. It seems preferable to regard **a s** gen. as a depression, dry or down which water merely seeps, and to restrict swallow-hole to a vertical shaft with a waterfall. In the broad sense, there are 60 000 ss on the plateau of Kentucky, USA. (iii) Volcanic s, in a large SHIELD VOLCANO or lava dome; as molten lava is removed from beneath, subsidence forms a depression, several hundreds of m deep and 1·5 km or more across. See pl. 62.

sinter Hard, crusty deposit of silica formed around a GEYSER or hot spring. S. alone signifies *siliceous s* (*geyserite*). Sometimes *calcareous s* (or *calc-sinter*) is used, but this is correctly TRAVERTINE.

Plate 62. The famous SINKHOLE of Gaping Gill, which penetrates deeply into the CARBONIFEROUS LIMESTONE of the Pennines, northern England. (*Eric Kay*)

sinuosity ratio An expression of the degree to which a stream channel winds or meanders, derived from the ratio between channel length and either valley length or length of the meander belt axis (found by the joining of meander CROSSOVER POINTS). Where the s r exceeds 1·5 the channel may be described as meandering; a s r of 4·0 or more denotes a strongly meandering channel.

Sinusoidal Projection See SANSON-FLAMSTEED P. [*f*].

siphon A passage in a cave-system, in section an inverted U, which will allow water to flow through the passage whenever the head in the cave behind rises above the level of the top of the s passage. If the cave below the s opens out to the surface, this may form an *intermittent spring*. S is often inaccurately used to signify a SUMP or TRAP.

[*f* INTERMITTENT SPRING]

sirocco, scirocco (It.) Hot S. or S.E. wind, sometimes oppressively humid, sometimes dry, which blows from the Sahara across N. Africa, Sicily and S. Italy, in advance of depressions moving E. through the Mediterranean, and though dry when leaving the desert, it may be very humid on reaching S. Italy.

site The smallest unit in D. L. Linton's hierarchy of MORPHOLOGICAL REGIONS.

Sixth Power Law A l relating to the ability of a stream to move particles of a certain size, formulated in 1842 by W. Hopkins. This ability (COMPETENCE) is said to be proportional to the 6th power of its velocity; i.e. if a stream flows twice as rapidly, it can carry particles 64 times ($=2^6$) as heavy, if 3 times as rapidly it can carry them 729 ($=3^6$) times as heavy. But it must be remembered that other factors modify this, partic. grain size. A critical velocity exists for each grain size at which it can be moved, though once in motion it may continue to be transported. Some fine-grained substances, e.g. mud, are cohesive and difficult to move, though once in motion they continue easily in suspension.

skeletal soil See AZONAL S.

skerry (derived from Swedish *skär*) Low hummocky island, often in a series offshore and parallel to the main trend of the coast. Consists sometimes of solid rock, sometimes of morainic material; e.g. off W. Norway and S.E. Sweden; Sule S. and Stack S. in S. Bank, 64 km W. of Orkneys. See *f* FIARD.

skerry-guard Inaccurate translation of Swedish and Norwegian terms as a line of skerries, forming a 'breakwater'. Correctly refers to calm water between the line of skerries and mainland; e.g. 'inner-lead' along W. Coast of Norway.

sky cover The amount of s covered or obscured by clouds. This may be considered as the amount completely hidden (*opaque s c*), or that just obscured but not completely hidden (*total s c*). S c is measured in UK on a scale of 0 (cloudless) to 8 (sky entirely covered). Also measured in 10ths of the sky covered in accordance with international practice.

slack (i) Depression between lines of coastal sand-dunes. (ii) Period of stand-still about high and low water, when the tide is neither flowing nor ebbing. (iii) Quiet part of a stream where the current is slight, as on the inside of a bend.

slaking (i) The crumbling of clay-rich sedimentary rocks into small pencil-like fragments when exposed to air, as a result of alternately absorbing and giving up moisture. (ii) Also used in a more gen. sense for any crumbling and disintegration of earth on drying-out after being saturated.

slash Area of swampy ground, esp. in S. and S.E. of the USA. Much of the Gulf Coast lowlands has been planted, esp. with s pine (*Pinus caribaea*), from which gum and turpentine are obtained. Other trees found in similar locations are bald cypress, longleaf pine, pond pine.

slate (i) A fine-grained metamorphic rock, formed from deposits of mud (clay-slate) subject to pressure as a result of earth-movements; splits easily into thin layers. It is also formed from deposits of fine-grained volcanic ash; e.g. Buttermere Ss. in the Borrowdale Volcanic Series, English Lake District. (ii) Occas. used for any easily split rock, though in strict geological usage only for rock which splits along lines of CLEAVAGE, and not along bedding-planes (as with shales). (iii) Used as a proper name in stratigraphical formations; e.g. Llanberis S., N. Wales (Cambrian); Skiddaw S., English Lake District (Ordovician); Delabole S. (Devonian).

sleet (i) (UK definition): mixture of snow and rain, or partially melted falling snow. (ii) (US definition): raindrops which have frozen and then partially melted.

slickenside A polished, sometimes finely fluted or striated, surface-plane of a FAULT, caused by friction and sometimes fusion by heat, as a result of movement along divisional planes of the fault; e.g. in Bunter Sandstone near Frodsham, Cheshire; in Chalk to W. of Lulworth Cove, Dorset.

slide (i) Mass of rock or earth fallen through gravity down a hillside, with a speed constant throughout the mass. (ii) The mark left on the hillside by such a fall. See ROCK-SLIDE.

sling psychrometer A WET- AND DRY-BULB THERMOMETER mounted on a frame fixed on a handle, such that the frame whirls round the handle when operated. This ventilates the wet-bulb thermometer, and readings of the temperatures enable the RELATIVE HUMIDITY of the atmosphere to be read off from tables.

slip (i) A form of landslide involving a mass of material, usually saturated with water, and moving *en bloc* as a constituent mass which commonly stays intact. (ii) In a FAULT, the actual rel. movement along the fault-plane, which may be only part of the total displacement (SHIFT) of the rocks. The component of movement may be in the direction of STRIKE of the fault-plane (*strike-s*), or in the direction of DIP of the fault-plane (*dip-s*).

slip-face The sheltered leeward side of a sand-dune, steeper than the windward. Eddy motion helps to maintain a slightly concave slope. Sand constantly blows up the windward side, and down the **s-f**, hence the dune advances, unless its movement is checked by vegetation, some prominent obstacle, or water.

slip-off slope A low, gently sloping spur projecting from the opp. (convex) side of a meander to the river-cliff in a valley. The stream has migrated outward down the **s-o s**, cutting into the opposite bank. [*f* MEANDER]

slope profile analysis The sub-division of surveyed slope profiles into component parts, each of which has certain form properties e.g. convex and concave elements, and rectilinear segments. The analysis may be statistically based, involving computation of degree of curvature, construction of percentage composition graphs (showing the relative importance, in terms of ground length, of segments and elements) etc.

slope length The actual length of a **s**, measured on its surface from the higher to the lower point, and not by its projection on to a plane (as on a map), which is the HORIZONTAL EQUIVALENT. [*f*]

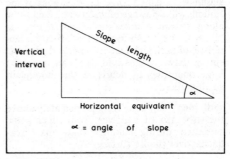

slope-profile line A straight line of traverse down a hill-slope, orthogonal to rectilinear contours. See PROFILE.

slope retreat The progressive back-wearing of the **s** profile during the course of erosion. 3 basic concepts have been advanced. W. M. Davis suggested that ss progressively decline in angle. W. Penck proposed *s replacement*, a form of parallel **r** in which **s** facets are replaced by others of different angle. L. C. King maintained that PARALLEL R OF SS as a whole

272 SLOPE

takes place, that they maintain their form almost until the end of the cycle.

slope unit A SEGMENT (rectilinear) or ELEMENT (convex or concave) of a slope profile. S us may be combined to give 'profile sequences' (e.g. convex-rectilinear-concave, convexo-concave etc.).

slough (i) Fanciful, dramatic or poetic name for an area of marsh, bog, mud or water-logged ground; e.g. 'S. of Despond'. (ii) In USA a backwater.

sludging Flow of a thawed surface sheet of mud over still frozen ground; syn. with SOLIFLUCTION. Common in early summer in high latitudes and at high altitudes.

slump, slumping A type of mass-movement involving actual shearing of the rocks, tearing away of a mass of material (*s-block*), with a distinct rotational movement on a curved concave-up plane, leaving a fresh scar on a hill-side. Partic. common when more massive rocks overlie clay or weak shale, along a sea-cliff or an escarpment; e.g. along S. coast of I. of Wight, and Kentish coast between Folkestone and Dover, where Chalk overlies Gault Clay; along Devon coast between Exmouth and Lyme Regis; along Cotswold scarp between Gloucester and Cheltenham, where Jurassic limestones slump down over clay.
[*f* EARTH-FLOW]

slurry A flow of very wet, highly mobile mud.

slushflow Defined by A. L. Washburn as the 'mudflowlike flowage of water-saturated snow along stream courses'; also referred to as 'slushers' or 'slush avalanches'. S is characteristic of periglacial environments during the spring thaw, when rapid flow can transport large quantities of debris to the mouths of small gullies.

slush zone The zone lying immediately above the FIRN LINE of a glacier, from which snow saturated during summer thaw may avalanche over the ice surface.

small circle Any c on the surface of the Earth, the plane of which does not pass through its centre (i.e. not a GREAT C). All lines of latitude, with exception of the Equator, are s cs.

smog Thick yellow RADIATION FOG over a town, where large quantities of soot act as nuclei for condensation, and sulphur dioxide adds to the acrid flavour. Coined in 1905, and became well known as a result of some very bad ss, such as those of London in winter, 1952. These are said to have resulted in 4000 deaths from bronchitis and pneumonia. This led to campaigns for smoke abatement, creation of smokeless zones, and use of smokeless fuels. Consid. improvement has been effected and ss are now rare.

snout The end of a valley-glacier, usually with a cave from which flows a melt-water stream. Sometimes the s is obscured by masses of morainic material. See pl. 63.

snow The type of precipitation when water-vapour condenses at a temperature below freezing-point, passing directly from gaseous to solid stage, and forming minute spicules of ice. These unite into crystals, either flat hexagonal plates or hexagonal prisms, revealing infinite variations in their patterns. These crystals aggregate into *s-flakes*. Where the lower atmosphere is sufficiently cool, they reach the ground without melting. S may be dry and powdery under low temperature conditions, as in Antarctica; about 750 mm of fresh s may be equivalent to 25 mm of rain. It may be wet and compact, so that only 100–150 mm of s will melt to form 25 mm of rain. 300 mm of new fallen snow on avge = 25 mm rainfall. For records in UK, s is melted and added to the precipitation total. A tall cylinder is placed on a rain-gauge to collect s, but it may become choked and readings are uncertain.

snow-avalanche A fall of a mass of s down a hill-side: (i) *wind-slab*, where s is crusted and compacted; (ii) *dry s-a*, usually of new s in winter; (iii) *wet s-a*, caused by a sudden spring thaw.

snow-bridge A mass of s, more or less compacted, spanning a CREVASSE or BERGSCHRUND.

snowdrift Bank of s drifted by wind to accumulate against obstacles, sometimes to great depths; this may block roads and railway-lines, unless protected by barriers, sheds or s-fences.

snow-field Area of permanent s accumulated in a basin-shaped hollow among mountains or on a plateau; e.g. Ewig Schneefeld near the Jungfrau in Bernese Oberland.

snow-line The lowest edge of a more or less continuous s-cover. (i) The *permanent s-l* is the level at which wastage of s by melting in summer fails to remove winter accumulation. This varies with latitude, altitude and aspect.

Plate 63. The steep face of the Mont Miné glacier, Switzerland. Note the SNOUT, the emergence of the subglacial meltwater stream, the 'thrust planes' in the ice, and the collapsed blocks of ice. (*R. J. Small*)

At the Poles it lies at sea-level, S. Greenland at 600 m, Norway at 1200–1500 m, the Alps about 2700 m, E. Africa at 4900 m. (ii) The *winter s-l* fluctuates from year to year, but is markedly lower than (i). There is no permanent s in Gt. Britain, though in winter in Highlands of Scotland s lies above 900 m for about 80 days.

snow-patch erosion See NIVATION.

socle (Fr.) Convenient name given by French geomorphologists to the granitic or crystalline basal rocks underlying an area of sedimentary rocks.

soffoni, soffione, suffione (It.) See FUMAROLE.

soft hail See GRAUPEL.

soil The thin surface-layer on the earth, comprising mineral particles formed by break-down of rocks, decayed organic material, living organisms, s-water and a s atmosphere. It has a physical structure, and chemical constituents. Ss may be classified into: (i) ZONAL, in broad latitudinal zones or belts; (ii) INTRA-ZONAL, resulting from special parent rocks (e.g. limestone), presence of salt (solonchaks) or of water (peat-ss); (iii) AZONAL, *skeletal* or *immature ss*, new materials on which soil-forming processes have not had sufficient time to work. See individual s-types.

soil climate The temperature and moisture conditions within a s.

soil erosion Removal of s by forces of nature more rapidly than various s-forming processes can replace it, partic. as a result of Man's ill-judged activities (over-grazing, burning and clearance of vegetation, unprotected fallow, DRY-FARMING, up-and-down slope ploughing). This causes a loss of agricultural land which Man can ill afford. Chief types: (i) *wind e* or DEFLATION; e.g. 'Dust-bowl' of S.W. USA; (ii) SHEET or SHEET-FLOOD E; (iii) RILL E; (iv) GULLY E. S e can be checked by maintenance of an effective vegetation cover, CONTOUR-PLOUGHING, ROTATION OF CROPS, terracing, composting, planting of COVER CROPS, creation of wind-breaks (trees, fences), pipe drainage to prevent gullying, damming of gullies or filling them with brushwood.

soil profile A vertical section of the s, from the surface down to bed-rock, showing individual HORIZONS, often of different

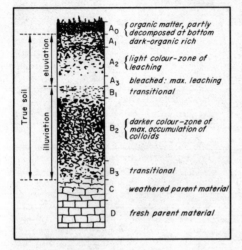

colours and textures, sometimes grading into each other, sometimes changing sharply. Each main s-type has its own p. A scientist may remove a complete p in a s-box for laboratory examination. [*f*]

soil moisture deficit The degree to which soil moisture content falls below FIELD CAPACITY. During late winter, after protracted rain and little or no evapotranspiration, zero s m d exists in Britain. Thereafter increasing s m d is calculated by the formula: cumulative rainfall minus evapotranspiration (which in summer exceeds rainfall).

solano (Sp.) E. or S.E. wind, hot and oppressive, which sometimes brings rain in summer to the coast of S.E. Spain.

solar constant The rate per unit area at which s radiation is received at the outer limit of the atmosphere; this is equal to approx. 1·94 langleys (1 l. = 1 calorie per cm^2) per minute, though values up to 2·04 have been recorded (hence c is a misnomer).

solar day The interval between 2 successive transits of the sun across a meridian; since the Earth's orbit is an ellipse and is inclined to the Equator, this varies slightly (see EQUATION OF TIME), and therefore *mean s d* of 24 hours is used. Ct. LUNAR DAY.

solarimeter An instrument for measuring intensity of solar radiation.

solar system The 9 PLANETS and the ASTEROIDS revolving in very nearly circular orbits around the sun in nearly the same plane.

solar year Period of time taken by the Earth to make 1 complete orbit around the sun = 365·2422 MEAN S DAYS.

sole (i) The ice-base of a glacier. (ii) The lowest THRUST-PLANE in an area of such intense compressional crustal movements that overthrusting has occurred: e.g. the lowest of the Caledonian thrust-planes in N.W. Highlands of Scotland, called the S.

solfatara (It.) A vent quietly emitting sulphurous gases, usually assoc. with approaching extinction of volcanic activity. So called after a small volcano in Phlegraean Fields, near Naples. Ct. FUMAROLE.

solid geology The g of the rocks underlying the layers of superficial deposits (DRIFT). The UK Geological Survey produces for many areas maps both of s g (as if drift were removed) and of drift g, in which case solid rocks are not shown The maps are not entirely consistent, as usually more extensive and thicker areas of alluvium are shown as an integral part of the geological picture even on a s g map, otherwise it would be unrealistic.

solifluction, solifluxion The downhill viscous flow of surface deposits saturated with water, esp. when released by thaw, over still frozen ground beneath. Formerly used syn. with soil creep (downward movement of material whether aided by water saturation or not). Hence *s slope*, *s sheet*, *s lobe*, *s stripe*. See pl. 59.

solonchak (Russian) INTRAZONAL SOIL in which soluble salts are present in considerable quantity. Of widespread occurrence wherever there is a sufficient degree of evaporation, both in hot deserts and in cooler continental interiors where summer heat allows seasonal evaporation. Strong salt solutions rise by capillarity and form a greyish surface crust, below which is a granular salt-impregnated HORIZON. E.g. Great Basin of Utah, Jordan Valley, around Caspian S.

solonetz (Russian) A saline soil in an area with appreciable rainfall, so that some salt in the surface layer is leached out, to be concentrated in a lower horizon.

solstice One of the 2 solstitial points on the plane of the ECLIPTIC, where the overhead midday sun is at its furthest declination (angular distance) from the Equator (approx. 23° 27′ N. and S.). The sun reaches the N. s (at Tropic of Cancer) about 21 June, the S. s (at Tropic of Capricorn) about 22 Dec. These are *summer* and *winter ss* respectively in the N. hemisphere. [*f*]

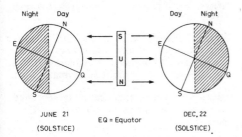

solum Used in USA for the Earth's soil layer.

solution (i) The state of a substance (*solute*) being dissolved in a liquid (*solvent*). This can act as a form of weathering; e.g. s of rock-salt in water. (ii) The process by which rain-water (acidulated with carbon dioxide derived from its passage through the atmosphere) acts as a dilute acid upon calcareous rocks (chalk and limestone). $CaCO_3$ is slowly dissolved and removed as calcium bicarbonate, $Ca(HCO_3)_2$. More correct to refer to this as CARBONATION-SOLUTION. Rivers carry a vast load in s; the Mississippi transports to the sea each year 136 million tons of matter in s (ct. 340 millions in suspension and 40 by SALTATION). It is estimated that each km^2 of the Earth's surface loses 20 tons of material annually by s.

solution collapse, subsidence In limestone country, sub-surface CARBONATION of rock fissures and underground water-courses may lead to subsidence, producing surface hollows.

solution pipe Cylindrical or cone-shaped hole, filled with debris, occurring partic. in chalk country, varying in size from small features to vertical shafts 120 m deep. The result of CARBONATION-SOLUTION concentrated along fissures and joints. A p is filled with material from overlying Tertiary rocks or superficial deposits such as gravel and sand. E.g. in the E. of plateau of Hesbaye, Belgium, are ps known as '*aard pijpen*' or '*orgel pijpen*', diameter 0·3 to 2·0 m, depth 1 m to 20 m.

Sonnenseite (Germ.) The sunny slope in a valley; syn. ADRET.

sonograph Graphical depiction of patterns of the seafloor produced by echo-sounding of reflected sound waves; now essential for oceanographers.

sop Vertical mass of iron-ore, usually occurring in Carboniferous Limestone, e.g. in W. Cumbria. Deposited from mineralized solutions derived from former overlying Triassic sandstones, which percolated along lines of weakness in the limestone.

sotch (Fr.) Term in Grands Causses (Central Massif) for a DOLINE.

sound (i) Area of water connecting 2 larger areas, too wide to be called strait; e.g. in W. Scotland, S. of Jura, Sleat. (ii) Lagoon along S. and S.E. coast of USA; e.g. Pamlico S., N. Carolina. (iii) Applied to inlet of the sea; e.g. Plymouth S. (Devon); Marlborough S. (New Zealand). (iv) 'The S'. separates Zealand (Denmark) from Sweden.

sounding (i) Depth of water in sea or lake, formerly obtained by 'the lead', a fine hemp line with a weight attached, then by a wire attached to a s-machine, now by sonic or ultrasonic echo-s. By the last, vibrations are transmitted through the water to the sea-floor, to return as an echo, electrically recorded. (ii) The actual depth of water thus obtained, normally expressed as being 'below chart datum', usually lowest possible tide-level, i.e. worst navigational conditions. Some port authorities (e.g. Mersey Docks and Harbour Company, Southampton Harbour Board) have their own Datum, to which all ss in waters under their control are related. Ss are marked on charts as 'spot-depths', by British Admiralty in fthms; submarine contours are interpolated.

source The point at which a river rises, or begins to flow as an identity: (i) at a spring; e.g. R. Churn (Thames headstream) at Seven Springs in Cotswolds; (ii) from a lake; e.g. R. Mississippi; (iii) from a glacier; e.g. R. Rhône; (iv) as a resurgence from a cavern; e.g. R. Axe from Wookey Hole in the Mendips; (v) from marsh, bog or swamp; e.g. Goredale Beck, Malham, Yorkshire. Many streams begin as tiny rills or trickles on a hillside, or as a seepage, and gradually grow into a distinctive flow of water.

Southeast Trades TRADE WINDS of the S. hemisphere.

Southerly Burster A strong, dry wind, bringing unusually low temperatures to New South Wales, Australia. An example of a 'polar outbreak', whereby cold air-masses 'burst' into normally milder areas, drawn N. behind a depression.

spalling Used in USA as syn. with EXFOLIATION.

spar A light-coloured mineral, used as a suffix: feld-, fluor-, calc-.

spate Sudden flood or downflush of water on a river, caused by intensive rain or sudden snow-melt higher up its valley.

spatter cone A large mass of lava ejected from a volcano, which 'spatters' and congeals as it hits the ground to form a small c, 3–6 m high; e.g. around Sunset Crater, Arizona, USA.

special scientific interest, site of Area in UK defined and listed because of its partic. floral, faunal or geological interest, protected in various ways; a local authority must consult the Nature Conservancy before granting any planning permission affecting it. 2600 sites are thus defined.

species, cloud A subdivision of C GENERA in terms of shape and type, with 14 forms; e.g. *castellanus* (castellated or turreted), *floccus* (tufted), *lenticularis* (lens-shaped), *nebulosus* (veiled).

specific gravity Ratio of the mass of a body to the mass of an equal volume of water at the temperature of its max. density (i.e. 3·945°C). S g is a rel. quantity, DENSITY an absolute quantity, though numerically the same.

specific heat Amount of heat (in calories) required to raise the temperature of 1 g of a partic. substance through 1°C. 1 calorie is required to raise 1 g of water from 0° to 1°C.

specific humidity Ratio of the weight of water-vapour in a 'parcel' of the atmosphere to the total weight of air (incl. water-vapour), in g of water-vapour per kilogram of air; e.g. very cold dry air may have a s h of only 0·2, while very humid warm air near the Equator may have s h of 15·0 to 18·0. Ct. MIXING RATIO.

speleology, spelaeology The science of cave-exploration. In USA known as 'spelunking', its exponents as 'spelunkers'.

speleothem A deposition feature of calcite on a cave roof. See HELICTITE, STALACTITE, STALAGMITE.

sphenochasm A triangular gap in the sea separating 2 continental blocks bounded by converging faults; e.g. Arabian S., Gulf of California, S. of Japan, Ligurian S. in W. Mediterranean.

sphere A solid figure, any point on the surface of which is equidistant from its centre. Area of surface $=4\pi r^2$; volume $=\frac{4}{3}\pi r^3$. Used commonly as a suffix: atmosphere, biosphere, cosmosphere, ecosphere, exosphere, heterosphere, homosphere, magnetosphere, planetosphere, stratosphere, troposphere.

spherical triangle A t on the surface of a sphere, bounded by arcs of 3 GREAT CIRCLES. Solving a s t by means of s trigonometry is a basic operation in geodetic surveying.

spheroid A nearly spherical body generated by the rotation of a sphere about its axis. See EARTH.

spheroidal weathering Swelling or expansion of outer shells of a rock by penetration of water, so forcing them to pull successively away and loosen. A form of chemical w similar to larger scale mechanical w (EXFOLIATION). Partic. noticeable in basalt and granite.

spillway See OVERFLOW CHANNEL.

spine, volcanic Solidified mass of viscous lava, either forced out by extrusion and congealed, or hardened in the pipe and exposed by denudation; e.g. the destroyed s of Mont Pelée, formed in 1902; Santiagito s in Guatemala 1922–4; Devil's Tower, Wyoming.

spinifex A plant with tufted, spiny leaves, growing in deserts of central Australia in large clumps separated by bare sand; sometimes called *porcupine grass*.

spit Long narrow accumulation of sand and shingle, with one end attached to the land, the other projecting into the sea or across the mouth of an estuary; e.g. Calshot S. at S. end of Southampton Water; Hurst Castle S. at W. end of Solent; Blakeney, Norfolk. [*f*]

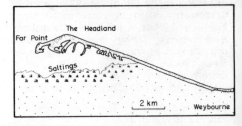

splash erosion See RAIN-DROP EROSION.

splaying crevasse A c which starts as a longitudinal feature and splays outwards towards the side of a glacier.

spot-height A precise point, its height above a given datum measured and indicated on a map, not necessarily on the ground; ct. BENCH-MARK, actually indicated on the ground.

spring (i) Natural flow of water from the Earth's surface, gen. issuing strongly, though at times it may just ooze or seep out. Its occurrence is rel. to the nature and relationship of rocks, esp. of permeable and impermeable strata, with the profile of surface relief. A s occurs where the WATER-TABLE intersects the surface. It may be permanent, INTERMITTENT, SCARP-FOOT, DIP-SLOPE, FAULT, JOINT, VAUCLUSIAN. (ii) The season between winter and summer; in the N. hemisphere it occurs astronomically between s or vernal EQUINOX (about 21 Mar.) and summer SOLSTICE (about 21 June); pop. March, April, May.

spring-line A l of ss indicating the level where the WATER-TABLE intersects the surface, as at the foot of an escarpment. Along such a line are *s-l villages*; e.g. along N. edge of S. Downs. While availability of water is important in settlement establishment, it is probably not the only factor, for such villages may be situated to take advantage of 2 types of land or soil.

spring-sapping Erosion around the issue-point of a strongly flowing stream, which removes material from around it, creating a small amphitheatre cut progressively back into the slope. An important agency in the development of an escarpment DRY VALLEY. See pl. 11.

spring tide A twice-monthly t of considerable amplitude (i.e. high high t and low low t), occurring when moon, sun and Earth are in the same straight line (SYZYGY), either in conjunction (new moon) or opposition (full moon), so that gravitational effects are complementary. S ts occur about every 14·75 days and are about 20% higher than mean ts. Ct. NEAP T.

spur Prominent projection of land from a mountain or a ridge. See INTERLOCKING S, TRUNCATED S.

squall A precise definition adopted in 1962; sudden incr. of wind-speed by at least 8 m s^{-1} (16 knots), rising to 11 m s^{-1} and lasting for at least 1 minute.

stable equilibrium State of the atmosphere where the ENVIRONMENTAL LAPSE-RATE of an air-mass is less than the DRY ADIABATIC LAPSE-RATE. If a surface pocket of air is displaced upwards, it will cool at the D A L-R, become colder than the surrounding air-mass, and will sink back to its orig. level. Rising motion cannot be sustained, and the atmosphere is in NEUTRAL EQUILIBRIUM. Conditions in an ANTICYCLONE and in subsiding air are usually s. Ct. UNSTABLE E.

stac An isolated mass of hard igneous rock, the result of marine erosion, in the St Kilda group W. of Outer Hebrides. Some have been worn down to or just below sea-level, others are upstanding, such as S. an Arnim (191 m, the highest in Britain), S. Lee (160 m).

stack Steep pillar of rock rising from the sea, formerly part of the land but isolated by wave action. Forms part of a cycle of marine erosion: cave—arch—s—stump—reef; e.g. Needles (at the W. end of the I. of Wight), Old Harry (I. of Purbeck), both of chalk; Old Man of Hoy (Orkneys) of Old Red Sandstone; Sule S. (on Skerry Bank, W. of Orkney) of gneiss, 40 m high. See STAC. See pl. 64. [*f*]

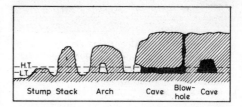

stadial moraine See RECESSIONAL M.

stage (i) In its restricted geomorphological sense, point of development reached in the course of a CYCLE OF EROSION. Characteristic ss are thought to be recognizable, viz. 'youth', 'maturity', 'old age'. Recent critics of Davisian geomorphology have urged that s should not be given this restricted meaning of a phase in a progressive and inevitable evolution of landforms in a cyclic manner, but a wider, more realistic implication of changes in causal relationships between structure and process with the passage of time. (ii) A subdivision of a SERIES in stratigraphy (4th order, corresponding to a unit such as a FORMATION), or in a mineral zone, or a zone of a partic. fauna; corresponding in time to an AGE. (iii) Division of the Pleistocene; i.e. glacial ss. (iv) In USA, the height of the surface of a stream, i.e. depth of water; see RATING CURVE [*f*].

stage-discharge curve Syn. with RATING C [*f*].

stalactite Pendant mass of calcite hanging vertically from the roof of a cave, deposited

278 STALACTITE

Plate 64. A prominent STACK formed by marine erosion of horizontally bedded chalk at Etretat, northern France. Note also the fine natural ARCH. (*Eric Kay*)

from drops of water containing calcium bicarbonate which have seeped through crevices and joints. Calcite is deposited partly because of evaporation, partly because some carbon dioxide in the water escapes and part of its dissolved calcium bicarbonate is changed back into calcium carbonate. In Ingleborough Cave, near Clapham, Yorkshire, rate of growth of a s was found to be 7·483 mm per annum, about 760 mm per century, much more rapid than is gen. believed.

stalagmite Similar mass to a STALACTITE, usually more stumpy, growing from the floor of a cave, deposited from drips from the roof. This forms below a stalactite, so that they may join to form a single pillar. In the Aven Armand, in Grands Causses, France, these pillars are 23–25 m high and 1 m or more in diameter.

stand Group of similar trees growing together; e.g. s of Douglas fir, teak. This allows economic forestry practice. In tropical rainforests species are scattered, and ss rarely occur, but are much more usual in mid-temperate forests.

standard atmosphere Avge condition of the **a**, defined in terms of its pressure and temperature for various altitudes, and used for calculation of aircraft performance, calibration of instruments, etc. Various **s** as are used by USA, USSR and others, though it is hoped to define one internationally acceptable, applicable to 60 km altitude.

standard hillslope A hillslope comprising four components: an upper convexity or 'waxing' element; a free face; a debris slope, rectilinear in profile; and a basal concavity or 'waning' element (alt. PEDIMENT). According to L. C. King **s hs** are formed by all types of slope process, and are thus independent of climatic control. A strong and massive bedrock is required to support all four components of the **s h**; in weak and incoherent rocks only the convex and concave elements may occur. [*f*]

```
WAXING
SLOPE
-----------
FREE
FACE
-----------
DEBRIS
SLOPE
-----------
PEDIMENT
(waning slope)
```

standard parallel A partic. **p** of latitude, selected for calculating and drawing a specif. projection (e.g. conical group), or for the horizontal axis of a grid-system.

Standard Time The mean time of a meridian centrally located over a country, and used for the whole area, instead of the inconvenience of APPARENT (LOCAL) TIME. In a large country each zone, of 15° longitude, has its Zone S. T.; e.g. USA has Eastern, Central, Mountain, Pacific, Yukon, Alaska-Hawaii and Bering zones, related to 75°, 90°, 105°, 120°, 135°, 150° and 165° W. meridians, and resp. 5, 6, 7, 8, 9, 10 and 11 hours slow on Greenwich. [*f, page 279*]

standing wave Syn. with LEE or stationary **w** in air or water.

star-dune Sand-dune shaped like a pyramid, often quite high (up to 90 m) and apparently

fairly permanent in position, with radiating ridges of sand from a central point.

static lapse-rate Syn. with ENVIRONMENTAL L-R. Ct. DYNAMIC L-R.

static rejuvenation R. of a river caused by re-stimulation of erosive activity by an incr. in precipitation due to climatic change, or to an incr. in volume due to CAPTURE. Ct. DYNAMIC R.

stationary wave Syn. with OSCILLATORY W in tide-formation.

Staublawine (Germ.) Avalanche of fine powder-snow.

steady time One of three categories of 'geomorphological time-scale' devised by S. A. Schumm and L. W. Lichty (see CYCLIC and GRADED TIME). S t refers to periods of short duration, embracing the formation of minor features of river channels, hillslopes etc.

'steady state' Syn. with DYNAMIC EQUILIBRIUM.

steam-coal A hard shiny c, intermediate in carbon content between BITUMINOUS coal and ANTHRACITE.

steam-fog F formed when cold air passes over the surface of much warmer water, from which moisture condenses into minute droplets; the water appears to 'steam'. In very low temperatures moisture is converted directly into ice-particles, hence ICE-FOG.

steering Used in meteorology for the directional effect of some atmospheric influence on another phenomenon; e.g. the s effect of high altitude STREAMLINES or of temperature differences on movement of surface depressions.

step-fault A series of parallel fs, each with a greater throw in the same direction, so producing a 'stepped slope'; e.g. W. slopes of Vosges and E. slopes of Black Forest, bounding the Rhine rift-valley.

[*f* RIFT-VALLEY]

STEREOGRAPHIC 279

steppe (Russian) Mid-latitude grassland extending across Eurasia from Ukraine to Manchuria, usually regarded as syn. with prairie. Some writers regard s as being somewhat drier than prairie.

steptoe A hill protruding from and surrounded by a widespread BASIC LAVA flow, which has otherwise blanketed the pre-existing land-surface. After Steptoe Butte, Oregon, USA.

Stereographic Projection A ZENITHAL P in which meridians and parallels are projected on to a tangent plane at a point on the opposite extremity of the diameter. Both parallels and meridians are circles. The **p** is CONFORMAL. Exaggeration increases outward symmetrically from the central point. It has one unique property: all circles on the globe appear as circles on the **p**. Used for maps of the world in two hemispheres, for star maps, and for maps used in geophysics (since problems in spherical trigonometry on it are easy to solve). It can be constructed in the polar, equatorial [*f*] and oblique cases. As for all zenithal ps, Great Circles passing through the centre are straight lines; it can be used for aeronautical maps centred on an airport, and the polar case for high latitude navigation. Graphically the polar case is easy to construct. The meridian circle is drawn to scale, with a line tangential to the Pole, and angles are projected on to the tangent line from the opp. side of the polar diameter; where these lines intersect the tangent line will be the radii of the individual parallels. Meridians are straight lines radiating from the centre (360° divided according to meridian interval). Trigonometrically, the radius of each parallel $= 2R . \tan \frac{1}{2}(90°-\text{latitude})$. [*f*]

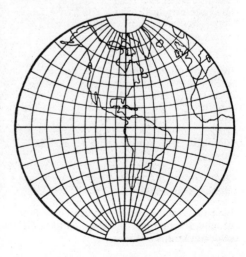

stereoscope An instrument with lenses or eye-pieces the approx. distance apart of human eyes (about 57 mm, 2·25 ins), through which a pair of photographs (*stereograms*) of the same landscape, taken from slightly different angles, are viewed. This gives a 3-dimensional effect (*stereomodel*). The simplest instrument comprises 2 lenses with folding legs; more complex patterns have binocular viewers, and large stereogrammetric plotters are used for accurate map-making from air photographs (*stereoplotting*).

stillstand A period of time during which the level of the sea rel. to that of the land remains virtually undisturbed.

stock Small intrusive mass, defined as having an upper surface extent of not more than 100 km^2; i.e., a small BATHOLITH. When more or less circular, known as a BOSS; e.g. the Cheviot, in Cheviot Hills.

Stone Age Given by archaeologists to the period approx. from the beginning of the Pleistocene to the BRONZE AGE, during which Early Man used stone implements or *artifacts*. Divided into EOLITHIC ('dawn' **s a**), PALAEOLITHIC ('old' **s a**), MESOLITHIC ('middle' **s a**) and NEOLITHIC ('new' **s a**).

stone line A layer of stones, frequently quartz pebbles, lying at the base of the upper horizon of tropical soil-profiles. The **s l** may represent a LAG gravel, formed by surface wash, then buried by a younger soil creeping down the slope or raised to the surface by termites.

stone-polygon, -stripe See PATTERNED GROUND. See pl. 65.

stoping, magmatic A process by which MAGMA penetrates overlying country-rock by shattering it, so that blocks sink into the magma (ENGULFMENT), allowing it to ascend. Once held to account for large-scale **m** penetration, but various objections, mainly geophysical, lead to the belief that penetration is small-scale, confined to upper parts of BATHOLITHS and around RING-DYKES.

storm (i) Any severe disturbance of the atmosphere, together with its effects, esp. at

Plate 65. Inactive STONE POLYGONS near Boulder, Colorado, USA. Note the angular fragments forming the polygon 'borders', and the vegetated areas of fine material forming the polygon centres. (*M. J. Clark*)

Plate 66. STORM BEACH with high ridge of pebbles, thrown up by powerful storm waves from the west, at Newgale, St Bride's Bay, Pembrokeshire, Wales. (*Eric Kay*)

sea: thunder-, rain-, snow-, hail, sand-, dust-. (ii) Force 11 (28·5–32·6 m s^{-1}) on BEAUFORT WIND SCALE. (iii) Many compound words: s-wave, s-tide, S-SURGE, S-BEACH.

storm-beach Accumulation of coarse material (shingle, cobbles, boulders), formed during exceptionally powerful storm-waves, usually above the foreshore, and well above the level reached by normal high spring-tides. See pl. 66.

storm-surge Rapid rise of the level of the sea above predicted tidal heights, whereby water is 'piled up' against the coast by strong onshore winds. This may result in breaching of coastal defences and widespread flooding. E.g. during 31 Jan.–1 Feb. 1953, s-ss developed along the coasts of E. England and the Netherlands, following an intense atmospheric depression crossing N. part of the N. Sea. This produced gale-force winds from N. and N.N.W., and caused a 'surge' or piling-up of water in the N. Sea basin, which narrows to the S. The vertical rise in the water-level above predicted high water was 2–2·5 m between the Wash and Strait of Dover, and 3–4 m along coast of the Netherlands. Widespread flooding, damage and loss of life occurred.

stoss (Germ.) Side of a prominent crag facing the direction from which movement of a glacier or ice-sheet occurred, striated and roughened by ice. The opp. side is the LEE; hence *s and lee relief*. See ROCHE MOUTONNÉE [*f*].

stow (i) Unit of smallest order in hierarchy of regional divisions devised by J. F. Unstead. (ii) Second order in D. L. Linton's system of MORPHOLOGICAL REGIONS.

strait, straits Narrow stretch of water linking two areas of sea; S. of Gibraltar, Dover, Magellan.

strand A fanciful name for the shore.

strandflat (Norwegian) A wave-cut platform off W. Norway, which now stands above sea-level as a result of ISOSTATIC recovery. It exceeds 50 km in width. Sub-aerial weathering (mainly frost action) has been so potent that cliffs have been rapidly worn back, and

282 STRANDFLAT

Plate 67. STRATOCUMULUS CLOUDS viewed from above. (*Aerofilms*)

powerful waves have quickly removed debris.

strath (Scottish) Long steep-sided flat-floored valley, wider than a GLEN. Commonly used as a prefix; e.g. Strathclyde, Strathmore, Strathpeffer. Occas. in N. England; e.g. Langstrath, Borrowdale, Cumb.

stratification (i) The accumulation of sedimentary rocks in layers or strata; hence *stratified*, vb. to *stratify*. (ii) Some authorities use the term more widely to indicate any rocks occurring in layers, incl. igneous intrusions (SILLS), but this is not advisable. *S index*: number of beds in a single formation, multiplied by 100, and divided by the total thickness in the formation; e.g. in a formation 50 ft thick with 7 distinct beds, s index = 14. See pl. 25, 64.

stratiform All sheet clouds, incl. STRATUS, CIRROSTRATUS, ALTOSTRATUS, STRATOCUMULUS, NIMBOSTRATUS.

stratigraphy A branch of geology dealing with order and succession of rock-strata, their occurrence, sequence, lithology, composition, fossils and correlation; basis of HISTORICAL GEOLOGY.

stratocumulus (*Sc*) A uniform heavy cloud, dark grey globular masses within a continuous sheet, occurring up to 2400 m, usually lower. See pl. 67.

stratosphere The layer of the atmosphere above the TROPOPAUSE, extending up to the *stratopause* which marks the base of the MESOSPHERE at about 50 km. The base height of the s avges 16 km at the Equator, 11 km at

50°N. and S., and 9 km at the Poles; this varies slightly with the season (rather higher in summer) and with gen. atmospheric conditions. The s contains little dust or water-vapour, but much OZONE. At the base of the s temperatures over the Equator vary during the year only from about −79°C to −90°C though over polar regions the seasonal difference is more marked (from about −40°C in summer to −79°C in winter). The temperature rises in the s from the tropopause to a max. of about 77°C at the stratopause, beyond which it falls again in the mesosphere.

strato-volcano Used in USA for a COMPOSITE VOLCANIC CONE.

stratum Layer of sedimentary rock (pl. *strata*). Usually the syn. *bed* is used in sing. strata in pl., but there is no fixed practice. Some geologists, esp. in USA, use s in a wide sense, so that it may be made up of a number of beds.

stratus (*St*) A grey, uniform cloud-sheet, often persistent, occurring in a thin sheet at any height up to about 2400 m. Not usually responsible for precipitation, other than a fine drizzle. Adj. *stratiform*.

stream A body of flowing water, covering all scales from a rill to a large river. Hence used to denote all characters of, processes and landforms resulting from ss.

stream-flood Coined by W. M. Davis to indicate a FLASH-FLOOD confined to a distinct channel, normally dry, esp. in arid and semi-arid lands. Ct. SHEETFLOOD.

streamline The direction of movement of all 'parcels' of air within an area, measured at a single moment in time, an instantaneous overall picture of air movement. Ct. TRAJECTORY, which follows a single 'parcel' of air over a period of time.

stream order A topological classification of ss according to their position in a drainage net. The scheme usually adopted is that of A. N. Strahler, modified from R. E. Horton. The s network can be thought of as a branching tree: those branches terminating at an outer point are designated *1st order ss*. The junction of 2 of these will produce a *2nd order s*, 2 *2nd* order ss will form a *3rd order s*, and so on until the 'root', or s mouth, is reached. A recent classification, proposed by R. L. Shreve, uses only 1st order links as an index of magnitude. See BIFURCATION RATIO, HORTON'S LAW. [*f*]

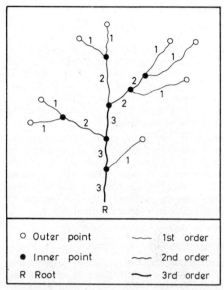

stress A force per unit area, obtained by dividing the total force by the area to which it is applied. If 2 forces act away from each other the stress is a *tension*; if they act towards each other it is a *compression*; if they act parallel it is a *shear*. When rocks are exposed to stress, they suffer strain.

striation, striae Scratches, grooves and scorings on ice-worn rocks, caused by dragging of rocks frozen into the base of a glacier by movement over the valley floor. Afford an indication of direction of ice-movement.

strike The direction along a rock stratum at right-angles to TRUE DIP. S is used as an adjective in connection with features developed along it; e.g. s-fault, -valley, -joint, -stream. [*f*]

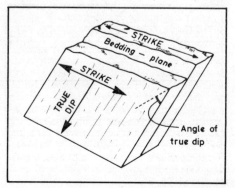

strike-fault A f whose direction is parallel to the strike of the strata it affects.

strike-slip fault See TEAR-F.

strike-valley V developed in rocks parallel to the strike of the strata; the v of a subsequent stream in a TRELLIS DRAINAGE PATTERN. E.g. Front Ranges of the Rockies, between Denver and Boulder (Colorado), where rocks dip E., s-vs trending from N. to S. have developed along the s of the outcrop of weaker shales between more resistant rocks (which form HOGBACKS). See pl. 17.

string bog A bog characterized by ridges of peat and other vegetal matter, separated by ill-drained depression, found mainly in sub-Arctic tundra regions, where s bs may cover 10% or more of the total area. S bs have been attributed to rucking of the bog surface by solifluction or differential frost heave.

striped ground See PATTERNED G.

stripped surface See STRUCTURAL SURFACE.

strombolian eruption So called after Stromboli, a volcano in the Lipari I. N. of Sicily. The crater contains molten lava, and gases can readily escape, with small-scale intermittent es at frequent intervals, but with no great accumulation of pressure which might cause a more violent e. A cloud constantly hangs over Stromboli, coloured pink by reflected light at night; hence its name 'lighthouse of the Mediterranean'.

structural surface A 'plane' surface formed by the erosional stripping of a weak stratum from an underlying resistant stratum, in such a way that the latter remains largely undissected. E.g. in S. England Salisbury Plain is a s s at 125–150 m, formed by the removal of unresistant Belemnite Chalk from the harder Echinoid Chalk, probably by periglacial solifluction during the Pleistocene. [*f*]

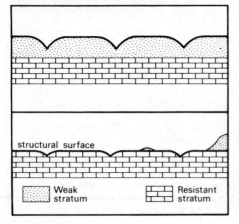

structure Arrangement and disposition of rocks in the Earth's crust. In some cases this implies more than just arrangement, including the nature of the rocks and the initial forms prior to erosion. In this sense used in W. M Davis's classic trilogy for the explanatory description of landscape: 's, process and stage'. It is preferable to restrict the term to actual ATTITUDE of rocks. Structural is used widely as an adj.: -basin, -feature, -terrace, -trap.

subaerial Occurring or forming on the surface of the Earth, as distinct from *subaqueous* or *submarine*.

sub-alluvial bench A rock surface, normally convex in profile (hence *s-a convex b*), formed beneath alluvial or scree deposits at the base of a retreating scarp. As the scarp is weathered, its lower part becomes progressively buried and so 'protected'. At each stage a greater amount of scarp recession is needed to provide material for the building up of the increasingly extensive slope foot deposits. Thus the **s-a b** develops its convex profile. [*f*]

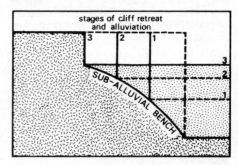

subarctic (i) Used with ref. to latitudes immediately S. of the Arctic Circle, and to phenomena occurring there. (ii) A group of climatic types defined by W. Köppen, with mean temperature of the warmest month above 10°C and mean temperature of the coldest month under −3°C; evaporation and precipitation are both low. They are denoted by letter *D*, subdivisions being *Dfc*, *Dfd*, *Dwc*, *Dwd*. (iii) An alt. name for PRE-BOREAL climatic phase.

Sub-Atlantic The climatic phase since 500 B.C., following SUB-BOREAL, and still in being. Conditions became milder and moister, and an alder-oak-elm-birch-beech flora gradually spread, with beech gen. dominant.

Sub-Boreal The period from about 2500 to 500 B.C., following the ATLANTIC climatic STAGE, with a renewal of cooler climatic conditions of before 5500 B.C., though rather drier. The dominant tree of the Atlantic phase (oak) gradually gave way to pine.

subclimax A persistent SERAL community which has been permanently prevented from reaching a state of CLIMAX VEGETATION by factors such as bad drainage, poor soil, or exposure to strong winds; e.g. in heathlands of W. Europe, as the Kempenland.

sub-consequent stream See SECONDARY CONSEQUENT S.

subduction zone In PLATE TECTONICS a boundary between converging crustal plates, usually indicated by a trench, island arc or line of volcanic activity.

sub-glacial chute A valley, gully or depression eroded by meltwater plunging down a hillslope beneath an ice-margin. Some s-g cs are fed by ice-marginal streams which penetrate crevasses. S-g cs are usually now dry, and tend to begin and end abruptly, depending on the former height of the ice and the base of crevasses initiating them; most are not more than 100 m long.

subglacial moraine Material carried by a glacier, frozen into the ice at or near its base.
[*f* MORAINE]

subhumid A climate transitional in precipitation received, lying between areas with a well distributed rainfall and those with near arid conditions. The vegetation consists of tall grass; i.e. too dry for much of the year for tree growth, but not so dry as to be desert or scrub. It is a matter both of seasonal distribution and total of precipitation.

sublimation Conversion of a solid to a vapour, or vice versa, without an intervening liquid stage. Specif. the formation of ice-crystals directly from water-vapour when condensation takes place at temperatures below freezing-point.

sublittoral The zone extending from low-tide level to the edge of the CONTINENTAL SHELF, syn. with shallow water zone. On it are deposited sands, muds and coral fragments.

submarine canyon Steep-sided trench on the floor of the sea which crosses the CONTINENTAL SHELF, sometimes continuing across the continental slope into deep water. It may be due to tidal scouring, to erosive activity of s TURBIDITY currents, to former river erosion when sea-level was lower than at present, to faulting, or to sapping action of powerful s springs which burst out on the sea-floor.

Some are obviously submerged valleys of rivers continuing over the continental shelf, but others cross the entire shelf and slope into deep water, and a few occur only near the outward margins, forming deep gorges cut into the slope. E.g. Fosse de Cap Breton in the Bay of Biscay [*f*, *below*]; off the mouth of Zaïre R.; off R. Hudson. [*f*]

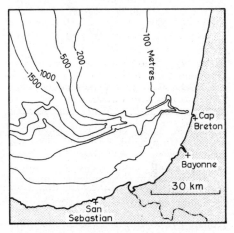

submarine ridge See MID-OCEAN R.

submerged coast A coastline which owes its form largely to a rel. rise of sea-level, covering the margins of the former land-surface. (i) *S upland coast*: (*a*) RIA; (*b*) FIORD; (*c*) DALMATIAN or longitudinal; (ii) *S lowland coast*: (*a*) broad shallow estuaries, with marsh and mud-flats uncovered at low tide, creeks and winding shallow inlets; e.g. S. Suffolk; Strangford Lough (N. Ireland) and Boston Harbor (Mass., USA), where a DRUMLIN coast is submerged; (*b*) FIARD; (*c*) FÖHRDE; (*d*) BODDEN. See also S FOREST.

submerged forest A layer of peat in which tree-stumps and roots are embedded, occurring between tide-levels or below present low tide, indicative of a positive change (rel. rise) of sea-level; e.g. revealed by excavations for dock basins at Southampton and Barry (Glamorgan). Peat, stumps and trunks can be seen at low tide along the coast near Formby (Lancs.), Wirral, near Harlech, and at Borth on the Dovey estuary. At Pentuan in Cornwall a layer of sediment containing oak stumps and roots lies 20 m below present sea-level.

subsequent stream A tributary to a CONSEQUENT S, which has developed its valley, mainly by headward erosion, along an outcrop of weak rocks; if outcrops occur more or less at right-angles to the consequent slope, the s s will join the consequent river at right-angles. See TRELLIS DRAINAGE. [*f*]

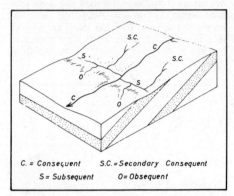

C. = Consequent   S.C. = Secondary Consequent
S = Subsequent   O = Obsequent

subsere A SERAL community which has developed to a certain level, but which has been prevented by extraneous factors from developing into CLIMAX VEGETATION. Only if the delaying factors are regarded as temporary is the community known as a s; otherwise the appropriate term is SUB-CLIMAX.

subsidence (i) The depression of part of the Earth's surface relative to its surroundings. This may be: (*a*) a large-scale crustal movement; e.g. a RIFT-VALLEY, a warped basin; (*b*) a small-scale local movement; e.g. the result of solution of salt deposits (e.g. in Cheshire; see MERE, FLASH); the collapse in a limestone area of the roof of a cave-system formed by carbonation-solution, leaving a hollow or gorge on the surface; e.g. Gordale Scar near Malham (Yorkshire); Cladagh valley, near Enniskillen (N. Ireland). (ii) Used in climatology for the slow settling of large masses of air, as in an ANTICYCLONE.

subsoil A transitional layer of partially decomposed rock underlying the top-soil and overlying the bed-rock. It forms the C-HORIZON in the SOIL PROFILE.

sub-surface wash Sub-surface moisture migrating laterally within the regolith, and transporting material in solution and fine particles to give 'lateral eluviation'.

subtropical App. to the latitudinal zones between the tropic of Cancer and about 40°N., and the tropic of Capricorn and about 40°S. It is used in respect of: (i) belts of atmospheric high pressure; (ii) climatic regions with no cold season; i.e. with no month below about 6°C (43°F). The term is also used gen. of lands, climates, vegetation outside the tropics and merging into the warm temperate zone.

succession (i) The order of beds of rock in time. (ii) A gradual sequence of changes or phases in vegetation over a period of time, even if the climate remains unaltered; hence *plant s*. This will proceed until some situation of equilibrium is attained, and a CLIMAX COMMUNITY is established.

sudd (Arabic) A compact mass of floating vegetation on the White Nile in the Sudan, which blocks the main channel, forcing it to split up into a maze of backwaters and shallow lakes.

suffosion A process of chemical and mechanical erosion occurring in limestone overlain by glacial and alluvial drift. Percolating water selectively removes by solution some drift contents and washes fines into voids in the limestone. This results in the formation of numerous small surface hollows ('drift dolines').

sugar limestone A crumbly soft material formed by the THERMAL METAMORPHISM of Carboniferous L; e.g. by an INTRUSION, as in the N. Pennines above and below the Great Whin Sill.

sugar loaf A steep-sided residual hill, with smooth rock-faces produced by massive EXFOLIATION. E.g. Sugar Loaf Mountain, Rio de Janeiro, Brazil. Cf. BORNHARDT.

sulphur An element obtained: (i) from iron pyrites (FeS_2), as in Spain; (ii) from deposits of natural s, sometimes in the neighbourhood of volcanic regions, also in vast beds in Texas and Louisiana associated with salt-domes, and in Sicily associated with Tertiary limestones and sandstones.

sumatra A line-squall wind (or 'linear disturbance') in the Malacca Straits, usually accompanied by thunderstorms, occurring very suddenly, gen. at night, and esp. during the period of the S.W. monsoon.

summer (i) The warmest season of the year, in ct. to winter. (ii) In the N. hemisphere, the period June, July and August; in the S. hemisphere, the period December, January and February. (iii) In astronomical usage, the period between the summer solstice (about 21 June) and the autumn equinox (about 22 Sept.) in the N. hemisphere, and between about 21 Dec. and 21 March in the S. hemisphere.

summit The highest point or level of a mountain, railway, road, canal.

summit plane A plane passing through a series of ACCORDANT SUMMITS indicating the existence of either: (i) a former PENEPLAIN; or

(ii) a regional balance in down-wasting, so that divides are reduced in level together (the GIPFELFLUR hypothesis).

sump A deep pool in a cave-system below the WATER-TABLE, the exit of which lies below the level of the water.

sun An incandescent, almost spherical body, the centre of the solar system, with a diameter of about 1 392 000 km (865 000 mi.), a mean surface temperature of about 5700°C, and a very high internal temperature. It rotates once in 24·5 days at its equator and the whole solar system moves through space at about 18·5 km (11.5 mi.) per second. Its mean distance from the Earth is 150 million km (92·9 million mi.). See APHELION, PERIHELION.

sunrise, sunset colours At these times of day, the light-waves of solar radiation have had a long passage through the atmosphere because of their low angle to the Earth's surface. When these light-waves encounter obstacles (such as air particles, water-vapour, dust, smoke or volcanic ash), the constituents are scattered, esp. at the short-wave (blue-violet) end of the spectrum, which results in the colours at the long-wave (red) end of the spectrum becoming increasingly dominant. 'Red sky at night shepherd's delight; red sky at morning shepherd's warning.' Cf. ALPINE GLOW.

sunshine An important climatic element, the duration of which is partly a function of latitude (i.e. the total hours of daylight and therefore of possible s), partly a function of daytime cloudiness. It is measured by a CAMPBELL-STOKES RECORDER, and tables of data can be given in terms of duration in hours per day, month or year, or as a percentage of the possible amount. Lines of equal mean duration of s (plotted for various stations) are *idohels*. The sunniest parts of the earth are the hot deserts; at Helwan, Egypt, the mean s for the year is 3668 hours, or 82% of the possible for the latitude. The lands around the Mediterranean Sea have about 90% of the possible in summer, though they are somewhat cloudier in winter. The equatorial and cool temperate latitudes have much less s; Valentia in S.W. Ireland has a mean of 1·3 hours of s per day, or 17%. The sunniest place in Gt. Britain is claimed to be Shanklin, I. of Wight.

supercooling This occurs when water remains liquid below 0°C (32°F). In clouds well below freezing-point, water droplets frequently remain unfrozen if undisturbed. Freezing may ultimately occur when these droplets come into contact with flying aircraft, causing 'icing' (which can be extremely dangerous unless provided for), and can freeze on telegraph wires. S may also lead to the growth of very large hail-stones.

superficial deposit Materials lying more or less loosely on the Earth's surface, formed independently of the rocks below and usually transported and deposited there by natural agencies: sand and loess blown by the wind, glacial drift, river-borne alluvium and gravels. Peat is also included, although it has developed *in situ*. A soil which has developed through weathering of the underlying rock is not regarded as a **s d**.

superglacial stream A s of melt-water flowing during a summer day on the surface of a glacier in a deeply cut runnel, until it vanishes down a crevasse. See MOULIN.

superimposed drainage, superimposition A pattern of **d** which orig. developed on a cover of rocks now removed. This pattern seems to bear no relation to the existing surface rocks; e.g. the radial **d** of the English Lake District; the rivers of E. Glamorgan which cross at right angles the varied Devonian and Carboniferous rocks of the S. Wales coalfield basin and its margins; the Meuse, which crosses the Ardennes transverse to the gen. trend of the structure; the rivers of the Hampshire Basin which flows gen. S. across the E. to W. folds of mid-Tertiary age. [*f*]

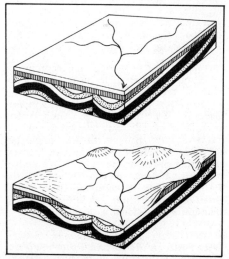

superimposed profile The construction of a diagram on which a series of **p**s spaced at regular intervals across a map of a piece of country is drawn. This may bring out such features as ACCORDANT SUMMITS and erosion platforms. [*f* PROFILE]

Superposition, Law of In sedimentary rocks (and also extrusive igneous rocks) an upper bed of rock is younger than a lower one, provided that earth-movements have not reversed the order.

supersaturation, supersaturated The state of a body of air with a relative humidity of more than 100%, with water-vapour sufficient to produce condensation, but in which this has not actually occurred. For condensation to take place, nuclei such as particles of dust, smoke, salt, pollen, even negative ions, are usually necessary, but not always available. Thus s is a fairly common atmospheric phenomenon.

surazo A cold wind experienced in S. Brazil, the result of an ANTICYCLONE in winter.

surf A mass of broken foaming water, formed when a long wave breaks on a gently shelving sandy beach.

surface detention The layer or film of water, on a slope surface, associated with OVERLAND FLOW (i.e. the water actually in transit). S d is normally less than 2 mm in depth, but increases with distance from the slope crest owing to 'downslope increment'.

surge, glacier In a g, the downstream passage of a bulge of ice at a considerably greater speed than the normal g flow. R. P. Sharp describes a s that has been watched for 15 years moving down one of the gs on the flanks of Mount Rainier in the Cascade Mtns, USA. See KINEMATIC WAVE.

survey (i) The measuring and recording of lines and angles so as to make an accurate map of part of the Earth's surface; this has become a very exact process. Hence *surveying*, *surveyor*. See GEODETIC S. The various operations include BASE-LINE measurement, TRIANGULATION and LEVELLING. Simple but effective methods of local s include compass traversing and plane-table ss. (ii) To examine, record and depict certain features of the Earth's surface, in cartographic, diagrammatic or written form; e.g. an industrial s, a land-use s, a regional s, a geological s, a CADASTRAL S.

suspension The holding-up of small particles of matter, transported by moving water, by turbulent upward eddies. The resistance of the water through which a particle is sinking by gravity helps to keep the finest material in s for a long time; e.g. if a jar of muddy water is left, it takes many hours for most material to sink to the bottom even though the water is still; coarse sand, however, sinks very rapidly. The R. Mississippi carries each year to the sea about 340 million tons of material in s, ct. 136 million tons in SOLUTION, 40 millions by SALTATION. Approx. 80 tons of solid matter in s are removed annually from each km^2 (200 tons per sq. mi.) of the Earth's surface. Hence the *suspended load* of a stream.

swale A long narrow depression on a beach, broadly parallel to the coastline, separating two ridges of shingle; syn. with LOW.

swallow-hole, swallet (i) A deep, vertical shaft, usually in limestone, down which a surface stream disappears as a waterfall; e.g. Gaping Gill, Ingleborough, Yorkshire, down which the Fell Beck pours as a waterfall 111 m (365 ft) high, see pl. 62; Hunt Pot, Pen-y-ghent, Yorkshire. Known also as a PONOR in KARST terminology. (ii) A hole in the bed of a stream down which water disappears; e.g. in the Mole Valley near Dorking Surrey; near N. Mimms, Hertfordshire.

swamp A gen. term applied to a permanently waterlogged area and its associated vegetation, commonly reeds; ct. MARSH, only temporarily inundated; and BOG, where the vegetation is partially decayed. A s is intermediate between a wholly aquatic habitat and a marsh.

swamp-forest A waterlogged area in which certain trees can grow; e.g. the cypress swamps of S.E. USA; the tropical freshwater swamps of E. Sumatra; MANGROVE (saltwater) swamps.

swash The mass of broken foaming water which rushes bodily up a beach as a wave breaks; syn. with *send*. [*f* LONGSHORE DRIFT]

S-wave See SHAKE-W. [*f* EARTHQUAKE]

sweep zone The max. vertical difference in surface profile of a beach during the period of observation.

swell (i) The regular undulating movement of waves in the open ocean, not breaking, and with an appreciable distance between crests; there may be as much as 12–15 m (40–50 ft) vertically between the crest and trough, and the longest s measured was 1128 m (3700 ft) horizontally between 2 successive crests. (ii) A long, gently sloping elevation rising from the sea-floor, though its summit is still far from the surface; e.g. the Hawaiian S., 1000 km (600 mi.) from W. to E. and 3000 km (1900 mi.) from N. to S.

swing, of pressure belts The seasonal migration of the atmospheric pressure-belts, N.-ward in the N. summer, S.-ward in the N. winter, following the overhead sun and the THERMAL EQUATOR, i.e. the zone of greatest solar heating and low pressure.

sword-dune A type of sand-d, syn. with SEIF-D.

syenite A group of coarse-grained igneous rocks, composed mainly of alkali-feldspar. One type resembles granite (though without quartz, or with the mica replaced by hornblende). It is usually classed as intermediate in chemical composition, and occurs in plutonic form.

syke, sike A small stream in the N. Pennines, specif. in Teesdale.

symbiosis In ecology a form of adaptation in which organisms derive benefit because of proximity and adaptation to other units, e.g. African 'tick birds' which feed on Cape Buffaloes.

synclinal valley A v formed by a downfold; e.g. the *vaux* (sing. *val*) of the Jura Mtns. S vs are not common when much erosion has taken place, because of the development of longitudinal streams, the destruction of the anticline, and thus the formation of synclinal peaks.

syncline A downfold in the Earth's crust, with strata dipping inward towards a central axis, caused by compressive forces. The s may be broad and shallow; e.g. the London Basin, the Paris Basin; while in strongly folded districts the LIMBS of the s may be much steeper; e.g. S. limb of the Hampshire Basin. Ct. ANTICLINE. See pl. 12. [*f*]

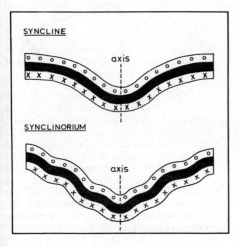

synclinorium A complex syncline upon which minor upfolds and downfolds are superimposed. [*f* SYNCLINE]

synoptic chart A c depicting meteorological conditions (isobars, winds and other features) at a moment of time. The construction of a s c is essential in weather forecasting. In Britain and USA such cs are published daily.

synoptic climatology The overall study on a regional basis of the condition of the atmosphere at a moment of time, primarily in connection with weather forecasting. Ct. DYNAMIC C.

synoptic map (i) A cartographic technique of presenting all the various factors involved in any problem on a single m, or by a series of overlays. E.g. for choosing the site of a new town, various ms of relief, slope, drainage, soil, accessibility, presence of minerals, liability to undue frost, fog and flooding, landscape value, etc. may be superimposed to give a 'cumulative picture'. Cf. SIEVE MAP. (ii) Used specif. for climatic charts, showing ISOBARS, temperatures, windforce and direction, etc.

system (i) A division of the succession of stratified rocks, equivalent to a period in geological time; e.g. Devonian, one of the six (seven in USA) divisions of the Palaeozoic. (ii) See GENERAL SYSTEMS THEORY.

systems analysis A search for generalizations based on the whole rather than on individual parts; a consideration of a set of objects and the functional and structural relationships and organizations linking these objects. It is not a replacement for analytical methods, but an alternative additional line of scientific enquiry. See GENERAL SYSTEMS THEORY, CLOSED SYSTEM, OPEN SYSTEM.

syzygy When the sun, moon and Earth are in the same line, either in conjunction or opposition.

tableland A level-topped upland; esp. where its edges are well defined; e.g. in S. Africa, terminating in the Drakensberg Mtns. Used loosely on various scales; e.g. for a MESA in Arizona, and for the Chinese T., the plateau of S. China.

tabular iceberg A large horizontal floating ice-mass in the S. Ocean, which has broken off from the edge of the Antarctic ice-barrier. These are often of great size; about 100 km lengths have been recorded.

tacheometer A THEODOLITE adapted for the measurement of distance, with which both the horizontal and vertical position of a point can be established by instrumental observations. Hence *tacheometry* or *tachymetry*, the making of a contained plan.

taconite A low-grade iron ore, consisting of ferruginous chert, which contains unleached haematite, magnetite, siderite and hydrous iron silicates. It has an iron content of 25% or less.

tafoni (Corsican patois) Hollowed out blocks of granite, as in the massif of Agriates in N. Corsica. Nightly dew-fall and daily heating lead to the capillary rise of mineral solutions, forming a surface 'varnish', or hard crust, behind which the rock-core gradually disintegrates as the 'cements' are removed. Some granite blocks may become virtually empty shells.

taiga (Russian) A needle-leaf evergreen forest, extending across the N. continents where they are at their broadest, from Scandinavia to the Pacific coast of the USSR, and from Alaska to Labrador and Newfoundland. In W. Europe they extend S. to 60° N., in E. Asia to about 50°N. The trees have to withstand very cold winters, short cool summers, a light summer rainfall, and winter precipitation in the form of snow. They grow slowly; their hard needle-shaped leaves reduce transpiration, and their conical structure helps their stability against wind and snow. Their timber is 'soft'. There are not many varieties of tree: pine (*Pinus*), spruce (*Picea*), fir (*Abies*), with larch (*Larix*), silver birch (*Betula pendula*), aspen (*Populus tremuloides*). In parts the **t** becomes boggy, as the MUSKEG in Canada. *Note:* Some authorities differentiate between the main coniferous belt (BOREAL FOREST), and the thinner more open **t** on the TUNDRA margins.

tail See CRAG-AND-TAIL.

tail-dune A sand-d on the lee-side of an obstacle, tapering gradually away, varying in length from 3 m (10 ft) to 0·8 km (0·5 mi.). [*f* DUNE]

talc Hydrous magnesium silicate, sometimes called *soapstone* because of its greasy or soapy feeling. It can be scratched with the finger-nail, and is given the number 1 on the Mohs' hardness scale. It occurs widely among weathered magnesium-rich basic igneous rocks such as peridotite, or in association with dolomite and marble.

talik A layer of unfrozen ground between seasonally frozen ground and the PERMAFROST TABLE. Alt. an unfrozen layer *within* the permafrost, resulting from local heat sources or circulating ground-water.

talus By most authorities, **t** is syn. with SCREE. Some would make a distinction, regarding **t** as a land-feature, scree as the ingredient (broken rock-debris) of which it is formed; others refer to the **t**-slope as specif. at the base of a cliff. See pl. 25, 60.

talwind A wind which blows upvalley, in ct. to a BERG WIND.

tank An artificial pool or small lake in India and Ceylon, used to store water for irrigation.

tarn A small lake among the mountains, usually on the floor of a CIRQUE-basin, specif. in the English Lake District; e.g. Sprinkling T., Red T., Bleaberry T.

tar-sand See OIL-S.

tautochrone 'A line joining together points of varying condition or value referring to a partic. moment or period of time; e.g. soil temperatures at varying depths at 16·00 hrs on April 1st.' (J. A. Taylor.)

taxonomy The scientific classification of features according to gen. principles and laws.

tear-fault A FAULT in which the displacement of the rocks is mainly horizontal along its line; usually regarded as syn. with STRIKE-SLIP FAULT. [*f*]

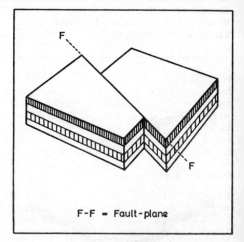

F-F = Fault-plane

tectogenesis The processes which collectively create an OROGENIC zone, incl. FOLDING, FAULTING, OVERTHRUSTING and VULCANICITY.

tectonic From a Gk. word, *tekton*, 'builder', app. to all internal forces which build-up or form the features of the crust, incl. both DIASTROPHISM and VULCANICITY. Hence the term is applied to the features which result; e.g. *t valley*.

tektite A small rounded stony mass, black to green in colour, with a glassy appearance; probably of meteoritic origin.

tellurometer A modern surveying instrument which can measure distance up to 64 km (40 mi.) to a very high degree of accuracy, by sending short-wave radio signals from the observing station to the other point.

temperate Introduced in classical times for one of the 3 temperature zones, the others being TORRID and FRIGID. In climatic terms, t is usually thought of as 'moderate', without any extremes, whereas in actual fact climates within the t zone may vary considerably and be far from moderate. Thus the expression MID-LATITUDE is often preferred. T is commonly used as a qualifying adj.; e.g. t forest, t grassland, t low-pressure belt. W. Köppen included the name t in 5 of his climatic types: *Cfa* (T rainy, with hot summers); *Cfb* (T rainy, with warm summers); *Cfe* (T rainy, with cool short summers); *Csa* (T rainy, with dry hot summers); *Csb* (T rainy, with dry warm summers).

temperature As a climate element, the degree of sensible heat or cold within the atmosphere. It is measured on various thermometric scales (see CENTIGRADE, CELSIUS, FAHRENHEIT, RÉAUMUR, ABSOLUTE). The highest shade t ever recorded in the world was $58.0°C$ ($136.4°F$), at Azizia in Tripoli and at St Louis Potosi, in Britain $38°C$ ($100.5°F$) at Tonbridge (Kent). The lowest shade t recorded is $-89°C$ ($-127°F$) in the Antarctic continent, in Britain $-27.2°C$ ($-17°F$) at Braemar.

temperature-humidity index An i constructed from various formulae, which affords an indication of the physiological effects of the weather on human comfort. One such i, used by the US Weather Bureau, is $0.4 \times$ (*dry bulb temperature + wet bulb temperature*) + 15. On this scale, 60–65 represents ideal conditions, while 80 is so uncomfortable that offices and factories may be closed.

temporary base-level A b-l in the course of a stream, such as a lake, or a resistant stratum, or (for a tributary) the level of the main stream which it joins. A dam is a man-made t b-l. These are impermanent, in ct. to the ocean, the ultimate b-l (though even that may change as a result of relative movements of land and sea); a lake may be filled up or its outlet downcut, a hard bar may retreat upstream.

tension A stress caused by 2 forces moving in opposite directions, in ct. to COMPRESSION. In the Earth's crust, t involves an extension of the strata, and results in JOINTS and NORMAL FAULTS. On a glacier tension results in the development of CREVASSES.

tepee butte A residual conical hill found in Arizona, esp. in the Painted Desert. The edge of a plateau, consisting of mostly rather soft, horizontal strata of sandstone, shale, bentonite and clay, has been rapidly eroded, but a layer of more resistant sandstone near the top forms a capping which helps to preserve isolated hills, with softer material resting on the hard stratum giving them a conical appearance. These are sometimes called similarly '*tent-hills*' in Australia.

tephigram A thermodynamic diagram on which are plotted temperature and dew-point data at different isobaric levels.

tephra Used more commonly in USA for volcanic dust, ash, cinders, scoria, pumice and bombs, ejected during an eruption in solid form.

tephrochronology A geological time-scale based on the dating of volcanic ash layers found around volcanoes and widespread as buried layers in soil horizons. Introduced by S. Thorarinsson, and used with success in Iceland where volcanic eruptions have been numerous.

Terai A zone of marshy jungle along the edge of the Himalayan foot-hills, at a height of about 1500 m (5000 ft).

terlough In W. Ireland, a shallow depression, with a sink-hole, containing water when the WATER-TABLE rises to the surface.

terminal moraine A crescentic mound of till deposited beyond the snout of a glacier or at the edge of an ice-sheet; syn. with *end-m*. A large t m indicates a prolonged stand-still of the ice; e.g. in N.W. Europe, a clearly defined line of t m, representing the max. advance of the ice-sheet during the 4th (*Weichsel*) glacial period, can be traced through Denmark, E. Germany and Poland as the Baltic Heights or Baltic End Moraine. At several points these ridges exceed 300 m in height. The t ms of former valley-glaciers often form natural dams, thus creating long narrow lakes; e.g. the English Lakes; the Italian Lakes.

[*f* MORAINE, ESKER]

terminal velocity The rate of fall attained by a particle moving through a fluid; it represents a balance between gravitational force and the resistance of the fluid (related to the size of the particle). This is an important factor in the transportation of solid matter in SUSPENSION in a stream current, where TURBULENCE is also involved, and in the development of rain-drops and HAIL.

termite mound A mound of soil, often pinnacle-like and reaching heights of 5 m or more, constructed by termite colonies in the tropics. T ms are geomorphologically significant; it has been calculated that in the African savannas termites carry to the ground surface the equivalent of a 5 cm layer every 1000 years. Lateral displacement of this material contributes to soil creep on slopes. See pl. 68.

Plate 68. A striking TERMITE MOUND, about 8 m high, in savanna woodland near Jebba, Niger valley, Nigeria. (*R. J. Small*)

terrace A shelf or bench, relatively flat and horizontal, sometimes slightly inclined. A *river t* lies along the side of a valley; it represents either: (*a*) the remains of a former FLOOD-PLAIN at a higher level dissected by renewed downcutting by a rejuvenated river, with a t at each side (*paired* t); or (*b*) the formation left by freely swinging MEANDERS cutting into a former flood-plain (*meander* t). River-ts are carefully identified (often named) and are correlated by their height, sequence and the deposits (esp. of gravels) on them. E.g. the Thames in London has three well-known pairs. (*a*) the present Flood-Plain T.; (*b*) the Taplow T.; (*c*) the Boyn Hill T., as well as a number of intermediate minor pairs. Near Oxford are: (*a*) the Summertown-Radley T. (on which the city stands); (*b*) the Wolvercote T.; (*c*) the Hanborough T., which is 30 m above the present river-level. See also KAME-T [*f*], ALLUVIAL T [*f*], MEANDER T [*f*].

terrace-gravel A deposit of **g** on a river-t. The orig. material deposited by a river consists of alluvium and **g**; the former is more easily removed when renewed erosion takes place, so that most older, higher terraces are covered with **t-g**s, which are often named and correlated.

terracette A small terrace, from a few cm to 0·3 m or so high, across the face of a slope on which SOIL-CREEP is prevalent. These are sometimes called 'sheep-tracks', though it is not likely that the features have been caused by animals treading out paths, though they may be used in this way once formed.

terrain The physical character of an area, its configuration (as in '*t studies*' and '*t intelligence*').

terrain model See RELIEF MODEL.

terrain-type map Partic. in USA, a m on which the land-surface is divided into a number of categories (or '*t-types*'), and shaded distinctively. Sometimes quantitative methods are used, in order to attain some degree of objectivity, and **t-t**s such as 'nearly flat plains', 'rolling and irregular plains', 'partially dissected tablelands' and 'hills' are numerically defined. The surface of a map is gridded with squares, for each of which a value is obtained, using such criteria as max. difference in elevation, slope and proportion of near-level land. Boundaries are drawn to include all areas within the same quantitatively determined categories.

terra rossa (It.) A reddish residual material, rich in iron hydroxides, the indissoluble residue of chemical weathering, which accumulates in depressions in limestone country, notably under semi-arid or summer drought (Mediterranean) conditions, as in Yugoslavia, S. Italy, Malta. They also develop, less markedly, on limestones in Britain.

terrestrial App. gen. to the land or Earth; e.g. t radiation, -deposits, -sphere, -magnetism.

terrigenous deposits Inorganic ds of sand and shingle derived from the denudation of the land, and laid down in the littoral zone of the sea-floor. One of the 13 categories of marine ds. Ct. NERITIC, PELAGIC.

Tertiary A division of geological time, gen. recognized as lasting from 70 million to about 1 million years ago, but about whose use and definition there is considerable confusion. (i) In Britain it is regarded by many as one of the 4 eras, with its corresponding groups of rocks, and with its 4 divisions (periods) of Eocene, Oligocene, Miocene and Pliocene. (ii) In USA it is regarded as a PERIOD, the earlier part of the Cainozoic era (the Quaternary being later), together with the corresponding system of rocks; this is divided into 5 epochs and the corresponding rock series: Paleocene, Eocene, Oligocene, Miocene, Pliocene. *Note:* This practice is followed in A. Holmes's revised time-scale (1959). (iii) Some European and American workers regard it as an era (with its corresponding rock group, but divide it into 2 periods (systems): Palaeogene and Neogene; this puts the Pleistocene and Holocene (or Recent) into the Neogene, and avoids the word Quaternary.

Tethys The GEOSYNCLINE which extended from W. to E. across what is now the Old World, between Laurasia and Gondwanaland.

Tetrahedral Theory Formulated by Lowthian Green in 1875. He envisaged the Earth as a 4-faced figure, contained by 4 triangles, with its apex at the S. Pole, its base at the N. Pole (representing the Arctic Ocean). This pattern corresponds with the greater part of the land-masses in the N. hemisphere, of the oceans in the S., the triangular tapering nature of the continents, and the antipodal arrangement of land and sea. A spheroidal tetrahedron would fit the concept of a shrinking globe. But there are serious geophysical objections, and the theory is no longer considered tenable.

texture, of soil The physical quality of a s, depending on the sizes of its individual constituent particles. It may be: (i) *coarse-grained*, with a gritty feel; i.e. a sandy s, with particles between 0·02 and 2·0 mm in diameter; (ii) *fine-grained*, with a sticky feel; i.e. clay s, with particles less than 0·002 mm in diameter; (iii) *intermediate*, with a silky feel; i.e. a silt s, with particles from 0·02 mm to 0·002 mm in diameter; (iv) a mixture of particles of many sizes; i.e. a loamy s.

texture ratio A ratio of valley density, derived by identifying from a detailed topographical map the basin contour with the largest number of crenulations (indentations marking valley courses), and dividing this number by the length of the basin perimeter.

thalassostatic A term used by F. E. Zeuner to describe river terraces resulting from aggradational episodes induced by high sea-levels during Pleistocene interglacials, and intervening degradational phases during periods of low sea-level under full glacial conditions.

thalweg, talweg (Germ.) Lit. a 'valley-way', but used to denote the longitudinal profile of a river.

thaw The physical change from snow and ice to water when the temperature rises above freezing-point, and hence is applied gen. to periods during which this happens; i.e. 'the t'.

thaw lake A lake formed by the thawing of frozen ground in permafrost region. T ls occur within ALAS valleys and collapsed PINGOS, and are a common feature of THERMOKARST.

theodolite A precise optical surveying instrument used for measuring angles, with both horizontal and vertical graduated arcs, a telescopic sight and a spirit-level, mounted on a tripod. Some ts are so accurate that with a vernier or micrometer microscope they enable fractions of a second of arc to be read.

thermal (noun) A vertically rising current or updraught of air in the atmosphere. On a sunny day, the sun's rays heat the Earth's surface, parts of which (e.g. bare sand, concrete) will heat more rapidly then others (e.g. woodland or pasture), causing conductional heating and absolute instability, and therefore the vertical currents will rise more rapidly. Both birds and glider-pilots look for ts to assist their ascent. A t may cause condensation, the formation of CUMULUS clouds, and possibly heavy convection rainfall and a thunderstorm. T is also used as an adj., app. to temperature gen.

thermal depression A small-scale, though sometimes intense, low pressure system, the result of local heating and convectional rising of air, responsible for thunder and heavy rainstorms, and in hot deserts for certain storms such as DUST-DEVILS, the SIMOOM of the Sahara, and the KARABURAN of the Tarim Basin.

thermal equator A line drawn round the Earth joining the point on each meridian with the highest average temperature. If this is done for each month, this line will move N. and S. with the apparent motion of the sun, though for the most part it lies N. of the equator because of the larger land-masses and therefore the greater heating in summer. An annual **t e** is sometimes defined, esp. in USA, as the line connecting places of the highest mean annual temperatures for their longitudes. Syn. with *heat e*.

thermal fracture The cracking or fissuring of rocks as a result of sudden temperature change. This is partic. likely: (i) where the rock is heterogeneous; i.e. containing various minerals with differing coefficients of expansion, so setting up strains; (ii) on the face of a steep crag which receives and loses the sun's rays rapidly; (iii) where there is a rapid drop of temperature after sundown. *Note:* Doubt has recently been cast on the efficacy of **t f**, or indeed whether it happens at all in the absence of moisture.

thermal (or temperature) efficiency index An empirical **i** devised by C. W. Thornthwaite, in conjunction with a PRECIPITATION EFFICIENCY I, as the basis for his first scheme of climatic classification. **T e** is defined for each month as $\frac{T-32}{4}$, where T=mean monthly temperature in °F.

thermal metamorphism M caused by a rise in temperature, usually as a result of the INTRUSION of a mass of molten igneous rock at a high temperature. It can produce a fusion or recrystallization of the minerals or grains in a rock; e.g. a coarse-grained sandstone changes into quartzite; limestone into marble.

thermal spring See HOT S.

thermograph A self-recording THERMOMETER, which contains a bimetallic strip in the shape of a coil; one end is fixed, the other actuates a pen which traces a continuous record on to a chart fixed to a rotating drum.

thermokarst The formation in periglacial environments of a highly irregular surface owing to collapses induced by the thawing of ground-ice masses. Pits and basins are so numerous that there is a resemblance to KARST (though no chemical weathering is involved, merely thermal changes). **T** forms include THAW LAKES, ALASES and collapsed PINGOS.

thermomeion An area with a high negative temperature ANOMALY.

thermometer An instrument for measuring temperature. One type consists of a glass tube graduated with a thermometric scale (see CENTIGRADE, FAHRENHEIT, RÉAUMUR, ABSOLUTE) in degrees, containing a column of mercury or alcohol, rising from a bulb of the same liquid, which expands or contracts with temperature change. Other types include metals which expand or contract at a known extent with temperature, or which possess varying resistance to the passage of electricity with temperature change.

thermopleion A region with a high positive temperature ANOMALY.

thermosphere Syn. with IONOSPHERE.

tholoid A steep-sided volcanic dome which has grown up inside the CRATER or CALDERA of a volcano as the result of the slow extrusion of viscous acid lava. E.g. a **t** grew within the caldera of Bezymianny, Kamchatka, USSR in 1956. A **t** developed in the crater of Mt Pelée in 1902, through which a SPINE was later forced.

thorn forest Dense thickets of thorny scrub, growing in semi-arid climates, such as in N.E. Brazil. See CAATINGA.

threshold A factor complicating the simple self-regulation of systems by NEGATIVE FEEDBACK and thus the maintenance of equilibrium states. When a **t** is crossed, irreversible changes may be set in motion e.g. the permanent destruction of vegetation cover by a major flood may initiate a wholly new run-off regime and a different texture of landscape dissection, even though geological and climatic 'controls' remain unchanged.

throw, of a fault The vertical change of level of strata or rocks as a result of their displacement by faulting. The blocks of rock on either side of the fault-line are referred to as upthrow or downthrow, relating to their relative displacement. The amount of throw may vary from a fraction of an inch to thousands of feet.

thrust A compressional force upon strata in a low-angle or a near-horizontal plane, the cause of an extreme RECUMBENT FOLD or of a very low-angled REVERSE FAULT. See T-FAULT, T-PLANE.

thrust-fault A REVERSE F of very low angle, in which the upper beds have been pushed far forward over the lower beds.

thrust-plane The p or surface of movement in a REVERSE FAULT, over which the upper strata are thrust, usually inclined at a very low angle; e.g. four main t-ps can be distinguished in the N.W. Highlands of Scotland: the Glencoul, Ben More, Moine and Sole (all of CALEDONIAN age); in the HERCYNIAN folding in the N. Ardennes of Belgium is a major t-p, the Grande Faille du Midi, over which rocks from the S. have been driven N. towards and partly over the Namur syncline; in Glacier National Park, Montana, the Precambrian 'Belt' Series has been driven E., as a result of the LARAMIDE folding, over the Lewis T-p, so that these rocks now lie over Cretaceous strata. [*f* OVERTHRUST]

thufur A surface hummock, up to 0·5 m high, found in Iceland and other periglacial and Alpine environments. The t may comprise a central core rock, raised by FROST HEAVE, but it may be composed wholly of mineral soil or turf. Features resembling ts are found at high altitudes in Britain (e.g. the grassed mounds at Coxe's Tor, Dartmoor), though whether these are active or fossil is uncertain.

Thule-Baffin moraine A type of ice-cored end-moraine developed at the margins of cold ice-sheets (e.g. in Baffin Island and the Thule area of Greenland). Debris brought to the ice-surface, possibly via shear-planes, restricts ablation in the zone approx. 100 m along the ice-front. Ablation of the unprotected ice away from this protected zone leaves the latter standing up as a moraine ridge. In time the ice core of this ridge may melt to give a low terminal moraine composed of debris only. [*f*]

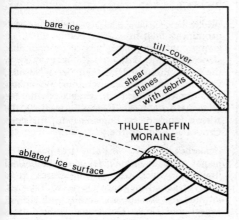

throughflow The lateral movement of water, below the ground surface and above the water-table. Max. flow may be concentrated in the soil, above relatively impermeable bedrock, and follow PERCOLINES. T, rather than OVERLAND FLOW, may be the most important type of water flow on hill-slopes in humid temperate climates.

thunder The result of the expending of electrical energy in a flash of lightning, which heats the gases in the atmosphere, causing their expansion and creating sound-waves.

thunderhead Graphic name, used esp. in USA, for the top of a towering CUMULONIMBUS cloud.

thunderstorm A storm in which intense heating causes convectional uprising of air, extreme conditions of INSTABILITY, CUMULONIMBUS clouds, heavy rain and/or hail, lightning and thunder. The exact mechanism of a t is still far from understood. *Frontal ts* are associated with the passage of a COLD FRONT. Other ts accompany OROGRAPHIC PRECIPITATION, esp. in the tropics where very warm and moist air-masses rise sharply over steep mountain ranges.

tidal current A movement of water set up in areas affected by the rise and fall of the tides. A distinction is sometimes made between the normal movement in and out of an estuary (T STREAM), and an hydraulic t c set up by differences of water-level at either end of a strait due to differing tidal régimes. The latter is the stricter, more limited, usage; e.g. in the Menai Straits high tide occurs at different times at either end, resulting in a powerful t c flowing through the Straits. The same phenomenon takes place in the Pentland Firth, in the N. of Scotland. In the Seymour Narrows, between Vancouver I. and the mainland of British Columbia, there can be as much as 4 m (13 ft) difference in water-level at either end, with resulting strong t cs.

tidal datum A d obtained from a long period of t records. It may serve as: (i) a d for national geodetic and topographical survey; e.g. the ORDNANCE D (O. D.) of Gt. Britain; (ii) a d for a port authority, or for an Admiralty chart, by indicating the lowest possible depth of water, i.e. the worst navigational conditions (CHART D), or a basis for the port tide-tables (*port d*).

tidal flat An area of sand or mud uncovered at low tide. See also MARSH.

tidal (tide-water) glacier A valley-g which reaches the sea, and there discharges bergs or floes; e.g. along the E. and W. coasts of Greenland.

tidal range The difference in the height of the water at low and high tide; this varies constantly, but a mean is given for most ports; e.g. about 3·7 m (12 ft) at Southampton, 5 m (17 ft) at Sheerness in the Thames estuary, 7 m (23 ft) at London Bridge, 9 m (30 ft) at Liverpool, 13 m (44 ft) at Avonmouth. In the open ocean, the **r** may be less than 1 m. The **t r** is at its max. at SPRING TIDES.

tidal stream The normal movement of water in and out of an estuary or other inlet, as a result of the alternate high and low water stages, known as the flood-**t** and the ebb-**t**. The former may be responsible for a BORE. This powerful movement of water, apart from its obvious importance to navigation, may have marked geomorphological effects; see SCOUR. Most flood-tides seem to flow strongly for about 3 hours before high tide, and ebb for about 3 hours after, with a period of slack water about high and low tide. But there are anomalies; e.g. the YOUNG FLOOD STAND at Southampton. A **t s** should not be confused with a T CURRENT.

tidal wave An inaccurate term for a TSUNAMI.

tide The periodic rise and fall in the level of the water in the oceans and seas, the result of the gravitational attraction of the sun and moon. The strength of this attraction varies directly as the masses of the sun and moon, and inversely as the square of their distances. The sun's mass is 26 million times that of the moon, but it is 380 times farther away; its **t**-producing force is 4/9 that of the moon. See also SPRING T, NEAP T, PERIGEAN T, APOGEAN T, OSCILLATORY WAVE THEORY OF TS. [*f*]

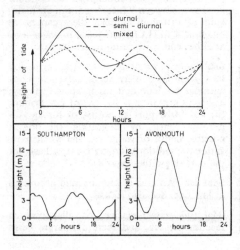

Tierra (Sp.) An altitudinal climax zone, distinguished specif. in Mexico and central America: (i) *T Caliente*, the hot tropical coastlands from sea-level to about 900 m; (ii) *T Templada*, the zone between about 900–1800 m, with monthly means of 18°–24°C (65°–75°F); (iii) *T Fria*, the zone above 1800 m, with monthly means of 12°–18°C (55°–65°F); (iv) *T Helada*, the zone of high peaks and permanently frozen or snow-covered terrain.

till An unsorted, heterogeneous mixture of rocks, clays and sands, carried within, upon or below a glacier or ice-sheet and deposited directly on melting, without any subsequent water transport. It is more usually called BOULDER-CLAY in Gt. Britain, though **t** is perhaps a better word as it does not imply any precise constitution. It is increasingly used as a proper name in Britain; e.g. the Cromer T., Lowestoft T., Gipping T., which have replaced former boulder-clay names. **T** is one of the two main types of DRIFT, the other being *stratified drift*, or FLUVIOGLACIAL deposits. *T plain*: an extensive area covered with a fairly level sheet of **t**; e.g. much of Illinois, Iowa and Indiana, though later stream erosion has produced modifications.

till fabric analysis Analysis of the direction and dip of elongated stones within till, to determine the former direction of ice flow (with which many stones become aligned during the depositional process). **T f a** has been employed to differentiate the LOWESTOFT and GIPPING TILLS in E. Anglia. In the former the ice-sheet moved predominantly from the west; in the latter it moved from the north and north-west.

tillite A compact and very ancient BOULDER-CLAY, deposited during one of the ice-ages earlier than that of the Quaternary. See DWYKA T.

tilt-block A crustal **b** left upstanding between prominent fault-lines, but bodily tilted, so giving contrasting steepness of its bounding slopes; e.g. the Basin-and-Range country in the Great Basin of Utah; the Sierra Nevada, California and the Grand Tetons, Wyoming; the Central Massif of France (with the Cévennes forming its steep S.E. margin); the Alston Block of N. England (with the steep Cross Fell Edge on the W.).

timberline (in Gt. Britain, **tree-line**) The height on a mountain-side, or the latitudinal limits, at which tree growth ceases. It is rarely an actual line, since it varies with species, latitude and climate, and locally with shelter, aspect and slope. Occasional stunted trees may survive beyond the gen. limit. The t in S.W. USA is at about 3500 m (11 500 ft), up to which grow aspen and Engelmann spruce, and above which is a zone of Arctic flora or TUNDRA. In this area the Douglas fir grows to about 2900 m (9500 ft), the pinyon pine to about 2100 m (7000 ft). Some authorities in USA define: (i) a *dry t*, a lower one, separating the forested slopes above from the semi-arid scrub below; and (ii) a *cold t*, an upper one, separating forested slopes from the Tundra.

time-distance curve, of an earthquake A c produced on a graph (vertical axis, time; horizontal axis, distance), showing the travel of an earthquake from its EPICENTRE. The time intervals between the arrival of the various waves increase with distance, and the curves help to determine the distance from the epicentre and the time of the earthquake. If curves from several stations are available, the position of the earthquake can be accurately determined.

time signal A world-wide radio-signal transmitted, e.g. from Greenwich, at frequent intervals, indicating an exact moment of time for the regulation of chronometers and thus for the calculation of longitude.

time-transgressive Used of the climatic sequence in post-glacial times; the S. boundary of each stage varies with latitude.

time zone A longitudinal division of 15°, or less if a country is small, within which the MEAN T of a meridian near the centre of the z is adopted as standard for the whole, hence STANDARD T [*f*].

tin A metal resistant to corrosion by air, water and most acids derived from its ore, *cassiterite* (tin oxide) (SnO_2), which occurs either as 'alluvial' (or 'placer') deposits which have weathered out from metallic veins in granite (e.g. in Cornwall, Malaya), or directly from lodes (as in Bolivia).

tind (Norwegian) Syn. with HORN, and used thus occas. in English. A common mountain name in Norway; e.g. Glittertind (2484 m, 8150 ft) in the Jotunheim.

tipping-bucket rain-gauge A type of self-recording R-G.

Tissot's indicatrix An application of the law of deformation, developed by M. A. Tissot in 1881, which enables the amount of angular deformation, or of areal exaggeration or reduction, to be determined for any point on a map projection.

tjaele (Swedish) Permanently frozen ground under PERIGLACIAL conditions, containing lenses of ice within the solidly frozen soil. English form = *taele*.

toadstone Sheets of dark-coloured basaltic rock occurring among the Carboniferous Limestone of Derbyshire. The t is impermeable, and thus may create a PERCHED WATER-TABLE among the limestone, of great importance in the siting of villages. The t also seems to be associated with mineral veins near its margins, though it contains no metallic ore itself. Hence one theory as to its name: *todstein* (Germ. for dead or worthless stone), the other being that its dark roughish surface resembles a toad's skin.

tombolo (It.) A bar of sand or shingle linking an island with the mainland; e.g. Chesil Beach joining the I. of Portland to the mainland beyond Weymouth; the island of Monte Argentario, on the W. coast of Italy between Leghorn and Rome, is linked to the mainland by 2 bars, T. della Gianetta, T. di Feniglio. [*f*]

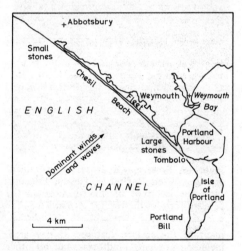

ton A measure of mass: (i) *avoirdupois* (*long t*) = 2240 lb. (ii) *metric* = 2204·6223 lb., or 0·98421 of an avoirdupois t. (= 1000 kilogrammes). (iii) *American* or *short t* = 2000 lb. (20 cwt. of 100 lb. each) = 0·907 of a metric t.

topographic science The collection of measurable data concerning the distribution of features on the Earth's surface, and of the phenomena related to them; i.e. the whole field of surveying, photogrammetry, cartography and certain aspects of geodesy.

topography Derived from the Gk. *topos*, a place. The description (or representation on a

Plate 69. Bowerman's Nose, a prominent granite TOR on Dartmoor, Devon, England. Note the influence of JOINT-PLANES on the detailed form of the tor. (*R. J. Small*)

map) of the surface features of any area, including not only landforms, but all other objects and aspects both of natural or human origin. Hence *topographic, topographical*, as applied to a map of fairly large scale (e.g. 1:63 360) showing **t**. The term is commonly but incorrectly used for relief features alone, even as syn. for relief; this should be strictly avoided. Some American authorities use the term *topographic map* as indicating a large-scale map of a small area, as distinct from a *chorographic map* of scale between 1:500 000 and 1:5 000 000, and a *global map*, on a smaller scale still.

topology A branch of geometrical mathematics concerned with order, contiguity and rel. position, rather than with actual distance and orientation. In much geographical research, topological relationships are expressed in terms of networks (*graph theory*). **T** is sometimes referred to as 'the rubber sheet geometry', since a pattern on such a sheet can be deformed yet points on it remain in the same order or relationship.

topset beds The fine material deposited on the surface of a delta, continuous with the landward alluvial plain. [*f* BOTTOMSET BEDS]

topsoil Used by a gardener or farmer for the top 'spit' or layer of mature soil; that part of the soil which is cultivated.

tor An isolated exposure of much jointed rock, e.g. granite, standing as a prominent castellated mass above the gen. surface of a plateau, notably in Cornwall and Devon; e.g. Haytor, Dartmoor, at 454 m (1490 ft). Ts (sometimes called KOPJES) are also common in many tropical landscapes. The origin of **t**s has been the subject of much controversy. D. L. Linton suggested that they result from sub-surface (rather than subaerial) rotting of granite, through the action of acidulated rain-water penetrating along the joints into the body of the granitic mass. The pattern of the **t** is controlled by the joints, which leave between them broadly rectangular 'core-stones'. Where the jointing is widely spaced, massive core-stones remain; where the jointing is close, there is more shattering and removal, forming depressions between the **t**s. This may have taken place in pre- or interglacial times. Then followed a period of exhumation, when the overlying weathered material and the fine-grained products of the rock-decay were removed by melt-water or SOLIFLUCTION (or in tropical lands wet-season surface wash following reduction of the vegetation cover and/or uplift), thus revealing the 'blockpile' character of the **t**. J. Palmer believes that PERIGLACIAL processes account for the formation of **t**s in Britain. See pl. 69. [*f*]

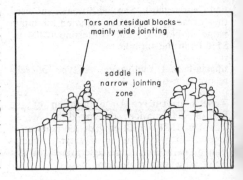

tornado (i) A counter-clockwise whirling storm (sometimes called a '*twister*' in USA), formed around an intensely low pressure system, with winds of great velocity (300 km h^{-1} plus), and often a dark funnel-shaped cloud. It occurs partic. in the Mississippi basin in spring and early summer. It is associated with a trough of low pressure, where cool air from the N. and warm, damp air from the Gulf of Mexico come into frontal contact (commonly along a SQUALL-LINE). Local heating contributes to a vortical uprush of air. It is very short-lived, lasting for only an hour or two, and is usually only a hundred metres or less across. Its destructive effects are limited in extent, but it can cut a swathe across a town. Not only are buildings destroyed by the winds, but the very low pressure at the centre causes them to collapse outwards. T warnings are put out by radio from a centre in Kansas. (ii) A name given to a squall associated with a thunderstorm and torrential rain on the coast of W. Africa. See pl. 70.

torrent A rapidly flowing stream in an upland area; hence *t tract*, or mountain tract, of a stream.

torrid Introduced in classical times for the warmest of the 3 latitudinal temperature zones they designated, meaning 'burning' or 'hot'. Hence *t zone*, the others being temperate and frigid.

Torridonian, Torridon Sandstone The upper of the 2 main Precambrian divisions of rocks in N.W. Scotland and the Hebrides. It is a stratified deposit, often horizontally bedded or with a gentle dip, including red feldspathic grits and sandstone. It is characterized by much false-bedding. The constituent grains are mainly rounded, and it appears to have been laid down in water. It rests unconformably on the denuded surface of the Lewisian Gneiss, the other main Precambrian division.

tower karst See TURMKARST.

trace element A minute quantity of an **e**, such as boron, manganese and iodine, which is usually present in the soil; any deficiency may have adverse effects upon the development of organic cell-structure in plants and animals. Its presence or absence can be detected, but quantitative determination demands refined techniques.

trachographic map Used by E. Raisz for a type of **m** showing ruggedness by means of perspective symbols, using 2 major elements: relative relief and average slope.

trachyte An extrusive, fine-grained igneous rock, composed mainly of alkali-feldspar, with crystals of orthoclase, plagioclase and hornblende; it has a distinctive flow-structure. In composition it is regarded as intermediate. Phonolite is a variety of **t**. Many of the puys of Auvergne, France, are of **t**; e.g. the Puy de Dôme; and there are extensive outcrops in the East African rift valley (Plateau **t**s).

tract (i) A unit of the 2nd smallest order of magnitude in the hierarchy of regional divisions, devised by J. F. Unstead in 1933. (ii) The 3rd order of unit in D. L. Linton's system of MORPHOLOGICAL REGIONS.

traction load, of a river Syn. with the BED-LOAD of a **r**.

Trade Winds Winds blowing from the Subtropical High-Pressure 'cells' towards the Equatorial Low from a N.E. and a S.E. direction in the N. and the S. hemispheres respectively, hence N.E. **T** and S.E. **T** winds. The name comes from the phrase 'to blow trade' (Lat. *trado*), i.e. in a constant direction,

Plate 70. A severe TORNADO passing north of Denver, Colorado. (*Popperphoto*)

and has nothing to do with commerce. The ws are noted for their constancy of force and direction, esp. over the E. side of the oceans, though there are interferences by pressure disturbances (esp. E. waves) near the W. sides of the oceans.

trajectory The path followed by a single 'parcel' of air moving during a period of time. Ct. STREAMLINE

tramontana (It., Sp.) A cold dry N. or N.E. wind in the W. Mediterranean basin; the name is applied commonly to any wind blowing down from the mountains, as in Italy and Spain. It occurs in Corsica in winter behind a depression, commonly bringing snow and bitter weather, and sometimes affects the Balearic Is.

transcurrent fault Syn. with TEAR-FAULT.

transfluence A type of watershed breaching, in which impeded ice escapes from its valley not by a lateral distributary crossing a col into a neighbouring valley (see DIFFLUENCE), but at the head of the valley across a major watershed e.g. in Scandinavia ice moved westwards during the Pleistocene from the Baltic regions across the main N.–S. watershed to the Atlantic coastlands of Norway.

transgression An extension of the sea over a former land-area as a result of a positive movement of sea-level; this may be caused either by a EUSTATIC rise of sea-level, or by an actual sinking of the land. E.g. the Flandrian T of post-glacial times, which created the S. part of the N. Sea, the Strait of Dover and the English Channel (a eustatic rise).

transgressive intrusion Syn. with DISCORDANT I.

transit In surveying and astronomy, the apparent passage of a heavenly body over the MERIDIAN.

transmission capacity The capacity of a soil to allow the passage of water. Sometimes the **t c** will be less than the INFILTRATION CAPACITY of the upper horizons. Thus although rainfall may not exceed infiltration capacity, because of the limited **t c** run-off will occur after saturation of the upper soil.

transpiration The loss of water-vapour from a plant through the minute pores (*stomata*) which cover the leaf-surface. See also EVAPOTRANSPIRATION.

transport (of sediment, etc.) In the whole process of denudation, **t** is that phase which is concerned with the actual movement of material by some natural agent: rivers, glaciers, ice-sheets, the wind, waves, tides and currents. It does not include mass-movement by gravity. The material transported is the LOAD, which itself acts as an eroding agent (abrasion), and suffers progressive diminution by impact both against other parts of the load and the surface over which it is being transported (attrition).

transportation slope A slope (i) developed to provide a gradient over which detritus from above can be transported by wash processes etc. (ii) in which transportational processes alone operate (the two are not mutually exclusive). In W. Penck's theories of slope development basal concavities (or wash slopes) are considered to arise as **t** ss beneath steeper parallel-retreating faces. Many regard PEDIMENTS as **t** ss, across which the fine products of granular disintegration can be washed, without significant erosion or deposition, by ephemeral flows.

transverse coast See DISCORDANT C.

Transverse Mercator Projection A case of the MERCATOR P in which the cylinder is tangential to the globe not along the equator, as in the normal case, but along a meridian; i.e. it has been turned transversely through 90°. The central meridian is divided truly. The **p** is used mainly for maps of small areas with the main dimensions from N. to S. It is used for all British Ordnance Survey maps, and as a basis of the NATIONAL GRID. It is also used for the American UTM GRID. The scale error increases away from the central meridian. Hence on the **T M P** for Gt. Britain the lines of correct scale are transferred 200 km E. and W. of the 2°W. central meridian, resulting in a negative error between them, a positive error outside them. The overall error is thus spread out, and in effect is halved. On this **p** the LOXODROMES are curved lines, not straight as in the normal case. It is also known as the *Gauss Conformal P*.

transverse valley A **v** which breaks across a ridge at right-angles; e.g. a *cluse* in the Jura Mtns.

trap (i) In a cave-system part of which is below the WATER-TABLE, a **t** is formed where the roof of the cave dips down below the water, but rises again above the water-level some distance further on. This involves diving, sometimes necessitating 'frog-man's' apparatus, and is one of the main hazards (and thrills) of speleology. (ii) A gen. name applied to dark-coloured basaltic rocks, notably occurring in DYKES and SILLS. (iii) A structural arrangement in the rocks favourable for the accumulation of oil and or natural gas. [*f*]

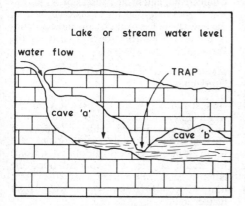

traverse A surveyed line. The easiest method is by using a magnetic compass, hence a *compass t*, entering the observations in a fieldbook. In such a survey method, the error is cumulative, and at the end may be considerable; this error is distributed proportionally along the t. A series of such lines (or 'legs') can be linked together in a circuit to make a *closed t*, from which a map of the whole area can be made.

travertine A crusty deposit of calcium carbonate formed from a strong solution around a hot spring; sometimes it is called *calc-sinter*, or calcareous tufa; e.g. the Mammoth Terraces, Yellowstone National Park, Wyoming, which are magnificent cascades of delicately coloured t.

tree-line The l or zone marking the limit of tree growth, both altitudinally on a mountain and latitudinally as regards distance from the Poles. It is known in USA as the TIMBERLINE.

trellis(ed) drainage A rectilinear pattern of d with CONSEQUENT, SUBSEQUENT, OBSEQUENT and SECONDARY CONSEQUENT streams, usually occurring in scarplands where outcrops of alternately more and less resistant rocks occur at right angles to the initial slope, and where adjustment to structure has occurred. [*f, opposite*]

tremor A minor earthquake with a low intensity; usually termed an *earth-t*.

trench An elongated trough or DEEP in the ocean floor; e.g. the Mariana T. off Guam in the W. Pacific.

trend line The gen. tendency or pattern of the structural ls in an area (e.g. folds, faults, dykes); akin to GRAIN.

trend surface map A m which isolates, measures and depicts quantitative components in any geographical pattern. The 2 methods which can be used are: (i) *filter mapping*, and (ii) *nested sampling*. In (i), where complete data are available, the area is covered with a grid, and the pattern is expressed as a ratio or value for each grid-sq. (e.g. forested/non-forested); isopleths are then interpolated. This gives a response s, resulting from both regional and local factors, which can be filtered to show local anomalies. (ii) Can be used when information is incomplete, as in a reconnaissance or exploratory survey, when a few equal-sized regions are selected at random, and then broken down, also at random, into smaller units. Values for each level can be determined using an appropriate type of variance analysis. Thus by either methods (i) or (ii), the variability of areal patterns can be broken down and sampled. T s mapping has been used to indicate the possible existence of a dissected surface and to reconstruct former EROSION SURFACES; however, this application has been criticized as misleading on the grounds that the 'residuals' *above* the t s commonly bear the remnants of the former surface, which does *not* coincide with the contours of the t s m.

triangle of error In finding the position of a point on a plan by RESECTION [*f*], using a plane-table, if three intersecting back-rays are drawn from known objects, a t of e will result. (i) If the observer's position is within the t formed by the 3 known points, the position on the plan will lie within the t of e. To fix this more exactly, the true position will be vertically away from each ray a distance proportional to its total length. (ii) If the observer's position is outside the t formed by

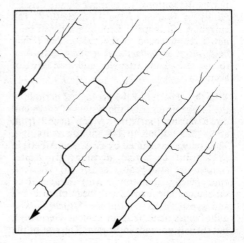

the 3 known points, the true position will be either to the right or to the left of all 3 rays, at a vertical distance from each proportional to the length of each. In each case, estimate the true position by eye, realign the planetable, and check; the **t of e** should have been eliminated, or at least be much smaller, in which case the operation is repeated.

triangular diagram The plotting of 3 related or associated aspects of some feature or item on **t** graph-paper, ascribing a max. value for each aspect to each apex of the triangle; e.g. 3 aspects of climate (pressure, temperature, humidity), population age (young, middle-aged, old), sediment (sand, clay, silt), hillslope analysis (% of slope-angle observations falling within summit, midslope and basal portions, as used by A. F. Pitty). [*f*]

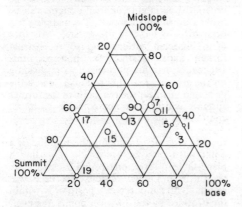

triangulation A system of triangles developed by angular measurement with a theodolite from a BASE-LINE, thus affording an accurate geodetic framework for a topographical survey. The principles of **t** are lost in antiquity, but the 1st national **t** was made by the Cassini family in France in the 18th century. The current O. S. maps in Britain are based on a new **t**, commenced in the 1930s. The **t** may be *primary*, *secondary* and *tertiary*, according to the size of the triangles and standards of accuracy.

Triassic The 1st of the geological periods of the Mesozoic era, and the system of rocks laid down during that time, which lasted from about 225 million to 180 million years ago. The system was named by F. A. von Alberti in 1834 from its 3-fold division: (i) *Bunter* (sandstones and pebble beds); (ii) *Muschelkalk* (shelly limestone); (iii) *Keuper* (red sandstones and marls, with beds of rock-salt and gypsum). In W. Europe *Muschelkalk* is quite widespread (e.g. in central Germany), but there is no trace in Britain. The rest of the **T** is strikingly developed in the Bristol Channel area, the W. Midlands, S.W. and N.E. Cumberland, and in small areas of N. Ireland The **T** rocks are difficult to distinguish from the underlying Permian system, and commonly the term New Red rocks is used for the Permo-Trias. See RHAETIC.

tributary A stream or river which joins a larger one.

'trigger action' In an AIR-MASS, any process which initiates the development of CONDITIONAL INSTABILITY. Forced uplift of a pocket of air (e.g. orographic uplift, or the rising of a warm air mass over a cold air mass at a front) will cause cooling and condensation: the latter process will release heat (the latent heat of condensation) which will trigger further uplift.

trim line A sharp change of vegetation cover, related to the advance and retreat of glaciers. In Alps **t** ls, separating areas of bare lateral moraine below and well-vegetated older moraine above, were formed by the advance of the glaciers during the 'Little Ice Age' and the major recession since 1850.

tropic One of the 2 parallels of latitude of approx. 23½°N. and S. (see CANCER, CAPRICORN). In the pl., broadly the zone between these 2 parallels, hence *tropical*, app. to this zone, as -forest, -grassland, -climates.

tropical air-mass An a-m which has originated within the sub-tropical ANTICYCLONE belt, either over the ocean (*tropical maritime*) (*Tm*, or *mT* in USA), or over a continental interior (*tropical continental*) *Tc* or *cT*).

tropical climates A group of climatic types occurring within the tropics, but of varying definition. (i) W. Köppen defines **t** cs as having an average temperature for each month above 18°C (64·4°F), thus no winter season, with a considerable precipitation, mostly of a convectional nature with a marked summer max. Köppen includes three main types: *Af*. **T** Rain-forest; *Am*. **T** Monsoon; *AW*. **T** with dry winter, hence SAVANNA. (ii) A. A. Miller's classification includes (*a*) T. Marine; (*b*) T. Marine (Monsoon); (*c*) T. Continental (summer rain); (*d*) T. Continental (Monsoon).

tropopause The discontinuity plane between the TROPOSPHERE and the STRATOSPHERE, marked by an abrupt change in the lapse-rate; it varies in height slightly with the seasons, but lies at about 18 km (11 mi.) above the equator, 9 km (5·5 mi.) at latitude 50°, and 6 km (4 mi.) at the Poles. Recent research shows that the **t** is not a single plane, but a series of overlapping planes.

tropophyte A plant, not XEROPHYTIC, which possesses various adaptations (e.g. leaf-shedding) to enable it to survive a period of seasonal adversity (cold or drought). It behaves as a XEROPHYTE at one period of the year, as a HYDROPHYTE at another.

troposphere The lower part of the atmosphere, from the surface of the earth to the TROPOPAUSE.

trough (i) An elongated TRENCH or DEEP in the ocean floor. (ii) A U-shaped valley, as a glacial **t**; e.g. Lauterbrunnen, Switzerland; Yosemite, California. These **t**s are of varied size and form, and have been usefully classified by D. L. Linton into four types: (*a*) *Alpine* where the **t**s occupy pre-glacial valley systems; (*b*) *Icelandic*, where plateau-accumulation of ice rather than high ground has led to discharge into peripheral valleys; (*c*) 5 sorts of *composite* types; and (*d*) *intrusive* **t**s, which have been formed by lowland ice intruding into and through uplands. (iii) A depression between the crests of two successive waves in the sea. (iv) A narrow elongated area of low atmospheric pressure between 2 areas of higher pressure.

trough-end A steep rock-wall forming the abrupt head of a glaciated valley; e.g. Warnscale Bottom, the head of the Buttermere valley, English Lake District; the **t-e** at the head of the Rottal, under the S.W. face of the Jungfrau, Switzerland.

true dip The max. **d** of a stratum, as ct. with APPARENT D. [*f* STRIKE].

true north The direction of the Geographical N. Pole along the meridian through the observer. Ct. MAGNETIC N, GRID N.

truncated spur A **s** which formerly projected into a pre-glacial valley, partially planed off by a glacier which moved down the valley; e.g. the **t** ss of Saddleback (Blencathra), a mountain lying E.N.E. of Keswick in the English Lake District. [*f, opposite*]

tsunami (Japanese) A large-scale seismic sea-wave (incorrectly known as a tidal wave), caused by an earthquake shock in the ocean floor. It may travel for long distances over the sea, then surge over the land margins, where it may cause considerable damage. In the open ocean the wave height may only be a foot or so, but the wave-length may be 80–190 km (50–120 mi.), travelling at 640–960 km h^{-1} (400–600 m.p.h.). On entering shallow water the wave steepens rapidly up to 15 m (50 ft) or more. The eruption of Krakatoa in 1883 caused a **t** to travel right round to the other side of the world. Another **t** off Japan in 1896 drowned 27 000 people and destroyed nearly 11 000 houses.

tufa (It.) A porous, cellular or spongy type of TRAVERTINE, deposited around the point of issue of a spring of calcareous water.

tuff A rock formed from a compacted or cemented mass of fine volcanic ash and dust, with particles less than 4 mm in diameter. Ct. BRECCIA. See also WELDED TUFF. See pl. 20.

tuffisite Fragmentary material produced by explosive volcanic activity, but deposited in the PIPE, rather than over the surface as TUFF. E.g. in the Swabian Jura, W. Germany.

tundra (Lapp) A zone between the N. latitudinal limit of trees and the polar regions of perpetual snow and ice, experiencing a climate with a brief summer above freezing-point (W. Köppen's *ET* type), in which the mean temperature of the warmest month is below 10°C (50°F), but above 0°C (32°F). It is characterized by PERMAFROST, a vegetation cover of dwarf shrubs, berried plants such as cloudberry in Norway and cranberry in Canada, herbaceous perennials, lichens (e.g. *Cladonia*) and mosses, with no trees other than a few stunted Arctic willow and birch, and with widespread marshland in summer. The pattern of vegetation in detail is very varied, in response to slight differences in slope, aspect and drainage.

tungsten A metal mostly obtained from the minerals wolframite and scheelite, commonly found in VEINS cutting through metamorphosed sedimentary rocks.

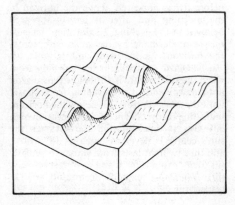

tunnelling The process of sub-surface washing out of soil along seepage lines or PERCOLINES. T may cause soil subsidence, collapse of the turf mat, and the initiation of surface depressions and gullies.

turbidity The churning up of sediment by water, forming a dense, heavy flow, the sediment remaining in suspension, esp. in the sea. This applies partic. to bottom currents on the ocean floors, which are believed to have important erosional effects on the continental slope, and also to currents formed when rapidly flowing, sediment laden, river waters enter a lake. See SUBMARINE CANYON.

turbulence (i) An irregular eddying flow, in ct. to a smooth laminar flow, partic. used in a meteorological context for the flow and mixing of air by this means. Though it is easy enough to say what it is in gen. terms, the mathematical definition and explanation of the phenomenon is a difficult problem. Instability due to the uneven heating of the Earth's surface may cause the rapid rise of 'pockets' of air. This may make an aircraft distinctly uncomfortable, as it abruptly rises and falls; see THERMAL. T can usually be recognized by the presence of CUMULONIMBUS clouds and thunderstorms, but '*clear-air*' t may occur, esp. in a JET-STREAM, which is difficult for a pilot to avoid. (ii) In a stream t contributes to the transport of material in suspension. (iii) T is a marked feature of an ocean DRIFT, which has an overall direction of water movement, but contains irregular speeds and directions within it.

turm karst (Germ.) A karst landscape dominated by vertical walled MOGOTES standing above alluviated plains e.g. in parts of S. China and Vietnam.

twilight The reflection of the sun's light before it has appeared above the horizon in the morning and after it has disappeared below it in the evening. Its duration depends on the angle made by the sun's path across the horizon, i.e. on date and latitude. (i) *Astronomical t* extends in the morning from when the sun's centre is 18° below the horizon until dawn, and in the evening from sunset until the sun's centre is 18° below the horizon, indicating theoretical perfect darkness. In high latitudes at times around the SOLSTICE the sun's centre is never as much as 18° below, and thus twilight lasts from sunset to sunrise. (ii) *Civil t* extends from and to 6° respectively. (iii) *Nautical t* extends from and to 12° respectively.

tychoplankton Minute organisms, animal and vegetable, that have been transported from the margins of lakes and ponds into those bodies of water by currents.

typhoon A small, intense, vortical tropical storm in the China Sea and along the margins of the W. Pacific Ocean, accompanied by winds of terrific force (160 km h^{-1}, 100 m.p.h. plus), torrential rain and thunderstorms. Cf. CYCLONE, HURRICANE.

ubac (Fr.) A hill-slope, esp. in the Alps, which faces N. or N.E., and so receives minimum light and warmth; cf. Germ. *Schattenseite*. It is usually left under forest, while settlements, terraced agriculture, and meadows (ALPS) occur on the S. and S.W.-facing slopes. Ct. ADRET.

Uinta structure A type of structure named after the Uinta Mtns in N.E. Utah, where a broad flattened anticlinal flexure was upraised in late Cretaceous times, and then extensively denuded, so that Precambrian rocks (sandstones and quartzites) are exposed over most of its surface. At the end of the Eocene, the worn-down mass was again bodily uplifted, with major faults flanking it to S. and N. The Uinta Mtns experienced further uplift during late Pliocene-Pleistocene times; their highest point rises to 4114 m (13 499 ft) in King's Peak.

ultrabasic rock An igneous r containing less than 45% silica, and more than 55% basic oxides, mainly ferromagnesian silicates, metallic oxides and sulphides; e.g. peridotite, consisting mainly of olivine.

ultraviolet rays That part of the solar RADIATION which lies just beyond the blue end of the visible spectrum, with a wave-length range from 4×10^{-5} cm to 5×10^{-7} cm. Much of the u light is absorbed by ozone molecules in the upper atmosphere, but some reaches the Earth's surface; the intensity is most marked in high mountains.

unconformity A break or gap in the continuity of the stratigraphical sequence, where the overlying rocks have been deposited on a surface produced by a long period of denudation. Strictly, an overlying stratum does not conform (i.e. is *unconformable*) to the dip and strike of the underlying strata.

See also NONCONFORMITY, DISCONFORMITY, ANGULAR U, which are various types of **u**. [*f*]

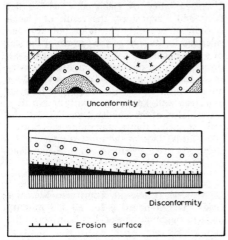

Unconformity

Disconformity

Erosion surface

underclay See SEAT-EARTH.

undercliff A mass of material at the base of a cliff, the result of falls due to weathering; this is partic. pronounced where chalk lies over clay. E.g. the U along the coast of the I. of Wight to the S.W. of Ventnor; along the Devon coast between Seaton and Lyme Regis, formed by the settlement of a mass of Chalk and U Greensand, estim. to weigh 8 million tons, on Christmas night 1839.

undercutting Partic.: (i) the erosive action of a river current as it impinges against its bank on the outside of a bend; hence undercut slopes; (ii) the erosion by sand-laden wind near the base of a rock in the desert; (iii) of coastal cliffs; e.g. the Dorset coast, where removal of the Lower Portlandian Sands has undercut the overlying Purbeckian Beds.

underfit App. to a stream: syn. with MISFIT.

underground stream In areas of rock characterized by joints and fissures, water flow below ground may be concentrated in a distinct channel. The courses of such **u** ss, from SINKHOLE to RESURGENCE, have been traced in a number of cases, by actual exploration, by putting fluorescein (green colouring) into the water, and more recently by the employment of lycopodium spores. These are frequently far more complicated than one would suspect from the surface; e.g. the R. Aire at Malham, Yorkshire. N. Casteret proved that the Garonne rises in Spain on the S. side of the main Pyrenean watershed, which it penetrates as an **u s**.

undertow An undercurrent flowing down the beach near the bottom of the water, the result of a back-flow of water piled up on the beach by a breaking wave.

unequal slopes, Law of A principle which states that where the opposing ss of a ridge are steep and gentle respectively, the former will be eroded more rapidly, and the ridgeline will therefore move back on that side. This process is involved in the recession of an ESCARPMENT and the development of a COL.

uniclinal Used by some authorities for beds of rock dipping uniformly and evenly in one direction; hence *u structure*. Others use 'monoclinal', but this is inadvisable, as the word MONOCLINE has specif. implications.

uniclinal shift If the cross-profile of a river valley is asymmetrical, esp. one which lies along the line of the STRIKE in an area of gently dipping rocks, the course of the river may tend to migrate down the DIP, and the escarpment forming the valley-side on the down-dip side will move sideways in a similar manner. [*f*]

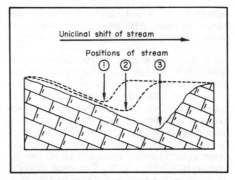

Uniclinal shift of stream

Positions of stream
① ② ③

uniformitarianism An important geological principle, established by J. Hutton in 1785, and formulated by C. Lyell in 1830, that processes and natural laws which existed in geological time are sensibly those that may be observed today. This is contrasted with CATASTROPHISM, which holds most physical phenomena to be the result of short-lived and exceptional events.

unloading See PRESSURE RELEASE.

unroofed anticline See BREACHED A.

unstable equilibrium The state of the atmosphere where the ENVIRONMENTAL LAPSE-RATE of an air mass (sometimes called the 'Environment Curve') is greater than the DRY ADIABATIC LAPSE-RATE. A surface pocket of unsaturated air, if heated, will rise, cool at the

DALR, and because it is still warmer than the surrounding air mass will continue to rise. Rising pockets of moist air will cool even more slowly (at the SATURATED ADIABATIC LAPSE-RATE) and will therefore tend to be even more unstable, producing large CUMULUS clouds, possibly causing heavy rainfall, hail and even thunderstorms. Vertical ascent will cease when it reaches the same temperature as the surrounding air, and it is then in NEUTRAL (or indifferent) E. Cf. CONDITIONAL INSTABILITY, and ct. STABLE E.

upfreezing The upward movement of stones, within finer soil, by frost action. Various mechanisms of u may include (i) expansion of the ground upwards upon freezing, raising stones and fines equally, but preferred sinking of fines on thawing; (ii) the collapse of fines into voids beneath the upthrust stones; (iii) the maintenance of ice 'pedestals' beneath upraised stones while surrounding fines thaw and collapse.

upland A gen. term for higher land, in contrast to lowland, with no specif. connotation.

upland plain A not very satisfactory term for a comparatively level area of land at some altitude; a high-lying planation surface.

UPS Grid (Universal Polar Stereographic Grid) The US military g for areas between 80°N. and the N. Pole, and 80°S. and the S. Pole, complementary to the UTM G. It is superimposed on a Polar STEREOGRAPHIC PROJECTION bounded by the parallel 80°, and divided into 100-km squares, with their sides parallel to a line joining 180° to 0° longitude (central meridian), crossed at right-angles by that joining 90°E. and W. The origin of the g is the Pole, the centre of the projection; the FALSE ORIGIN is transferred 2 million m W. and 2 million m S. in the N. Zone, 2 million m W. and 2 million m N. in the S. Zone.

Urstromtal, pl. **-täler** (Germ.) Lit. an 'ancient stream-valley'. A broad shallow trough or depression, eroded by a melt-water stream flowing along the front of the continental ice-sheets which lay at various stages across N.W. Europe. As the ice-sheets retreated, so a series of U was successively formed; 5 main lines can be traced across Germany and Poland. The post-glacial rivers (Elbe, Oder, Vistula) flow in a general direction from S.E. to N.W. across the N. European Plain to the North Sea. They occupy various sections of the U, then continue sections of their N.W. courses, which helps to explain the frequent 'elbows' in their courses. Some of the U are dry, others have facilitated the construction of sections of Germany's '*Mittelland*' Canal.

U-shaped valley A glaciated v, with a flat floor and steep sides, the result of glacial erosion of not only the floor but also the sides (up to the level of the surface of the ice) of a pre-glacial river v. The v is straightened, projecting spurs are planed off, high tributary vs are left hanging, and rock-steps in the floor are common. The v ends abruptly at its head in a steep wall, known as a TROUGH-END. E.g. the radial vs of the English Lake District; the Scottish glens; the vs containing the Norwegian fjords. Two of the most spectacular U-shaped vs are Lauterbrunnen in Switzerland and Yosemite in California. See pl. 71.
[*f* ALP]

UTM Grid (Universal Transverse Mercator Grid) The standard g for all US military maps throughout the world. It is drawn on the TRANSVERSE MERCATOR PROJECTION between 80°N. and 80°S. (see UPSG). The world is divided into 60 grid-zones, each 6° of longitude (with $\frac{1}{2}$° overlap into adjacent zones on each side). Each zone is numbered, 1 to 60, beginning at 180°W. and moving E.-ward. Each latitudinal division of 8° is lettered from C (80°S. and 72°S.) to X (72°N. to 80°N.). Hence a quadrilateral is known by a figure and letter; e.g. 4Q is between longitude 162°W. and 156°W., 16°N. and 24°N. A 2nd degree of reference is given in respect of squares of 100 km, indicated by 2 other letters. The 3rd degree of reference is within each grid-zone, with 6 figures; i.e. a full world reference could be 4QXC264839. A km g is drawn within each zone, based on the central meridian and the equator. The FALSE ORIGIN for each zone is 500 km W. of the central meridian on the equator for the N. hemisphere, 10 000 km S. of this for the S. hemisphere.

uvala (Serbo-Croat) A hollow in limestone country, larger than a DOLINE (and commonly formed by the coalescence of several of these), but smaller than a POLJE. Its floor is more uneven than that of a polje.

vadose water 'Wandering' w moving through permeable rock above the WATER-TABLE. Ct. GROUND W.

val, pl. **vaux** (Fr.) A valley; specif. a longitudinal valley in the Jura Mtns. [*f* CLUSE]

vale A somewhat poetical name for a valley, used commonly in place-names; e.g. V. of York, Evesham, Lorton (English Lake District), Clwyd (N. Wales).

Plate 71. A glaciated U-SHAPED VALLEY (Strath More) in Wester Ross, Scotland. The flat floor of the valley has been emphasized by AGGRADATION in the POST-GLACIAL period. (*Eric Kay*)

valley An elongated depression sloping towards the sea or an inland drainage basin, usually though not always occupied by a river. See LONGITUDINAL, TRANSVERSE, RIFT-, U-SHAPED, HANGING, and V-SHAPED VS.

valley-glacier A tongue of ice moving downward and outward from a FIRN-field, along a pre-existing **v**; it is sometimes called an alpine or mountain **g**. See GLACIER.

valley train A line of fluvioglacial outwash material deposited in a **v** by melt-water from glacier-snout.

valley wind See ANABATIC, KATABATIC W.

vallon de gélivation (Fr.) A small valley formed not by stream erosion, but by the widening of lines of structural weakness (e.g. close-set joints) by frost disintegration and evacuation of debris by solifluction. Some chalk dry valleys in S. England may be **v de g**s, since their long and cross-profiles show little evidence of occupation by former streams.

valloni (It.) A long narrow gulf or channel in the Adriatic Sea between the islands of the Dalmatian coast of Yugoslavia; syn. with *canali*. They were formed by a rise of sea-level in an area of mountains and valleys parallel to the coast, thus turning the valleys into gulfs of the sea.

vanadium A metal widely dispersed as a mineral-ore in many igneous and sedimentary rocks, though in small quantities. Most is used in steel-alloys; it helps to make the metal less prone to fatigue under stress.

Van't Hoff's rule The law stating the rate of increase of chemical reactions with rise of temperature. In broad terms the increase is 2–3 times for every 10°C rise of temperature.

vapour-pressure The **p** exerted by the water-**v** present in the atmosphere. Ct. the **v-p** in England (7–15 mb.) and near the equator (30 mb.). The max. **v-p** at any temperature obviously occurs when the air is saturated. *V-tension* is syn., though obsolescent.

vapour trail See CONDENSATION T.

variation, magnetic See MAGNETIC DECLINATION.

Variscan App. to the mountain-building movements of late Palaeozoic (Carbo-Permian) times. Sometimes the name is given in central Europe to the equivalent of the ARMORICAN in the W.; sometimes it is regarded as syn. with HERCYNIAN and includes all the European ranges of this age, hence *Variscides*. The name was first used in 1888 by E. Suess, after the Germanic tribe of the *Varisci*. It seems preferable to regard the V uplands as the E. representatives of the Hercynian, the Armorican as the W. ones.

varve A distinctive banded layer of silt and sand, deposited annually in lakes ponded near the margins of ice-sheets; the coarser material, lighter in colour, settles first during summer melting, the finer darker deposits in winter. Each band of light and dark material represents one **v**. Thus it is possible to count the number of vs and so find the number of years involved in the formation of the **v** deposit; as **v** characteristics are fairly distinctive and recognizable in relative thickness for each individual year's sedimentation, correlations over quite wide areas are possible, affording a contribution to the glacial chronology. The concept was worked out by the Baron G. de Geer in Sweden, and first published in 1910.

Vauclusian spring The resurgence or reappearance of an underground stream, called after the Fontaine de Vaucluse in the lower Rhône valley. It occurs commonly in limestone country, where water wears subterranean ramifications, finally issuing from the limestone at its base. E.g. the R. Axe, issuing from Wookey Hole, Mendips, Somerset.

vector analysis A statistical method of analysing changes in the amount and direction of movement. A **v** is a quantity which has direction as well as magnitude (e.g. a wind, a tidal current), and can be denoted by a line drawn in stages from an original to a subsequent position. E.g. a diagram of a tidal current may comprise a continuous series of lines (a *v traverse*), each representing the hourly observation drawn on a correct bearing, and each proportional in length to the mean velocity during that hour. Another simple form is a WIND-ROSE.

veering, of wind (i) A change of direction of the w in a clockwise direction; e.g. from N.E. to E., to S.E. Ct. BACKING. (ii) In USA the term is used as above in the N. hemisphere, but in the reverse (i.e. anticlockwise) in the S. This is not gen. usage, which implies a change in a clockwise direction in both hemispheres.

vegetation The living mantle of plants (*flora*) which covers much of the land-surface, forming an important aspect of the physical environment. Hence **v** regions, NATURAL V, SEMI-NATURAL V, cultivated **v**, **v** types. V is controlled by various groups of factors: (i) *climatic* (temperature, moisture, wind, light); (ii) *edaphic* (soil conditions); (iii) *physiographic* (relief, aspect and drainage); (iv) *biotic* (the effects of organisms); and (v) *human* (burning, felling of trees, drainage, irrigation, grazing by domestic animals).

vein A fissure or a crack in a mass of rock containing minerals deposited in crystalline form, frequently associated with metallic ores. Syn. with LODE, the miners' term.

veld (Afrikaans) An area of open grassland on the African plateau, of great variety, including the *High* (above 1500 m, 5000 ft), *Middle* (1500–900 m, 5000–3000 ft) and *Low v* (below 900 m, 3000 ft), *Grass v*, *Bush v*, and *Sand v* (semi-arid).

vent An opening in the surface of the Earth's crust, through which material is forced during a volcanic eruption.

ventifact A boulder, stone or pebble, worn, polished and faceted by windblown sand in the desert. See DREIKANTER, EINKANTER.

verglas A thin layer of clear, hard, smooth ice which forms over rocks, the result of a hard frost following rain or snow-melt; this adds considerably to climbing difficulties.

vernier An auxiliary scale on a ruler, barometer, theodolite, etc., which enables readings to be taken to an additional significant figure. E.g. if a scale is graduated in 10ths of an inch, the **v s** is a length of 9/10 of an in., divided into 10 equal parts. The **v** is slid until its zero is opposite the quantity to be measured. By observing which division of the **v s** is exactly opposite a division on the main scale, the second decimal place can be read off.

vertical exaggeration (V. E.) A deliberate increase in the **v** scale of a SECTION, in comparison with the horizontal scale, in order to make the section clearly perceptible. E.g. if the horizontal scale is 1 in. to 1 mi. (5280 ft) and the vertical scale is 1 in. to 1000 ft, the **v e** is 5·28 times.

vertical interval The difference in **v** height between 2 points; see GRADIENT. A 1° rise in a HORIZONTAL EQUIVALENT of 100 ft involves a **v i** of 1·74 ft.   [*f* SLOPE-LENGTH]

vertical photograph An aerial **p** taken from as **v** a position as possible; the ct. is with an *oblique p*

vertical temperature gradient See LAPSE-RATE.

vesicular, vesicule (noun) Applied to the

texture of a rock containing many small cavities (vesicles), the results of the presence of bubbles of steam or gas in molten rock (LAVA) as it cooled. See AMYGDALE.

virga The trailing shreds of cloud under the low dark line of a passing FRONT.

virgation A bunching and divergence of fold-ranges from a central 'knot'; e.g. the Pamirs in central Asia, Pasco in the Andes.

viscous App. to a fluid, specif. LAVA with a high melting-point, a stiff, pasty consistency and rich in silica; it solidifies rapidly, does not flow far, and builds up a high, steep-sided cone.

visibility The distance which an observer can see, depending on: (i) his height above sea-level (see HORIZON), with which is involved the curvature of the Earth's surface; (ii) the amount of DEAD GROUND; (iii) the clarity of the atmosphere; see FOG, MIST, HAZE; (iv) the time of day or night.

vitrain Thin horizontal bands of strata in bright glossy coal (from Fr. *vitre*, glass), fracturing easily at right-angles to the bedding.

volcano Popularly a conical hill or mountain, built up by the ejection of material from a vent; better, a volcanic peak. The adj. *volcanic* is correctly used to signify all types of extrusive igneous activity, as opposed to PLUTONIC; the noun *volcanicity* is also sometimes used. The word's highest extinct v is Aconcagua (6959 m, 22 834 ft) in the Andes; the highest active v is Guayatiri (6060 m, 19 882 ft), also in the Andes, which erupted in 1959. See under specif. volcanic feature: AGGLOMERATE, ASH, BOMB, CALDERA, CINDERS, CRATER, DOME, DUST, ERUPTION, PIPE, PLUG, SPINE, THOLOID, TUFF. See pl. 24.

volume A measure of bulk or space. In Gt. Britain, USA, etc., different scales are used for solids and liquids.

Solids
1728 cu ins. = 1 cu ft
1 cu in. = 16·387 cm^2
27 cu ft = 1 cu yd

Liquids
4 gills = 1 pint (0·5682 ltre)
2 pints = 1 quart
4 quarts = 1 Imperial gallon (4·546 litres)
2 gallons = 1 peck
4 peck = 1 bushel
8 bushels = 1 quarter

Metric
1000 mm^3 = 1 cm^3
1000 cm^3 = very nearly 1 litre

Conversions
1 m^3 = 35·315 cu ft
1 litre = 0·22 gal.
1000 litres = 1 m^3

V-shaped valley Gen. a valley eroded by a river, as opposed to most U-shaped glaciated valleys. The angle of the V depends on: (i) the rate of river incision (a rapid uplift will promote a narrow V); (ii) the resistance of the rocks to the weathering of the containing slopes; (iii) climate (a humid climate will favour rapid mass movements and an open V); (iv) the stage of the cycle of river erosion, whether young (a steep V), or mature (a wider, more open V). In old age the V is replaced by a broad, almost level valley, bounded by low bluffs which may be far back from the river itself.

vulcanicity, vulcanism The processes by which solid, liquid or gaseous materials are forced into the Earth's crust and/or escape on to the surface. This includes igneous activity gen., besides that popularly associated with volcanoes.

wadi (Arabic) A steep-sided rocky ravine in a desert or semi-desert area, usually streamless, but sometimes containing a torrent for a short time after heavy rain. E.g. in Arabia. See pl. 20, 72.

wake-dune A sand-d which may occur on the lee side of a larger **d**, trailing away in the direction of wind movement. [*f* DUNE]

Wallace's Line A l between S.E. Asia and Australia, demarcating the distinctive Asiatic and Australian flora and fauna. [*f*]

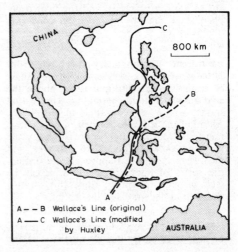

310 WALL

Plate 72. A WADI system in the Jebel Ram, Jordan. Note the extensive ALLUVIUM, deposited by streamfloods, and the pattern of dry braided channels. (*Aerofilms*)

wall-sided glacier A valley-g which flows out on to a plain, so forming a steep-sided tongue of ice unconfined by a valley.

waning slope Used by W. Penck and A. Wood for the gentle s of finer material (sometimes called the '*wash slope*'), accumulated at the foot of a CONSTANT s. These low-angle ss are said to become more and more dominant as erosion proceeds, and the final peneplain stage may be regarded as consisting of coalescing w ss of progressively gentler angles. [*f* FREE FACE]

warm front The boundary-zone at the front of the w sector of a DEPRESSION, where a mass of w air is overriding and rising above the cold air which it is overtaking. The frontal surface is at a very low angle of only $\frac{1}{2}°-1°$. Ahead of the w f, a broad belt of continuous rain falls from a heavily overcast sky, while the wind BACKS before the w f arrives, and then veers. As the w f approaches, a distinctive series of thickening cloudforms can be seen: cirrus, cirrostratus, altostratus, stratus, and then nimbostratus as the w f passes. [*f* DEPRESSION]

'**warm glacier**' A g where the ice-mass is at or near 0°C, partly due to warming by the percolation of melt-water produced by conduction heating at the surface. Thus a '**w g**' may approach 0°C throughout its mass in summer, though in winter superficial cooling will produce a very cold crust. These physical facts play an important part in g motion. Called a '*temperate g*' by H. W. Ahlman. Ct. 'COLD GLACIER'.

warm occlusion An o where the overtaking cold air is not as cold as the air-mass in front. [*f* OCCLUSION]

warm sector A bulge or 'bay' of w air in the S. part of a depression (in the N. hemisphere), its E. edge being a w FRONT, its W. edge a COLD FRONT. Within the w s over Britain, the airflow is between S. to S.W., with a complete cloud-cover, and continuous drizzling rain, which gradually becomes heavier and concentrated in showers as the cold front approaches. [*f* DEPRESSION]

warp, warpland Sediment deposited in a tidal estuary; e.g. on the shores of the Solway Firth, Cumberland.

warping (i) A gentle deformation of the crust over a considerable area. Though it may involve a vertical movement of only a few ft, it may have important results on the surface of an uplifted peneplain, or along a coast. The Central Massif of France shows several warped surfaces intersecting each other at low angles. There has been w in the Gt. Lakes region of N. America during uplift. One type of w is the result of ISOSTATIC depression (*down-w*) or uplift (*up-w*), the latter esp. after the melting of an ice-sheet; e.g. in Scandinavia. (ii) The deposition by flooding of a layer of mud and alluvium over the low-lying land adjacent to an estuary. Hence WARP, WARP-LAND.

warren An area of waste land, often consisting of sand-dunes, in former times reserved for breeding game, esp. rabbits. E.g. Dawlish W., near the mouth of the R. Exe, Devon.

Warthe A stage of the Quaternary glaciation in N.W. Europe, the moraines of which can be traced in the Fläming (to the S.W. of Berlin) and in Poland (esp. Upper Silesia). Some authorities contend that the W was an early stage of the 4th (Weichsel) main glaciation, but the consensus of opinion is that it represents a temporary halt and slight readvance during the retreat of the ice-sheets towards the end of the 3rd (Saale) glaciation.

wash (i) Fine material moved down a slope, esp. where there is little vegetation to fix and hold it; sometimes called *downwash*. (ii) The movement of water up a beach after a wave breaks; ct. BACKWASH. (iii) An area of tidal sand- and mud-banks. (iv) A term applied in the S.W. of USA to a shallow, dry, stream-channel in the desert.

washboard moraine A series of closely spaced and approx. parallel small morainic ridges. Mechanisms of formation may include: (i) concentration of till at the base of thrust planes within the ice; (ii) the pushing forward of moraine below a steep ice-front with each winter advance; (ii) the squeezing out of sub-glacial till into an overhanging crevice formed each season at the base of an ice-cliff terminating in a lake.

washland Embanked low-lying lands bordering a river or estuary, deliberately allowed to flood so as to cope with high water-level in the river; e.g. the land between the Old and New Bedford Levels in the Fen District. W can be used for grazing at some times of the year.

washout (i) The result of a sudden concentrated downpour of rain, causing extensive scouring, the sweeping away of bridges, and the undermining of the river-banks. (ii) A gap in a coalseam, usually filled with sandstone; it represents concentrated stream erosion during or after the formation of the coal-seam; a stream similar to a distributary in a delta formed a channel, which was later filled with sand.

wash trap A trough sunk into a slope surface in such a way that sediment washed from above is trapped. If the area 'feeding' the trap is known, the rate of surface erosion by rainwash can be calculated. The use of w ts has shown that slope retreat by wash processes alone is less than 0·01 mm/year in cool temperate environments, but may be 10 times as high in Mediterranean climates where the summit drought restricts vegetation.

wastage (i) The loss of ice in a glacier or ice-sheet; a better term than shrinkage. It is sometimes regarded as syn. with ABLATION. (ii) A gen. term for the denudation of the Earth's surface.

waste mantle See REGOLITH.

waterfall A steep fall of river water, where its course is markedly and suddenly interrupted. This may be the result of: (i) a transverse bar of resistant rock across the river's course, interrupting its progress to a graded profile; e.g. the Nile cataracts, Niagara, Kaieteur Falls (Guyana), Gibbon Falls (Yellowstone National Park, Wyoming); (ii) a sharp well-defined edge to a plateau; e.g. Aughrabies Falls in S.W. Africa, where the Orange R. crosses the edge of the African plateau; (iii) faulting, later forming a FAULT-LINE SCARP; e.g. Gordale Scar, Malham, Yorkshire; Victoria Falls, due to faulting in part; (iv) the presence of a deep glaciated valley, with HANGING tributary valleys; e.g. Yosemite Falls, California; the falls (e.g. Staubbach) of the Lauterbrunnen valley, Switzerland; (v) along the edge of a cliffed coast; e.g. Litter Water on the Devon coast near Hartland.

Some Major Falls

	m	(ft)
Angel F., Venezuela	979	3212
Sutherland Fall, New Zealand	580	1904
Ribbon F., Yosemite, USA	491	1612
Upper Yosemite F., USA	436	1430
Uitshi, Guyana	366	1200
Staubbach, Switzerland	264	866
Vettisfoss, Norway	260	852
Kaieteur, Guyana	251	822
Victoria, Africa	110	360

Note: (*a*) The Yosemite Falls include the Upper Falls (436 m, 1430 ft), the intermediate cascades (248 m, 815 ft), and the Lower Falls (98 m, 320 ft); total: 782 m, 2565 ft. (*b*) The Victoria Falls on the Zambezi R. are 1372 m

312 WATERFALL

Plate 73. The Iguacu Falls, southern Brazil. Note the stepped nature of the falls, related to the horizontal geological structure. (*Popperphoto*)

(4500 ft) in width. (*c*) Niagara Falls descend 51 m (168 ft), with 2 distinct falls, the Canadian or Horseshoe Falls with a frontage of 853 m (2800 ft), separated by Goat I. from the American Falls with a frontage of 323 m (1060 ft); these falls are receding upstream at an average rate of 0·67 m (2·2 ft) a year. See pl. 73. [*f*]

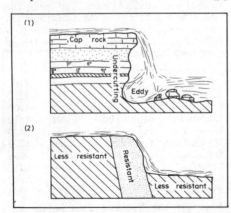

water-gap A low-level valley across a ridge, through which flows a river; e.g. the Wey g, near Guildford, the Mole g near Dorking, through the N. Downs; the Ouse g near Lewes through the S. Downs; the Goring g, where the R. Thames flows between the Berkshire Downs and the Chilterns.

water hemisphere Half of the globe, more or less centred on New Zealand, which contains only 1/7 of the land surface, as opposed to the land **h**.

waterhole A hole containing water, found esp. in the hot deserts or savanna lands, sometimes in the bed of an intermittent stream, used by animals and men.

water-meadow An area of **m** on the flood-plain of a river, which is naturally or artificially flooded, so stimulating grass growth. It is found in chalk country adjacent to a stream, where flooding early in the year produces an early crop of grass. Some w-ms are intersected with channels which can be flooded from a river; e.g. along the Itchen flood-plain S. of Winchester.

watershed (i) The line separating head-streams which flow to different river systems; it may be sharply defined (the crest of a ridge), or indeterminate (in a low undulating area).

In USA this is equivalent to a DIVIDE. The term *water-parting* is used in both Britain and USA. (ii) In America, the whole 'gathering-ground' of a single river-system, equivalent to a DRAINAGE BASIN. The term is thus used with 2 quite different meanings. [*f*]

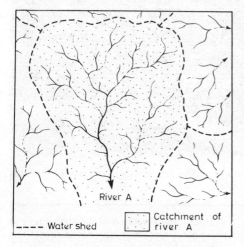

waterspout An intense, though small-scale, rapidly moving, low pressure system, similar to a TORNADO, but over the sea. From the low-lying base of a CUMULONIMBUS cloud, a whirling cone or funnel of cloud elongates until it touches the surface of the sea. The water-drops are derived both from condensation caused by cooling within its vortex and from water picked up from the agitated surface of the sea.

water-table The upper surface of the zone of saturation in permeable rocks; this level varies seasonally with the amount of percolation. Where it intersects the land surface, springs, seepages, marshes or lakes may occur. The w-t in a permeable AQUIFER roughly follows the surface profile of the ground, with its gradients somewhat flattened out. The slope of the w-t is inversely proportional to the permeability of the aquifer. The situation becomes more complicated, however, in rocks where flow is concentrated along faults and fissures, when no simple w-t may exist. See also PERCHED W-T. [*f, opposite*]

Watten (Germ.) Tidal marshes lying between the mainland coasts of Denmark, W. Germany and the Netherlands and the protective line of the offshore N. and E. Frisian Islands. At low tide a maze of creeks and channels separates the sheets of mud. Syn. with *wadden* in Netherlands.

WAVE 313

wave, ocean An oscillation of water particles. In the open o, the particles describe a circle in a vertical plane as the w passes; each moves slightly forward on the crest, and returns almost to its orig. position in the trough. The oscillation is caused by the friction of wind upon the surface of the water. Its size is determined by the wind speed and duration, and by the length of FETCH. The nature of the w is defined by its: (*a*) *height*, the distance from crest to trough, which may be as much as 12–15 m (40–50 ft), the record instrumentally measured being 21 m (70 ft) in 1961; (*b*) *length*, the distance between two successive crests (the longest measured is 1128 m, 3700 ft); (*c*) *period*, the time taken for the w-form to move the distance of one w-length; (*d*) *velocity*, the speed of the forward movement of an individual crest; (*e*) *steepness*, the ratio of its height to its length, which increases as the w enters shallow water until it reaches 1:7, when it breaks (hence a BREAKER); (*f*) *energy*, dependent on all the preceding features, including fetch; the average pressure exerted by a large w in winter on an exposed coast is nearly 11 tons per m² (a ton per sq. ft) (e.g. on the coast of W. Ireland), and during a storm this may be 3 times as great. A w is an erosive agent through: (i) hydraulic action; (ii) corrasion; (iii) attrition; (iv) solution. The greatest depth at which sediment is disturbed on the sea-floor is the *w-base*. Deposition is carried out by constructive ws, transport by ws breaking obliquely, hence LONGSHORE DRIFT. See also SWASH, BACKWASH, DOMINANT WS.

wave-built terrace A feature of marine deposition beyond the W-CUT BENCH [*f*].

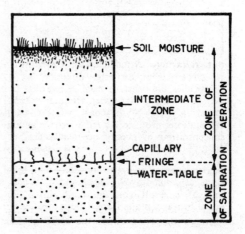

314 WAVE

Plate 74. An extensive WAVE-CUT rock PLATFORM in nearly vertical beds of OLD RED SANDSTONE, Manorbier Bay, Pembrokeshire, Wales. (*Eric Kay*)

wave-cut bench A feature of marine erosion at the base of sea-cliffs; this can develop into a w-c platform. See pl. 74. [*f*]

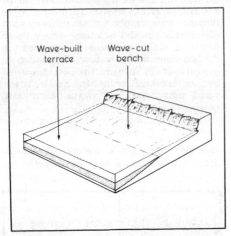

wave refraction A tendency for a w-front to be turned from its orig. direction as it approaches the coast. It may be retarded by the shallowing of the water. Oblique ws tend to turn parallel to the coast, while coastal indentations produce curved w-fronts and a concentration of energy on headlands. See pl. 4. [*f, opposite*]

waxing slope The upper convexity on a hillslope, as defined by W. Penck and A. Wood, thought to result from the weathering of an initial break of slope from 2 sides at once.
[*f* FREE FACE]

weather The condition of the atmosphere at any place at a specif. time, or for a short time, with respect to the various elements (temperature, sunshine, wind, clouds, fog, precipitation). This is an hour to hour, or day to day, condition. Ct. CLIMATE.

weathering The disintegration and decay of rock, so producing *in situ* a mantle of waste, depending on: (*a*) the nature of the rock; (*b*) the relief; and (*c*) the potency of the climatic elements in operation. W may be: (i) *mechanical* or *physical*: frost action, temperature change; (ii) *chemical*: SOLUTION, CARBONATION, HYDROLYSIS, OXIDATION, HYDRATION; (iii) *biological*: the presence of moss and lichen,

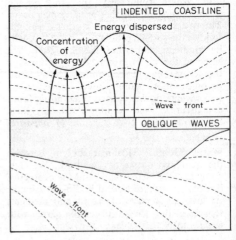

tree-roots, worms, moles, rabbits; this is not strictly w, but it assists, both mechanically and physically. The work of wind and rain (except for the latter providing lubrication of material and water which may freeze in cracks) are not included in w, since they involve transport of material, and are part of EROSION.

weathering front See BASAL SURFACE OF WEATHERING.

weather type A generalized type of synoptic pressure pattern, with its associated set of characteristic w conditions. See ANALOGUE.

wedge, of high pressure A region of high atmospheric p between 2 depressions, narrower than a RIDGE, bringing a brief spell of fine weather.

Weichsel The final main phase of the Quaternary Glaciation in N.W. Europe, corresponding to the Würm in the Alpine Foreland, and to the Wisconsin in N. America. It is divided into 3 main stages:

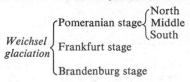

weir A dam across a river for raising and maintaining the level of the water, so controlling the current, keeping a navigable depth, or providing a head for a mill-wheel.

welded tuff A mass of hot volcanic ash that has fused to form such compact rocks as RHYOLITE.

Wentworth scale A geometric s of factor 2 of the size of particles of sediment, published by C. K. Wentworth in 1922. It ranges from particles of clay (0·004 mm diameter), silt, sand, granule, pebble, cobble to a boulder (exceeding 256 mm in diameter).

Westerlies The air-flow from the Subtropical High Pressure 'cell' to the Temperate Low Pressure zone between 35°N. and 65°N., and 35°S. and 65°S., blowing from the S.W. in the N. hemisphere, from the N.W. in the S. hemisphere. The W strengthen with altitude, and are locally concentrated into JET STREAMS.

wet adiabatic lapse-rate See SATURATED A L-R.

wet-bulb temperature The t recorded on a thermometer which has its bulb surrounded by a moist muslin bag, thus lowering the temperature by loss of latent heat through evaporation. With the aid of a dry-bulb thermometer and a set of tables, RELATIVE HUMIDITY can be ascertained.

wet-day In British meteorology, a day (24 hours, commencing at 09·00 hours) with at least 1·0 mm (0·04 in.) of rainfall.

wet spell A duration of at least 15 successive WET-DAYS.

wetted perimeter In hydrological studies, the actual length of the line of cross-sectional contact between the water in a river and its bed or channel.

whaleback (i) A rounded elongated mass of rock, commonly granite, shaped by moving ice; (e.g. in the Canadian Shield), or comprising an exposed part of the basal surface of weathering (see RUWARE). (ii) A smooth elongated mass of sand in a hot desert.

whinstone A quarryman's term in N. England for dolerite; hence the Great Whin Sill.

whirlpool (i) A violent circular eddy in the sea produced by a powerful tidal current flowing through an irregular channel, or by the meeting of 2 currents; e.g. the Maelstrom in the Lofoten Is. (ii) A similar phenomenon produced at the base of a large waterfall; e.g. Niarara Falls. Another w has developed 5 km (3 mi.) down the gorge of the Niagara R. where it bends at right-angles; the current thus impinges against the N. side of the gorge, swirls violently round, and flows off N.E. This w is apparently cut into the S. end of the drift-filled and abandoned interglacial gorge. See pl. 75.

whirlwind A rapidly rotating column of air, produced by local heating and convectional uprising. See CYCLONE, DUST-DEVIL, TORNADO, WATERSPOUT.

white box approach A type of 'systems approach', in which an attempt is made to identify as many storages, flows etc. as possible, in order to gain a very detailed understanding of the internal structure of the system (ct. BLACK BOX and GREY BOX APPROACHES).

white-out Under BLIZZARD conditions with a total snow-cover, it is extremely difficult to find one's direction; the impression is of being swathed in a white opacity.

wilderness Used in CONSERVATION, indicating an area left untouched in a natural state, with no human control or interference. Ct. NATURE RESERVE.

Plate 75. A clockwise WHIRLPOOL south of the equator. (*Popperphoto*)

williwaw A sailor's term for a sudden squall, esp. in the ROARING FORTIES.

willy-willy An intense tropical storm originating off the coast of N.W. Australia, sometimes crossing on to the land; a type of tropical CYCLONE.

wilting point A measure of soil moisture used by botanists and agriculturalists. It indicates the amount of water in the soil, below which plants will be unable to obtain further supplies, and will therefore wilt.

wind A horizontal current of air, varying from 'light air' to 'hurricane'. Ws can have a vertical movement, but this seldom happens at the Earth's surface. See BEAUFORT SCALE.

wind-chill index An **i** of physiological significance in cold climates, obtained from a formula involving temperature and wind-force.

wind-gap A g or notch in a ridge of hills, without a river flowing through, usually (though not always) at a somewhat higher level than a WATER-G. These are partic. identified as COLS in escarpments through which former consequent streams were thought to flow before CAPTURE or 'beheading' by scarp retreat and by scarp-foot streams. E.g. Clayton, Pyecombe and Saddlescombe cols in the S. Downs.

[*f* CAPTURE, RIVER]

'window' See FENSTER [*f*].

wind-rose A diagram with radiating rays drawn proportional in length to the mean percentage frequency of ws from each cardinal direction. The rays may be subdivided to show the frequency of various w strengths associated with the partic. direction, each direction of ray being of different width [as *f*]. An octagonal w-r can be constructed, with each side representing one of the 8 cardinal directions, and the 12 monthly frequencies of w from each of these directions are plotted as columns. Calms may be given as a percentage in the centre of the circle or octagon. [*f, page 317*]

wind-shadow A 'dead air space' in the lee of an obstacle in the path of the w, though it is rarely calm there, but the scene of eddying. This is an important fact in the creation of a sand-dune and in determining its shape.

wind-slab A sheet of snow which is hard-packed by the w, requiring great care by a ski-er, since it is liable to avalanche *en masse*.

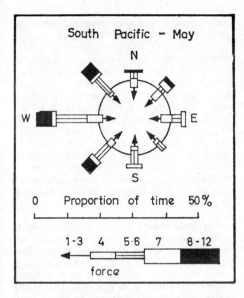

phases combined. Various stages and interstadial intervals have been suggested:

Wisconsin glaciation
- Mankato glaciation
- Valders glaciation
- Two Creeks interstadial
- Cary glaciation
- Tazewell glaciation
- Peorian interstadial
- Iowan glaciation (formerly regarded as a separate stage)

Sangamon (Third) interglacial

The glacial relief is very varied; it includes numerous well-defined terminal and recessional moraines, KAME-AND-KETTLE features (esp. in Wisconsin State), and thick clay-TILLS.

wold An open rolling chalk upland, used more specif. as an element in a proper name; e.g. Lincoln Ws, Yorkshire Ws. The term is sometimes used for hills of other rocks; e.g. Cotswolds (Jurassic).

Wolstonian The penultimate glacial period in Britain, equivalent to the SAALE glaciation of N.W. Europe, and represented typically by glacial tills at Wolston, Warwickshire.

woolsack A large, rounded, unweathered boulder within a mass of RESIDUAL DEBRIS. Also referred to as a CORESTONE or 'tor boulder', since exposure of ws may form some tors, esp. in tropical environments.

wrench-fault A nearly vertical STRIKE-SLIP FAULT, with the horizontal displacement on a large scale; e.g. San Andreas Fault, California; the Alpine Fault of the S. Island of New Zealand.

Würm The name given by A. Penck and E. Brückner to the last of the series of 4 periods of fluvio-glacial deposition during the Quaternary glaciation; see GÜNZ, MINDEL, RISS. This concept was subsequently widely applied to glacial periods in general, though the latest research indicates that the pattern of glacial and interglacial periods is not a simple 4-fold one. The W is equivalent to the Weichsel in N.W. Europe, and probably to the Wisconsin in N. America. The W is divided into 3 stages, known as W I, II, III.   [*f* MINDEL]

Winkel's Tripel Projection A development from LAMBERT'S ZENITHAL EQUAL AREA P, used by J. Bartholomew for world maps of climatology, vegetation and population because of its distributional merits.

winter (i) In gen. terms, the coldest season of the year, in ct. to summer. (ii) In the N. hemisphere the period Dec., Jan., Feb., in the S. hemisphere the period June, July, August. (iii) In astronomical terms, the period between the winter solstice (about 22 Dec.) and the spring equinox (21 March) in the N. hemisphere, and between about 21 June and 22 Sept. in the S. hemisphere.

winterbourne A stream breaking out in the floor of a dry valley only after a prolonged period of rainfall. This forms a common place-name element in chalk country; e.g. Winterbourne Abbas, -Strickland, -Zelstone and many more in Dorset. Syn. with BOURNE.

Wisconsin The final main phase of the QUATERNARY glaciation in N. America, corresponding to the Würm in the Alpine Foreland and to the Weichsel in N.W. Europe. The chronology of the W. is complex, but it has been closely studied because of its wide distribution, as indicated by the extensive deposits of boulder-clay and moraine, esp. in the area S.W. of the Great Lakes, and between Lake Michigan, Huron and Erie. The area covered in N. America was greater than that of all the other glacial

xenocryst A crystal in an igneous rock which is foreign to the rock in which it now occurs.

xenolith A metamorphosed piece of foreign rock occurring near the margin of a BATHOLITH within the solidified granite, usually derived from the invaded country rock.

xerophyte, xerophilous, xerophytic One of a category of plants adapted to withstand dry conditions, seasonal or perennial, in various ways: long roots, small, hard, glossy leaves; thick bark; thorns; various water storage devices. C. TROPOPHYTE.

xerosere (alt. **xerarch**) Applied in a botanical context to a plant succession (SERE), developing under markedly dry conditions. It may comprise either a rock sere (*lithosere*) or a sand sere (*psammosere*).

xerothermic index An **i** of the relationship of drought to plant-growth.

yard The British standard unit of length. 1 yd = 0·914399 m. 1 m = 1·09361 yds.

yardang A sharp keel-like crest or ridge of rock, separated from a parallel neighbour by a shallow groove or furrow. It is the result of differential erosion in a desert by the scouring effect of sand-laden winds. The ridges may be as much as 6 m (20 ft) high and 37 m (120 ft) wide.

yazoo A DEFERRED JUNCTION [*f*] of a tributary, called after the R. Yazoo which joins the lower Mississippi.

year A measure of time related to the revolution of a heavenly body around the sun, specif. the Earth. The exact concept differs: (i) *Sidereal Y*: the time taken by the Earth to make 1 complete revolution in its orbit with reference to the stars, 365·2564 mean solar days, or 365 days, 6 hours, 9 minutes, 9·54 seconds. (ii) *Tropical Y*: the average time taken by the Earth to make 1 complete revolution in its orbit with reference to the vernal equinox, as indicated by the First Point of Aries (see RIGHT ASCENSION), at present 365·2422 solar days, or 365 days, 5 hours, 48 minutes, 45·51 seconds; this diminishes by about 5 seconds in a millennium. It is also known as the *Equinoctial, Astronomical, Nature* or *Solar Y*. (iii) *Civil Y* (or *Gregorian Calendar Y*): a period of 365 mean solar days of 24 hours; to compensate for the extra 0·2422 solar days, every 4th is a Leap Y with 366 days (one extra in Feb.). This is still not quite right (0·01321478 days too long, or approx. 0·04 days every 4 years, and 1 day per century of 25 leap-years), so the last year of a century is only a leap y when the first 2 figures are divisible by 4; i.e. 1600 A.D. and 2000 A.D. are, but 1700, 1800, 1900 are not; (iv) *Anomalistic Y*: the time between 2 successive PERIHELIONS, 365·25964 solar days.

Younger Drift The younger tills deposited after the last interglacial period, which in part overlie the older glacial tills which have been weathered and eroded (OLDER D). In Europe the **Y D** is defined as the product of the Weichsel (or Würm) glacial advance, the Older **D** as that of the earlier ones. In Britain the same distinction applies; the S. limit of the **Y D** can be traced from the coast of N. Norfolk through the Vale of York, the W Midlands and S. Wales to the mouth of the Shannon in S.W. Ireland. The Older **D** and **Y D** are of course much subdivided.

young flood stand A period of slack-water interrupting the normal tidal rise, though not the ebb; e.g. in Southampton Water. This occurs about 1½ hours after low water (near mean tide level), and lasts for nearly 2 hours before the tidal rise is resumed. This is the result of a double entrance to the Solent, which is not in phase with the tidal rise and fall outside the I of Wight. A hydraulic gradient prevails at certain times between the Needles and Spithead entrances; at spring high water, the heights of the tide there are 1·07 m and 2·0 m respectively. This causes a flow from E. to W., the result of this 'head'. The reverse applies at spring low water when the level at the Needles is 0·9 m above that at Spithead. These interruptions to the normal tidal flow produce a pause in the normal tidal rise until the hydraulic 'head' has been wiped out. Added to the stand at high water, this gives 7 hours of slack in each 24.

'young' mountains Fold-mountains created during the last great period of folding (e.g. the Alps), by contrast with earlier ones (the 'old' fold ms of HERCYNIAN and CALEDONIAN age).

youth The 1st stage in the cycle of landform development; the orig. structure is still the dominant feature of the relief. Slopes are steep, gradients irregular, and denudation processes rapid (e.g. a 'young' river). Recent criticism of the indiscriminate use of this evocative terminology has been put forward; using the human life cycle as a simile can be carried too far. It cannot be assumed that the youth stage is necessarily the most active in terms of erosion, nor that 'old age' streams are the most sluggish. See CYCLE OF EROSION.

zawn A narrow rocky inlet in a cliffed coast, specif. in Cornwall.

zenith The point in the heavens (i.e. on the CELESTIAL SPHERE) vertically above the observer.

zenithal (or **azimuthal**) **projection** A class of **p** in which the globe is projected on to a plane touching at the Pole (*polar z*), at the equator (*equatorial z*), or anywhere between (*oblique z*). All bearings are true from the centre of the projection. See GNOMONIC, Z EQUAL AREA, Z EQUIDISTANT, STEREOGRAPHIC, ORTHOGRAPHIC, LAMBERT'S Z EQUAL AREA PS.

Zenithal Equal Area Projection (also called AZIMUTHAL EQUAL AREA) When drawn as the polar case (with the Pole at the centre), it has concentric parallels, with the area between each the same as on the globe to scale, of radius $= \sqrt{2R} (R - R \sin latitude)$, where R is the scale of the globe. Meridians are straight lines radiating out from the Pole. In the other cases, both meridians and parallels are curves. It forms a useful **p** for representing a single hemisphere. The oblique and equatorial cases are most conveniently constructed from tables.

Zenithal (Azimuthal) Equidistant Projection The only **p** in which all points are the true distance and true direction from its centre. In the polar case, the meridians are radii from the centre of the **p**, spaced at the desired angular intervals. Parallels are concentric circles, their true scale distances apart; i.e. $2\pi R . \frac{x}{360}$, where x is the parallel interval in degrees. The equatorial case is rarely used. The oblique case is used esp. for a map centred on an important city or an airport, since every place in the world is in the correct direction and at the correct distance, although shape becomes very distorted away from the centre. The emblem of the United Nations Organization is drawn on this **p**.

[*f* POLAR PROJECTION]

zero curtain Defined by S. W. Muller as 'the zone immediately above the permafrost where zero temperature (0°C) lasts a considerable period of time (as long as 115 days a year) during freezing and thawing of overlying ground'. Thawing of the **z c**, or the lowering of its temperature below 0°C, is prevented by the latent heat of ice freezing i.e. during a period of freezing this latent heat cannot easily escape, esp. if moisture content is high at the permafrost table.

Zeuge (pl. **Zeugen**) (Germ.) A tabular mass of resistant rock up to 30 m high, standing out from softer underlying rocks because of its protective capping. It is produced by differential erosion in a desert through the scouring effect of sand-laden winds. [*f*]

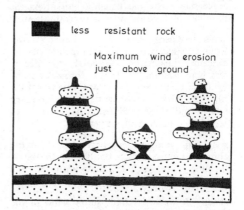

zinc A metal mainly obtained from *zincblende* (ZnS) (the mineral *sphalerite*), and *calamine* ($ZnCO_3$), an ore which often occurs in association with lead and silver, now commonly known in Britain as SMITHSONITE.

zincblende A major ore of ZINC, the mineral *sphalerite*, ZnS.

Zodiac An imaginary belt or zone in the heavens, within which are the apparent courses of the sun and the planets (except Venus and Pluto), bounded by 2 lines parallel to the ECLIPTIC, and 8° away on either side. In astrology, the **Z** is divided into 12 equal portions of 30°, their names derived from constellations which at the time of the Greeks lay in that partic. portion of the sky, but now, owing to the PRECESSION OF THE EQUINOXES (about 50 seconds a year), lie some distance (about 30°) to the W. The Signs begin from the First Point of Aries (see RIGHT ASCENSION), where the plane of the ecliptic intersects the CELESTIAL EQUATOR at the time of vernal equinox, about 21 March. The Signs are Aries, Taurus, Gemini, Cancer Leo, Virgo, Libra, Scorpio, Sagittarius, Capricorn, Aquarius and Pisces. Each Sign now lies in the next constellation to the W.; e.g. the First Point of Aries is in Pisces, which is astrologically awkward.

zodiacal light A cone of faint **l**, extending upward from a base below the horizon, best seen after sunset in spring and before sunrise in autumn (in the N. hemisphere), esp. on a clear moonless night. It seems to slant upward to the left in the evenings, to the right in the mornings. It probably consists of a cloud of rarefied dust particles or gas molecules, extending outward from the sun. In the tropics the z l sometimes appears as a band right round the sky.

zonal flow The movement of air along the parallels of latitude, in ct. to MERIDIONAL F.

zonal inselberg Used by J. Budel (Germ. Zonaler Inselberg auf Randpediment) to describe inselbergs marginal to deeply weathered plains. Z is are formed not by exhumation of the BASAL SURFACE OF WEATHERING (ct. SHIELD INSELBERGS), but by scarp retreat and allied pedimentation, as in L. C. King's PEDIPLANATION CYCLE.

zonal soil A s-type largely resulting from the climatic factors which contribute to the s-forming processes. The 2 main groups are the PEDOCALS and the PEDALFERS.

zonda A warm, humid, sultry wind in the Argentine, blowing from the N. in front of a low-pressure system. The name is also given to a FÖHN-type wind blowing down the E. slopes of the Andes in the same country.

zone (i) Used gen., even vaguely, for various belts (esp. of climate and soil) of a latitudinal character, or for any region defined between specif. limits with special characters (e.g. plants, animals). (ii) In geology, a thickness of rock with a partic. fossil-assemblage. The word z is used in conjunction with many other terms, esp. within rocks or the soil: z of aeration, capillarity, discharge, eluviation, illuviation, flow, fracture, saturation.

zone standard time The mean time within a longitudinal zone of 15°; see STANDARD TIME [f].

zoogeography The study of the geographical distribution of wild animals.

zooplankton A collective name for the minute marine animal life floating in shallow seas, esp. where warm and cold currents meet, and where cold water upwells. The main groups of z are *Radiolaria*, *Foraminifera* and *Copepoda*. Ct. PHYTOPLANKTON, similar vegetable organisms, though gen. much smaller.